128
132-133
134-135
136-137
138
142-143
144
90
92

jacaranda
jaconline.com.au

Society & Environment ATLAS

Contents

MAPS ARE SHOWN IN AQUA

Introduction iv

SKILLS

What is a map? 2–3
Title 2
Scale 2
Key 3
Direction 3

Finding your way 4–5
Grid references 4
Central Canberra 4
Using an atlas 5
Latitude and longitude 5

Using direction 6–7
Direction on maps 6
Sea World — Gold Coast, Queensland 6–7
Antarctica 7

Understanding scale 8–9
Using scale 9
Using scale to measure distances 9

Different types of maps 10–11
Political maps 10
Physical maps 10
Topographic maps 10
Mud maps 10
Finding your way with a mud map 10
Tourist maps 10
Tourist map of Carnarvon district, W.A. 10
Choropleth maps 11
Satellite images 11
Theme maps 11
Synoptic charts 11
Flow maps 11
Flow map showing visitors to Australia 11

Language of maps 12–13

Drawing contours and cross-sections 14–15
Contours 14
Cross-sections 15
Contour map of Uluru 15

Using topographic maps 16–17
Topographic map extract of Busselton area, Western Australia 16–17
Area and grid references 17

Aerial photographs 18–19
Interpreting aerial photographs 18
Change over time 18

Landsat images 20–21
Satellites 20

Primary and secondary sources 22–23
Lusitania — Primary sources 22
Lusitania — Secondary sources 23
Route of the *Lusitania* 23

Evaluating sources 24–25
Propaganda 24
Cartoons 25

Examining the evidence 26–27

Constructing timelines 28–29
BC and AD 28

Drawing line and bar graphs 30–31
Line graphs 30
Bar graphs 30
Climographs 31
Population pyramids 31

Drawing pie and other graphs 32–33
Pie graphs 32
Scatter graphs 33
Pictographs 33

Testing your skills 34–36
Tracking the thylacine 34
Sightings of the thylacine 34
Topographic map extract of Togari area, north-west Tasmania 36

AUSTRALIA

AUSTRALIA 38–41
Settlement, history and culture 38–39
Case study: Eureka rebellion 39
Landforms and land use 40
National Parks 41
Kakadu 41

Queensland 42–43
Aboriginal language groups 43

Brisbane 44–47
Change over time 44–45
Satellite image 46
Social and economic patterns 47

New South Wales 48–49
Aboriginal language groups 49

Sydney 50–53
Change over time 50–51
Satellite image 52
Social and economic patterns 53

Victoria 54–55
Aboriginal language groups 55

Melbourne 56–59
Change over time 56–57
Satellite image 58
Social and economic patterns 59

Tasmania 60–61
Aboriginal language groups 61

Hobart 62–65
Change over time 62–63

Satellite image 64
Social and economic patterns 65

South Australia 66–67
Aboriginal language groups 67

Adelaide 68–71
Change over time 68–69
Satellite image 70
Social and economic patterns 71

Western Australia 72–73
Aboriginal language groups 73

Perth 74–77
Change over time 74–75
Satellite image 76
Social and economic patterns 77

Canberra 78–79
National capital 78–79
Suburbs of Canberra 78

Northern Territory 80–81
Aboriginal language groups 81

Darwin 82
Change over time 82

THE WORLD

PACIFIC 84–91
Countries and dependencies 84–85
Earthquakes and volcanoes 86–87
Deaths caused by volcanoes and earthquakes 86
How do earthquakes occur? 87
Case study: Kobe earthquake 87

New Guinea 88–89
Mineral development in Papua New Guinea 88

New Zealand 90
Case study: Maori population 91
Maori Wars 91

ANTARCTICA 92–95
Research 92
Human endeavour 93
Working at Edgeworth David Station 93
Explorers' routes 93
Natural environment 94–95
Case study: Antarctic food web 95

ASIA 96–109
Settlement 96
Ancient cultures 97
Landforms and land use 98–99
Transect 99
Case study: Himalayas 99
South-East Asia 100–101
East Asia 102–103
Korea and Japan 104–105
Case study: Economic growth of South Korea 105
South Asia 106–107
Middle East 108
Case study: Ancient Sumer 109
Ancient Sumer 109

EUROPE 110–119
Settlement 110

Ancient cultures 111
Landforms and land use 112–113
Transect 113
Case study: Mt Vesuvius 113
North West Europe 114–115
Southern Europe 116–117
Commonwealth of Independent States 118–119

AFRICA 120–127
Settlement 120
Ancient cultures 121
Landforms and land use 122–123
Transect 123
Case study: Libyan Desert 123
Northern Africa 124–125
Southern Africa 126
Case study: Rwandan refugee crisis 127
Rwandan refugees 127
Mountain gorillas 127

NORTH AMERICA 128–135
Settlement 128
Ancient cultures 129
Landforms and land use 130
Transect 131
Case study: San Francisco Bay 131
Canada 132–133
United States of America 134–135
Central America and the Caribbean 136–137

SOUTH AMERICA 138–144
Settlement 138
Ancient cultures 39
Landforms and land use 140
Transect 141
Case study: Amazon Basin 141
Northern South America 142–143
Southern South America 144

REFERENCE

The political world 146–147
World countries and capital cities 146–147
The changing earth 148–149
Continental plates 148
World landforms and tectonic plates 148–149
Mountain building 148
Continental drift 149
Volcanoes 150–151
Ash or cinder cone 150
Fissure volcano 150
Composite cone 150
Shield volcano 150
Volcanic eruption 151
Australian volcanic activity 151
River systems 152–153
The life of a river 152
Waterfalls 152
Oxbow lakes 152–153
Murray–Darling Basin 153
Deltas 153
Desert systems 154–155
World deserts 154
Wind processes in deserts 154
Water processes in deserts 155
The Simpson Desert 155
Australian deserts 155

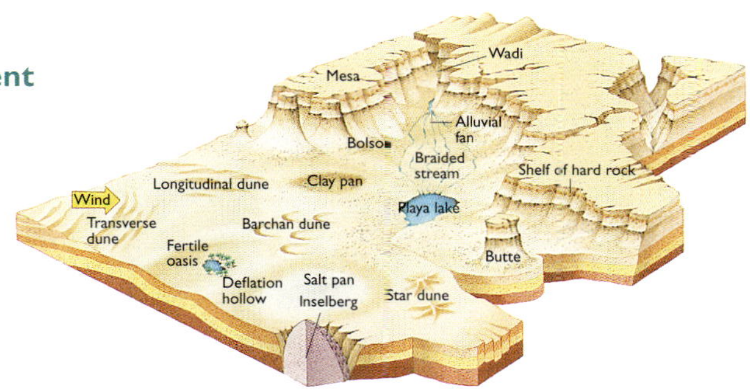

Wadi
Mesa
Alluvial fan
Bolson
Braided stream
Shelf of hard rock
Longitudinal dune
Clay pan
Wind
Transverse dune
Barchan dune
Playa lake
Fertile oasis
Butte
Deflation hollow
Salt pan
Inselberg
Star dune

Coastal systems 156–157
Headland system 156
Coral reef system 156
People and the coastal system 156–157
Beach-dune system 157
Port Campbell 157
World climate 158–159
World records 158
World average annual rainfall 158
World average temperatures — January 159
World average temperatures — July 159
Australia's climate 160–161
El Nino 160
Climate zones 160
Seasonal climatic patterns 161

Weather systems 162–163
Cold fronts 162
Synoptic chart 162
Cyclones 163
Major Australian cyclones since 1970 163
The natural world 164–165
Biomes 164
World biomes 164
Environments in danger 166–167
World land degradation 166
Endangered species 167
Endangered animals of Shark Bay, Western Australia 167
Revealing the past 168–169
The skill of the archaeologist 168
Early humans 168–169
World heritage 170–171
What is worth keeping? 170
World Heritage sites 170–171
Australia's World Heritage Sites 170
Indigenous people of the world 172–173
Tibetans 172
Tuareg 172
Inuit 173
Torres Strait islanders 173
Kayapo 173
Exploring, discovering, meeting 174–175
World exploration 174
First contact in the Pacific 175
A goblin comes to Whitianga 175
Cultural groups 176–177
Australia 176
World languages 177
World religions 177
The world at war 178–179
World War I 178
World War II 178
Australians at war 179
Pacific campaigns in World War II 179
People and resources 180–181
Agricultural land use in Australia 180
Mining and manufacturing sites in Australia 180
Resources and manufacturing 180
Transport and location factors 180
Raw materials 181
Manufacturing industries 181
Labour resources 181
An unequal world 182–183
Feast and famine 182
Wealth and health 182–183
The haves and have nots 182–183
Education and literacy 183
Cities of the world 184–185
World's largest cities 184
The location of towns and cities 185
Transport routes 185
Resources 185
Defence 185
World of work 186–187
Slavery 186
Informal work 186
Women and work 187
Statistics 188–195
Index 196–219
Acknowledgements 229

Introduction

The new *Jacaranda Society and Environment Atlas* introduces approaches and topic areas never before seen in a school atlas. The range of resources provided for History, Geography, Commerce and other Social Sciences ensures that this atlas will be more widely used than current Geography-based atlases. Topics and case studies have been selected from those currently popular in classrooms across Australia, as well as the new study areas suggested in the Studies of Society and Environment curriculum profile.

The 'Australia' and 'World' sections provide the most comprehensive map coverage of any Australian school atlas. In addition, they introduce innovative History, Culture, Resources and Systems perspectives within a regional context. The 'Skills' pages give an extensive introduction to the broad range of processes used by students to investigate, communicate and participate in Studies of Society and Environment (SOSE). The unique 'Reference' section provides teachers with an array of high-quality resources to assist in the development of key concepts in the SOSE strands of Time, Continuity and Change; Place and Space; Culture; Resources and Systems.

The *Jacaranda Society and Environment Atlas* provides strong coverage of National Curriculum SOSE outcomes at levels, 4, 5 and 6, with consolidation of lower levels as well as extension to higher levels. The chart below gives a broad interpretation of how the four sections of the atlas contribute to this coverage.

The separate *Jacaranda Society and Environment Atlas Worksheets* provide activity sheets that are specifically designed to assist teachers to implement new SOSE courses. Each worksheet outlines the relevant SOSE outcomes, and cross-references activities to the appropriate resource in the *Jacaranda Society and Environment Atlas*.

STRANDS		SKILLS	AUSTRALIA	THE WORLD	REFERENCE
Time, Continuity and Change	Understanding the Past		■	■	■
	Time and Change		■	■	■
	Interpretations and Perspectives	■			■
Place and Space	Features of Places	■	■	■	■
	People and Places	■	■	■	■
	Care of Places			■	■
Culture	Aboriginal and Torres Strait Islander Cultures		■		■
	Cultural Cohesion and Diversity			■	■
	Personal, Group and Cultural Identity		■		■
Resources	Use of Resources		■	■	■
	People and Work		■		■
	Management and Enterprise	■			
Systems	Natural Systems	■	■	■	■
	Political and Legal Systems		■		
	Economic Systems				■
Processes	Investigation	■			
	Communication	■			
	Participation	■			

SKILLS

2 WHAT IS A MAP?

A map is a simplified plan of a place seen from directly above. Things look different when viewed from directly above and these different shapes and patterns are what you see on maps. The photographs below show the Melbourne Cricket Ground from the ground and from the air. A map of the Melbourne Cricket Ground appears on the opposite page.

Title

The title will tell you the place on earth the map is representing. It may also give you a clue about what the map is particularly trying to show — Australia: Landforms and land use, Sydney: Social and economic patterns, World: Countries and dependencies.

Scale

Scale shows how distances on the map relate to distances in the real world. Maps show big areas on the earth's surface on a small sheet of paper. On the map of the Melbourne Cricket Ground the map maker, or **cartographer**, has been able to represent an entire cricket ground on less than one page of a book. The shapes of the features remain the same, but they are drawn smaller by using scale.

The actual scale of a particular map is shown using words, a line, or a ratio. The map of the Melbourne Cricket Ground has a **line scale**. Each centimetre on the map represents 25 metres in the real world. We can tell that the playing surface is approximately 175 metres long (east–west) by measuring this feature and comparing it with the scale.

Aerial photograph of the Melbourne Cricket Ground (*Inset:* Exterior of the Melbourne Cricket Ground)

MELBOURNE CRICKET GROUND MAP

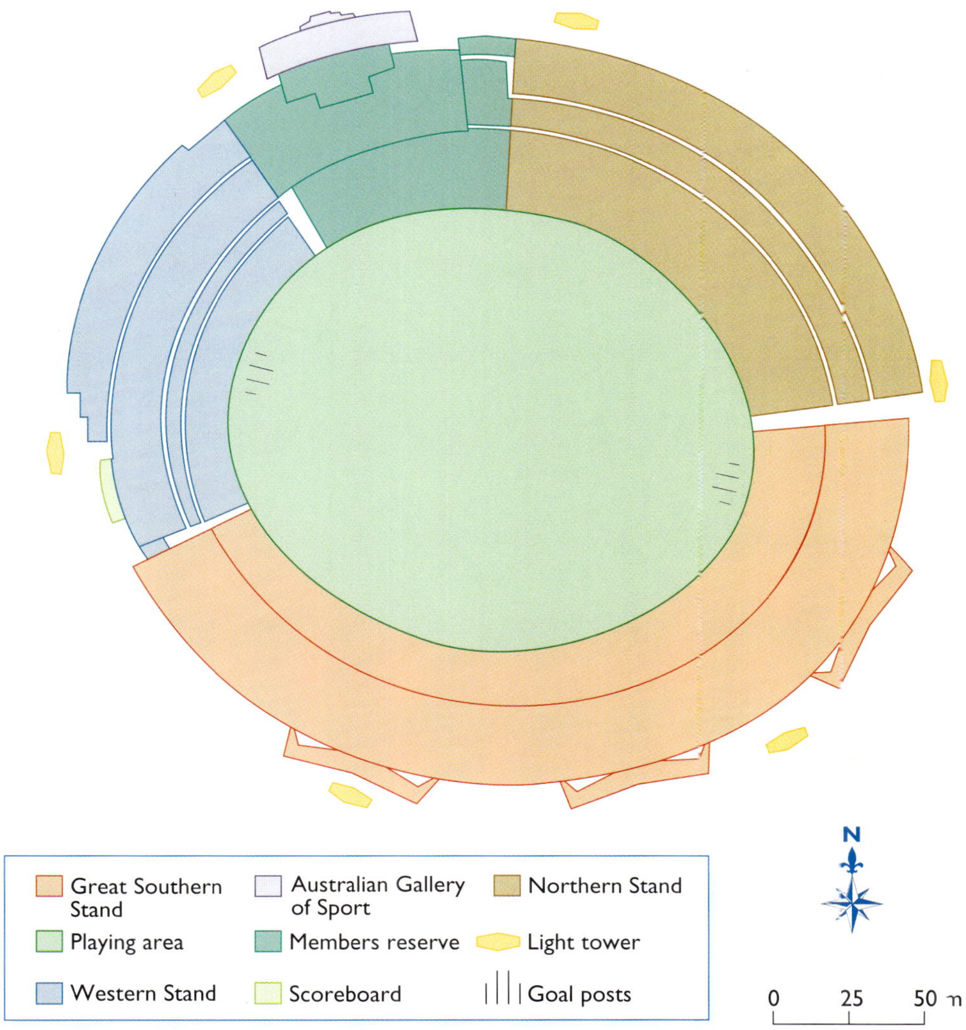

Key

A **key** or **legend** shows the signs which have been used to draw the map and explains their meaning. Symbols and colours are used to show different features. The importance of a feature might be shown by the size of the symbol or the size, style or thickness of the print. Cartographers try to use symbols which are clear in their meaning, for example an aeroplane represents an airport. A key or legend is necessary when the symbols are not clear, or to explain the colours used on a map.

When symbols are used, it is easy to see the pattern or distribution of a feature on a map, for example where the toilets and telephones are located in a shopping centre. Common features are usually shown using conventional signs, but some maps use different symbols. Always check the key.

Direction

Direction on maps is shown using an arrow pointing north. Most maps will have north at the top of the page, but always make sure to look for the north pointer, for this is not always the case. Some maps in this atlas have more than one north arrow to help teach you about the earth's curved surface.

ACTIVITIES

1. Make a list of the things that can be seen in the ground level photograph of the Melbourne Cricket Ground.
2. Make another list of those features that can be seen only in the aerial photograph.
3. Find one feature that can be seen in both photographs. Draw two sketches of the feature — one from the ground level and the other from the aerial view.
4. The map is drawn from the aerial view. How is the map of the Melbourne Cricket Ground similar to the aerial photograph?
5. What are the differences between the aerial photograph and the map?
6. Why do you think we use maps rather than aerial photographs to find our way around?
7. Why is the key so important to the use of the map?
8. The playing area in the aerial photograph is marked out for Australian Rules Football. Use the scale to work out how far a player standing on the semi-circular arc would be from the goal posts.
9. The rectangle at either end of the ground is the space in which the defensive team kicks when a point is scored. How long is it?
10. In which season was the aerial photograph taken and at what time of the day? Explain your answer.
11. Find out what each of the symbols shown below represents. Record your guess and then turn to the page to check your answer.

   Page 85 — Page 97

   Page 85 — Page 97

   Page 95 — Page 99

12. Which symbols were the easiest to recognise? Why?
13. Create your own map symbols to represent the following:

   telephone — church

   dangerous dog — car park.

4 FINDING YOUR WAY

Grid references

We all use maps to find our way. Road maps show us the way from town to town and street directories help us find our way within a town or city. Most maps have some sort of reference system to help locate features on the map.

The simplest and most common reference system is the **alphanumeric grid**. The map is divided into columns from top to bottom and rows from side to side. Columns and rows are named by numbers and letters. In an alphanumeric reference the column letter is given first, followed by the row number, for example A1.

The alphanumeric grid system is most common in atlases and on street and road maps. It tells you which grid square to look in to find a particular feature.

Look at the map of Central Canberra to find City Hill at grid reference D4. Find the letter D and the number 4. Draw a straight line with your finger down from the letter D and across from the number 4. The square (or grid) where the lines meet is grid reference D4, where you will find City Hill.

An index for the map has the main features of the area arranged alphabetically. The grid reference is found beside each feature. If the index is for more than one map, it will also tell you the map or page number to look at.

Index of Central Canberra's tourist sites

Aspen Island (Carillon) **F8**
Australian War Memorial **H4**
High Court of Australia **E8**
Jewish Memorial Centre **D12**
The Lodge **A11**
Australian National Gallery **F9**
Australian National Library **D8**
National Science and Technology Centre **D8**
Parliament House (new) **C11**
Parliament House (old) **D9**
Captain Cook Memorial Water Jet **D7**

CENTRAL CANBERRA

Using an atlas

If you are trying to locate a place in an atlas you can refer to the **table of contents** or to the **index**. The table of contents, which is found in the first few pages of the atlas, will help you locate broad areas such as continents or groups of countries.

If you want to find a specific place, you should look at the index at the end of the atlas. The index will tell you which map to refer to and the grid location of the place. Often the index gives the latitude and longitude as well.

Ahvaz **108 C4** 31.17N 48.43E
Ailinglapalap, island **84 L6**
Ain Beida **116 E2** 35.44N 7.22E
Ain Dalla **125 F4** 27.20N 27.23E
Ain M'Lila **116 E2** 36.02N 6.30E
Ain Oussera **116 D2** 35.24N 2.54E
Ain Salah **124 D4** 27.12N 2.29E
Ain Temouchent **116 C2** 35.18N 1.09W
Air, mountains **124 D3**
Airbangis **100 A3** 0.09N 99.25E
Aisega **89 E2** 5.44S 148.22E
Aitape **88 D3** 3.10S 142.17E
Aitutaki, island **85 P4**

Latitude and longitude

Basic grid systems are good for finding the general location of a feature quickly. **Latitude** and **longitude** help you to pinpoint places with far greater accuracy. This grid system is used in atlases and on globes or on maps used for long distance navigation.

Lines running east–west across the map are called **parallels of latitude**. They are measured in degrees north and south of the **Equator** (0 degrees latitude). These include:

- the Tropic of Capricorn (23.5 degrees south)
- the Tropic of Cancer (23.5 degrees north).
- the Arctic Circle (66.5 degrees north)
- the Antarctic Circle (66.5 degrees south).

Lines running north/south down the map are called **meridians of longitude**. They are measured in degrees east and west of Greenwich (0 degrees longitude). These include:

- the International Date Line (180 degrees east or west)
- the **prime meridian** (0 degrees longitude).

Latitude is always given before longitude. To give accurate locations of exact points, each degree is divided into 60 minutes (′) and each minute is divided into 60 seconds (″). This system gives very accurate locations. For example, Ain Salah is located 27° 12′ north of the Equator, and 2° 29′ east of the prime meridian. The index of this atlas shows Ain Salah's exact location as 27.12N 2.29E.

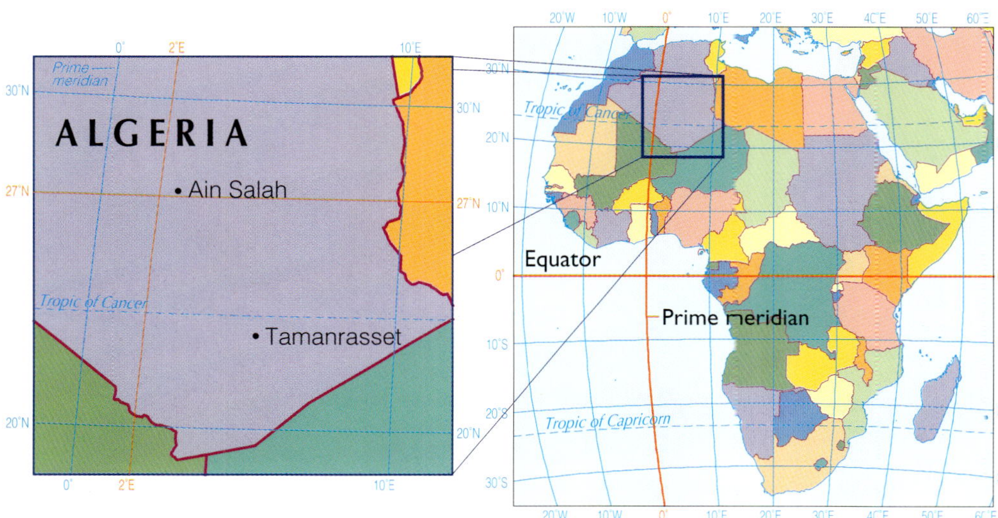

ACTIVITIES

Use the map and index of Canberra to answer the following.

1. Find the High Court of Australia and the Australian National Gallery. In which suburb are they found?
2. What shape is the road surrounding new Parliament House?
3. What shape do the roads surrounding old Parliament House form?
4. What features are found at the following grid references?
   A4, C12, E4, E13, G3
5. Use the contents of this atlas to find the page number for the following maps.
   - Tasmania: Settlement
   - Brisbane: Change over time
   - South-East Asia: Settlement and landforms
   - Korea and Japan: Settlement and landforms
   - United States of America: Settlement and landforms
6. Use the index to give the map page, grid reference and latitude/longitude of the following cities.
   - Port Moresby
   - Tokyo
   - Mexico City
   - Darwin
   - Dacca
7. Find the following mountain names using the page number, grid reference and latitude/longitude. Check your answers in the index.

48 E1	36.27S	148.16E
102 C2	27.59N	86.56E
104 I4	35.23N	138.42E
126 D4	3.02S	37.20E
134 B3	46.12N	122.11W

8. Estimate the latitude and longitude of the following towns shown on the map of Tasmania on page 60. Check your answers in the index.
   Wynyard C3
   Franklin E1
   Launceston E3
   Port Arthur E1
   Hobart E2

6 USING DIRECTION

Direction on maps

Geographers use the points of a compass to give directions and to work out the location of certain features, such as those shown on the diagram of Sea World. Words such as left, right, up, down, top and bottom can be confusing and inaccurate. Using **compass points** is more accurate because, in whichever direction you are facing, the compass direction always remains the same.

The **cardinal points** on a compass are north, south, east and west, and the sub-cardinal points are north-east, south-east, south-west and north-west. For greater accuracy, a **bearing** (angle) is given in degrees and minutes measured clockwise from north. Sixty minutes equals one degree.

True north is the actual direction of earth's geographical north pole, while magnetic north is the direction in which the magnetic needle of a compass points. Magnetic north and south poles move due to the spinning of the earth. The map of Antarctica shows how the south magnetic pole has moved over time.

Points of the compass

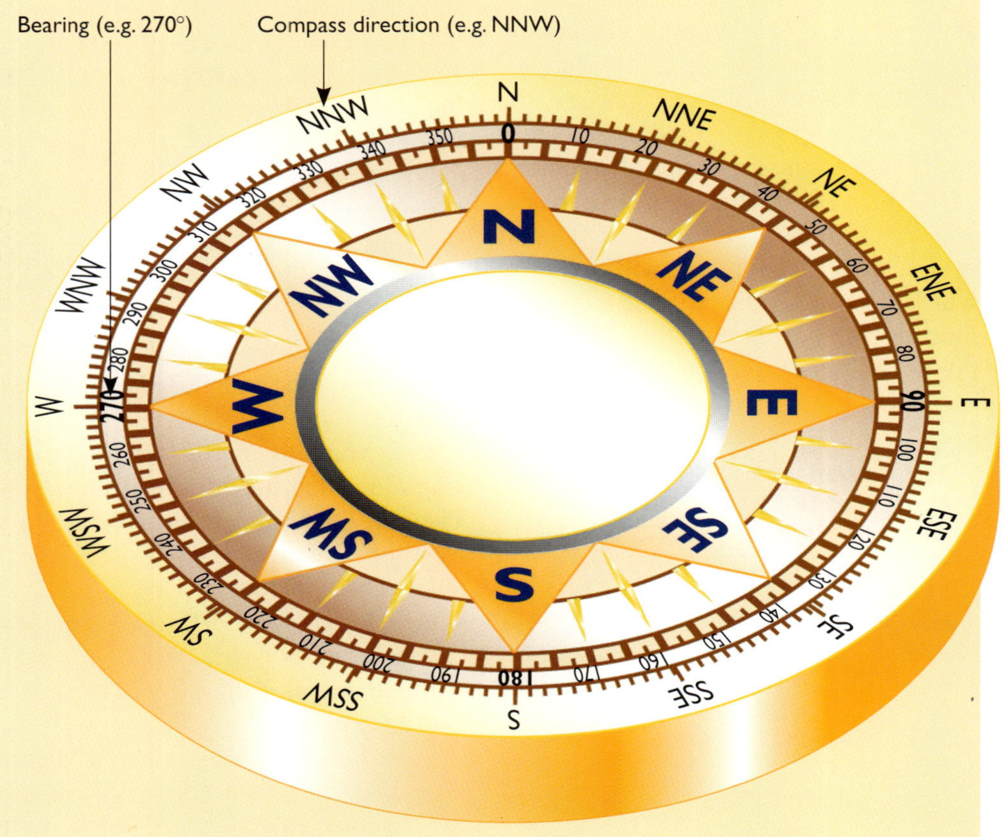

SEA WORLD — GOLD COAST, QUEENSLAND

1. Entrance
2. The World of the Sea Aquarium
3. Sky High Skyway
4. The Wondrous World of Dolphins
5. Grand Central Train Station
6. Sea Lions at Play
7. Thrill Seeker Rollercoaster
8. Viking's Revenge Flume Ride
9. Water slides and pool
10. Corkscrew Rollercoaster
11. Bermuda Triangle
12. Pirate Ship
13. Sea World Monorail

ANTARCTICA

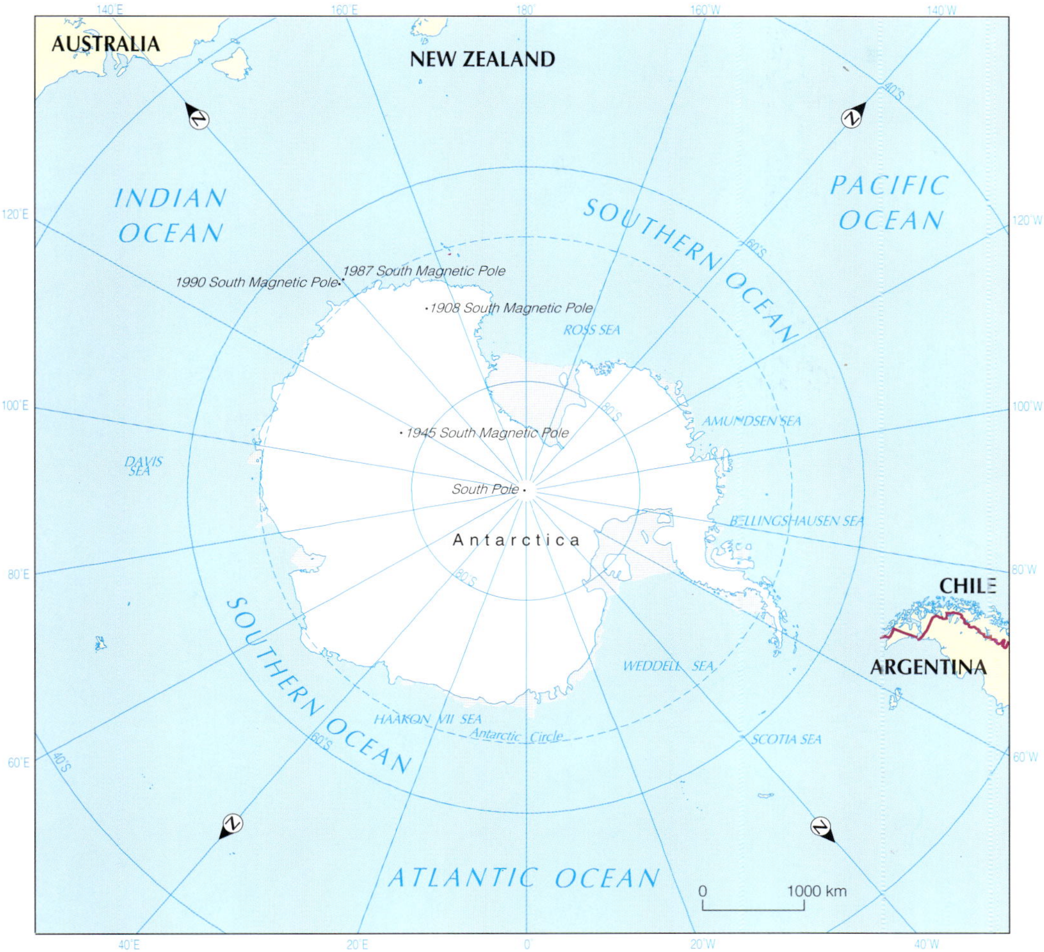

ACTIVITIES

1. What are the cardinal points on a compass?
2. What is north's compass bearing?
3. What are the bearings of the other cardinal points?
4. In which direction would you travel to reach the following countries from the south pole?
   - Australia
   - Chile
   - New Zealand
5. Why do you think there are four north points shown on the map of Antarctica?
6. How many different locations of the south magnetic pole are shown?
7. What is the difference between the south magnetic pole and the geographical south pole?
8. Complete the following sentences using compass directions:
   (a) The Wondrous World of Dolphins is located to the ________ of the Sea Lions at Play attraction.
   (b) The Bermuda Triangle is found to the ________ of the Thrill Seeker Rollercoaster.
   (c) The water slides and pool are located in the ________ corner of Sea World.
9. Find the entrance to Sea World. Once inside, what direction would you walk to visit the World of the Sea Aquarium?
10. In which general direction is the monorail train on the diagram travelling?
11. Find the steam train in the north-east corner of the diagram. Use directions to describe its journey back to the Grand Central Train Station.
12. Look at the map on page 111. In which direction did Caesar travel to invade Britain?
13. Look at the satellite image on page 131. In which direction would you be travelling if you drove
    (a) across the Golden Gate Bridge into San Francisco?
    (b) across the bridge from San Francisco to Oakland?

8 UNDERSTANDING SCALE

Aerial photograph of Flemington Racecourse, Melbourne, where the Melbourne Cup is run each November

An aerial photograph or a map of Flemington Racecourse is a scaled down version of the real thing, just as model cars are scale models of real cars and kindergarten chairs and tables are scale models of adult furniture.

Using scale

The scale on an aerial photograph or a map tells us how much the real area has been scaled down to fit on the map. **Large-scale** maps show detailed information about a small area. A map of your bedroom is a large-scale map and it shows a lot of detail. **Small-scale** maps show broad patterns and large areas such as global weather patterns.

There are three basic ways of showing scale on a map.

1. Scale statement or written scale

This statement says how much one centimetre on the map represents on the ground, usually in kilometres, for example, one centimetre represents one kilometre. Sometimes a map may say "one centimetre to five kilometres", but it should never say "one centimetre equals five kilometres" or "one centimetre measures five kilometres" as these statements are not true.

2. Line scale or linear scale

A line scale is a numbered line showing how many metres or kilometres are represented by a given distance on the map. This is a visual way of showing scale and is easy to understand and to use. On the large aerial photograph of Flemington Racecourse, the line shows that 1 centimetre on the photograph represents 80 metres on the ground.

3. Representative fraction (R.F.) or ratio

This is a way of showing scale as numbers: 1:1000 or 1/1000 both mean that one unit on the map (mm, cm) represents 1000 of the same units on the ground. For example 1:100 000 means that one centimetre on the map represents one hundred thousand centimetres, or one kilometre on the ground. When a map uses this type of scale, convert the ratio into words to make it clearer.

Using scale to measure distances

The easiest way to calculate distance on a map is to measure the distance between the two points in centimetres. For example, if you measure 10 centimetres between town A and town B and one centimetre on this map represents 5 kilometres on the ground, the distance between towns A and B is 50 kilometres (10 × 5).

Where a line scale is supplied, place the edge of a piece of paper along the distance to be measured. Mark in the starting and finishing points on the paper. Hold your piece of paper against the linear scale to work out the actual distance between the points.

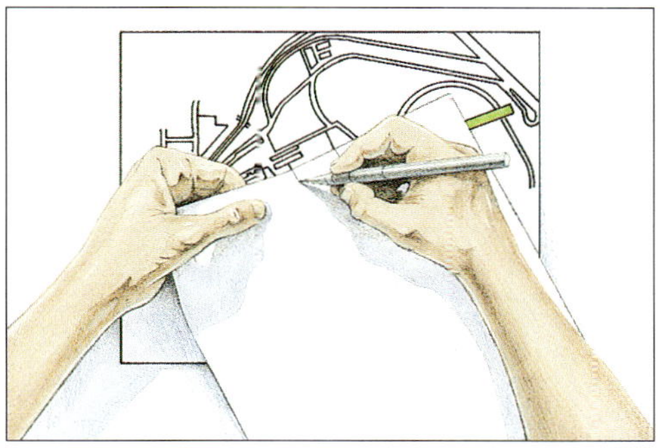

Measuring straight distances with a scale

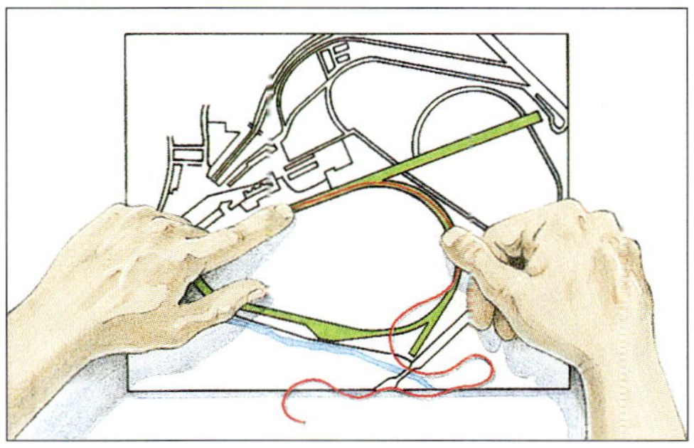

Measuring curved distances with a scale

Sometimes you may need to measure a distance which is not straight, for example a road or river. In this case you can use a piece of string or cotton to follow the shape of the feature. When you have completed the measurement, straighten the string along a ruler or line scale to determine the distance. Finally, convert the distance using the scale to find out the actual distance on the ground.

ACTIVITIES

1. Express the scale of the aerial photograph of the Flemington Racecourse
   (a) in words
   (b) as a ratio.
2. Show the following written scales as line scales.
   (a) One centimetre on the map represents five kilometres on the ground.
   (b) One millimetre on the map represents ten kilometres on the ground.
   (c) Ten centimetres on the map represents one hundred kilometres on the ground.
3. Horse races at Flemington are run in an anticlockwise direction around the outer green track. 1200-metre races are run in a straight line. Longer races involve curves. Use a piece of string and a ruler to work out where the following races start (work backwards from the finish line) and mark them on a sketch map of the Flemington Racecourse.
   1200 m 1600 m 2000 m
   2400 m 2600 m 3200 m
4. Why do jockeys usually try to keep horses close to the inside rail of the track and not the outside rail?
5. Check the following maps and images in the atlas and write each scale as a statement, a line scale and a ratio.
   (a) North west Europe, page 115
   (b) Ancient Sumer, page 109
   (c) Mt Vesuvius, page 113.
6. Look at page 127. How far did refugees living in Kigali travel to reach Goma?
7. What is the length of the transect line shown on page 123? If you walked from point A to point B, what is the highest point you would climb to? What type of farming takes place there?
8. Use the map of Australia on page 38 to answer the following questions. How far is it in a straight line from:
   (a) Brisbane to Perth
   (b) Sydney to Melbourne
   (c) Adelaide to Darwin?

10 DIFFERENT TYPES OF MAPS

Maps are designed for many different purposes. They are an important part of many jobs, including surveying, meteorology, archaeology, geology, building, architecture and engineering. Many of us use maps in our daily lives without actually realising it.

The types of maps shown on these pages represent a range of maps that you are likely to meet at school and in the community.

Political maps

These are the maps in an atlas (or on a wall map) that show different political regions in different colours and their borders. The most common political map is a world map (see pages 146–47). Another example is the Australian political map on page 38, which divides the country of Australia into political units (States).

Physical maps

The physical landforms of a region are shown on these maps. Political boundaries are often not shown on physical maps, which show patterns of mountain ranges, rivers and other physical features.

Variations of colour and shading are used to show height on physical maps, such as the gradation of shades of brown used in this atlas. A map showing the world's physical features is shown on pages 148–49.

Topographic maps

"Topo" maps show relief or height using contours. They also show physical features such as forests, rivers and lakes and cultural features such as roads, railways and settlements. This type of map is useful when bushwalking, planning roads and checking the steepness of slopes. Pages 16 and 17 look at topographic maps in greater detail.

Mud maps

Also known as mental maps or sketch maps, these are rough maps often drawn quickly to explain how to find a particular feature. They are often sketched on a piece of paper or drawn in the dirt (hence the name mud map). This type of map does not show correct scale. Mud maps typically show the streets to take to get to a particular location. Some companies use mud maps in advertisements to show where they are located.

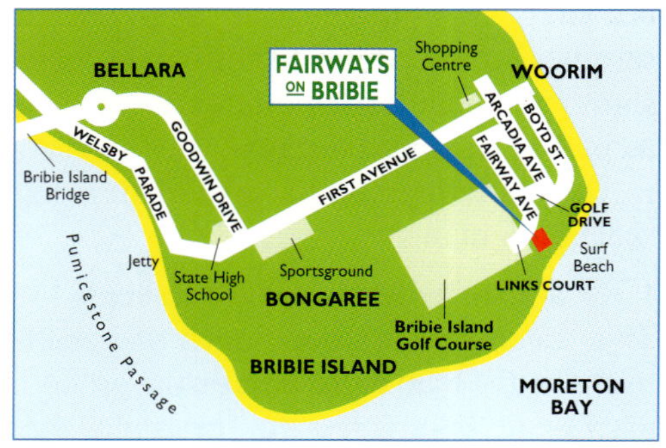

Finding your way with a mud map

Tourist maps

These maps show settlements, tourist attractions, road distances, street names and often accommodation and service stations. Tourist maps are not as detailed as maps in a street directory, but they serve a similar function for people travelling long distances.

Tourist map of Carnarvon district, W.A.

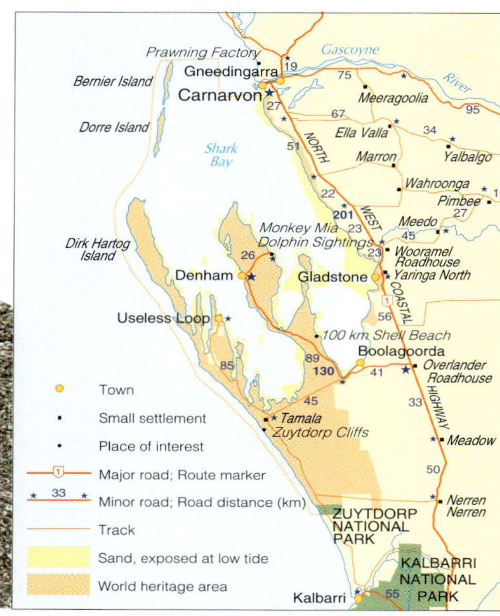

Choropleth maps

Choropleth maps use dark and light shades of the same colour to show a pattern. The darker shades show "the most" and the lighter shades show "the least". A range of choropleth maps can be found for each capital city of Australia, for example, Sydney on page 53.

Satellite images

These are not strictly maps, but they serve the same purpose. False colours are used to highlight particular features on the images relayed to earth from satellites. Satellite images are particularly useful for showing surface temperatures, cloud patterns, smoke and ash from bushfires or volcanoes, large-scale flooding, vegetation patterns and geological and settlement patterns. Pages 20–21 look at satellite images in greater detail.

Theme maps

A wide range of themes can be shown in map form. For example, historical sites or events, invasions and empires are themes that can be explored using maps. Vegetation maps, Olympic sites and resources maps are all examples of theme maps. A map of World Heritage sites appears on pages 170–71.

Synoptic charts

Synoptic charts, or **weather maps**, are seen daily on our television screens and in our newspapers. These maps show the movement of weather systems (pressure cells, fronts, winds). They show patterns of temperature, rainfall and wind. They are also used to predict rainfall and temperature for the near future by tracking approaching weather systems. Page 162 explains how to read a synoptic chart.

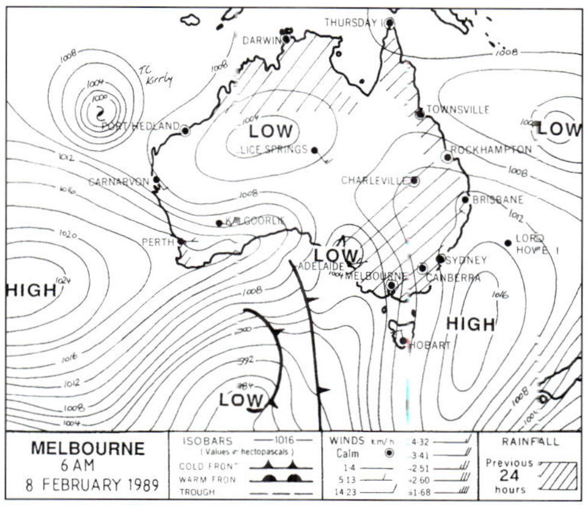

A synoptic chart can be found in the daily paper

Flow maps

These maps show the movement of things like traffic, trade and people (migration). Usually different-sized lines are used to show amount or volume.

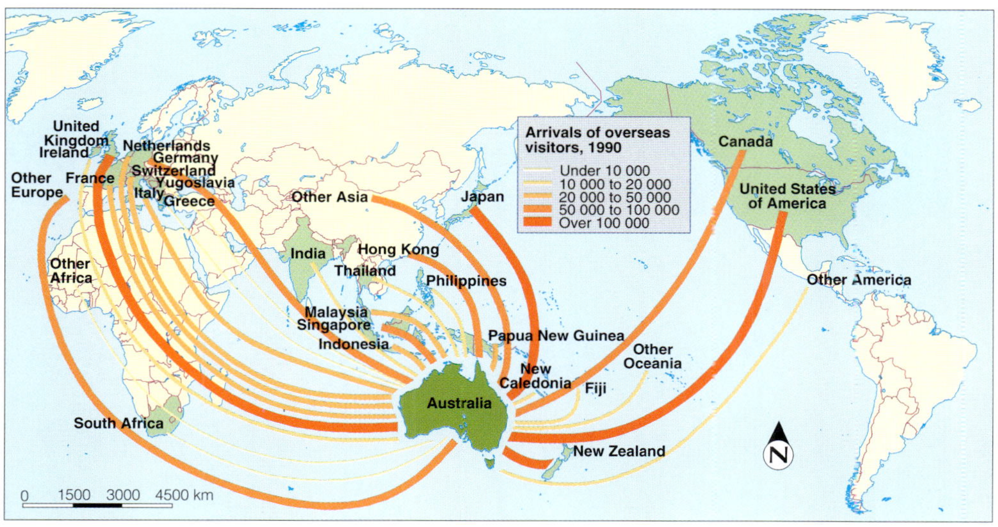

A flow map showing visitors to Australia, 1990

ACTIVITIES

1. Which type of map would be the best for each of the following activities?
   (a) Planning a trip from Melbourne to Brisbane
   (b) Checking the size of the Roman empire
   (c) Planning a hiking trip through bushland
   (d) Checking the path of a cyclone moving towards your State
   (e) Describing the amount of trade between Australia and other countries
   (f) Searching for quartz outcrops to go gold prospecting
   (g) Researching electoral boundaries
   (h) Describing to a friend how to find your house
2. Which types of maps do you use most often, and why?
3. Look at the mud map of Bribie Island. Which features are shown on this map to attract people to live on Bribie Island? List three features that you would not show on this type of advertisement.
4. Look at the tourist map of the Carnarvon area of Western Australia. What is the distance by road between:
   (a) Ella Valla and Yalbalgo
   (b) Carnarvon and Overlander Roadhouse
   (c) Carnarvon and Yaringa North?
5. What features other than road distances suggest the map has been designed for tourists?
6. Can the physical map on page 98 also be described as a choropleth map? Explain.
7. Find three examples of theme maps in the atlas.
8. The lines on a synoptic chart join places of equal air pressure. What do the lines on a topographic map join?
9. Look at the flow map. Where do most overseas tourists come from?
10. What other patterns could be shown as a flow map?

12 LANGUAGE OF MAPS

Archipelago — a group of islands made up of the crests of an underwater mountain range
Atoll — a coral reef in the shape of a ring or horseshoe that encloses a lagoon
Bay — a sheltered part of a sea or lake formed by a curve in its shore. A bay is smaller than a gulf.
Beach — the sandy or pebbly land at the edge of a sea, lake or river
Breakwater — a structure built to break the force of the waves or stop the erosion of a beach
Canal — a waterway constructed by people for transport or irrigation
Canyon — a deep, narrow valley with steep sides. A canyon is longer and wider than a gorge.
Cape — land which extends into the ocean, river, or lake; larger than a point.
Cirque — a bowl-shaped depression formed by snow and ice. It is found on mountains in glaciated areas.
Cliff — a high, steep, rocky slope formed by erosion or earth movements
Delta — a flat area of deposited material at the mouth of a river which divides the river into small streams
Desert — an arid area of land that receives less than 250 millimetres of rainfall each year
Drainage basin — the area drained by a river and its tributaries
Escarpment — an extended ridge of high land often with a steep slope on one side and a gentle slope on the other
Glacier — a large mass of ice which moves slowly through a valley
Gulf — a sheltered part of a sea or lake formed by a large curve in its shore. A gulf is larger than a bay.
Harbour — a body of water sheltered by natural or artificial barriers and deep enough to moor ships
Hill — a raised part of the earth's surface. Hills are smaller than mountains.
Iceberg — a large floating piece of ice that has broken away from an ice shelf or a glacier

Island — a piece of land completely surrounded by water
Isthmus — a narrow strip of land with water on both sides that connects two larger pieces of land
Lagoon — a shallow area of water, partially or completely separated from the sea by a bar, spit or reef
Lake — a large area of water surrounded by land. In arid areas the water in the lake can completely evaporate to form a salt lake.
Meander — the loops in a river caused by fast water eroding the outside of the bend and slower water depositing material on the inside of the bend
Mesa — a flat-topped hill with steep sides found in deserts
Mountain — land that is much higher than its surroundings; larger than a hill
Mountain range — most mountains are arranged in groups known as ranges
Oasis — an area of the desert with ground water at or near the surface

Ocean — large areas of salt water which surround the continents
Peninsula — a piece of land almost surrounded by water but still connected to the mainland
Plateau — an extensive area of flat land which is higher than the land around it
Point — land which extends into the ocean, river, or lake. A point is smaller than a cape.
Reef — an offshore ridge of rocks, coral or sand
River — a stream of fresh water which flows in a natural channel into a larger river, lake, or sea
Sound — a long, wide ocean inlet that normally runs parallel to the coast
Spit — a narrow deposit of sand and other material which extends into a body of water such as the sea, a lake or river

Strait — a narrow passage of water connecting two larger bodies of water
Tributary — a stream or river which contributes to another (usually larger) river
Valley — low land between hills or mountains that usually has a river flowing through it
Volcano — a mountain with a crater at the top which throws out molten rock, steam and ashes when active
Waterfall — a steep fall of water from a height which cascades over hard rock onto a section of eroded land below

Watershed — high land between two drainage basins. Water flows in different directions on either side.

ACTIVITIES

1. Name five coastal landforms.
2. Name five terms associated with rivers.
3. What do the terms in each of the following groups have in common?
   (a) iceberg, glacier, cirque
   (b) drainage basin, watershed, river, tributary
   (c) gulf, bay, lagoon, sound
   (d) escarpment, plateau, mesa
4. Complete the following paragraph using terms from the diagram.
   When rain falls high in the __________ it forms small streams known as __________ , which in turn flow into larger __________ . These __________ move quickly on the steep slopes, and erode the land to form __________ . When the __________ reaches the flatter land it begins to __________ from side to side. In its final stages the __________ slows even further and deposits its load near the mouth to form a __________ .
5. Match the following features (using the page and grid reference) with the type of landform.

Christmas (**84 F4**)	Mountains
Goodenough (**94 G2**)	Archipelago
Tien (**98 E3**)	Sea
Mergui (**100 A4**)	Plateau
Arabian (**106 A3**)	Island
Baltic (**114 H4**)	Cape
Central Siberian (**119 J5**)	Peninsula
Kalahari (**126 B2**)	Desert

6. Choose three landform features that you would expect to find in the following places:
   (a) Antarctica
   (b) Central Australia.
7. Look at the Antarctic cross-section on page 94. What two ice features cover the landforms below?
8. Use the Geo Facts information on pages 38, 98, 112, 122 and 130 to prepare pictographs (see page 33) of:
   • the highest mountains
   • the longest rivers
   in each continent.

14 DRAWING CONTOURS AND

Contours

Contour lines are one way you can show height on a flat piece of paper. They are numbered lines which join places of equal height above sea level on a map. Contours show the height, shape and location of hills and mountains. The difference in height between one contour line and the next is always the same. This is known as the **vertical interval** or **contour interval**.

On any contour map, evenly spaced contours mean an even slope, close contours mean a steeper slope and widely spaced contours mean a gentle slope or flat area.

Spot heights are also given to show the actual height of particular locations on a map. The highest points of a region are usually shown with spot heights.

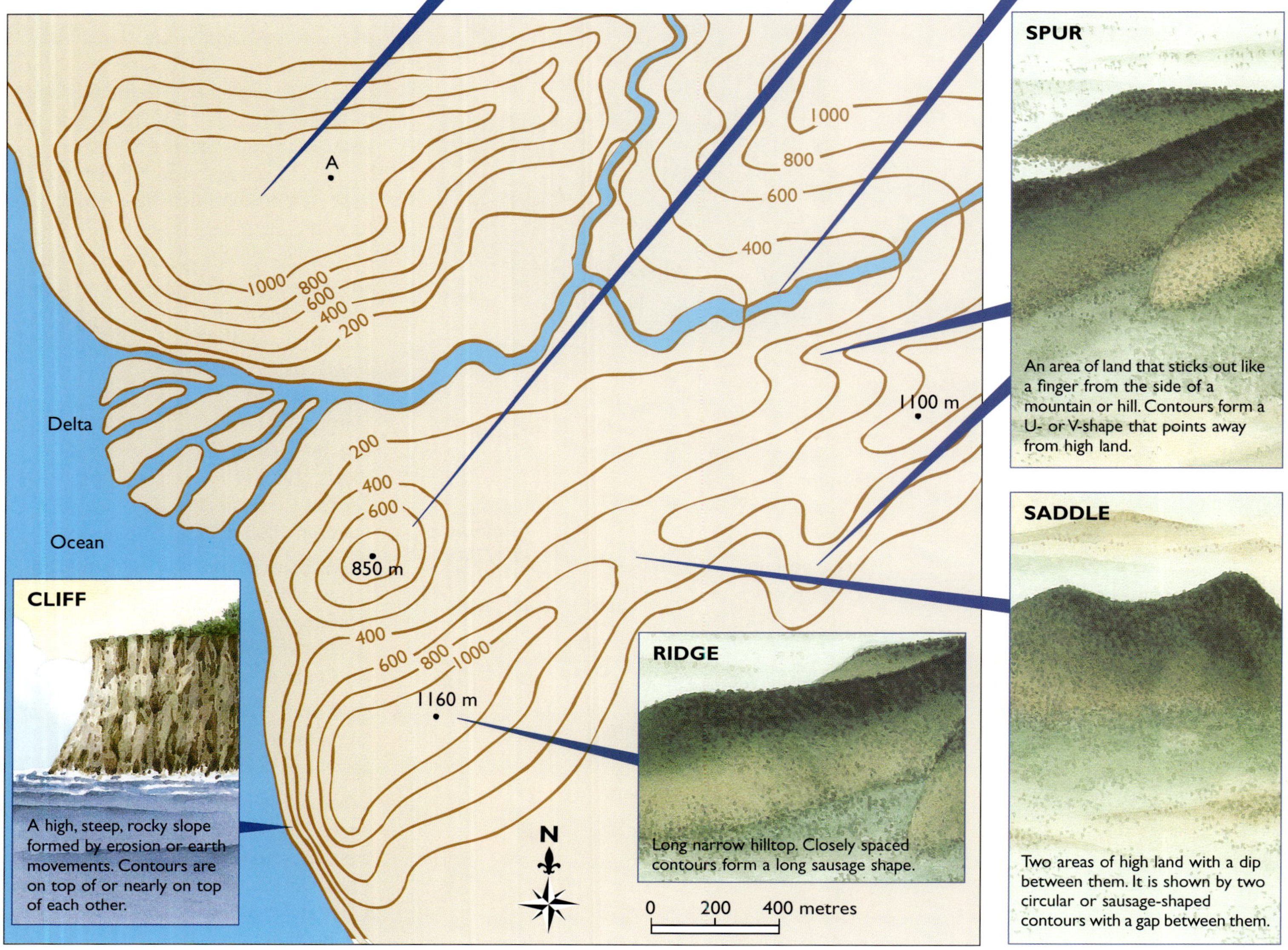

CROSS-SECTIONS

Cross-sections

A cross-section shows the shape of a feature from the side, as if it had been sliced with a knife. Contour lines are used to construct a cross-section.

Cross-section of Uluru from point A to point B

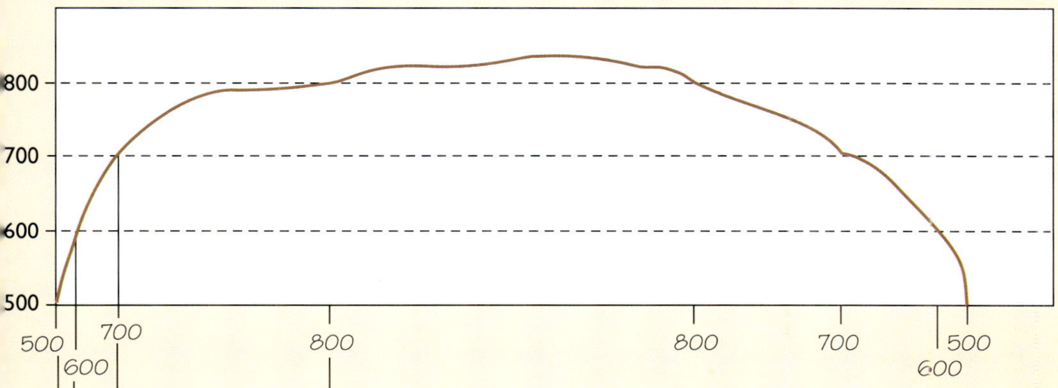

Horizontal scale = 1 cm : 285 metres
This is the scale which is shown on the map or aerial photograph.

Vertical scale = 1 cm : 100 metres
This is the scale you have chosen for your cross-section.

Vertical exaggeration = 2.85
This is the number of times you are exaggerating the horizontal scale. No exaggeration would have Uluru looking as flat as a pancake. Too much would make it look like Mount Everest.

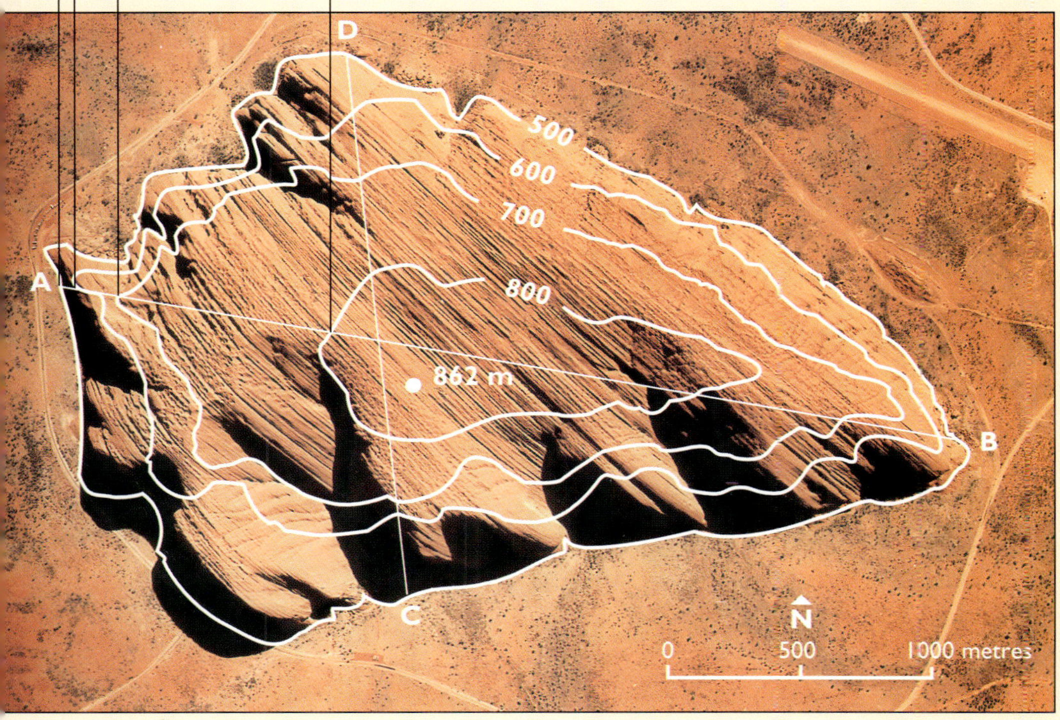

ACTIVITIES

1. What vertical interval is used on the contour map?
2. How high is point A on the plateau?
3. In what direction does the river flow?
4. Which type of land feature is associated with the highest point on the map?
5. Dangerous rocks are found offshore. Where would you place a lighthouse to warn passing ships? Why did you select this location?
6. What is the association between the contour shapes of spurs and valleys? Draw a diagram to explain your answer.
7. What is the height of the land at points A, B, C and D on the contour map of Uluru?
8. Uluru rises 348 metres above the surrounding plain. Why is the highest point on the map 862 metres?
9. Which side of Uluru is the steepest? How can you tell?
10. To draw your own cross-section of Uluru from points C to D, follow these instructions.
    (a) Place the straight edge of a piece of clean paper along the line between points C and D. Mark each contour line at the point it touches the edge of your paper. Against each stroke indicate the height of the contour line it stands for.
    (b) On a sheet of graph paper or blank paper draw a suitable vertical scale for your cross-section. In this case use the same vertical scale as the example, 1 centimetre represents 100 metres.
    (c) Place the marked edge of the paper at the base of the vertical scale, with the point "C" directly on the vertical scale, and mark the appropriate height directly above the stroke with a dot. Join the dots with a smooth line.
    (d) Add a title and a statement of vertical and horizontal scale.

16 USING TOPOGRAPHIC MAPS

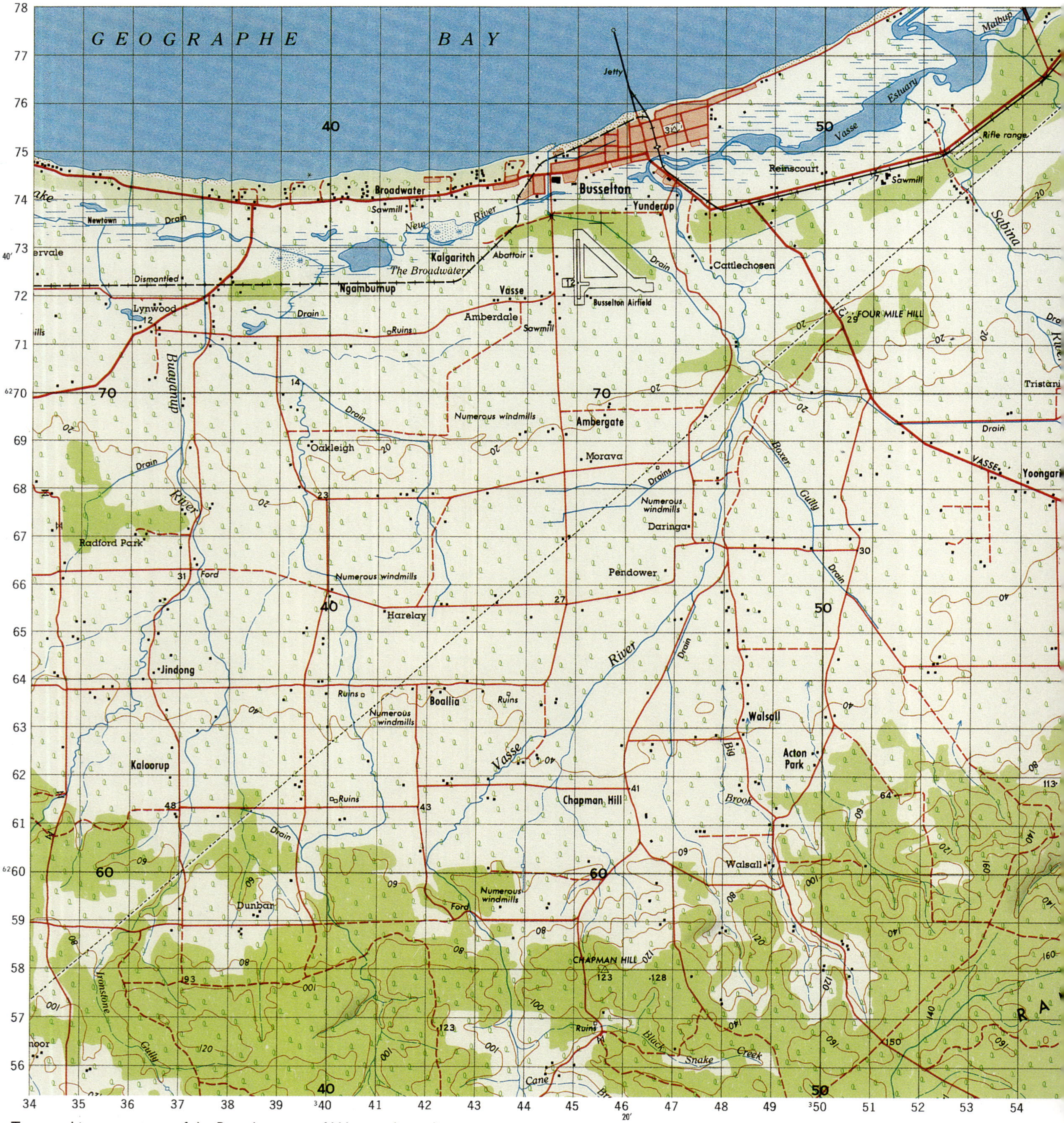

Topographic map extract of the Busselton area of Western Australia

Topographic maps show **relief** or height using contours. They also show **physical features** such as forests, rivers and lakes and **cultural features** such as roads, railways and settlements. This type of map is useful when bushwalking, planning roads and checking slope steepness. The key on topographic maps uses conventional signs to show the main features in the landscape.

Area and grid references

A one-kilometre-square grid is overprinted on a topographic map. These grid lines are numbered with two-digit numbers on the map's margins. Lines which run up and down the map (north/south) are called **eastings**. The numbers increase as you move east. Lines which run across the map (east/west) are called **northings**. The numbers increase as you move north. In a grid reference the eastings are given first, then the northings.

Four-figure area references tell you the grid square in which to find a feature. The letters AR are placed in front of a four-figure reference to show area reference, for example AR4569 contains the town of Ambergate. **Six-figure grid references** pinpoint an exact spot in the grid square. The third and sixth figures represent one tenth of the distance between two grid numbers. These divisions are not written on the map, so they must be estimated. The letters GR are used in front of a six-figure grid reference. There are no spaces between the digits in 4- or 6-figure references, for example Ambergate's grid reference is GR456695.

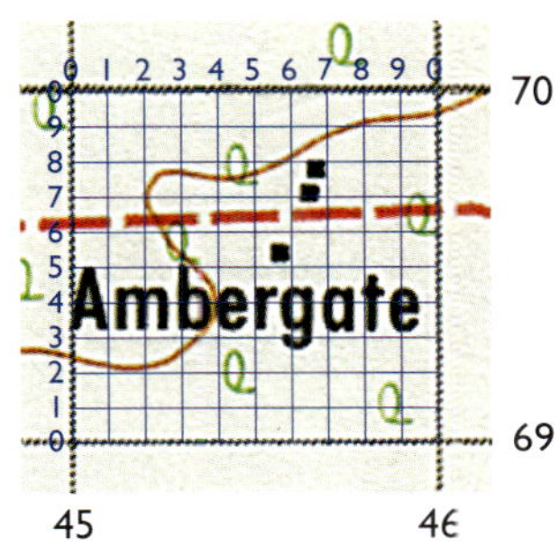

ACTIVITIES

1. What is the highest point on the map? Describe its location using an area reference.
2. What type of vegetation is found on the higher slopes?
3. Complete this description.
   The Vasse River begins in the Whicher Range. It travels in a __________ easterly direction, supplying farms with water from a system of __________. At Busselton the river widens into an __________ before flowing out to sea .
4. Use the key to help you discover what features are found at the following grid references.
   GR457775 GR376644
   GR407728 GR450720
5. Give grid references for the following settlements.
   Yunderup Vasse
   Jindong Walsall
6. How do you think the land on the map is used? What clues are on the map?

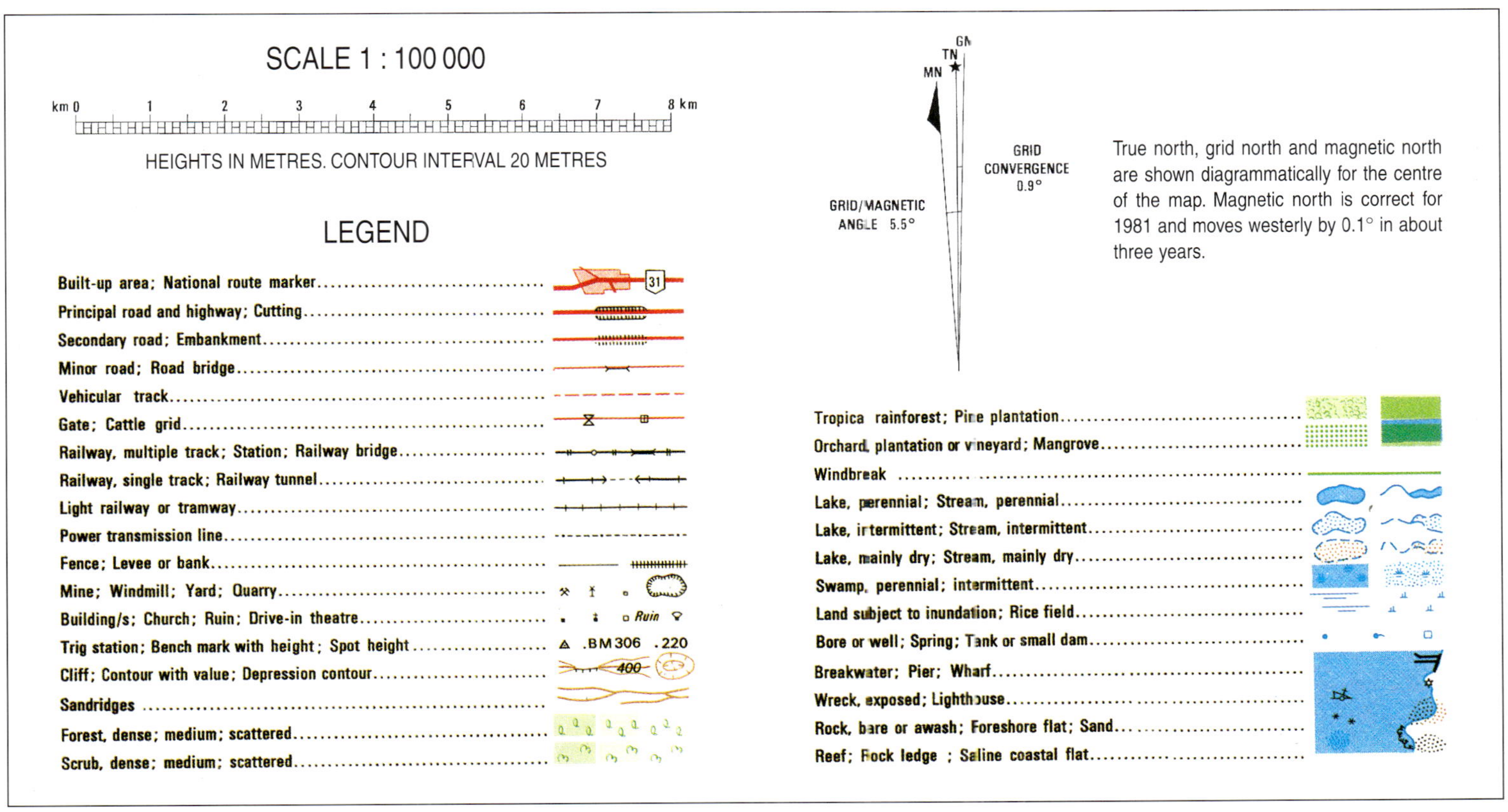

Key to topographic map extract

18 AERIAL PHOTOGRAPHS

Photographs taken from aeroplanes are a key part of map making. **Vertical** or **plan view aerial photographs** are taken with the camera lens pointing directly down. Photographs taken in this way give you a good idea of shapes and patterns that are commonly found on maps.

Oblique aerial photographs are taken with the camera lens at an oblique angle to the ground. They look like the view from a plane window. These photographs have a foreground in which features appear large, and a background where they appear small. There is no one correct scale on oblique aerial photographs.

Interpreting aerial photographs

- Check whether photograph is vertical or oblique. If vertical, you can use the scale for the entire photograph.
- Check time of year (season) and day (time).
- Identify major patterns — vegetation, landforms, drainage, transport, settlement.
- Identify main features — rivers, roads, forests, buildings, airstrips.
- Identify areas of human activity — agriculture, industry.
- Try to explain relationships between features — for example rivers and trees.
- Do landforms influence vegetation, transport, agriculture or settlement?
- Where do humans have an impact on the environment?

Change over time

Aerial photographs can also be used to show change over time. The development of Hamilton Island, part of the Whitsunday Group off Queensland, is clearly shown in the two photographs on page 19. The island's grazing land was transformed in the early 1980s into one of the most elaborate island resorts in Australia.

Vertical aerial photograph of Sydney

Oblique aerial photograph of Sydney

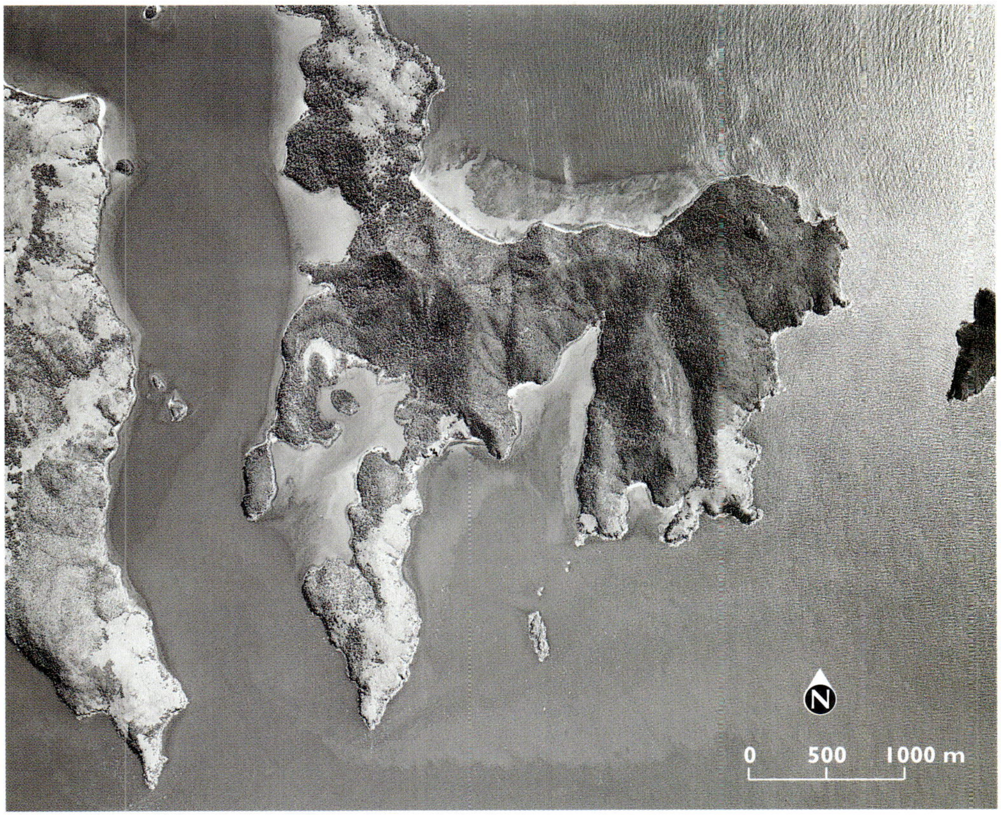

Hamilton Island, 1975

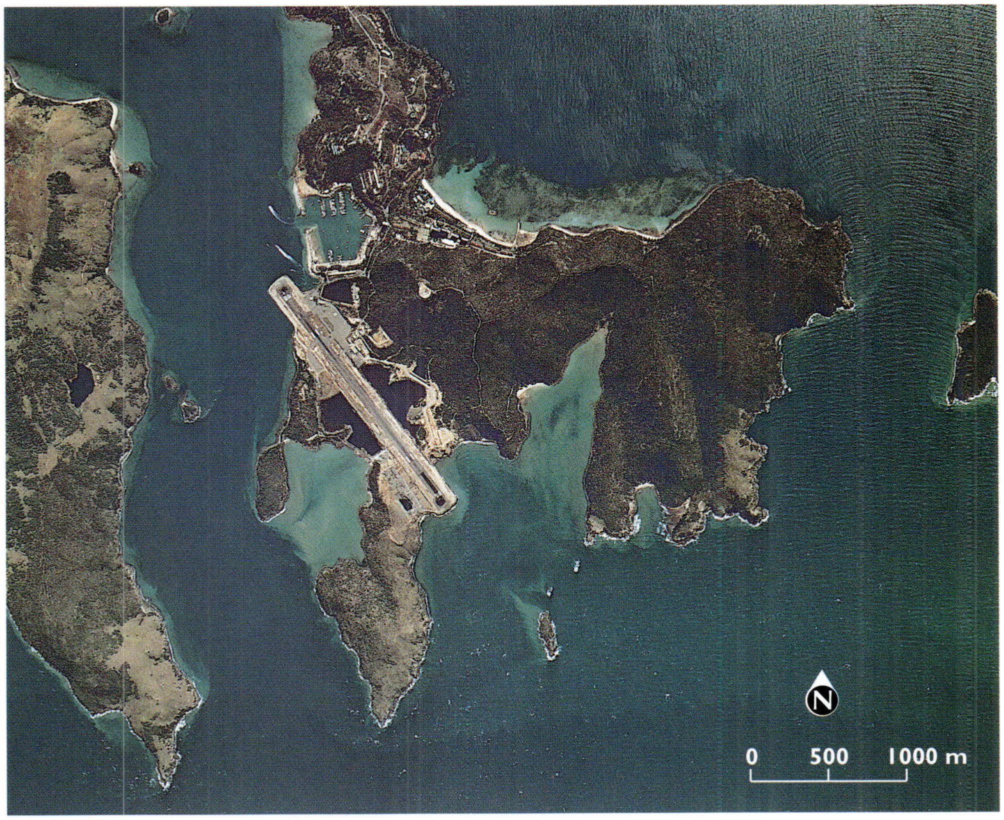

Hamilton Island, 1995

ACTIVITIES

1. Compare the vertical and oblique views of Sydney. Make a list of the advantages and disadvantages of each type of photograph.
2. Make two sketches of the Sydney Harbour Bridge — one from the vertical view and the other from the oblique view. Which sketch looks more like the real thing?
3. Is the city, or Central Business District, easier to recognise in a vertical or oblique aerial view? Why?
4. Which type of aerial photograph shows the entire Central Business District? Why can you see the complete area in these types of aerial photographs?
5. Would a vertical or oblique view be the most use to a cartographer (map maker)? Why?
6. On which side of Hamilton Island has most development taken place? Why do you think this area was chosen?
7. How was Hamilton Island used in 1975?
8. Give two examples of changes to the natural environment as shown in 1975.
9. Which part of Hamilton Island has undergone the least amount of change? Why do you think this is the case? What changes have taken place in this area?
10. Hamilton Island is the only resort in the world with its own jet airport. How long is the runway? What type of land was it built on?
11. Look at the vertical aerial photograph of Adelaide on page 68. What surrounds the Central Business District?
12. Compare the recent aerial view of Adelaide to the original plan for the city. Which features on the original plan have remained?
13. Compare the two oblique views of Brisbane shown on page 44. Describe two features that remain the same, and two features that have changed over time.

20 LANDSAT IMAGES

Satellites

Landsat images are beamed from **satellites** to earth, where they are transformed into pictures. **Satellite images** show a much larger area of the earth's surface than photographs taken from an aircraft. **Landsat imagery** is used by cartographers and geographers to discover broad patterns on the earth's surface.

Landsat images are picked up in Australia by a huge parabolic dish antenna in Alice Springs. Images are digitally recorded on tape and are transferred daily to Belconnen in Canberra for processing. Images are also received and processed in the USA.

Different features can be highlighted in the satellite scenes by selecting only some of the bands of light reflected from earth. Landsat images can give a true or false colour image of an area. Some typical false colour images are:

- dark blue-black — represents water because it absorbs light energy.
- bright red or pink — represents crops and other vegetation, usually in high rainfall or irrigated areas.
- red brown — shows forests.
- bright pale blue or green — shows shallow water.
- grey, grey-blue or purple — represents urban areas.

In the satellite image opposite, the Mississippi River runs from the north-west to the south-east. It is joined by the Missouri River in the north-eastern corner of the image. The pink and green strips of colour show the pattern of farms which rely on the rivers for water. The city of St Louis is found on the Mississippi River in the south of the scene, and its suburbs spread north.

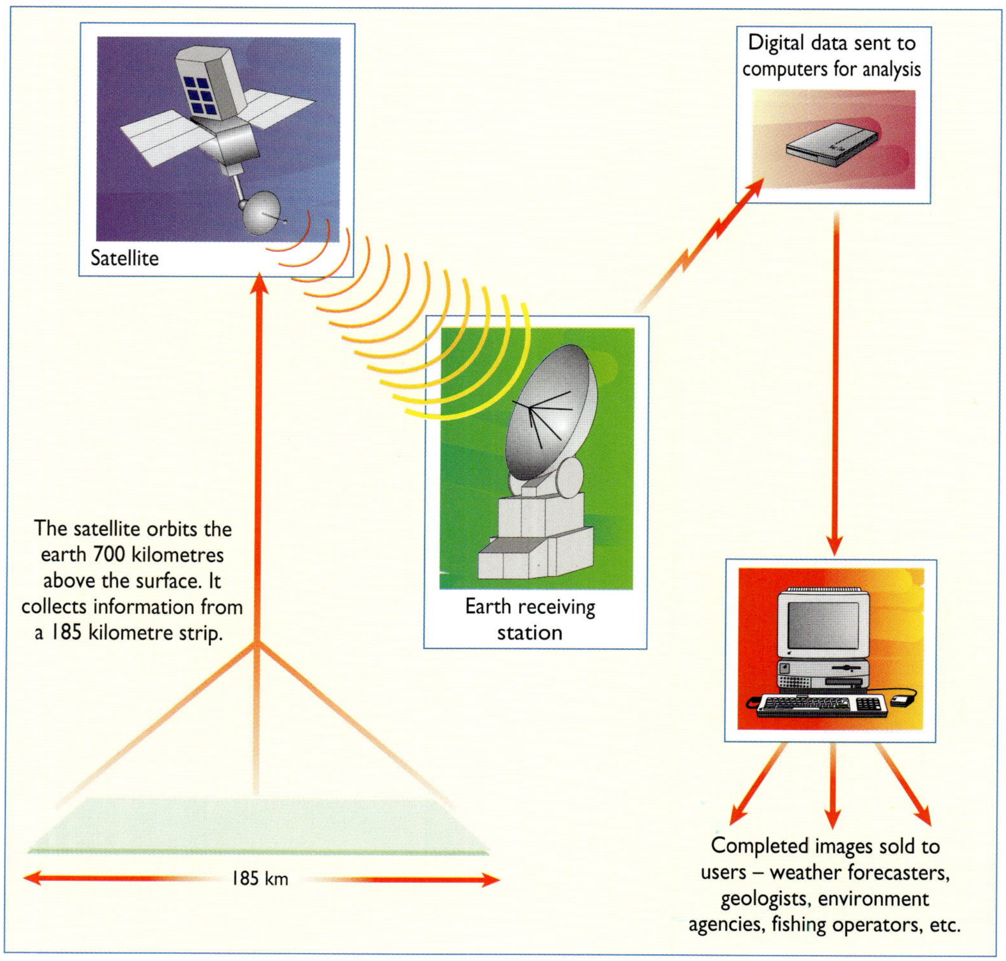

Remote sensing

Satellite image of the Mississippi River delta

ACTIVITIES

1. How do satellites send information to earth?
2. What size area do the satellite images cover?
3. Why are satellite images not called "photographs"?
4. Why do you think "false colours" are used to show features such as crops?
5. Is the satellite image of the Mississippi River a true or false colour image? Give an example to justify your answer.
6. Water generally shows up as black or dark blue. What do you think the straight black feature just south of the Mississippi/Missouri junction is?
7. What features are found on either side of the Mississippi along its path? Why are they located here?
8. Find the following features in the urban area of St Louis. Describe the shape and colour of each.
   - central city
   - suburbs
   - roads
   - golf courses
   - airport
9. Explain the red circles on the Landsat image on page 123.
10. Describe the area of the Amazon rainforest that is being cleared in the satellite image on page 141.
11. What do you think the white colour represents in the Landsat image on page 99? If deep water appears black because it absorbs light energy, suggest why the area shown might appear as white.
12. Compare the St Louis satellite image with the vertical aerial photograph of Sydney on page 18.
    (a) Measure the height and width of each image and convert the measurements to kilometres using the scale. Multiply the width by the height to obtain the area in square kilometres.Which type of image covers the greatest area?
    (b) Does a satellite image or a vertical aerial photograph show features more clearly? Give examples.

22 PRIMARY AND SECONDARY

In May 1915, a torpedo from the German submarine *U-20* sank the British Cunard ocean liner *Lusitania* with the loss of 1195 passengers and crew. Among the dead were 123 Americans. This was the first step in involving the United States of America in World War I. The mystery surrounding the sinking of the *Lusitania* and other important historical events can be explored by studying evidence.

Information about the past can be based on primary or secondary sources of evidence.

Primary sources come from the time the event occurred. A primary source can be a useful eyewitness account of an event, but it may only provide one viewpoint. Examples of written primary sources include diaries, contemporary maps and documents and gravestones. Non-written primary sources such as photographs, clothes, tools, houses, contemporary paintings and spoken sources are also useful tools when reconstructing the past.

Secondary sources are produced after the event they describe. These sources are an interpretation of what happened. Secondary sources such as books, maps, and artists' reconstructions are generally based on many hours of researching primary resources.

Lusitania — Primary sources

Oral evidence

Saw the Submarine 1000 Yards Off and Watched Torpedo as It Struck Ship

Queenstown, Saturday, May 8, 3:18 A.M.

A sharp lookout for submarines was kept aboard the *Lusitania* as she approached the Irish coast, according to Ernest Cowper, a Toronto newspaper man, who was among the survivors landed at Queenstown.

"As we neared the coast of Ireland," said Mr Cowper, "we all joined in the lookout, for a possible attack by a submarine was the sole topic of conversation.

"I was chatting with a friend at the rail about 2 o'clock when suddenly I caught a glimpse of the conning tower of a submarine about a thousand yards distant. I immediately called my friend's attention to it. Immediately we both saw the track of a torpedo followed almost instantly by an explosion. Portions of splintered hull were sent flying into the air, and then another torpedo struck. The ship began to list to starboard.

"The crew at once proceeded to get the passengers into boats in an orderly, prompt, and efficient manner. Miss Helen Smith appealed to me to save her. I placed her in a boat and saw her safely away. I got into one of the last boats to leave."

New York Times, 8 May 1915

Physical evidence

A graveyard service for some of the victims of the *Lusitania*

The rusted remains of part of the promenade deck of the *Lusitania*

Lusitania — Secondary sources

Written sources

The day before the *Lusitania* was due to sail from New York, a submarine slipped out of the port of Emden on Germany's north coast and headed north-west. The figure 20 on the vessel's bow identified her as *U-20*, under the command of Lieutenant Walther Schwieger. *U-20* was embarking on a lengthy patrol, one that would take her right around the British Isles before turning up at her eventual station in the Irish Sea off Liverpool. In all, the patrol would involve a trip of some three thousand miles.

Troop transports were prime targets for U-boats during this period of the war. A huge Allied expeditionary force had been assembled to attack Turkey's Gallipoli peninsula, with the eventual taking of Constantinople (now Istanbul) as its strategic objective. It was a bold plan that, if successful, might knock Turkey out of the war and take some of the military pressure off Russia. The mission of *U-20* and other U-boats was to cut off the supply of troops and material to the Dardanelles.

R. D. Ballard 1995, *Exploring the Lusitania*, New York: Orion Publishing

For years there was speculation that the great Cunarder's mission might not have been as innocent as the British wanted everyone to believe. In defiance of maritime law, some claimed the luxury liner was being employed to carry high explosives purchased in America for use by the Allies at the front. Not only that, it was rumored that a monstrously cynical conspiracy had precipitated the disaster. The British *wanted* the ship sunk — reasoning that the violent and unprovoked deaths of prominent Americans would bring the United States into the war on the Allied side — and Winston Churchill, the wily First Lord of the Admiralty, was behind it.

R. D. Ballard 1995, *Exploring the Lusitania*, New York: Orion Publishing

Map

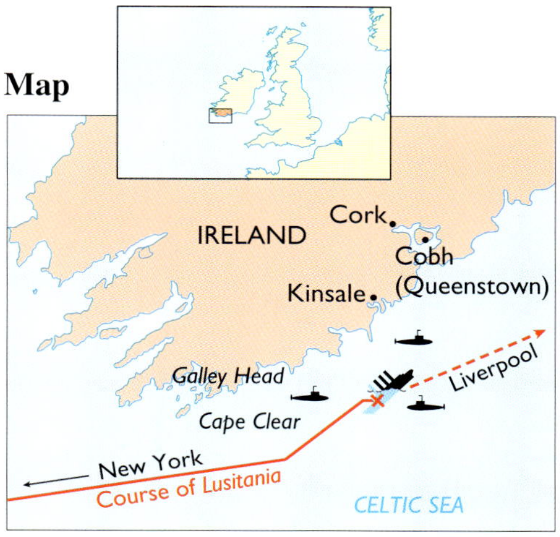

Route of the *Lusitania*

ACTIVITIES

1. Why are the remains of the *Lusitania* a primary source, when the wreck was only recently found?
2. Use the secondary sources to explain Ernest Cowper's primary evidence:
   (a) Why were passengers looking for submarines?
   (b) What does "list to starboard" mean?

Artist's reconstruction

Artist's reconstruction of the sinking of the *Lusitania*

EVALUATING SOURCES

Propaganda

The sinking of the *Lusitania* was used by the Allied forces and Germany in a propaganda war. Propaganda is the deliberate advertising and spreading of ideas which are intended to influence what people think. It often happens during wars and election campaigns. Enemies are made out to be evil and brutal, fears are exploited and truth is often forgotten.

Propaganda is the presentation of only one view, one side of an argument and one philosophy. Other opinions are ignored or forgotten. Propaganda has been used during war to motivate people to make an extra effort.

This Allied recruiting poster uses the sinking of the *Lusitania* to encourage people to increase their war effort.

The medal struck by the Germans to commemorate the sinking of the *Lusitania*. The Germans used it as propaganda against the British, accusing them of using the ship as a "blockade runner". The British made thousands of copies of the medal and reproduced it on posters.

Cartoons

Cartoons can be used to make social or political comment about current issues. They usually are intended to interest, entertain, stimulate discussion and amuse readers, but they are not always funny.

A cartoon published shortly after the sinking of the *Lusitania* portrayed the crew of the *U-20* submarine laughing and pointing at the drowning passengers. In fact, the *U-20* did not surface after the attack.

—*From The Herald, New York*

[The Notice on the Bulletin Board is the German Embassy's advertisement giving warning that travellers who sailed on ships of Great Britain or her Allies entering the War Zone did so at their own risk.]

Cartoonists use the following devices to communicate their message, effectively.

- Simple, clear, stylised images
- Brief, clear, witty captions
- Characters for whom readers can feel emotion, for example sorrow or hate
- Caricatures — figures with facial features or body types exaggerated
- Stereotypes of professions, races and personalities so that readers quickly recognise them, for example cowboys always have six shooters, ten gallon hats and spurs
- Images and symbols which are familiar, for example hearts to show love, doves to show peace.

When looking at a cartoon, you should ask the following questions.

- *What is the main idea being communicated?*
- *Where does the cartoonist appear to stand on this issue?*
- *What are the geographical or historical implications?*

ACTIVITIES

1. Look at the recruitment poster on page 24.
   (a) What devices do you think the artist has used to make the woman look innocent and heroic?
   (b) What can be seen in the background?
   (c) Do you feel this would have inspired the Allies? Why or why not?
2. "The Announcer" is a cartoon which was drawn shortly after the sinking of the *Lusitania*. The Imperial German Embassy had just announced that any British or Allied ship within the war zone would sail at its own risk.
   (a) Look at the cartoon of the German Kaiser (emperor). How do his black hat and cape make him appear?
   (b) What do you think the symbol of the skull and cross-bones with a German helmet on means?
   (c) What emotion do you feel towards the character?

EXAMINING THE EVIDENCE

Historians find out about the past by bringing together a number of pieces of evidence. In attempting to uncover some of the mystery surrounding the sinking of the *Lusitania*, historians have spent countless hours of research time examining primary and secondary evidence. Eighty years after the event, many of the questions still remain unanswered.

- Why was there a second explosion?
- Was the liner carrying munitions for the British?

1 Oliver Bernard watched the torpedo trail come towards the ship, closed his eyes and grabbed the railing. The rail shook as the torpedo hit. "Well, that wasn't too bad", he thought. A few seconds later a second explosion rocked the ship.

2 The red hot boilers may have exploded when they came in contact with the cold seawater. None of the survivors from the three operating boiler rooms reported an explosion.

Robert Ballard's expedition noted coal scattered over the bottom of the sea. Highly combustible coal was carried all along the inside of the hull of the *Lusitania*.

An artist's reconstruction

3 The *U-20*'s torpedo hit just below the *Lusitania*'s bridge. Lieutenant Schwieger recorded in his war diary that it would have been impossible for him to fire a second torpedo at "this crowd of people struggling to save their lives".

5 Deckhand Leslie Morton saw what he thought were two torpedoes. What he probably saw was the trail of a single torpedo and a trail of bubbles as it broke the surface.

4 Underwater explorer Robert Ballard explored the wreck of the *Lusitania*. "We were able to inspect the entire exposed area of the magazine and it was clearly undamaged. If it held munitions, they were not the cause of the secondary explosion."

ACTIVITIES

1. To determine which is the most likely cause of the second explosion, we must look closely at the evidence for each of the following alternatives.
   - There were two torpedoes.
   - The boilers exploded.
   - Munitions which were in the ship exploded.
   - The coal stored in the ship's hull was ignited and then exploded.

   (a) Look at the primary oral sources of Oliver Bernard and Leslie Morton, as well as the oral source of Ernest Cowper on page 22. Now look at the primary source of Lieutenant Schwieger's war diary. Did Leslie Morton see two torpedoes? Did the commander of the *U-20* which sunk the *Lusitania* report that he fired two torpedoes? Do you think this is a likely solution?

   (b) Do the survivors from the boiler rooms state there was an explosion in any of these rooms? Do you think this is a likely solution?

   (c) Look at the secondary source on page 23 which discusses the rumour that the *Lusitania* was carrying munitions on board. Now look at the secondary source of explorer Robert Ballard who inspected the area where these munitions were believed to be kept. Is there evidence to support the theory that the munitions exploded? Do you think this is a likely solution?

   (d) Look at the secondary evidence provided by Robert Ballard concerning the coal found on the seabed. Does this support the theory that the coal which the *Lusitania* was carrying probably ignited causing the second explosion? Do you think this is a likely solution?

   (e) Now that you have looked at the alternatives and the primary and secondary evidence, what do you think is the most likely cause of the second explosion on board the *Lusitania*?

28 CONSTRUCTING TIMELINES

Timelines record important events and give them a "time location". These events may form a timeline of one person's life, or they may be of world importance. Timelines may record long-term events over thousands or millions of years, for example a geological timeline, or may record short-term events, for example events in one day.

Events on a timeline are shown in correct time or chronological order. Accurate timelines have a definite scale where a certain length on the timeline represents a given number of hours, days or years. For example, you can draw a timeline of your own life.

BC and AD

We are close to the year 2000. Most English-speaking countries use the birth of Jesus Christ as their starting point to count modern time. Years before that event are called BC, meaning before Christ and years after are called AD or *anno domini* which means "in the year of our Lord". Sometimes time is marked on timelines as so many years BP or before present.

In recent times the letters BCE, meaning "before the common era", and CE, meaning the "the common era", have been used instead of BC and AD. This form of dating is thought more relevant to our multicultural society.

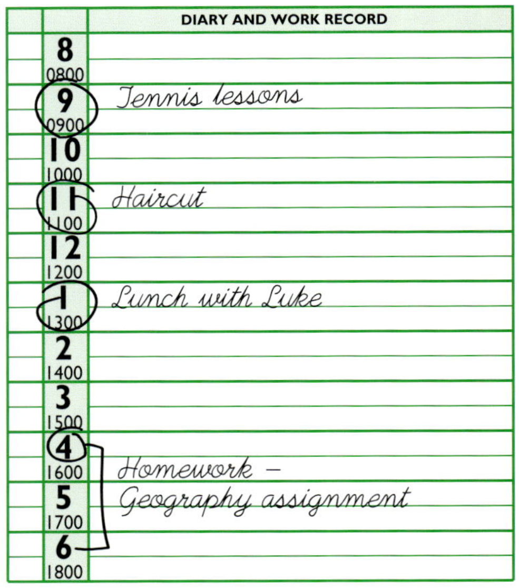

Ancient Rome

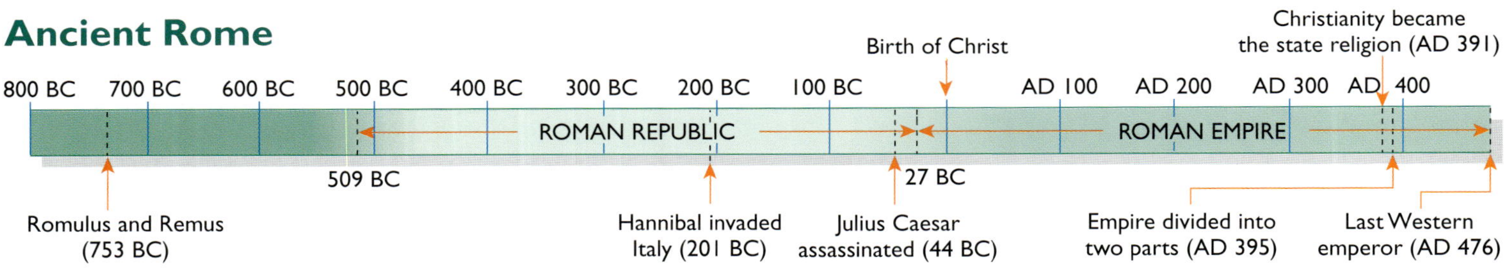

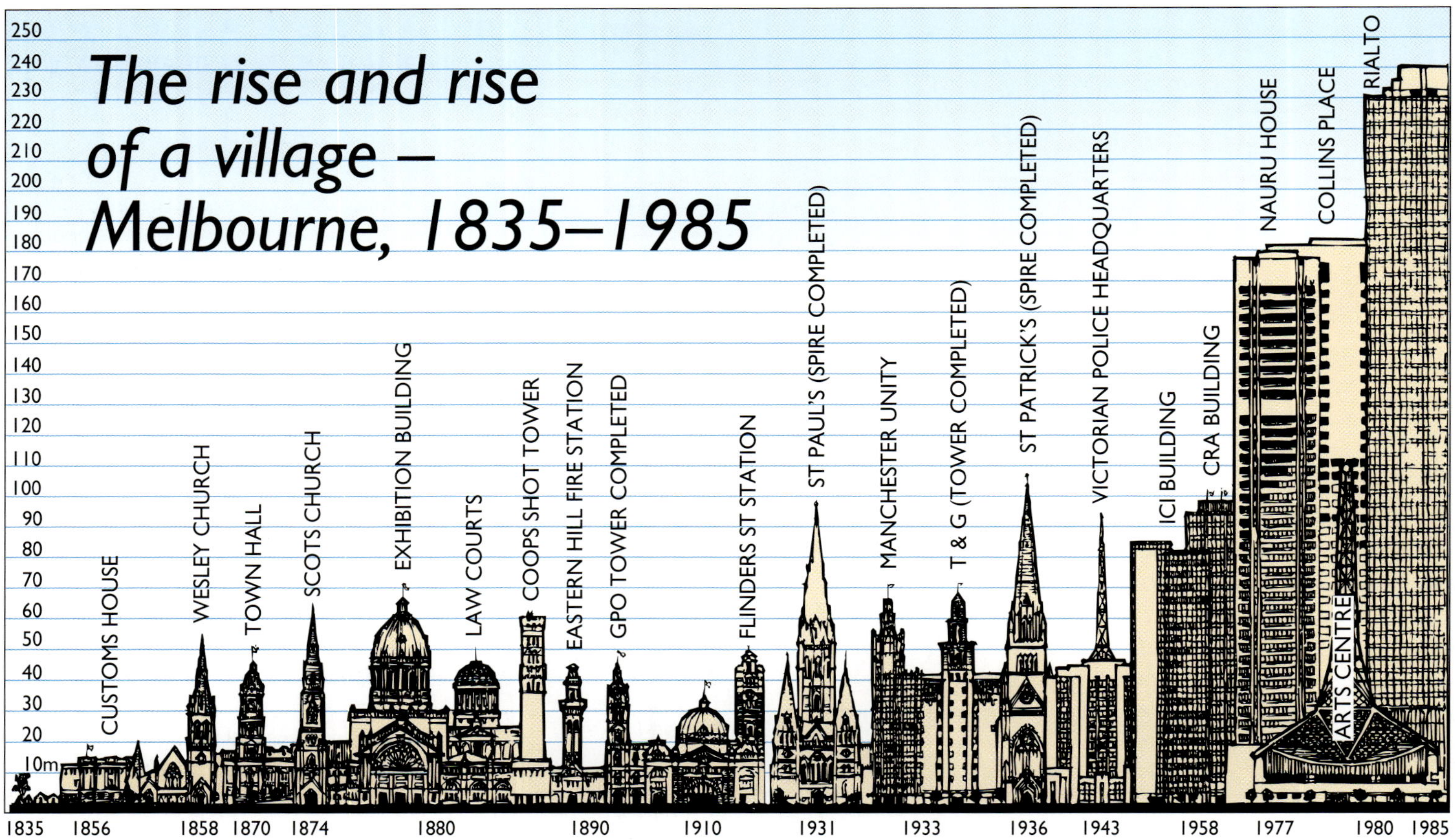

SPACE RACE

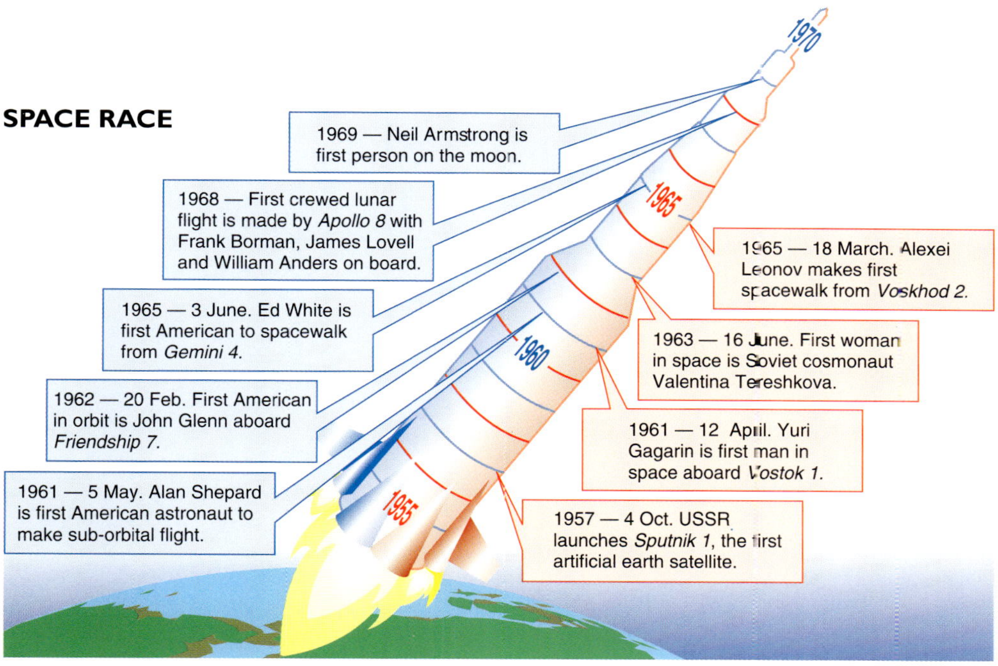

GEOLOGICAL TIMESCALE

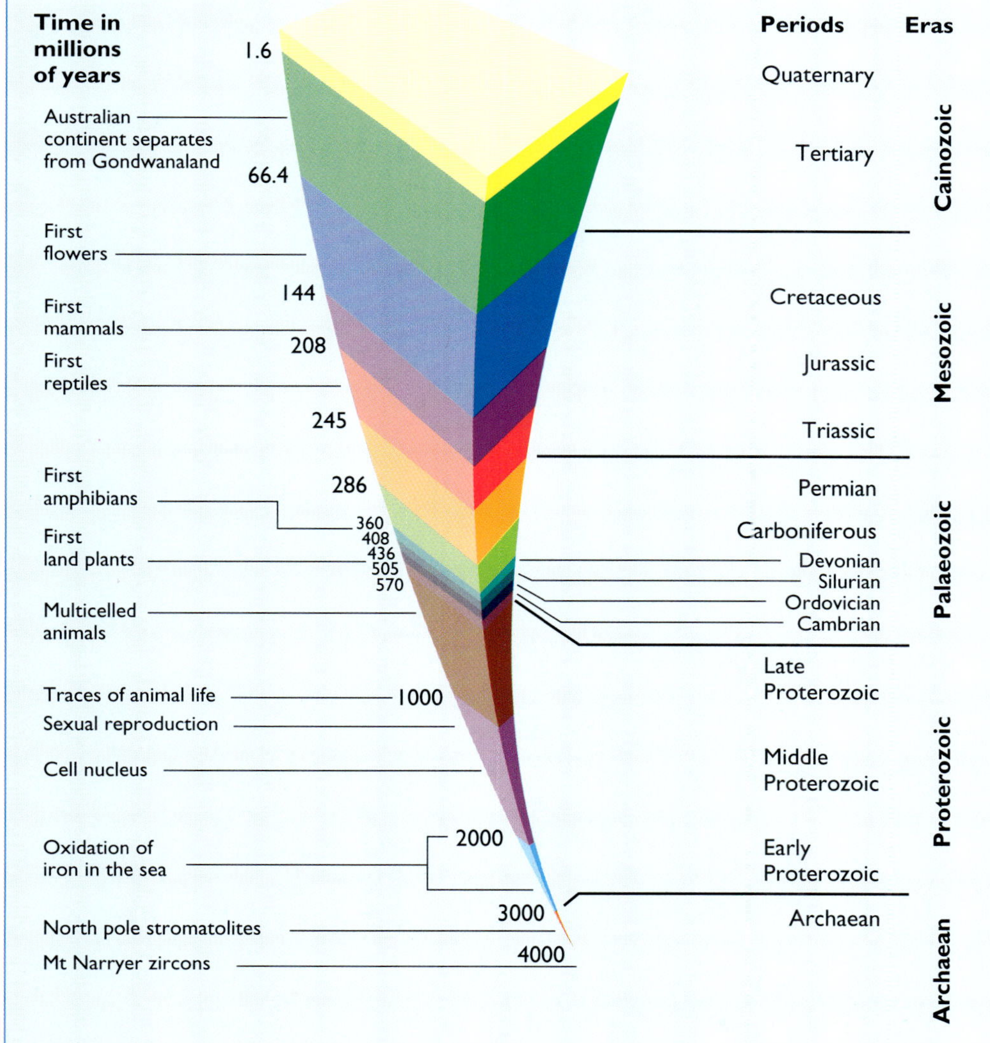

ACTIVITIES

1. Which of the five timelines shown covers the longest period of time? Which covers the shortest?
2. Give an example of an accurate timeline with a scale and of one where the dates are not evenly spaced.
3. How many years were there between the launching of *Sputnik 1* and the landing of the first person on the moon? Do you think there was a "space race" between the former USSR (now C.I.S.) and the United States of America?
4. Construct your own personal timeline by following these steps.
   (a) Collect the important events in your life and put a year beside each event. Examples will include year of birth, changes of address, births/deaths within your family, major holidays. (Try to include about 10 events.)
   (b) Arrange the events in time order (chronological order), from first to last.
   (c) Draw the line to a definite scale and mark on the base years, for example 1 centimetre might represent 1 year.
   (d) Mark the events on the timeline.
   (e) Now mark on the eras or periods of your education, for example at home, kindergarten, pre-school, primary school, secondary school. These eras should not overlap.
   (f) Mark on your membership of groups, clubs and activities, for example, swimming club, judo, cricket, scouts or craft. These will overlap.
   (g) Give your timeline a title.
5. Look at the timeline on ancient Asian cultures on page 97. This shows the Chinese dynasties. Which lasted the longest? How long did it last?
6. Refer to information on pages 168–69. Construct a timeline to show the major developments in the evolution of humans.

30 DRAWING LINE AND BAR

Line graphs

Line graphs show information as a series of points that are joined up to form a line. The line shows a trend or change over time.

Line graphs can be constructed by hand or through a **spreadsheet** program on your computer.

1. Select a variable to plot on the horizontal axis. On many of the line graphs you draw it will be units of time.
2. Select a variable to plot on the vertical axis. This will be the variable that changes over time.
3. Draw the vertical and horizontal axes of your graph, ensuring the scale will help show a clear pattern.
4. Plot a series of dots at the points where the variables meet.
5. Name the vertical and horizontal axes and give the graph a title.
6. Join the dots with a freehand line. Do not use a ruler. Use a dashed line to join points that represent estimates.

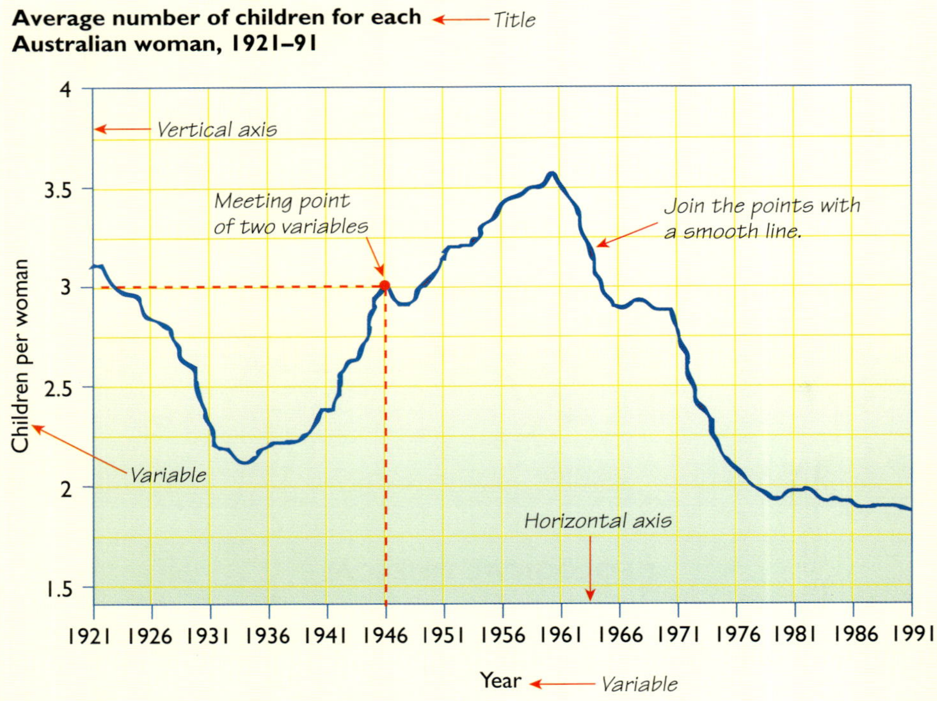

Source: Australian Bureau of Statistics

Bar graphs

Bar graphs present information in the form of a bar or column. They are useful for comparing quantities. Computer spreadsheets give various options to present bar graphs.

To construct your own bar graph follow these steps.

1. Select the variable you are comparing on the horizontal axis, for example year, country.
2. Select the quantity variable to plot on the vertical axis, for example tonnes, people.
3. Decide upon the width and spacing of the bars to be located along the horizontal axis. These should be uniform.
4. Draw the vertical and horizontal axes of your graph, ensuring the scale will help show a clear pattern.
5. Name the vertical and horizontal axes and give the graph a title.
6. Mark the meeting place of each pair of variables with a dot and complete each bar.

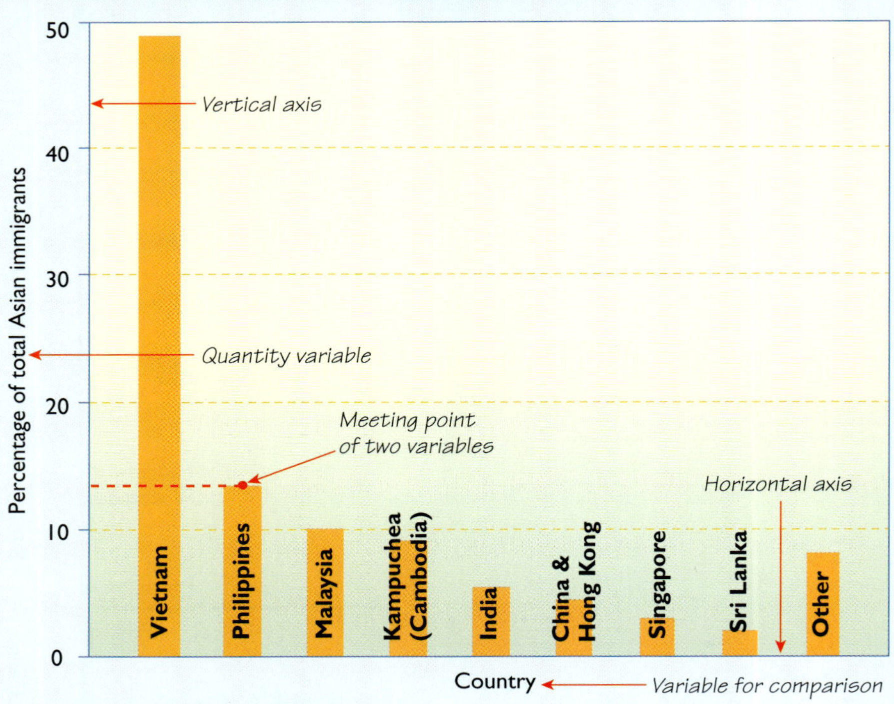

Source: Australian Bureau of Statistics

Climographs

Climographs are a combination of a bar and a line graph used to show the climate of a place over a 12-month period. The line graph section of a climograph always shows average monthly temperature and the bar graph represents average monthly rainfall.

There are a few extra steps required in drawing and reading climographs.

1. Prepare two vertical scales. The right-hand scale represents average rainfall in millimetres. Draw this scale first and use it to prepare the rainfall bar graph. It is useful to colour the bar graph green or blue to remind you that it is rainfall. The left-hand scale represents temperature in degrees Celsius. Use this scale to draw the temperature line graph. A red or orange line will help identify the variable of temperature.
2. Name the vertical and horizontal axes and give the graph a title.
3. Be careful to read from the correct scale when studying climographs.

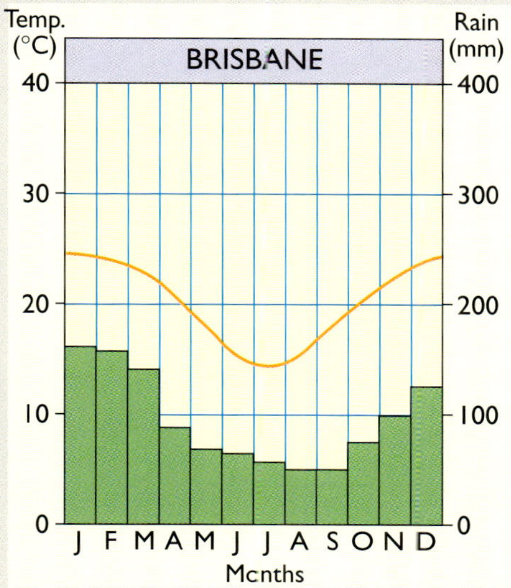

Population pyramids

Population or age/sex pyramids are horizontal bars that show the proportion of population in various age groups in a region. They are simply two bar graphs which are drawn on either side of a vertical axis. Population pyramids are used to show the characteristics of the population of a region. They show the percentage of the total population in each age group and the percentage of total males and total females in that age group.

1. The vertical axis shows the various age groups.
2. The horizontal axis is divided into males on the left-hand side and females on the right-hand side.
3. The scale on the horizontal axis shows the percentage of the total population, beginning in the centre and extending left for males, and to the right for females.
4. Name the vertical and horizontal axes and give the graph a title.
5. When reading population pyramids, first look at the overall shape. Young populations are shown by a wide base which tapers to almost nothing, like a pyramid. Old populations are shaped more like a vase with a small base, larger middle, and small top. Look for any bars that stick out and think about why this section of the population might be different.

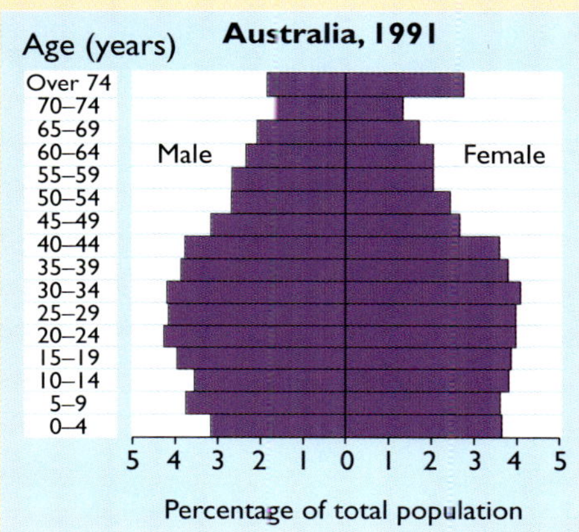

ACTIVITIES

1. Use the following figures to draw your own line graph. Remember to use a dashed line to show estimates.

World population (billions)

Year	Population
1800	0.7
1850	1.2
1900	1.5
1950	2.4
2000	6.2
2050	9.6

2. Construct a simple bar graph using the figures below.

Time spent by Aborigines of Fish Creek, Arnhem Land, collecting food

Day	Hours
1	7.0
2	3.0
3	0.0
4	7.0
5	6.5
6	0.5
7	4.0

3. Construct a climograph for Melbourne using the following statistics. Use the same scale as the Brisbane example and compare your results.

	J	F	M	A	M	J	J	A	S	O	N	D
Average temp (°C)	20	20	18	15	12	10	10	11	12	14	16	18
Average rainfall (mm)	48	50	54	59	57	50	49	50	59	68	59	58

4. Build a population pyramid for Botswana using the statistics below. Is Botswana an old or young population?

Age (years)	Male (% of total population)	Female (% of total population)
0–4	8.6	8.9
5–9	7.6	7.8
10–14	6.5	6.5
15–19	4.9	4.8
20–24	4.3	4.1
25–29	3.7	3.7
30–34	2.9	3.0
35–39	2.2	2.9
40–44	1.8	2.0
45–49	1.3	1.5
50–54	1.1	1.5
55–59	1.1	1.2
60–64	1.1	1.0
65–69	1.0	0.8
70–74	0.8	0.3
Over 74	0.5	0.6

32 DRAWING PIE AND OTHER

Pie graphs

In a pie graph the slices or sectors represent the size of the different parts which make up the whole. The circle of 360 degrees represents the total, or 100 per cent, of whatever is being looked at. The size of the slices is easily seen. The individual parts or components should be presented in order, from largest to smallest, to make it easier to interpret the figures.

Steps in drawing a pie graph

Equipment needed

Set of statistics, a compass, a protractor, a pencil, a blank sheet of paper, colouring pencils.

1. Order the statistics from largest to smallest. If you have an "other" category, put it last. Make sure you know what the total of all the parts adds up to.
2. Convert the figures to a percentage of the whole, that is, divide the part's figure by the total figure and multiply by 100.
3. Convert the percentage to degrees of a circle by multiplying by 3.6. 100% of the circle = 360 degrees, so 1% of the circle = 3.6 degrees. This gives the angle of each slice of the pie. Study the following example.

Population of Australia's largest cities, 1994

City	Population	%	Degrees
Sydney	3 656 500	20.5	74
Melbourne	3 080 900	17.3	62
Brisbane	1 301 700	7.3	26
Perth	1 193 100	6.7	24
Adelaide	1 049 800	5.9	21
Canberra	310 000	1.7	6
Other	7 220 000	40.6	147
Total	17 800 000	100.0	360

4. Use the compass to draw a circle. Don't make it too small or it will be hard to draw and read. Make the radius at least 3 centimetres. Put a pencil dot on the middle of your circle.
5. Draw a straight line from the centre of the circle up to the 12 o'clock position.
6. Use the protractor to mark the first and largest slice, working clockwise.
7. Place the 0 degrees line on your protractor along the line you have just drawn. Now mark in the second largest group. Use the protractor to mark each of the other slices in descending size, marking the "other" category last.
8. Label each slice, give the graph a clear title and colour each segment so that it is easy to read.

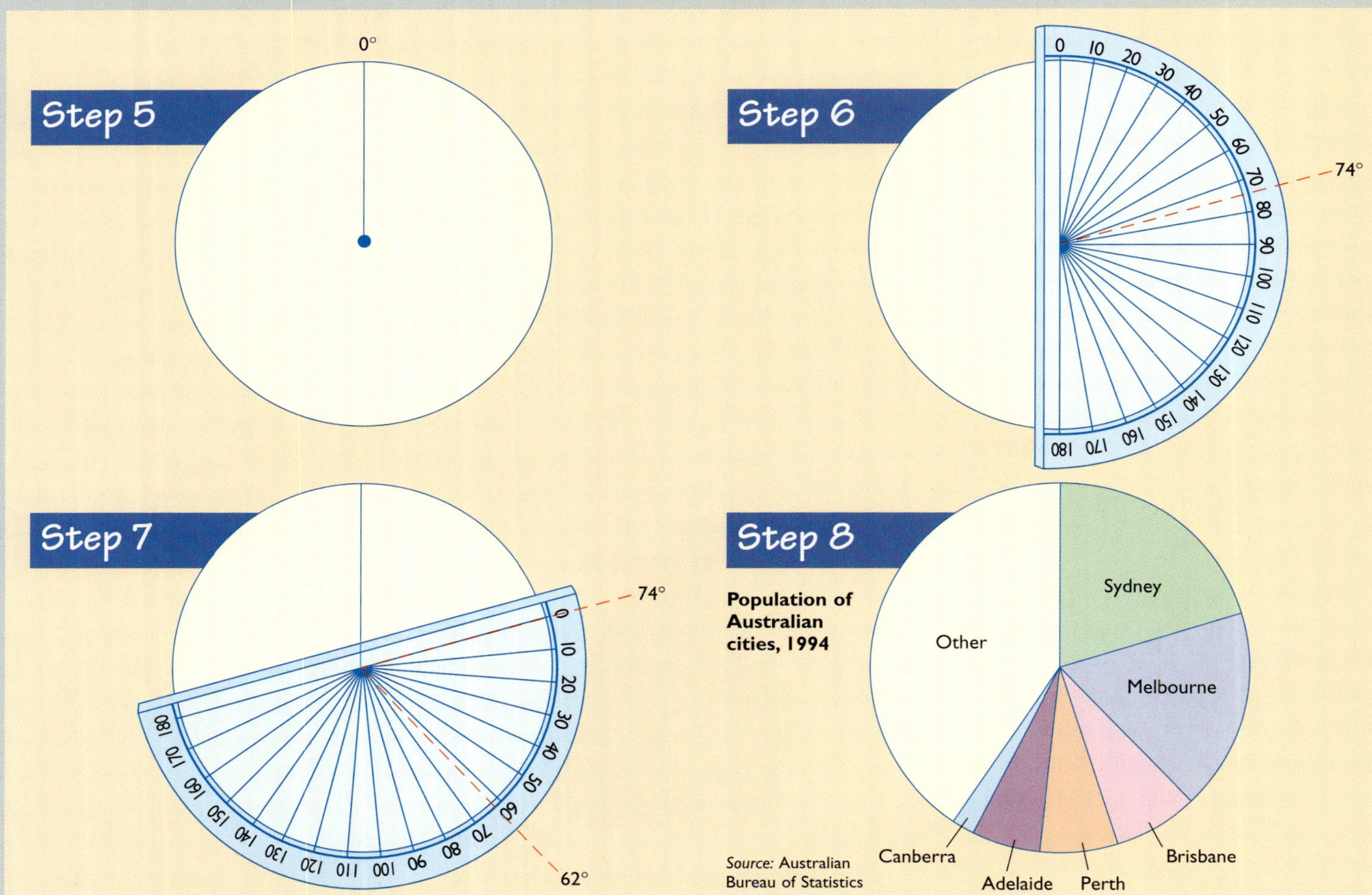

Scatter graphs

Scatter graphs are used to see if information about two different things is related. For example, the size of a city's population may be related to the number of schools it has.

The data for the two variables is plotted for a number of locations. The graph shows the population of the towns on the vertical axis and number of schools in each town on the other axis. Every location is plotted as a dot showing the population and number of schools in each town.

Once all the dots are plotted, we can look for a relationship. If the dots seem to form a line from bottom left to top right, there is a positive relationship between the variables. If the dots seem to form a line from top left to bottom right, the data has a negative relationship.

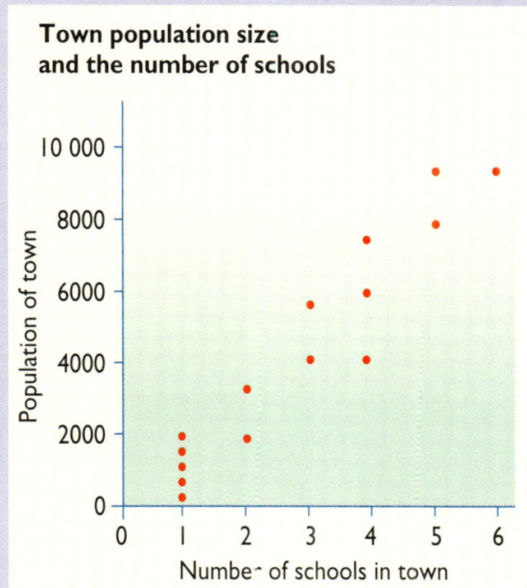

Pictographs

Pictographs use pictures to help you to understand, instantly, the content of the graph. Wheat production for various years might be represented by bags of wheat, each representing a quantity of wheat, for example, 1000 tonnes. Water consumption might be shown by buckets or dripping taps. Sometimes the size of the picture varies, for example, a big bucket means lots of water used and a small bucket means only a small amount of water used.

ACTIVITIES

1. Draw two pie graphs to show the pattern of Australia's imports and exports in 1995.

   Imports
   Chemicals — 11%
   Manufactured goods — 22%
   Food — 2%
   Raw materials — 6%
   Fuel — 3%
   Machinery and transport — 48%
   Other — 8%

   Exports
   Food and livestock — 24%
   Other — 1.5%
   Coal and fuel minerals — 32%
   Non-foods, including minerals and timber — 33%
   Machinery — 8%
   Chemicals — 1.5%

Look at the scatter graph.

2. Is there a positive or negative relationship between a town's population size and the number of schools it has? Explain what this means.
3. Why don't towns with only one school seem to fit the pattern?
4. Prepare your own scatter graph using the world statistics on pages 188–97 to compare the infant mortality rate with the number of persons per doctor. Select 10 countries to show a range of data.
5. Is there a positive or negative relationship between infant mortality rate and persons per doctor? Why might this be?
6. Try to draw your own pictogram to show differences between car ownership in Australia (8 million), Mozambique (100 000) and Brazil (14 million).
7. Select three categories from the statistics on pages 188–97. Prepare a pictogram for each category to illustrate the range of data.

34 TESTING YOUR SKILLS

Tracking the thylacine

The thylacine or "Tasmanian Tiger" was once common in mainland Australia. Competition from the dingo meant it was eventually confined to Tasmania. Here, where there were no dingoes, it survived in habitats in the north-east, north-west, the central plateau and the valleys of the south-east. European settlement in the 1800s meant the introduction of sheep. Unfortunately, the thylacine was very fond of mutton. A bounty was paid for any thylacine captured dead or alive.

This spelt the end for the thylacine. Or did it? Persistent reports of sightings and eyewitness accounts over the past 60 years suggest that the thylacine may not be extinct. Large sums of money and energy have been spent trying to prove its continued existence in Tasmania's wilderness. Despite these efforts, no conclusive evidence has been found. For many, however, the case is not yet closed.

SOURCE 1

1825	Sheep established in open grasslands of eastern Tasmania
1835	Van Diemen's Land Company introduces tiger bounty
1888–1912	2184 thylacines killed under government bounty scheme
1933	Last "wild" thylacine shot
1936	Last thylacine in captivity dies in Hobart Zoo
1938	Expedition to Jane River takes plaster casts of tracks
1945–46	Naturalist David Fleay searches in Jane River area
1963–64	Dr Eric Guiler searches an area bounded by Sandy Cape, Woolnorth and Balfour
1966	Hair samples from a "tiger lair" near Whyte River not conclusively identified as thylacine hair
1968–72	Intensive searches investigate many sightings
1979–80	World Wildlife Fund sponsors search team
1982	Sighting by Hans Naarding 10 km north of Arthur River
1984	Peter Wright searches Lake Adelaide area near Mole Creek after sightings
1985–96	Private searches continue with no conclusive evidence of thylacine's existence

SOURCE 2

Last thylacine in captivity in Hobart Zoo

SOURCE 3

Sightings of the thylacine

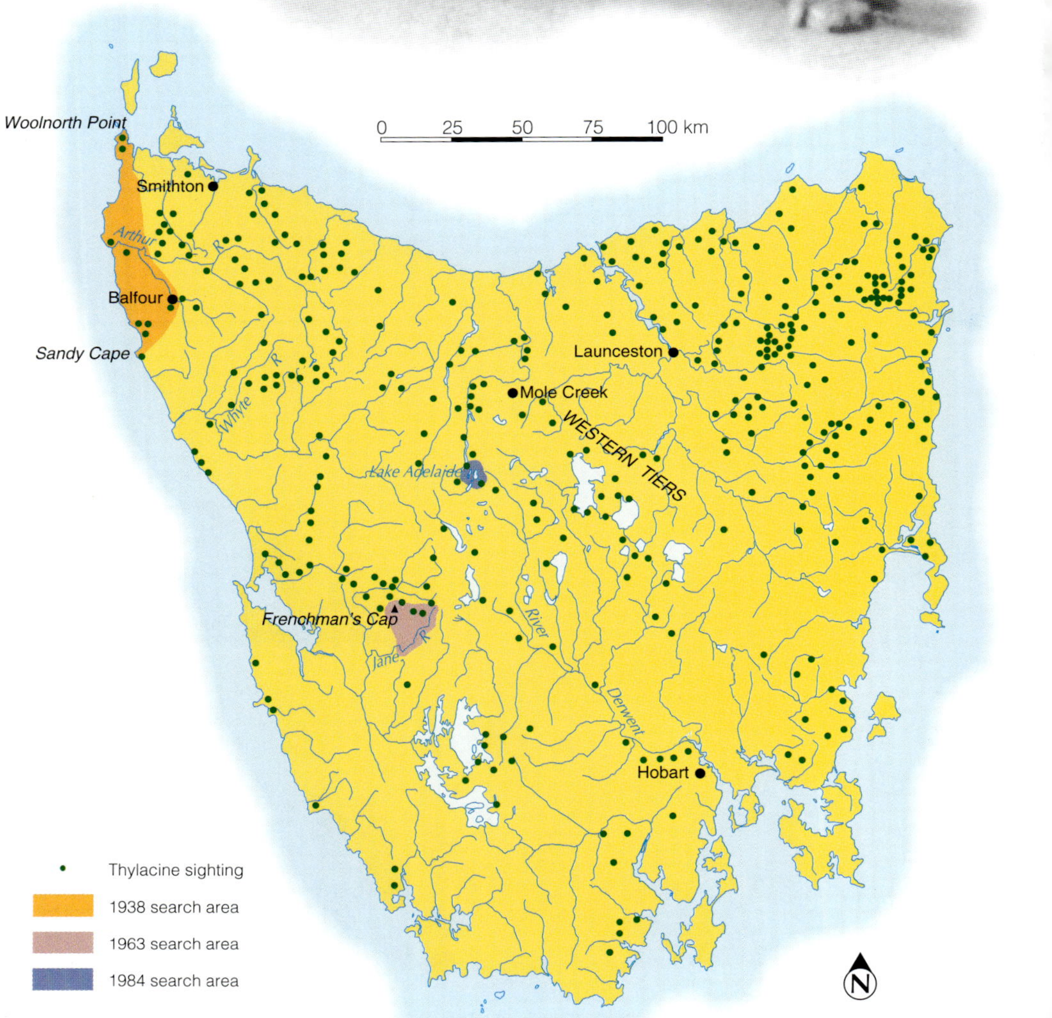

In 1938, a team of six searched for the thylacine in the Jane River area, east and south-east of Frenchman's Cap. Plaster casts were taken from footprints in a number of places over the eleven days of the expedition. The team even saw footprints around their camp, but an actual sighting of a thylacine eluded them. The tracks they saw were very different from those a dog would make. The team believed the number and distribution of the tracks they found confirmed that "the animal was fairly common and well-distributed".

SOURCE 4

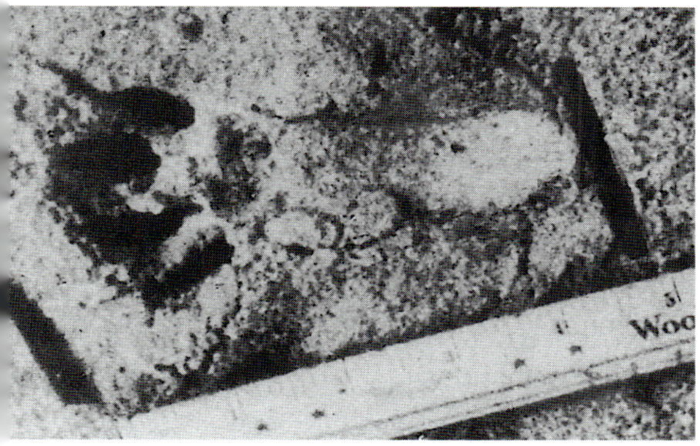

A plaster cast was made of this thylacine footprint, which was about 125 centimetres (5 inches) long.

SOURCE 5

Sighting of "tiger" by Hans Naarding, March 1982

"I had gone to sleep in the back of my vehicle which was parked at a road junction in a remote forested area in the north-west of the state. It was raining heavily. At 2 a.m. I awoke and, out of habit, scanned the surrounds with a spotlight.

As I swept the light-beam around, it came to rest on a large thylacine, standing side-on some 6 to 7 metres distant. My camera bag was out of immediate reach so I decided to examine the animal carefully before risking movement. It was an adult male in excellent condition with 12 black stripes on a sandy coat. Eye reflection was pale yellow. It moved only once, opening its jaw and showing its teeth. After several minutes of observation I attempted to reach my camera bag but in doing so I disturbed the animal and it moved away into the undergrowth. Leaving the vehicle and moving to where the animal disappeared, I noted a strong scent. Despite an intensive search, no further trace of the animal could be found."

SOURCE 6

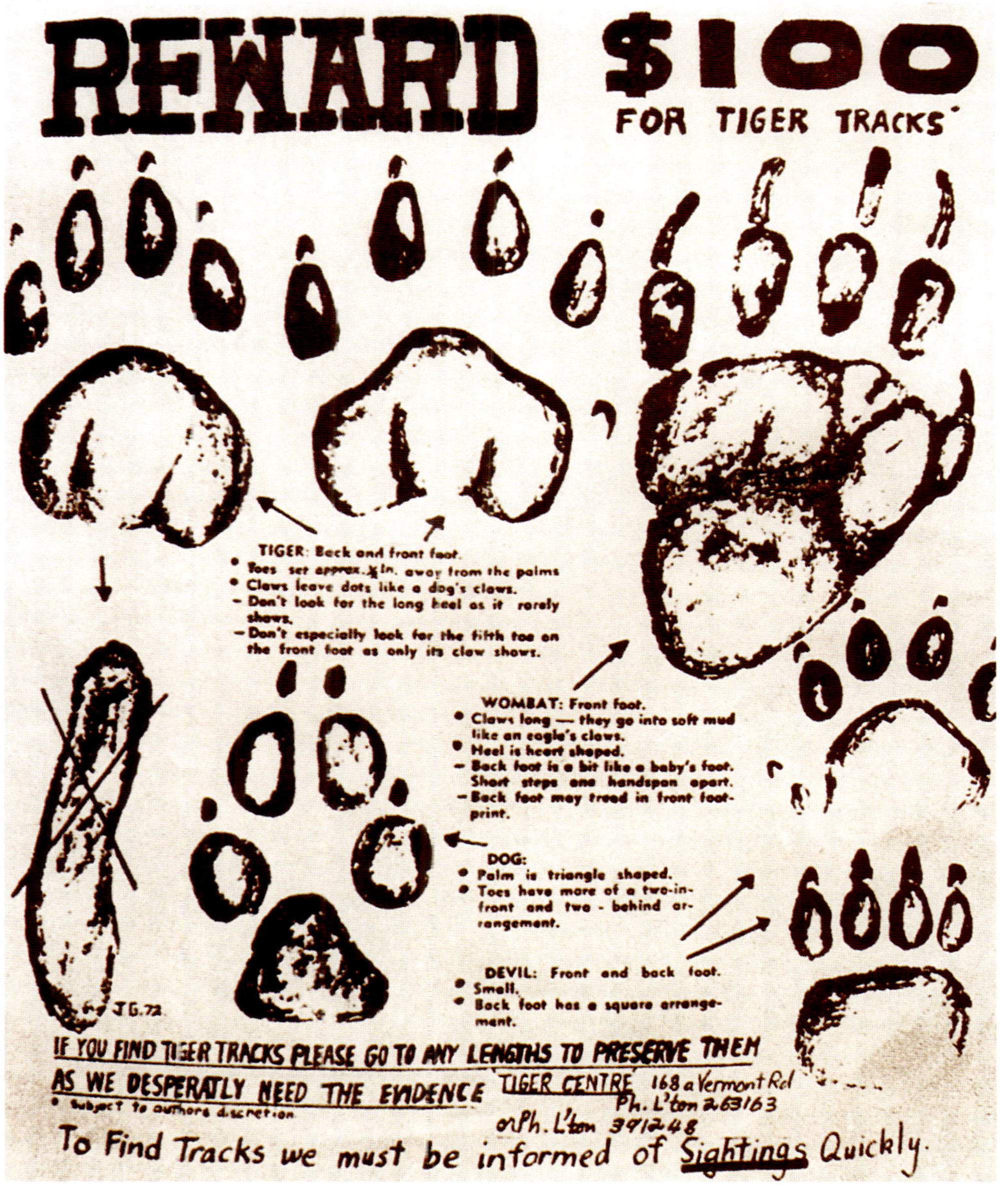

Reward poster showing tracks of various animals

SOURCE 7

Oblique aerial photograph of the area where Hans Naarding sighted the tiger

36 TESTING YOUR SKILLS

SOURCE 8

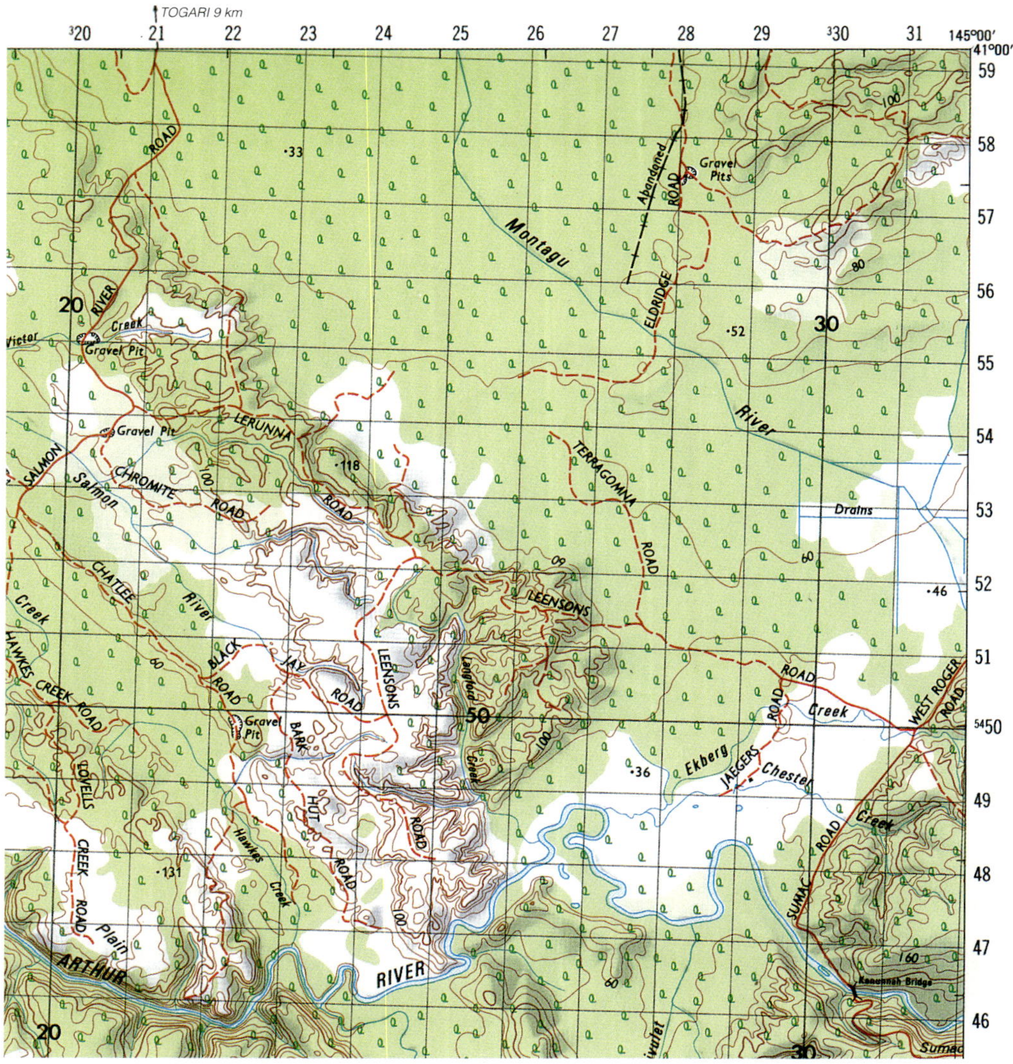

ACTIVITIES

Find the appropriate source, and answer the following questions.

1. When did the last known thylacine die?
2. Why did the numbers of thylacines decrease so quickly?
3. Write a description from the photograph of the last thylacine.
4. Compare your description to that of Hans Naarding in source 5. Do you think he could have seen a thylacine?
5. Why do you think a reward was offered for evidence of "tiger tracks"?
6. Find the 1938 search area in source 3. What type of landscape do you imagine it to be?
7. Use direction to describe the areas of Tasmania where most thylacine sightings have taken place.
8. Can you dispute any of Hans Naarding's primary evidence based on the time of his sighting and the weather conditions? Explain.
9. If Naarding's thylacine did exist, why do you think searches could find no trace of it?
10. The photograph in source 7 is located at GR208573. Use the key to describe the vegetation in this area.
11. If you were a thylacine, which area of the topographic map would you hide in? Give the grid reference, and suggest reasons for your choice.

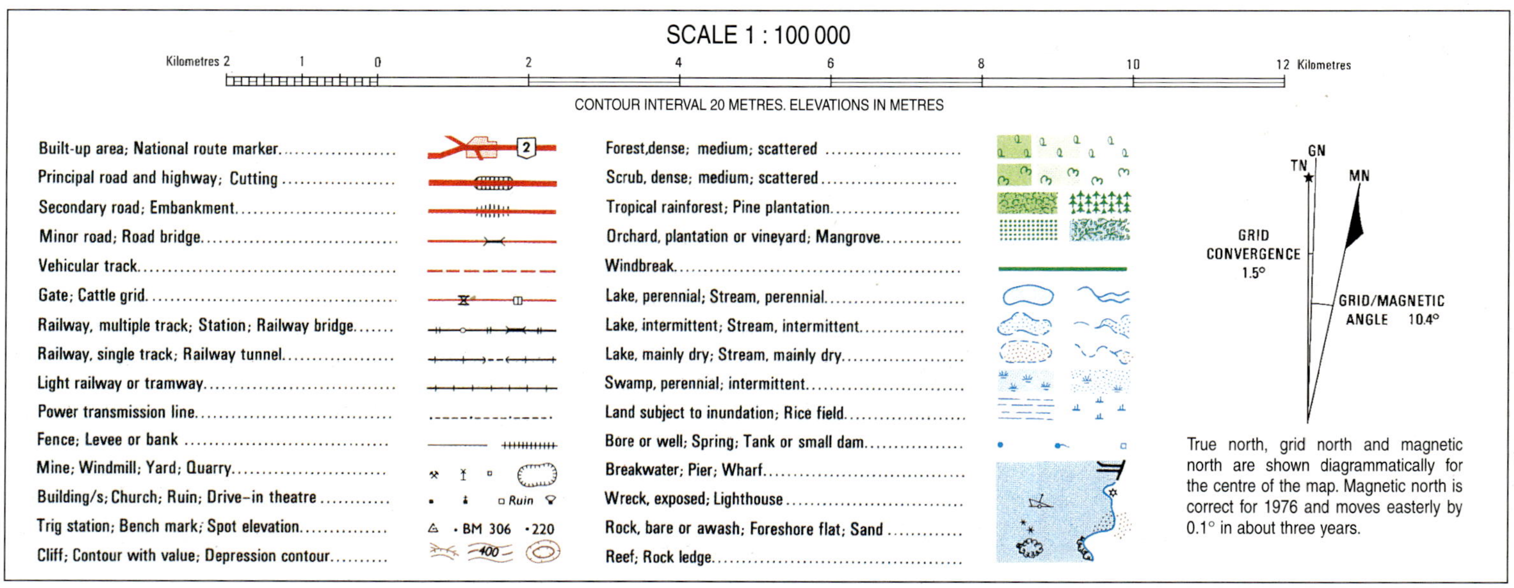

Topographic map extract of area around Togari in north-west Tasmania where Hans Naarding sighted the thylacine

AUSTRALIA

AUSTRALIA

Settlement, history and culture

Key

Built environment

AUSTRALIA	Country name
CANBERRA	Country capital underlined
(line)	Country border
(line)	State/territory border
Victoria	State/territory name

■ PERTH	More than 1 000 000 people
no example	500 000 to 1 000 000 people
● GEELONG	100 000 to 500 000 people
• Cairns	50 000 to 100 000 people
• Port Hedland	10 000 to 50 000 people
· Lock	Less than 10 000 people

Natural environment

(symbol)	River
(symbol)	Lake
King Island	Island

Scale

1 : 25 000 000

0 250 500 750 km

1 centimetre on the map represents 250 kilometres on the ground.

Geo Facts

Largest state (by area) Western Australia ▶ 2 526 000 sq km

Largest state (by pop.) N.S.W. ▶ 5 902 000

Largest city (by pop.) Sydney ▶ 3 600 000

GNP per head $17 065

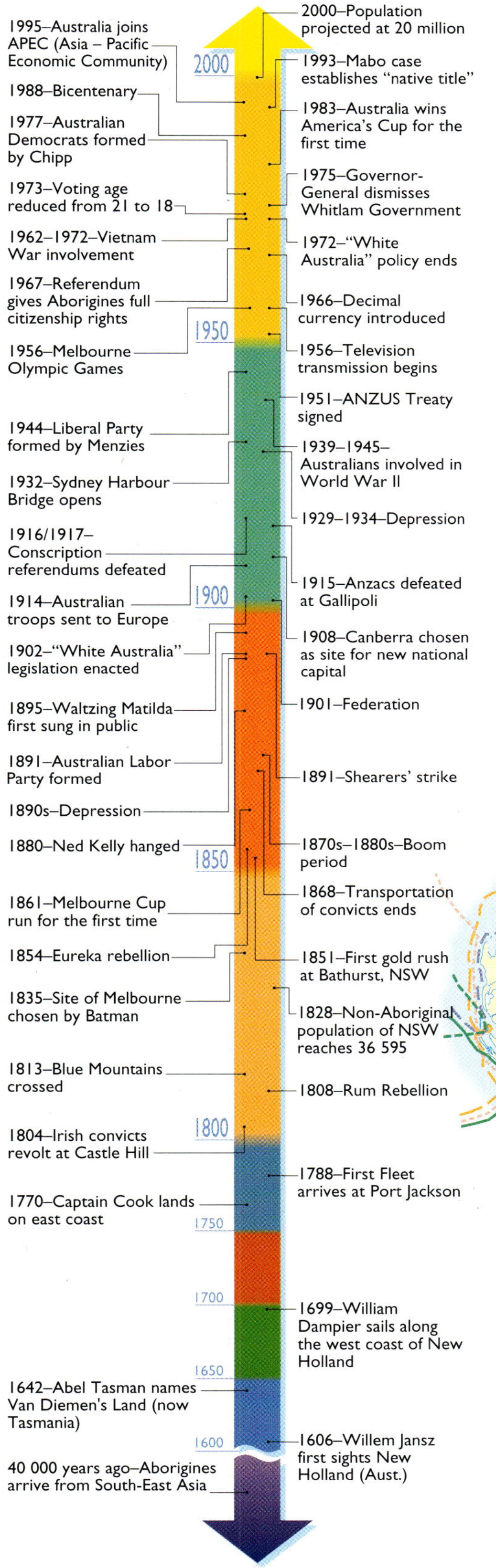

CASE STUDY
Eureka rebellion

The Eureka rebellion occurred in 1854 on the Ballarat goldfields in Victoria. Goldminers were angry about high licence fees and licence hunts by police. Led by Peter Lalor, the diggers (miners) built a stockade at Eureka and clashed with government troops. Five soldiers and at least 30 miners were killed.

Although the diggers were defeated, the licensing system was reformed and democratic rights were introduced to the goldfields. The Eureka flag continues to be a symbol of the fight for justice and equality, as can be seen in this photograph of a protest by truck drivers in 1988. The flag has even been proposed as a national flag should Australia become a republic.

40 AUSTRALIA

Landforms and land use

Key

Built environment

State/territory border

Natural environment

River

Lake

Roti — Island

Mt Ossa 1617 m — Mountain and height in metres

Scale

1 : 25 000 000

0 250 500 750 1000 km

1 centimetre on the map represents 250 kilometres on the ground.

Natural environment

- Land over 4000 metres
- Land from 2000 to 4000 metres
- Land from 1000 to 2000 metres
- Land from 500 to 1000 metres
- Land from 200 to 500 metres
- Land from sea level to 200 metres
- Land below sea level
- Sea from sea level to 200 metres
- Sea from 200 to 2000 metres
- Sea from 2000 to 4000 metres
- Sea from 4000 to 8000 metres
- Sea over 8000 metres

Geo Facts

Highest mountain
Kosciuszko, N.S.W. ▶ 2228 m

Highest average annual rainfall
Babinda, Qld ▶ 4537 mm

Lowest average annual rainfall
Troudaninna, S.A. ▶ 105 mm

Largest desert
Great Victoria, W.A.–S.A. ▶ 300 000 sq km

Longest river
Murray–Darling ▶ 3750 km

National Parks

Kakadu

Aborigines have continuously occupied Kakadu for more than 20 000 years. Their knowledge of the seasons and the land are recorded in rock art galleries and through Dreaming stories:

> "During the Dreaming the crocodile-man Gumangan and the plover-man Birik-birik argued over fire-sticks.
> Grabbing the fire-sticks which Gumangan was about to dip into the river, Birik-birik ran into the hills. Since then, the crocodile has always lived near water and the plover in the hills or on the plains. Today, the Aborigines fear the crocodile, but love the plover who saved fire for them."

Today the traditional Aboriginal owners of Kakadu, Australia's largest national park, lease 20 000 square kilometres to the Australian Nature Conservation Agency.

Reserves of more than 5000 hectares

- National park
- Marine park
- Other reserve

World Heritage areas

1. Shark Bay, Western Australia
2. Kakadu National Park
3. Uluru–Kata Tjuta National Park
4. Australian Fossil Mammal Sites (Riversleigh/Naracoorte)
5. Tasmanian Wilderness
6. Willandra Lakes Region
7. Central Eastern Rainforest Reserves (Australia)(50 separate sites)
8. Wet Tropics of Queensland
9. Fraser Island
10. Great Barrier Reef
11. Lord Howe Island Group
12. The Greater Blue Mountains Area
13. Purnululu National Park

QUEENSLAND

Settlement

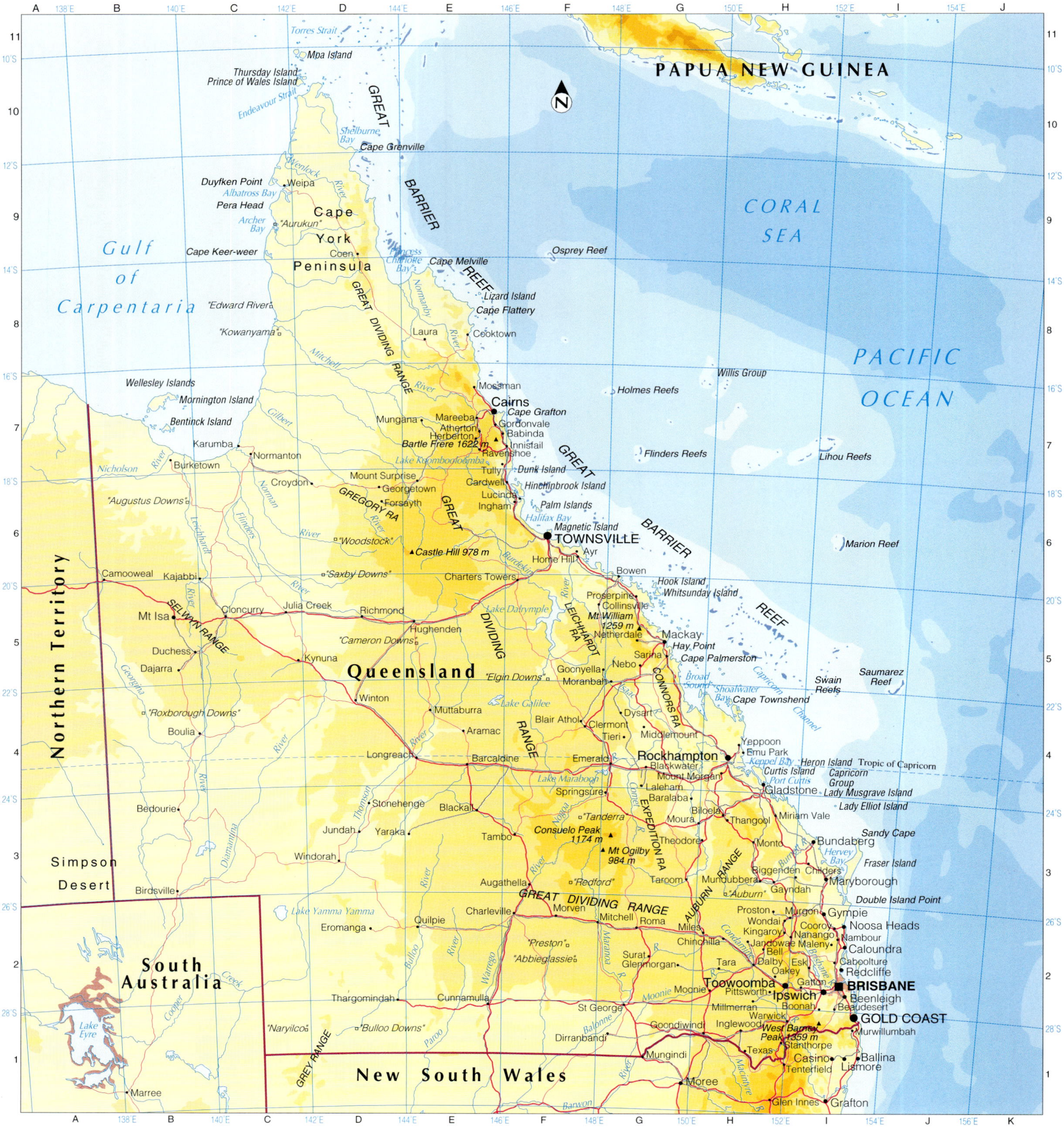

Aboriginal language groups

Eddie Mabo

Koiki (Eddie) Mabo (1937–92) was born on Mer (Murray Island) in the Torres Strait. He lived in the Torres Strait islands until 1959 when he moved to Townsville. He became a leader of the local Torres Strait islands community and helped establish the Aboriginal and Islander Health Service. Eddie also performed and taught Meriam song and dance.

In 1982 Eddie and four other Torres Strait islanders started legal proceedings to establish traditional ownership of their land on Mer. This became known as the Mabo Case. It was not until 1992 (4 months after Eddie died of cancer) that the High Court of Australia agreed that the Mer people had owned their land prior to annexation by Queensland. Eddie was named Australian of the Year in 1993 by the *Australian* newspaper.

Key

Built environment

PAPUA NEW GUINEA	Country name
	State/territory border
Queensland	State/territory name
	Major road
	Minor road
	Railway
	Urban area
■ BRISBANE	More than 1 000 000 people
no example	500 000 to 1 000 000 people
● TOWNSVILLE	100 000 to 500 000 people
● Cairns	50 000 to 100 000 people
• Gympie	10 000 to 50 000 people
• Quilpie	Less than 10 000 people
▫ *'Tanderra'*	Homestead/mission

Natural environment

	River
	Lake
Dunk Island	Island
▲ *Mt William 1259 m*	Mountain and height in metres
	Land over 4000 metres
	Land from 2000 to 4000 metres
	Land from 1000 to 2000 metres
	Land from 500 to 1000 metres
	Land from 200 to 500 metres
	Land from sea level to 200 metres
	Land below sea level
	Sea from sea level to 200 metres
	Sea from 200 to 2000 metres
	Sea from 2000 to 4000 metres
	Sea from 4000 to 8000 metres
	Sea over 8000 metres

Scale

1 : 10 500 000

0 100 200 300 400 km

1 centimetre on the map represents 105 kilometres on the ground.

Bindal	Aboriginal language group
- - - -	Aboriginal language group boundary
	Regional boundary

0 150 300 km

N

44 BRISBANE

Change over time

Brisbane, 1888

Brisbane today. What has changed? What remains the same?

Where is the CBD?

Brisbane is located on the banks of the Brisbane River and is Australia's fastest-growing capital city.

The Central Business District (CBD) of Brisbane is now centred around the Queen Street Mall. Since its establishment as a convict settlement on the north bank of the river, Brisbane has had three centres — South Brisbane, Fortitude Valley ("the Valley") and the Queen Street Mall area.

South Brisbane became the town centre when it became the terminus of the railway from New South Wales in 1891.

The Story Bridge, pictured in the background of the aerial photograph of Brisbane today, was built in 1940 in an attempt to move the business centre to "the Valley" (top left) where McWhirters and T. C. Beirne had big stores. Several "sky-scraper" projects and the opening of the Queen Street Mall in 1982 shifted the centre back to the original site chosen by John Oxley in 1824. South Brisbane today is a cultural and recreational centre, while the Brisbane City Council plans to rejuvenate the Valley as a retail, commercial and residential district.

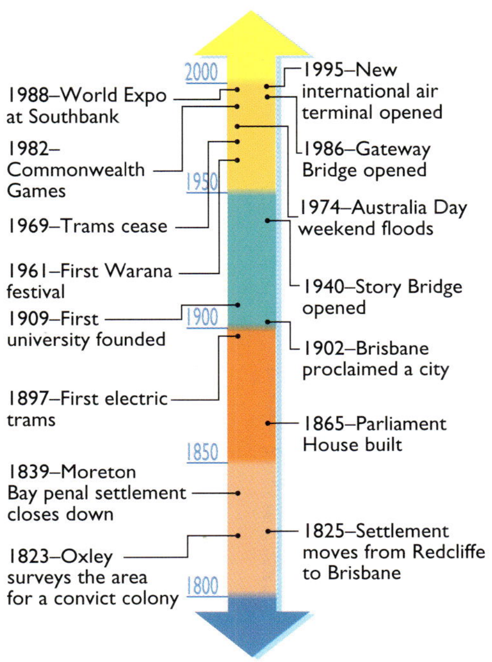

"Central station 7 a.m. (The morning rush) 1931"
Painting by Vida Lahey

Central station, 1996. The skyline has changed but there are still some clues linking this photograph to the painting. Can you find them?

46 BRISBANE

Satellite image

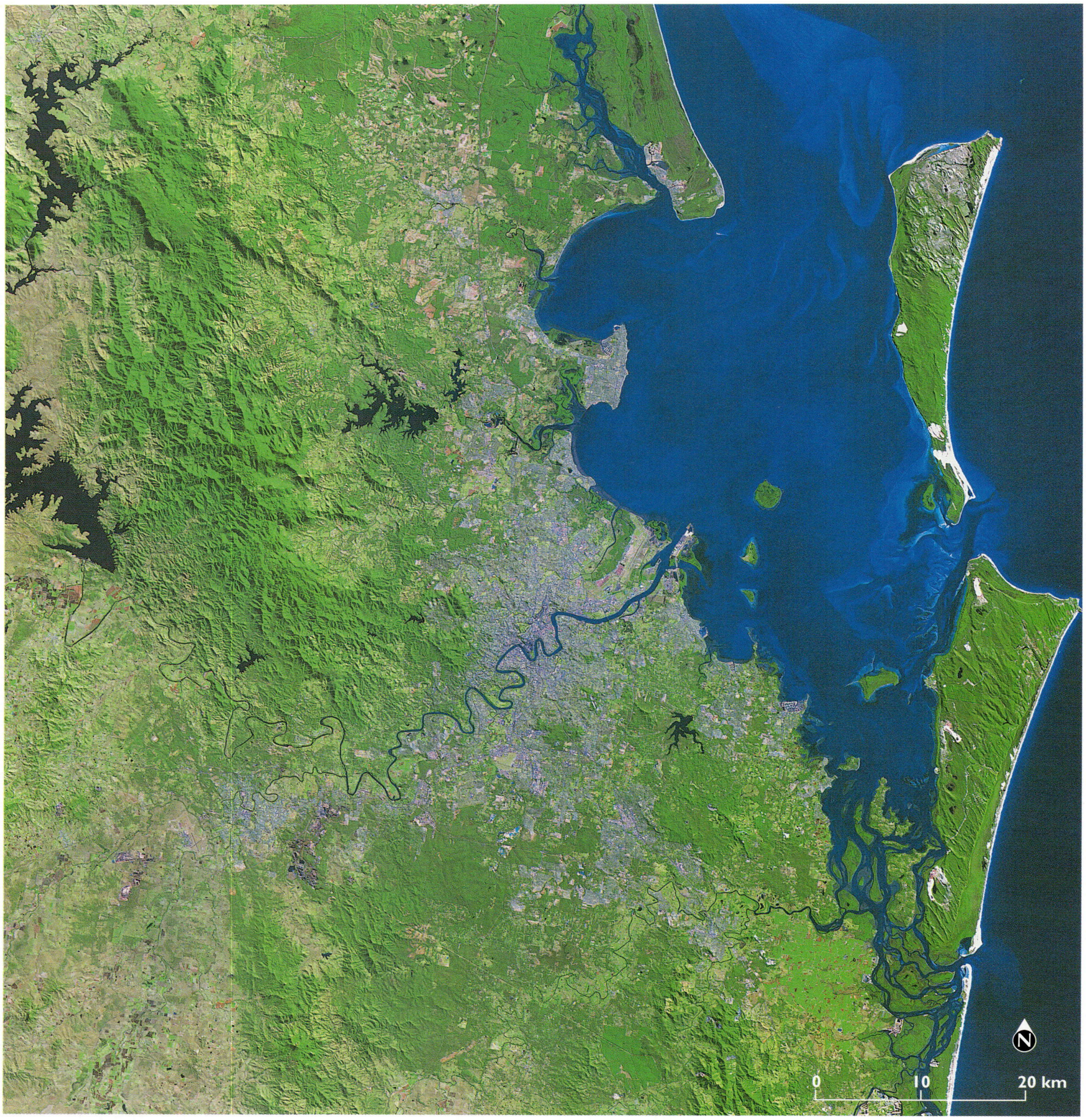

The urban area of Brisbane (shown in blue-pink) follows the path of the Brisbane River to the west, the Pacific Highway to the south (towards the Gold Coast), and the Bruce Highway to the north (towards the Sunshine Coast). Farmland is located to the west of the ranges, and shows as light green-brown on this satellite image.

Social and economic patterns

POPULATION CHANGE

The total population grew by 13.2% from 1986 to 1991. The greatest increases were in the outer northern and southern suburbs.

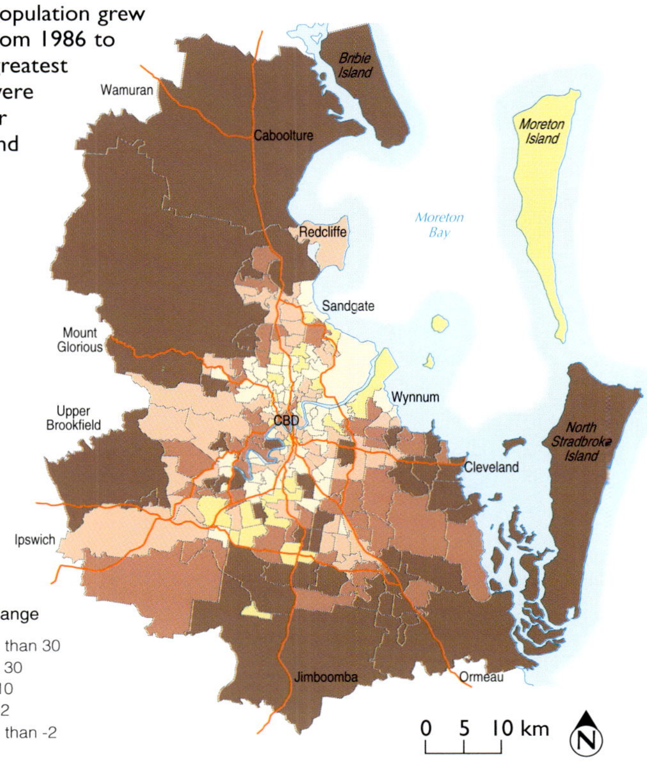

Percentage change

- More than 30
- 10 to 30
- 2 to 10
- -2 to 2
- More than -2

PEOPLE BORN OVERSEAS

In 1991 20% of Brisbane's population was born overseas. More live in the southern suburbs than in suburbs to the north of the Brisbane River.

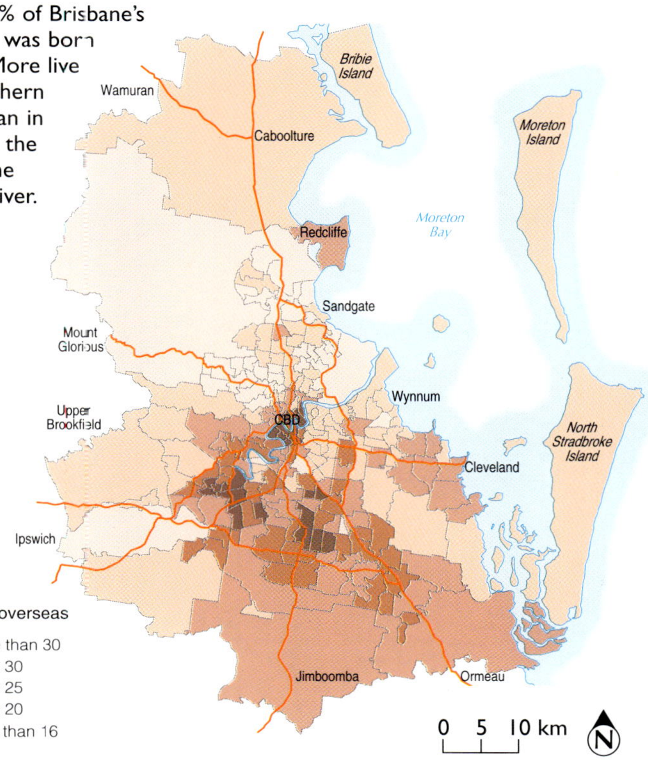

Per cent born overseas

- More than 30
- 25 to 30
- 20 to 25
- 16 to 20
- Less than 16

PEOPLE WITH UNIVERSITY QUALIFICATIONS

The inner western suburbs, near the University of Queensland, show the highest concentration of people with university qualifications.

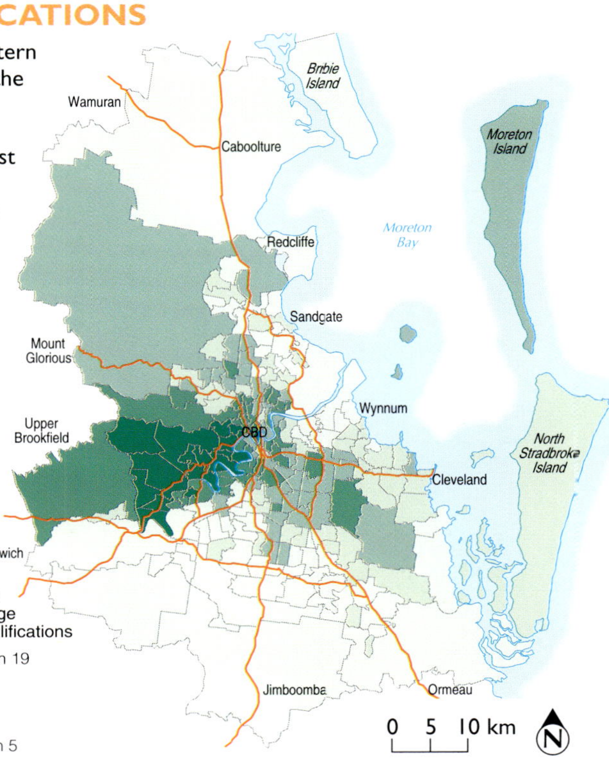

Per cent of people over 15 years of age with university qualifications

- More than 19
- 12 to 19
- 8 to 12
- 5 to 8
- Less than 5

PEOPLE WITH INCOMES OVER $50 000

Only 2.8% of people over 15 years of age in Brisbane earned more than $50 000 in 1991. Many of these people live in the leafy western suburbs of Brisbane.

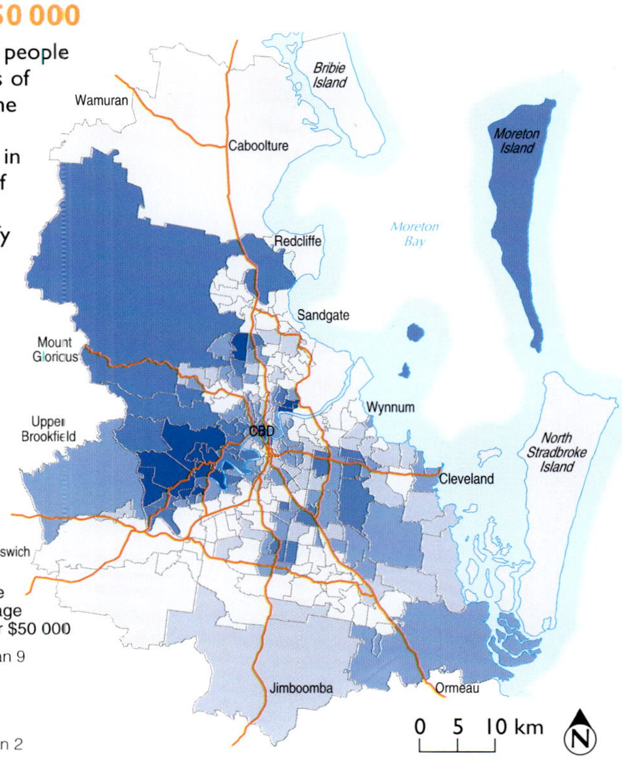

Per cent of people over 15 years of age with incomes over $50 000

- More than 9
- 5 to 9
- 3 to 5
- 2 to 3
- Less than 2

NEW SOUTH WALES

Settlement

The Three Sisters in the Blue Mountains, west of Sydney

Aboriginal language groups

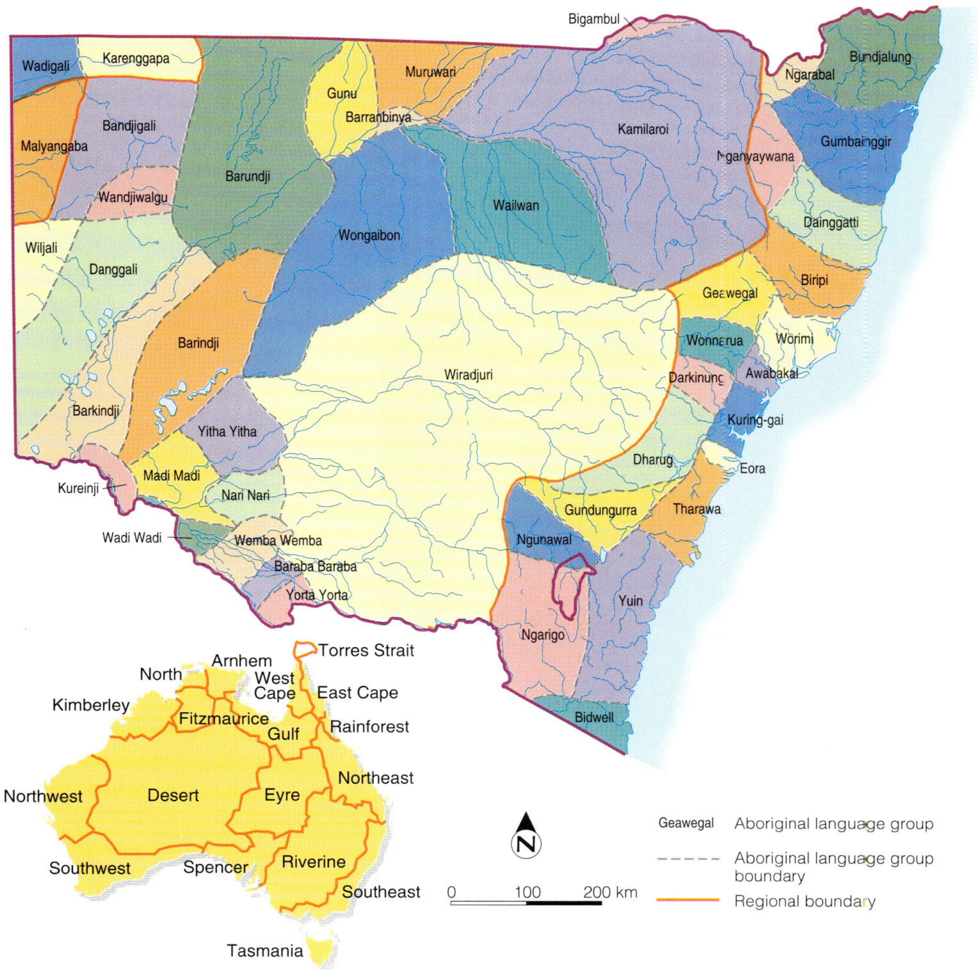

Key

Built environment

(line)	State/territory border
New South Wales	State/territory name
(line)	Major road
(line)	Minor road
(line)	Railway
(shape)	Urban area
■ **SYDNEY**	More than 1 000 000 people
no example	500 000 to 1 000 000 people
● NEWCASTLE	100 000 to 500 000 people
● Bendigo	50 000 to 100 000 people
• Moree	10 000 to 50 000 people
• Wilcannia	Less than 10 000 people
▫ *'Tonga'*	Homestead/mission

Scale

1 : 6 500 000

0 100 200 300 km

1 centimetre on the map represents 65 kilometres on the ground.

Natural environment

(line)	River
(shape)	Lake
▲ *Mt Myra 1128 m*	Mountain and height in metres
	Land over 4000 metres
	Land from 2000 to 4000 metres
	Land from 1000 to 2000 metres
	Land from 500 to 1000 metres
	Land from 200 to 500 metres
	Land from sea level to 200 metres
	Land below sea level
	Sea from sea level to 200 metres
	Sea from 200 to 2000 metres
	Sea from 2000 to 4000 metres
	Sea from 4000 to 8000 metres
	Sea over 8000 metres

Lorraine Mafi-Williams

Lorraine Mafi-Williams was born in 1940 at Purfleet, near Taree in northern New South Wales. As a very young Aboriginal girl Lorraine was forcibly removed from her parents, as often happened, and did not meet them again until she was 15. In 1967 she moved to Sydney and worked with the Aboriginal Family Education Centre.

Lorraine (*top right*) has written children's stories based on Aboriginal Dreaming stories, and short stories and poems. She has been involved with the Aboriginal Black Theatre Art and Culture Centre at Redfern. She took courses in film making, has worked as a production assistant and has appeared in several films and plays.

In 1988 she was given a grant to make her own film called *Eelemarni* based on the life of a warrior of her Gidabal (Bundjalung) people. Lorraine Mafi-Williams lives at Byron Bay and is still writing.

50 SYDNEY

Change over time

Harbour city

The harbour has been the focus of Sydney life since 1788 when Captain Arthur Phillip established Australia's first European settlement at Sydney Cove. It provided a secure anchorage for the British fleet and later was surrounded with forts and observation posts.

In its early years, Sydney became a Pacific port for the pork, sandalwood, seal and whale trades, and for trade with China. It became a major commercial and tourist port in the twentieth century. In 1932 the Sydney Harbour Bridge was opened, linking the north and south sides of the harbour. In World War II, nets were built to try to prevent Japanese submarines sneaking into the harbour.

Today, properties with harbour views sell for a very high price. At weekends the harbour becomes a patchwork of coloured sails. In 2000, Sydney will become the centre of world attention when the city hosts the Olympic Games. Because of their harbour locations, the Opera House, Darling Harbour, Taronga Zoo and the bridge attract tourists from everywhere.

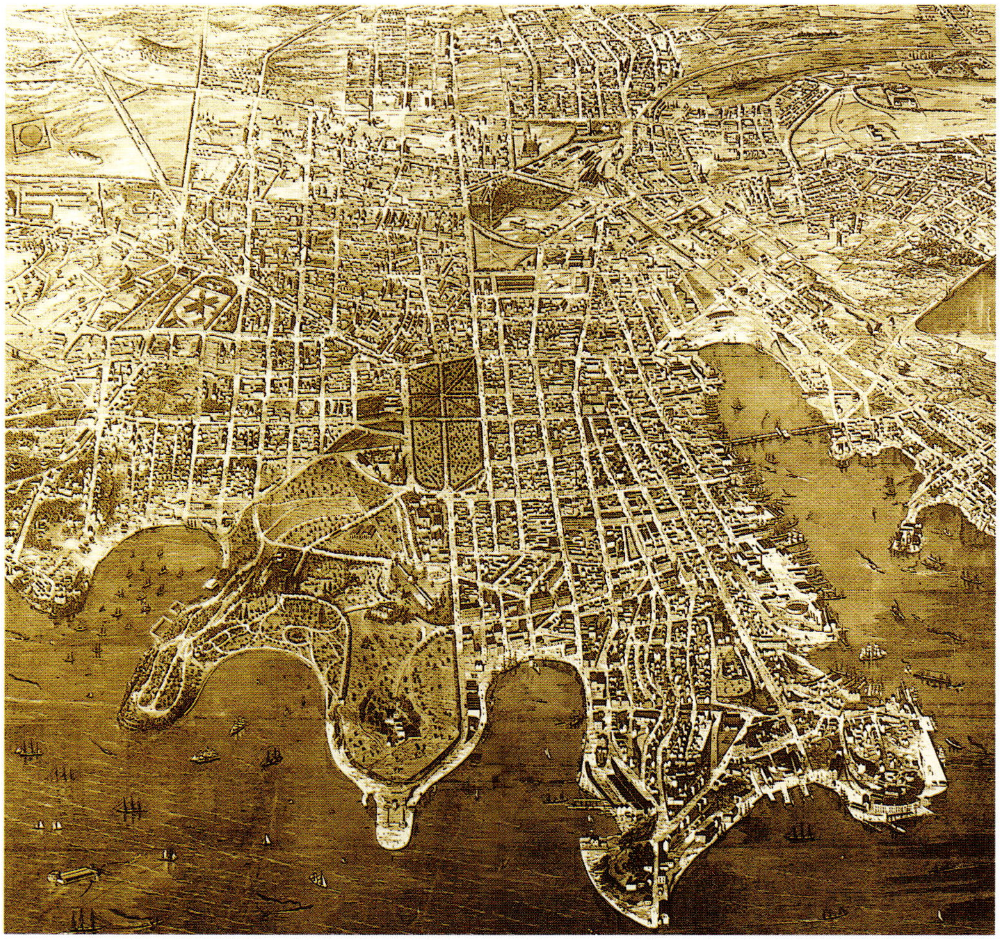

Sydney, 1879

Sydney today. What changes to the waterfront have taken place? Identify areas around the foreshores where the land use has changed very little.

Sydney in the 1920s, looking south along George Street, from the King Street corner

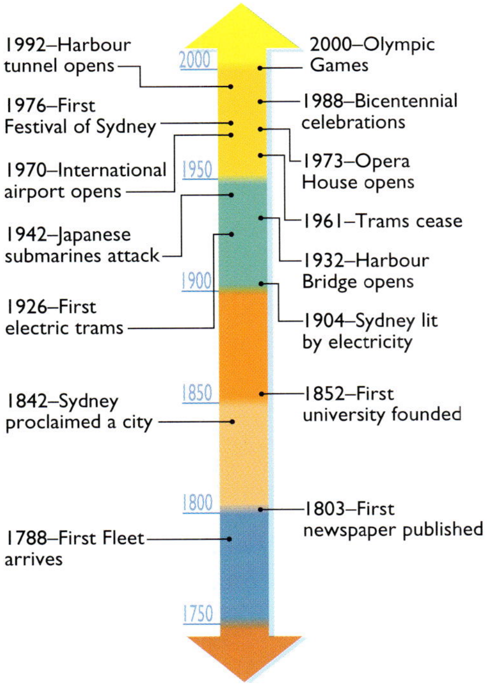

This modern photograph is taken from the same corner. Identify as many changes as you can. Has anything stayed the same?

52 SYDNEY

Satellite image

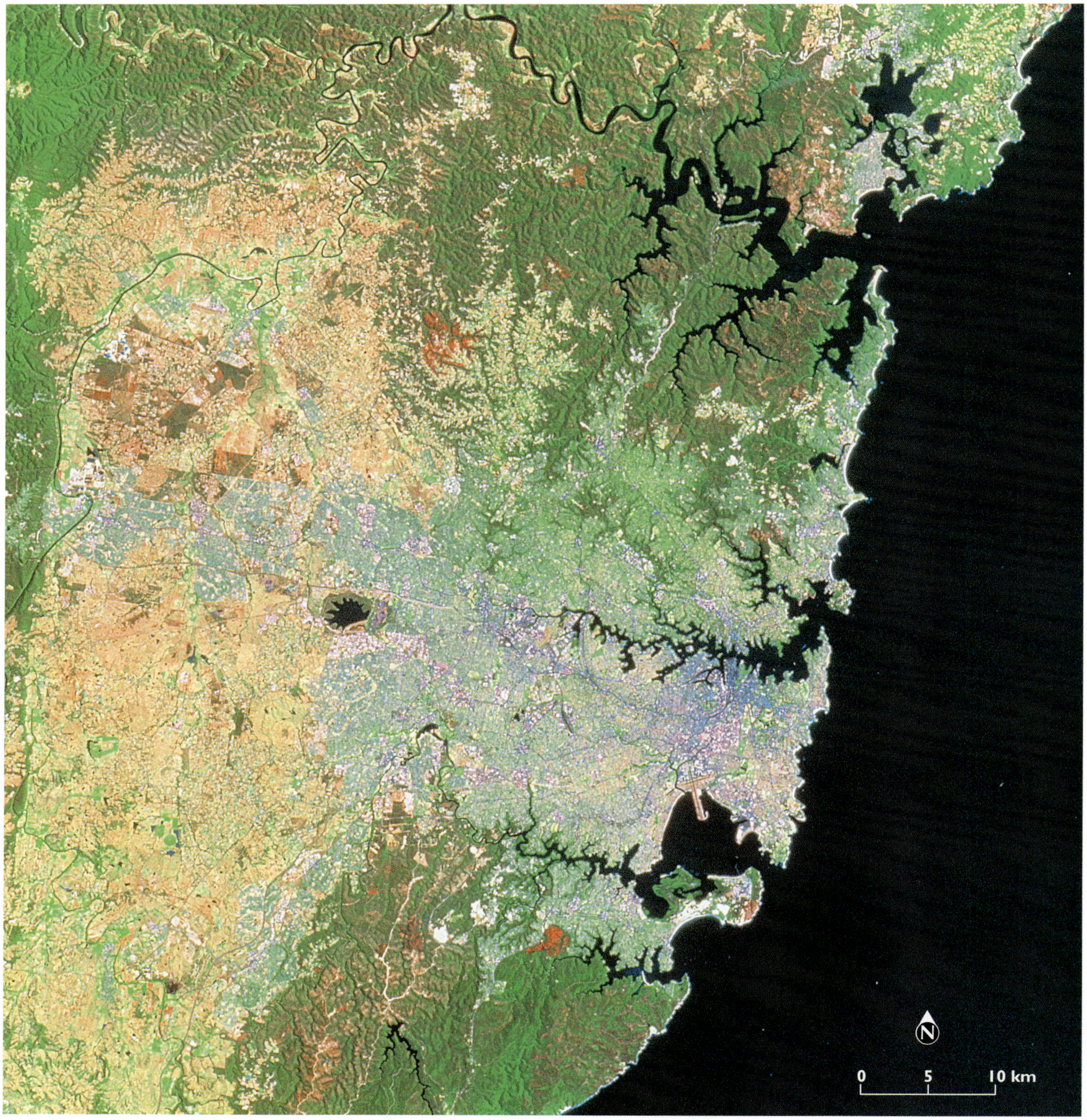

The sprawling suburbs of Sydney show up as the blue and purple area. They follow the Parramatta River westward through farmland (yellow-brown) to the Blue Mountains National Park (bright green) at the far west of the satellite image. Sydney's Kingsford Smith airport shows up as a brown cross on the northern point of Botany Bay.

Social and economic patterns

POPULATION CHANGE

Sydney's total population increased by 3.6% from 1986 to 1991. The main population growth was in the outer western and south-western areas.

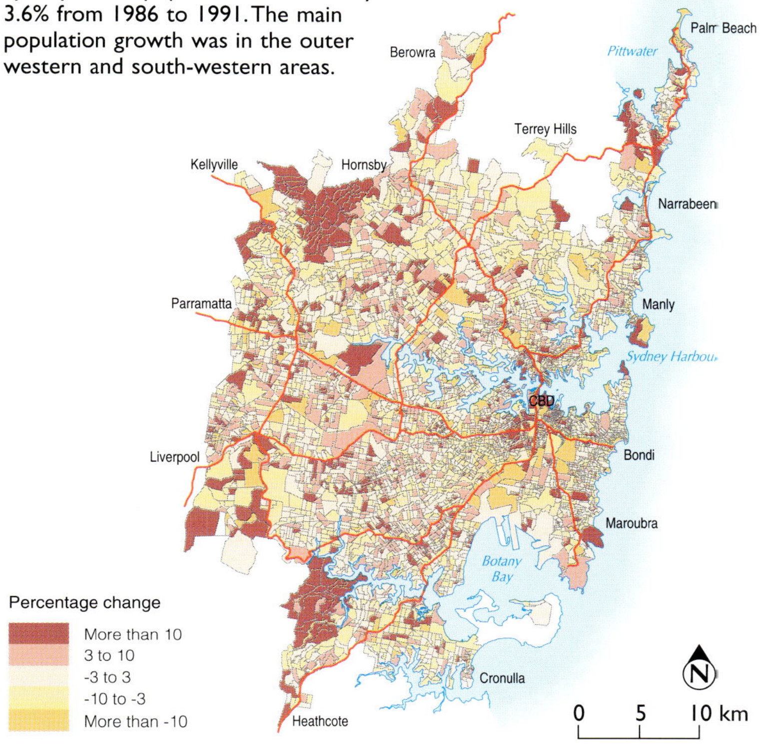

PEOPLE BORN OVERSEAS

In 1991, 32% of Sydney's population was born overseas. Most of these people live in the southern, south-western and western suburbs.

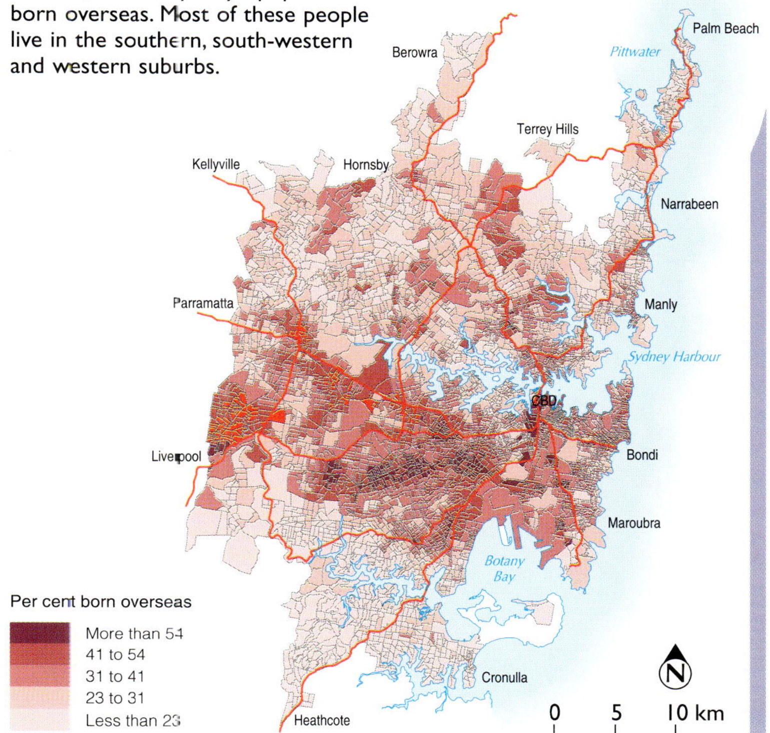

PEOPLE WITH UNIVERSITY QUALIFICATIONS

Areas with large numbers of university qualified people are located in the northern and eastern suburbs of Sydney.

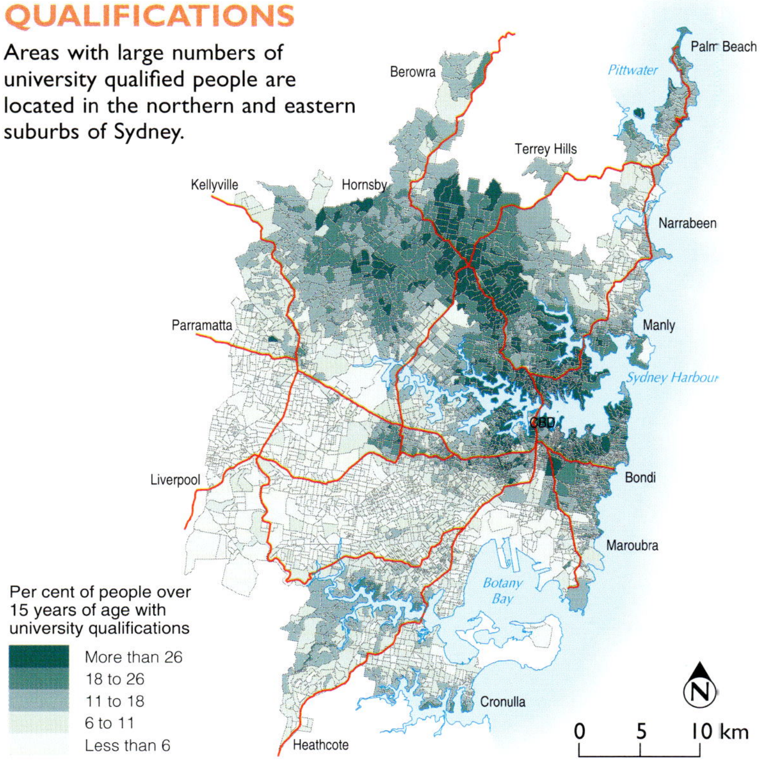

PEOPLE WITH INCOMES OVER $70 000

Many of the people with high incomes live around Sydney Harbour and in the northern suburbs.

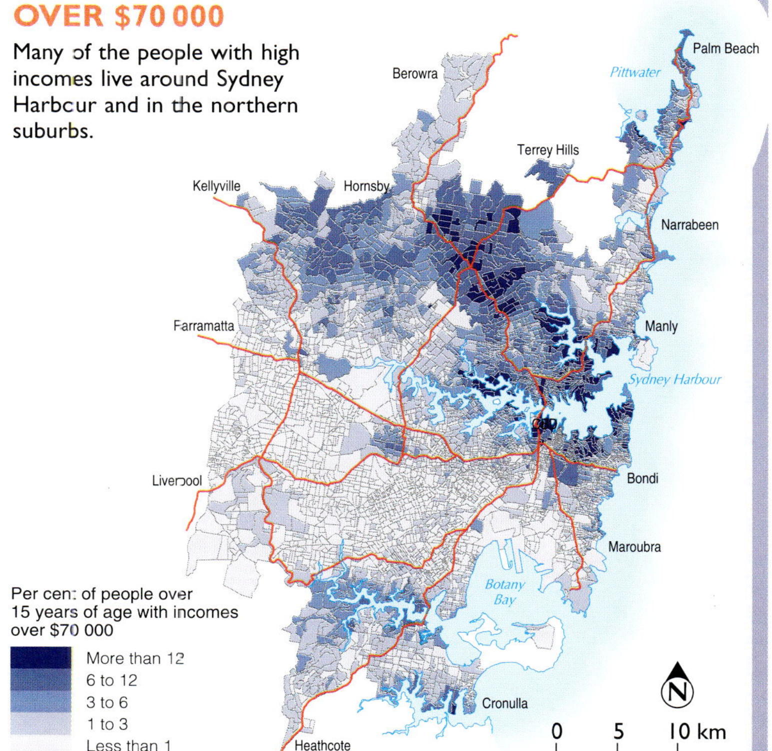

VICTORIA

Settlement

Aboriginal language groups

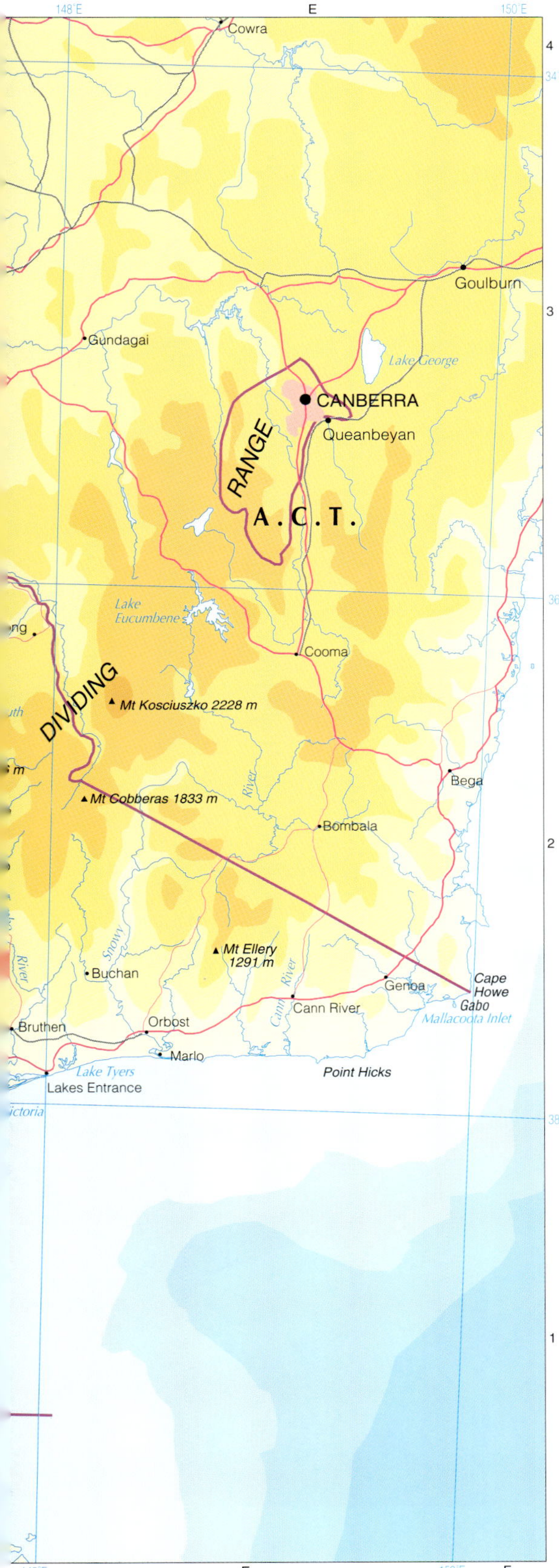

Lionel Rose

Lionel Rose, one of Australia's greatest fighters, was born in 1948 at Jackson's Creek, near Drouin in Victoria. He became Australian flyweight boxing champion in 1963. As a professional boxer he was bantamweight champion from 1966 to 1969. He won the world bantamweight title in 1968 by defeating Fighting Harada in Tokyo. He was voted Australian of the Year in 1968 and awarded an MBE. He had 53 fights, won 42 and lost 11. Lionel Rose became an inspiration to the whole nation and an idol to Aboriginal people.

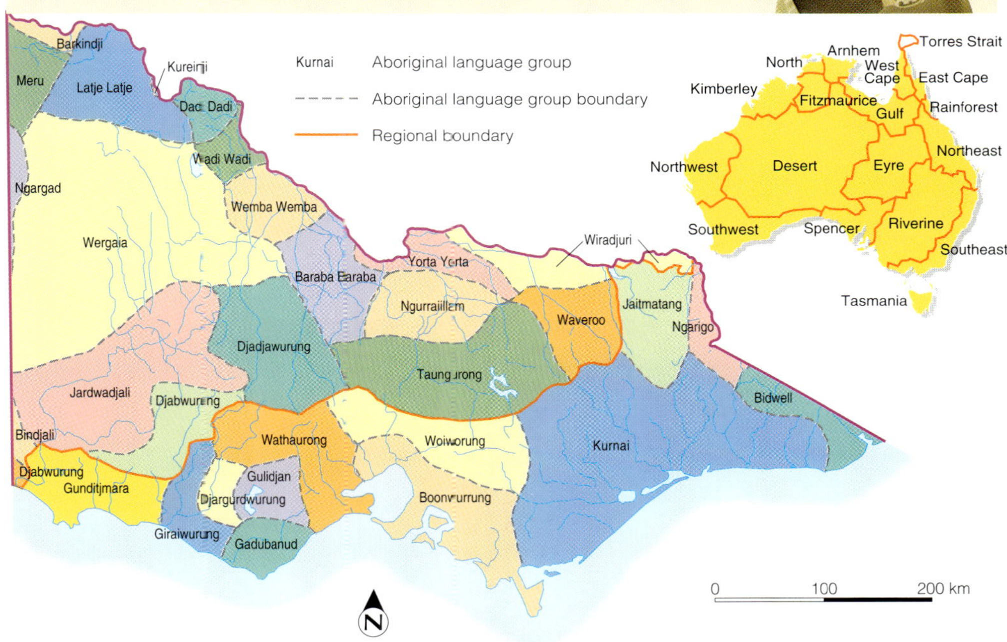

Key

Built environment

(line)	State/territory border
Victoria	State/territory name
(line)	Major road
(line)	Minor road
(line)	Railway
(area)	Urban area
■ MELBOURNE	More than 1 000 000 people
no example	500 000 to 1 000 000 people
● GEELONG	100 000 to 500 000 people
• Bendigo	50 000 to 100 000 people
• Portland	10 000 to 50 000 people
• Orbost	Less than 10 000 people

Scale

1 : 3 000 000

0 30 60 90 120 km

1 centimetre on the map represents 30 kilometres on the ground.

Natural environment

(line)	River
(shape)	Lake
Gabo	Island
▲ Mt Ellery 1291 m	Mountain and height in metres

- Land over 4000 metres
- Land from 2000 to 4000 metres
- Land from 1000 to 2000 metres
- Land from 500 to 1000 metres
- Land from 200 to 500 metres
- Land from sea level to 200 metres
- Land below sea level
- Sea from sea level to 200 metres
- Sea from 200 to 2000 metres
- Sea from 2000 to 4000 metres
- Sea from 4000 to 8000 metres
- Sea over 8000 metres

56 MELBOURNE

Change over time

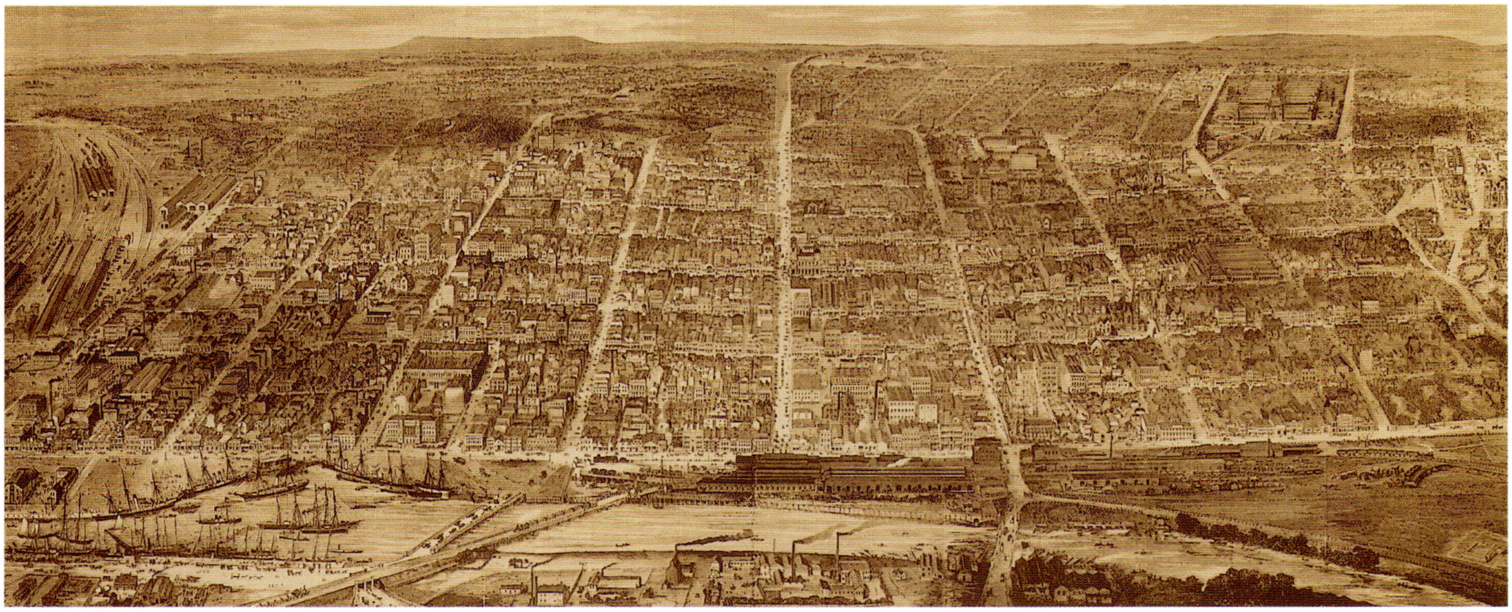

Melbourne, 1880

Melbourne today. The Yarra River is in the foreground of this photograph. Look closely and describe how the use of the river and its foreshores has changed since 1880.

Boom city

Melbourne was founded in 1835. From 1835–1850, it was a small service town for the expanding pastoral industry. However, after the gold rushes the population grew rapidly, stone and brick buildings went up everywhere and trade and commerce boomed.

In the 1870s and 1880s, a land boom occurred as suburbs were laid out. By 1880 Melbourne dominated Australian banking, manufacturing and shipping. It had excellent theatres, sporting facilities, libraries and exhibitions. However, the cost was high. Poor sanitation and poor building regulations led some to call it "Marvellous Smellbourne".

Modern highlights include the 1956 Olympics, the annual Melbourne Cup and AFL Grand Final, and the Australian Tennis Open. Although no longer Australia's largest city, it remains a major commercial and cultural centre and contains the head offices of many of Australia's largest companies.

Australian Rules Football at the Melbourne Cricket Ground (MCG), 1879. This was the first time electricity was used commercially in Victoria.

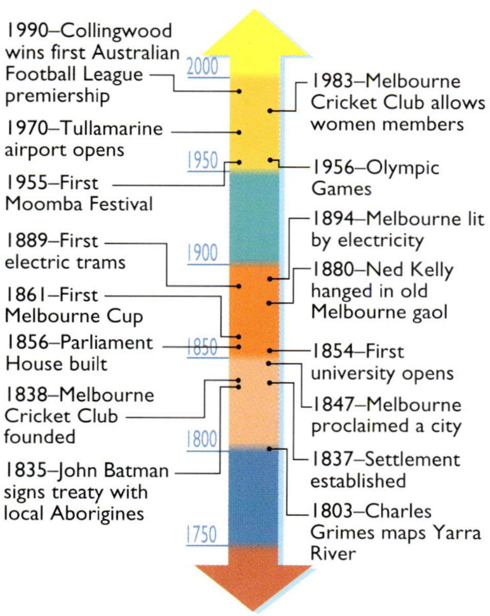

Night football and cricket at the MCG are very popular today. Describe how the MCG has changed.

MELBOURNE
Satellite image

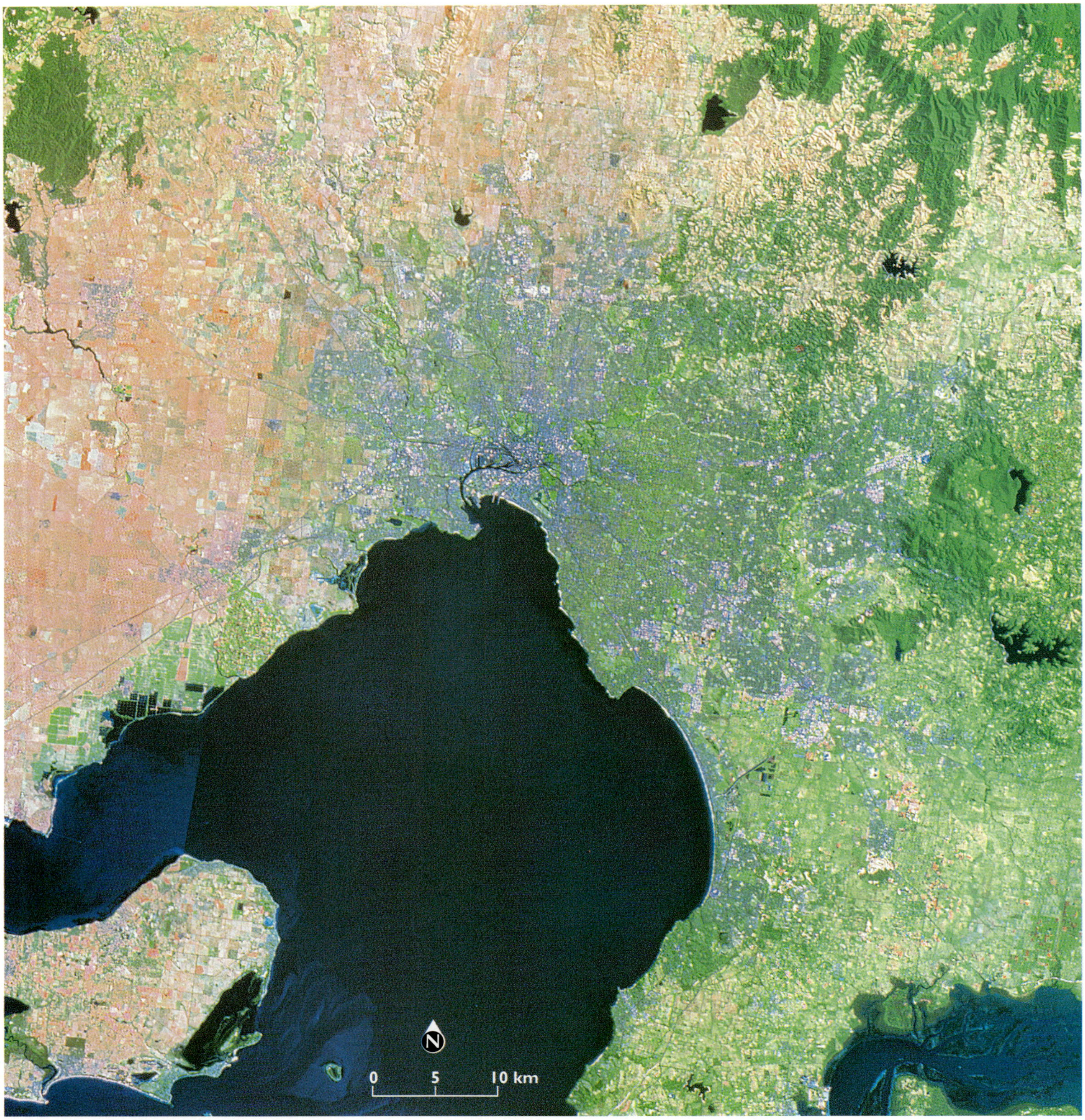

The urban area of Melbourne is the blue colour surrounding the head of Port Phillip Bay. Forest and park reserves form green corridors along the rivers and creeks of the northern and eastern suburbs. The pink-brown region to the north and west of Melbourne indicates farmland.

Social and economic patterns

POPULATION CHANGE

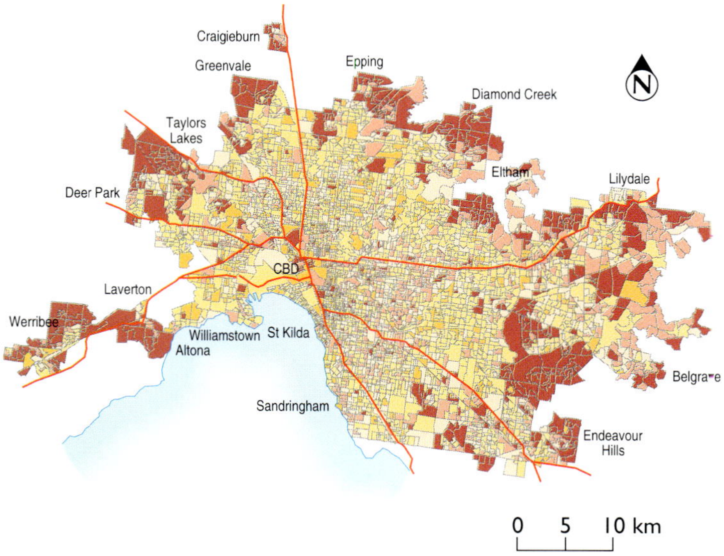

Percentage change

- More than 10
- 2 to 10
- -2 to 2
- -10 to -2
- More than -10

Between 1986 and 1991 the fastest-growing areas were the new outer suburbs, with population declines in many inner and middle suburbs.

PEOPLE BORN OVERSEAS

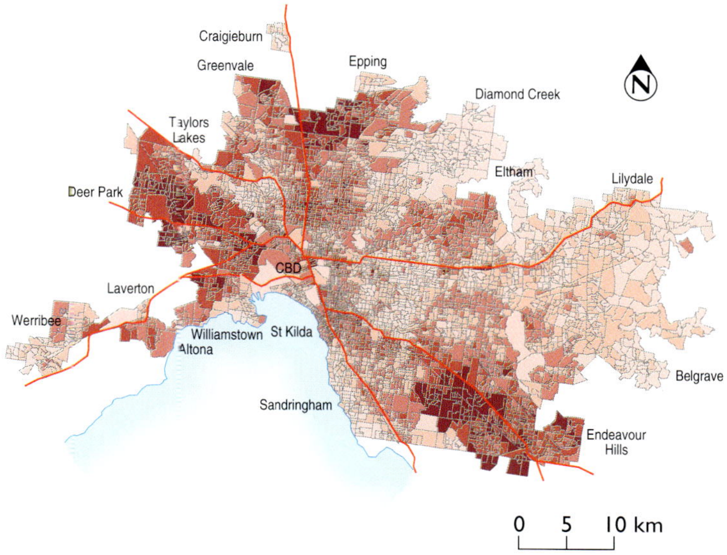

Per cent born overseas

- More than 48
- 37 to 48
- 28 to 37
- 21 to 28
- Less than 21

30% of Melbourne's population was born overseas. The highest concentrations are in the western, northern and outer south-eastern suburbs.

PEOPLE WITH UNIVERSITY QUALIFICATIONS

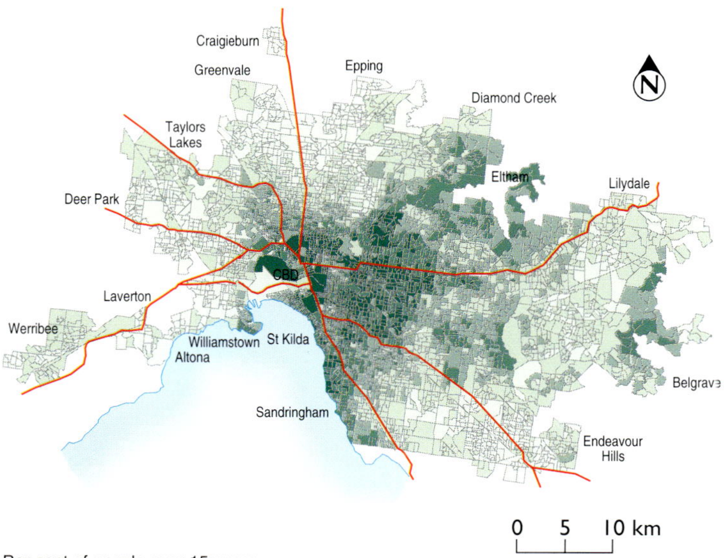

Per cent of people over 15 years of age with university qualifications

- More than 25
- 17 to 25
- 11 to 17
- 5 to 11
- Less than 5

People with university qualifications tend to live in the eastern suburbs, as well as the inner northern suburbs near Melbourne University.

PEOPLE WITH INCOMES OVER $70 000

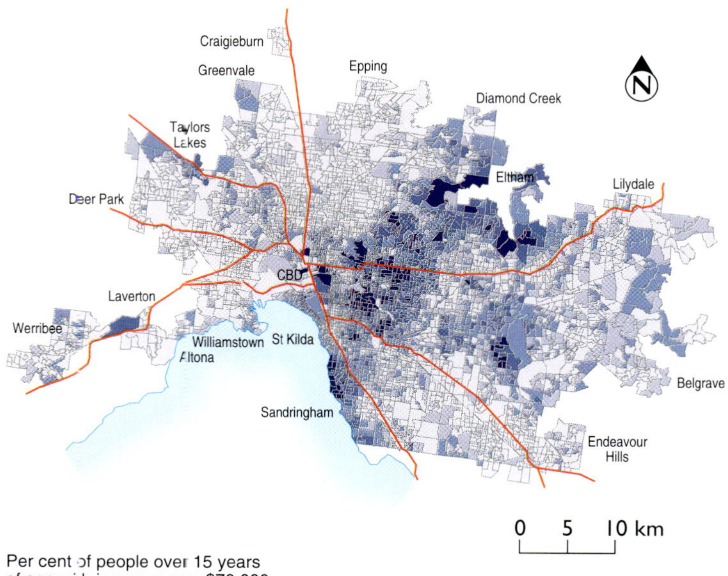

Per cent of people over 15 years of age with incomes over $70 000

- More than 11
- 5 to 11
- 2 to 5
- 1 to 2
- Less than 1

Areas of high income earners are mainly located in Melbourne's north-eastern and eastern suburbs and south-eastern bayside suburbs.

TASMANIA

Settlement

Aboriginal language groups

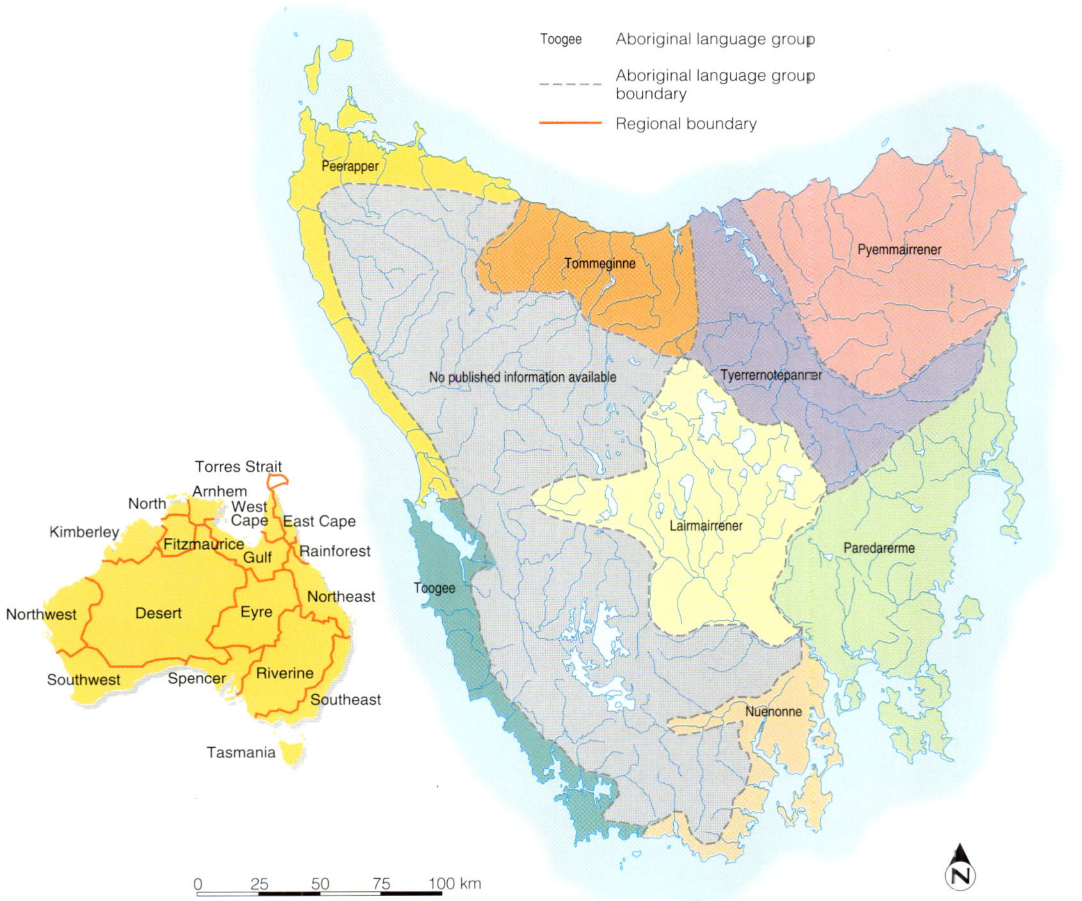

Key

Built environment

	State/territory border
Tasmania	State/territory name
	Major road
	Minor road
	Railway
	Urban area
no example	More than 1 000 000 people
no example	500 000 to 1 000 000 people
● HOBART	100 000 to 500 000 people
• Launceston	50 000 to 100 000 people
• Burnie	10 000 to 50 000 people
· Hastings	Less than 10 000 people

Scale

1 : 2 600 000

0 25 50 75 100 km

1 centimetre on the map represents 26 kilometres on the ground.

Natural environment

	River
	Lake
Pedra Branca	Island
▲ *Quoin Mtn 900 m*	Mountain and height in metres
	Land over 4000 metres
	Land from 2000 to 4000 metres
	Land from 1000 to 2000 metres
	Land from 500 to 1000 metres
	Land from 200 to 500 metres
	Land from sea level to 200 metres
	Land below sea level
	Sea from sea level to 200 metres
	Sea from 200 to 2000 metres
	Sea from 2000 to 4000 metres
	Sea from 4000 to 8000 metres
	Sea over 8000 metres

Aboriginal past

Tasmania has been isolated from the rest of Australia for 12 000 years. Aborigines inhabited all regions including the mountainous western areas. People traded over wide areas and large ceremonial gatherings were common. By the 1830s, most Aborigines in Tasmania had either been captured or shot by Europeans, or had died of disease. The survivors were forced to move to Flinders Island in Bass Strait where many died. The remaining Aborigines were taken to Oyster Cove south of Hobart. Truganini (pictured below) who died in 1876, aged 71, is often referred to as the last Tasmanian Aborigine. In fact, descendants of the original people live in Tasmania today — over 8000 in the last census.

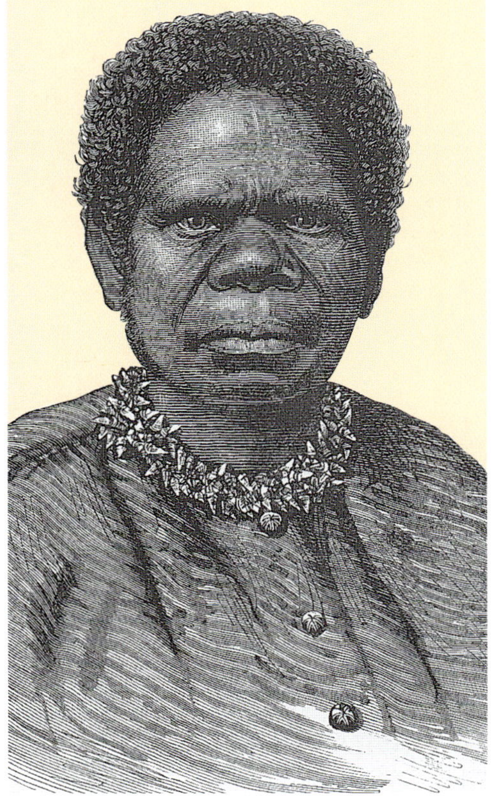

62 HOBART

Change over time

Hobart, 1879

Hobart today. What has changed? What remains the same?

Port town

Hobart is situated on the Derwent River about 20 kilometres from the coast. From 1803 to 1850 it developed as a service town for an expanding wheat–sheep industry. Whalers also used Storm Bay and the Derwent River from the beginning of settlement.

As a port town, Hobart was linked more to the sea than to the hinterland and more to London than to Sydney. European voyages into the Pacific usually stopped in Hobart for supplies and repairs. The export of grain and wool, as well as timber, hides, skins and meat, meant the wharves were always busy. Constitution Dock opened in 1850 and shipbuilding was a major industry.

By the 1830s Hobart was one of the great whaling ports of the world, servicing 35 whaling stations in Tasmania by 1841. Between 1818 and 1841, 16000 free settlers arrived and invested in new farms, factories, abattoirs, mills and wharves. After the gold rushes on the mainland, Hobart's growth slowed.

Today it is the centre of the rich agricultural areas around Tasmania's south-east. The city has great historical and geographical appeal and attracts the many tourists who visit Australia's "Island State".

Part of the Hobart waterfront in the early twentieth century

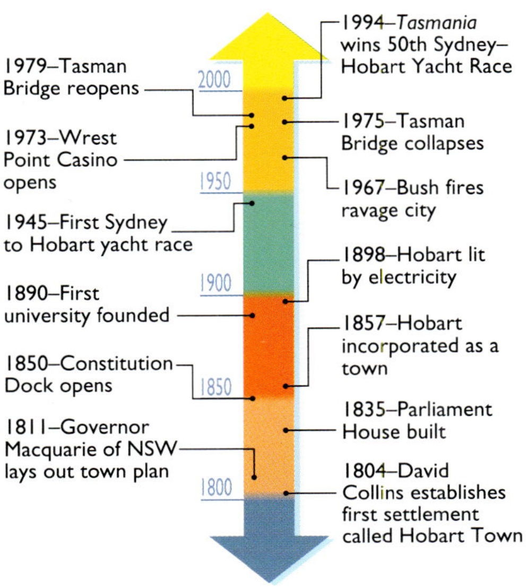

The same area today. In what ways has harbour use changed?

HOBART

Satellite image

The urban area of Hobart is the light-blue region shown in the centre of the image. Hobart is located on either side of the Derwent River, with Mount Wellington in the forested area to the west of the CBD. Farmland to the east of the Derwent River shows as pink on the satellite image.

Social and economic patterns

POPULATION CHANGE

Between 1986 and 1991, population showed the greatest increase in the Kingston area in Hobart's south and in the newer hillside suburbs overlooking the Derwent River. It has decreased through the older western shore suburbs.

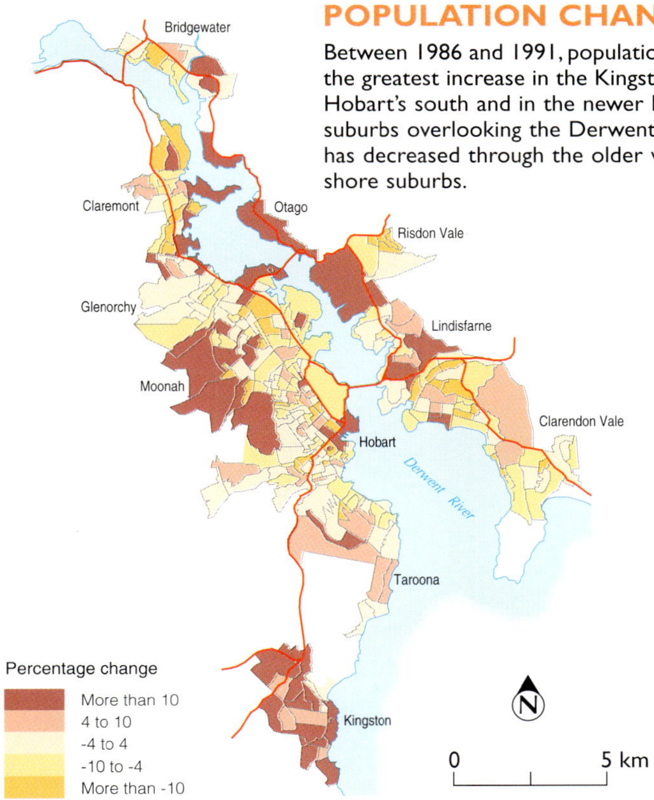

PEOPLE BORN OVERSEAS

Less than 15% of Hobart's population was born overseas. Most live on the western shores of the Derwent River. The eastern shore has very low percentages of overseas-born people.

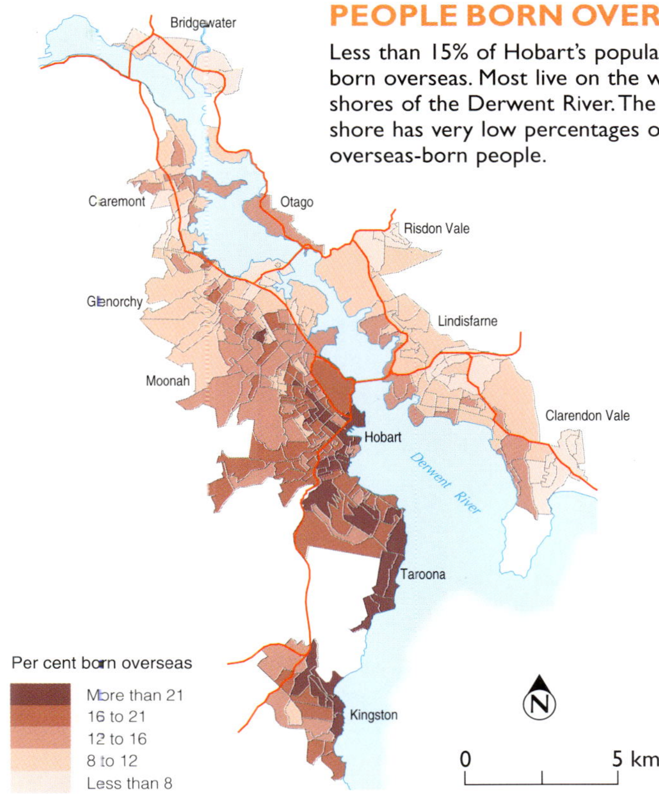

PEOPLE WITH UNIVERSITY QUALIFICATIONS

A number of inner Hobart rental areas and the south-western suburbs have high proportions of people who have university qualifications.

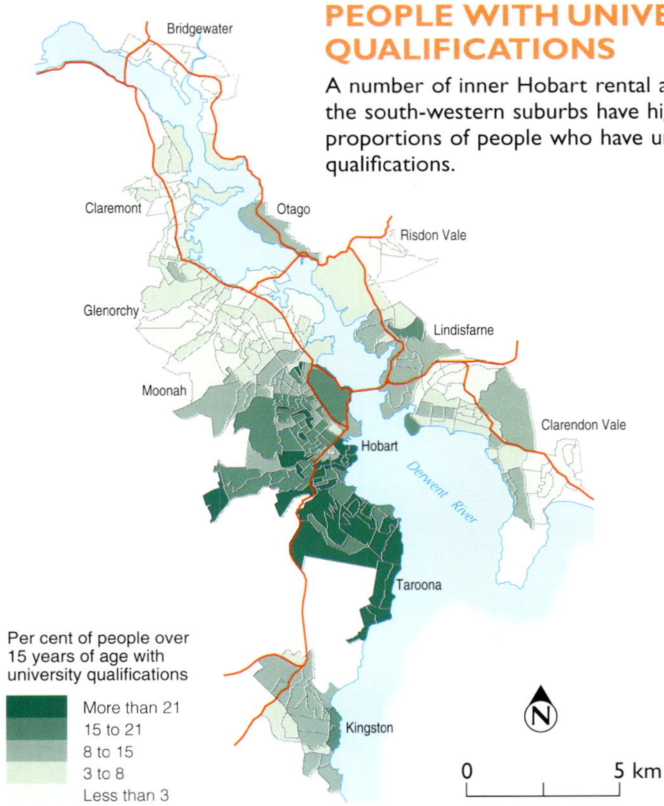

PEOPLE WITH INCOMES OVER $50 000

Many people with high incomes live around the western shore of the Derwent River.

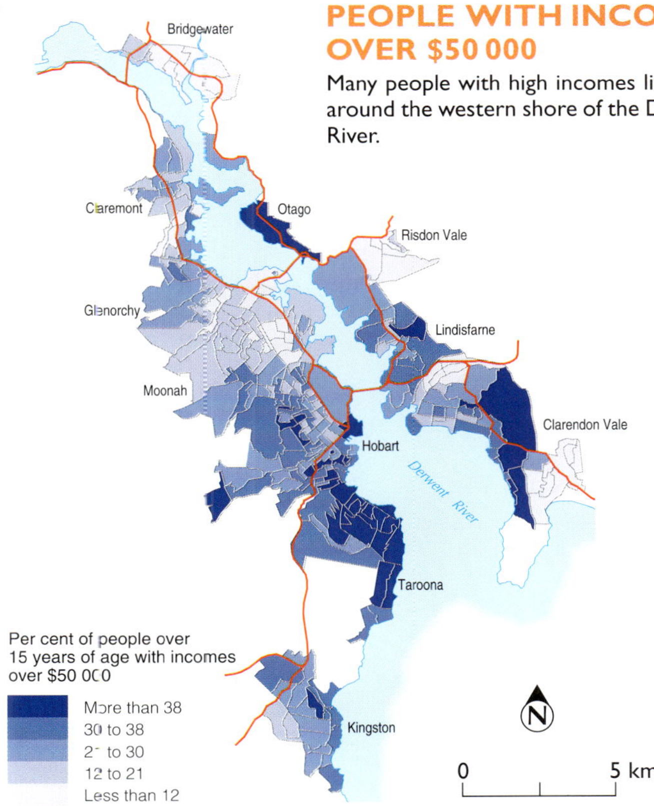

SOUTH AUSTRALIA

Settlement

Aboriginal language groups

Ben Murray

Ben Murray was born in 1893 near Frome Creek, east of Marree in northern South Australia. He learned the Thirrari language from his grandmother, and, at the time of his death in 1994, was its only speaker. He also learnt to read and write in German, English and Dieri at a German mission in the Birdsville area.

In 1915, Ben joined the Australian Army and went to Gallipoli where he narrowly escaped death and was taken prisoner. After the war he worked in the pastoral industry.

Ben Murray contributed to the study of the Dieri and Wangkangurru languages. He alone made possible the recording of the Thirrari language. Murray's historical knowledge of the Lake Eyre Basin area was unrivalled.

Key

Built environment

(line)	State/territory border
Victoria	State/territory name
(line)	Major road
(line)	Minor road
(line)	Railway
(area)	Urban area
■ **ADELAIDE**	More than 1 000 000 people
no example	500 000 to 1 000 000 people
no example	100 000 to 500 000 people
no example	50 000 to 100 000 people
• Port Augusta	10 000 to 50 000 people
• Cook	Less than 10 000 people
▫ *'Innamincka'*	Homestead/mission

Natural environment

(line)	River
(area)	Lake
Wardang	Island
▲ *Mt Lofty 725 m*	Mountain and height in metres
	Land over 4000 metres
	Land from 2000 to 4000 metres
	Land from 1000 to 2000 metres
	Land from 500 to 1000 metres
	Land from 200 to 500 metres
	Land from sea level to 200 metres
	Land below sea level
	Sea from sea level to 200 metres
	Sea from 200 to 2000 metres
	Sea from 2000 to 4000 metres
	Sea from 4000 to 8000 metres
	Sea over 8000 metres

Scale

1 : 7 000 000

0 100 200 300 km

1 centimetre on the map represents 70 kilometres on the ground.

ADELAIDE

Change over time

A planned city

Adelaide began as a planned city in 1836–37. It was laid out on a level, coastal plain astride the Torrens River, 10 kilometres from the coast. Areas were set aside for residential, commercial, recreational and government use, with parks separating these zones. Suburbs like Colonel Light Gardens were carefully laid out in 1924.

A huge, but well-planned, suburban expansion occurred in the 1950s to cater for post-war migration. In 1976 Rundle Street was closed to create a mall and parts of the city such as Victoria Square, North Adelaide and Port Adelaide were declared protected "heritage areas".

The idea of Adelaide as a planned city continued when it won the race to establish Australia's first Multi-Function Polis (MFP). However, the development of this scientific, educational and commercial zone in the northern suburbs has been slow.

Every alternate year, many people visit the city to participate in the Adelaide Arts Festival. Its art exhibitions, concerts and dance and theatre performances make it Australia's foremost arts festival.

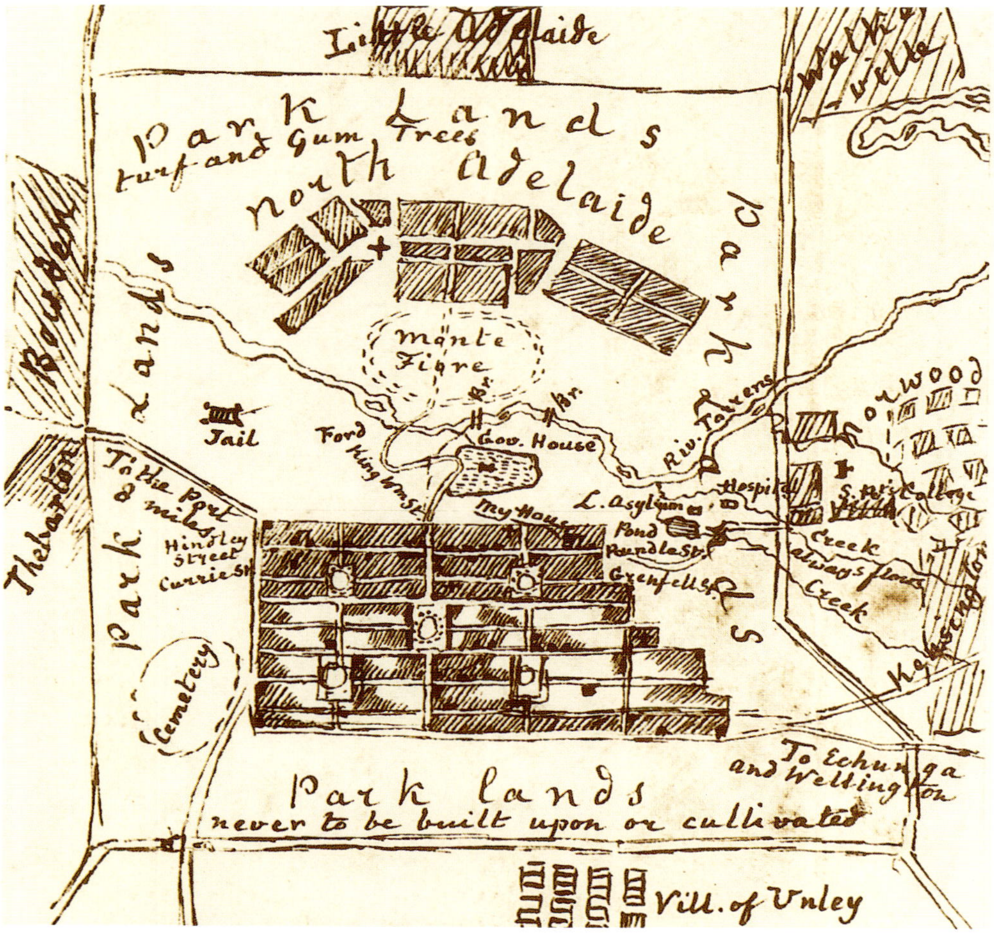

Map of Adelaide, 1853 from the letters of Charles Hastings Barton, an early resident

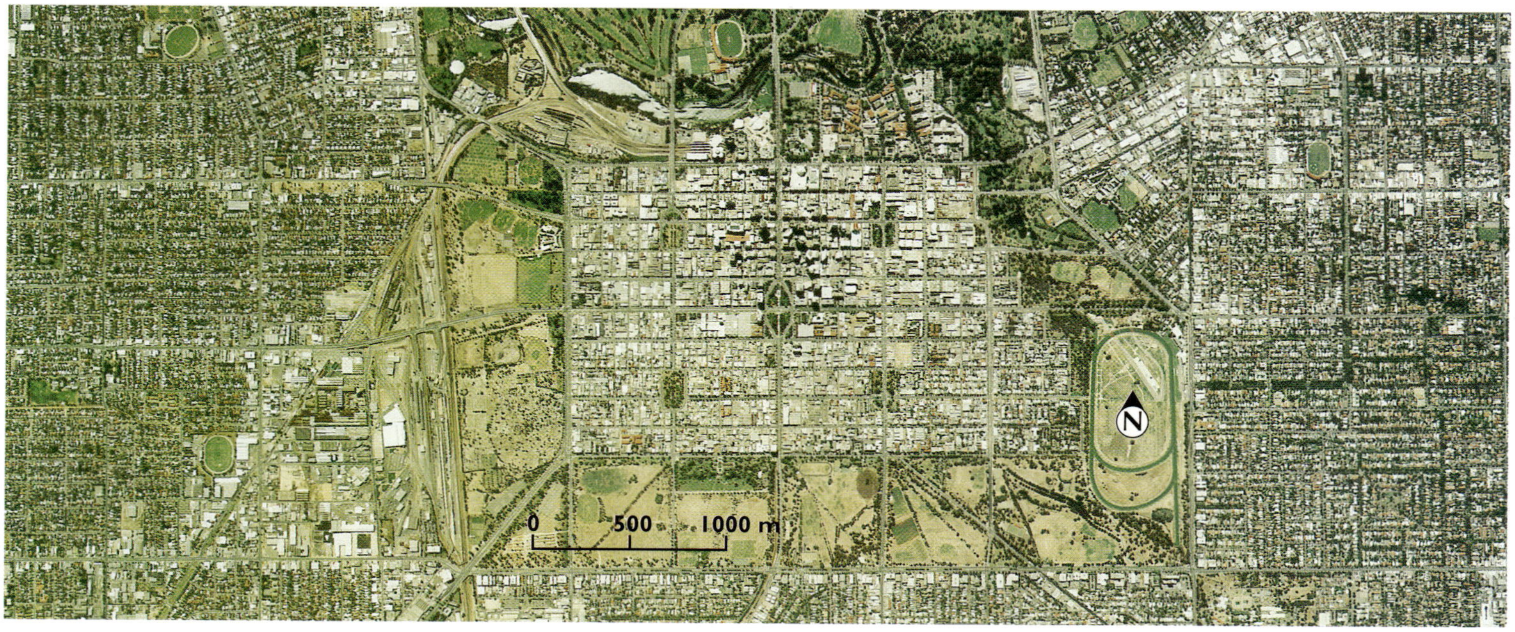

Compare the map with the modern photograph and identify any common features. Try to find the location of Barton's house on the photograph.

In the early twentieth century, Glenelg Beach in Adelaide was a very popular venue for sailing regattas.

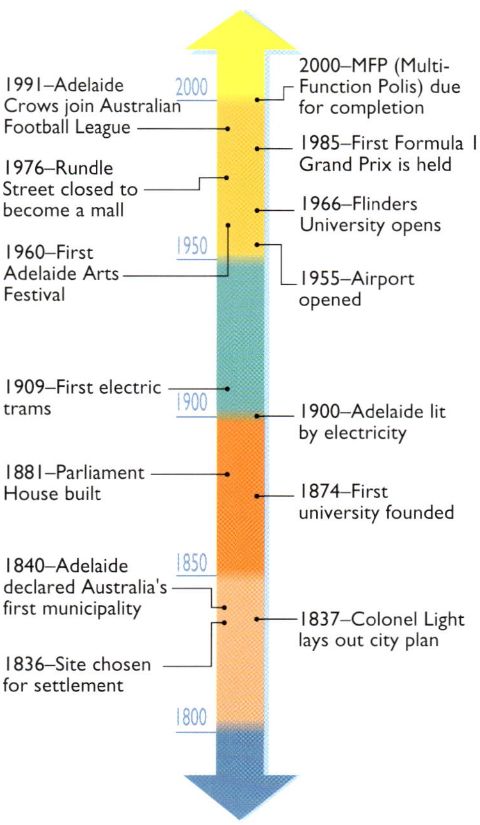

Today, Glenelg Beach is used mainly for swimming. What remains the same in the two photographs?

ADELAIDE

Satellite image

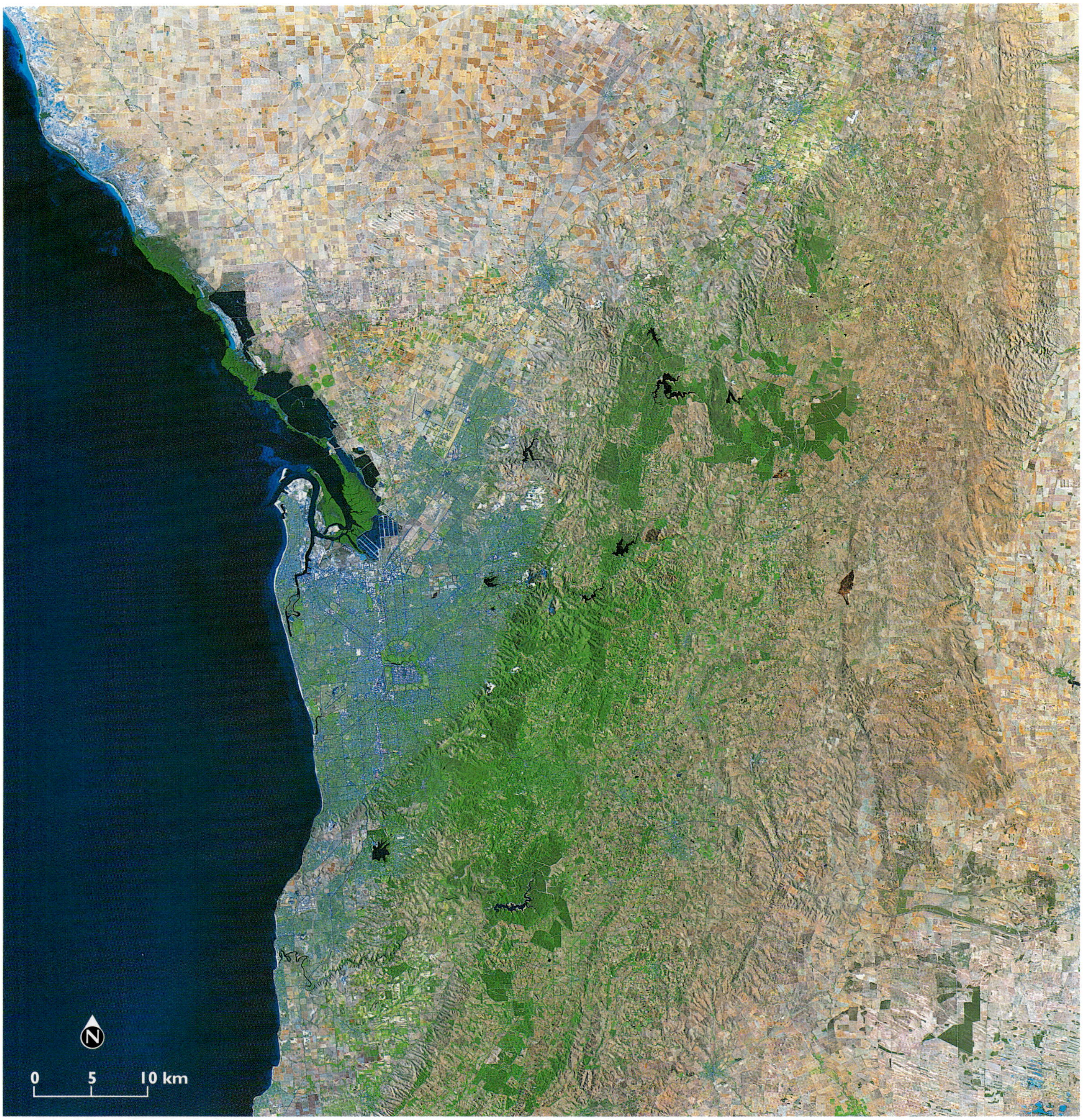

Adelaide's urban area, shown as turquoise in this satellite image, is wedged between St Vincent Gulf to the west and the Mount Lofty Ranges (green) to the east. The Central Business District can be clearly seen as it is surrounded by gardens and parklands (green). Farmland shows up as pink, brown, and light blue rectangles on this image.

Social and economic patterns

POPULATION CHANGE

Adelaide's strongest population growth between 1986 and 1991 was in the newly established suburbs in the outer southern and northern areas. The hill suburbs around Bridgewater continue to attract more people.

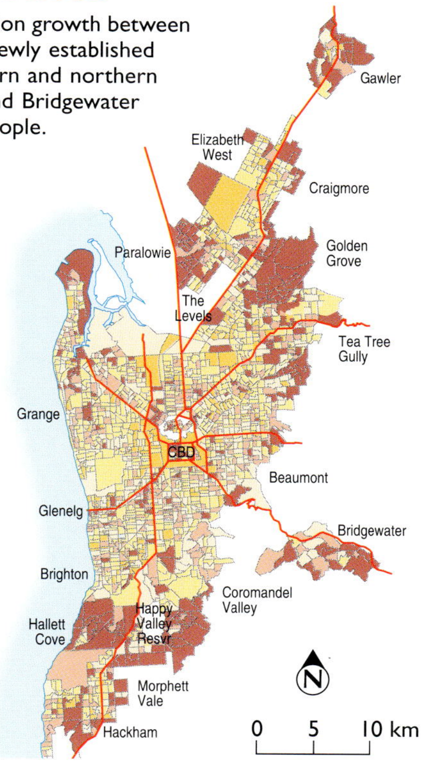

Percentage change

- More than 10
- 2 to 10
- -2 to 2
- -10 to -2
- More than -10

PEOPLE BORN OVERSEAS

More than 26% of Adelaide's population was born overseas. The north-western suburbs contain mainly South-East Asian-born migrants. The outer northern suburbs such as Elizabeth were settled by British migrants in the 1950s and 1960s.

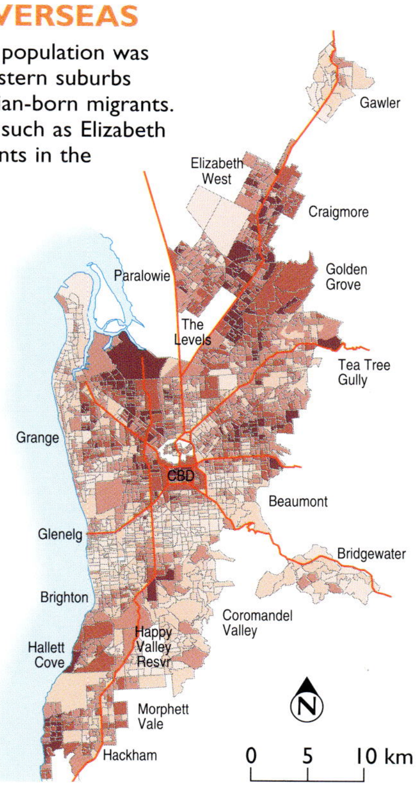

Per cent born overseas

- More than 37
- 30 to 37
- 24 to 30
- 19 to 24
- Less than 19

PEOPLE WITH UNIVERSITY QUALIFICATIONS

Adelaide's south-eastern suburbs contain the highest percentage of people with university qualifications. The inner northern suburbs, near the University of Adelaide, also record high percentages.

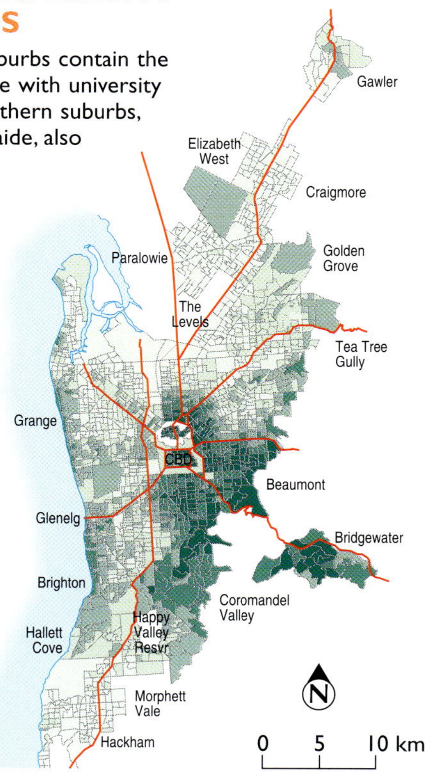

Per cent of people over 15 years of age with university qualifications

- More than 21
- 13 to 21
- 7 to 13
- 4 to 7
- Less than 4

PEOPLE WITH INCOMES OVER $50 000

Many high-income earners live in Adelaide's south-eastern suburbs and the hill suburbs around Bridgewater. The inner northern and some coastal suburbs also record larger percentages of high-income earners.

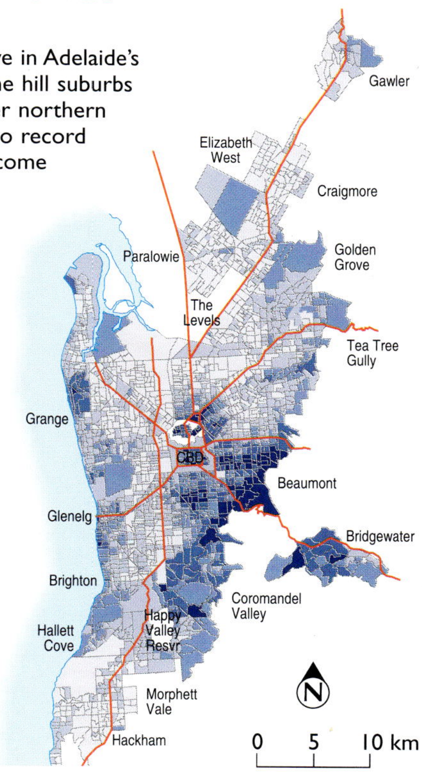

Per cent of people over 15 years of age with incomes over $50 000

- More than 11
- 6 to 11
- 3 to 6
- 1 to 3
- Less than 1

72

WESTERN AUSTRALIA

Settlement

Aboriginal language groups

Rover Thomas

Rover Thomas was born in the early 1920s at Yalda Soak near Well 33 on the Canning Stock Route in Western Australia, of Kukatja descent. As part of a cultural revival, Rover Thomas and a relative introduced a new ceremony, the Kuril Kuril, into the Wasmun community in the Kimberley region. In 1983 the ceremony was performed at the Aboriginal Festival in Perth with images painted by Thomas.

Rover Thomas became a highly acclaimed artist.

Maya — Aboriginal language group

Aboriginal language group boundary

Regional boundary

No published information available

Key

Built environment

State/territory border

South Australia — State/territory name

Major road

Minor road

Railway

Urban area

■ PERTH — More than 1 000 000 people

no example — 500 000 to 1 000 000 people

● WANNEROO — 100 000 to 500 000 people

no example — 50 000 to 100 000 people

• Kalgoorlie — 10 000 to 50 000 people

• Menzies — Less than 10 000 people

▫ 'Glenroy' — Homestead/mission

Natural environment

River

Lake

Barrow Island — Island

Mt Meharry 1249 m — Mountain and height in metres

Land over 4000 metres

Land from 2000 to 4000 metres

Land from 1000 to 2000 metres

Land from 500 to 1000 metres

Land from 200 to 500 metres

Land from sea level to 200 metres

Land below sea level

Sea from sea level to 200 metres

Sea from 200 to 2000 metres

Sea from 2000 to 4000 metres

Sea from 4000 to 8000 metres

Sea over 8000 metres

Scale

1 : 11 500 000

0 100 200 300 400 km

1 centimetre on the map represents 115 kilometres on the ground.

PERTH

Change over time

Modern city

Perth is situated on the banks of the Swan River, 18 kilometres from the port of Fremantle at the river's mouth.

Perth's rapid growth (from a population of 500 000 in 1966 to 1.2 million in 1996) is largely due to the discovery of extensive mineral deposits, especially iron ore in the Pilbara region of Western Australia.

In the late 1960s, this increasing population led to increased business activity and the city centre experienced much change and growth. Many of Perth's older buildings were replaced by tall office blocks and hotel towers, transforming the skyline. The CBD is restricted to a small area because of the river to the south, the Mitchell Freeway to the west and the suburban railway network to the east. Today, although a small city by world population standards, the CBD displays the high-rise development normally associated with much larger cities.

Map of Perth, 1838 by John Arrowsmith, London geographer and map maker

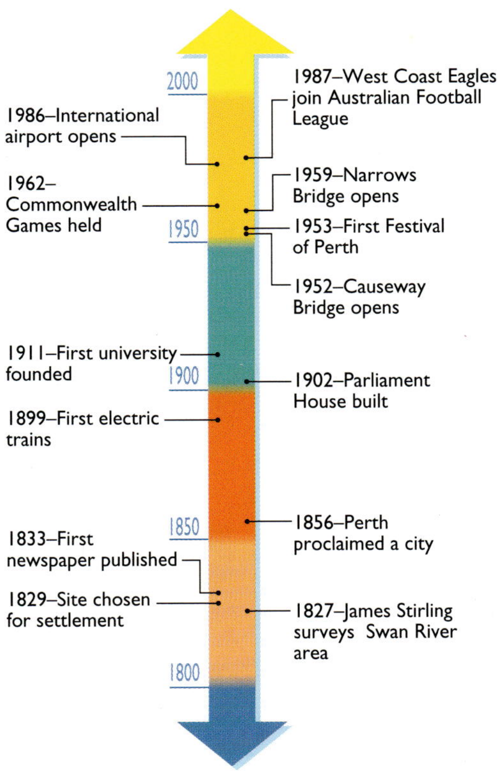

Perth today. What has changed? What has stayed the same?

Horace Samson
Perth 1847
watercolour, gouache and pen and ink
27.5 × 40.3 cm
Collection Art Gallery of Western Australia, presented by Mr D. Rannard

Perth today. This photograph is taken from Kings Park. Compare it with the painting above. What has happened to Perth since 1847?

76 PERTH

Satellite image

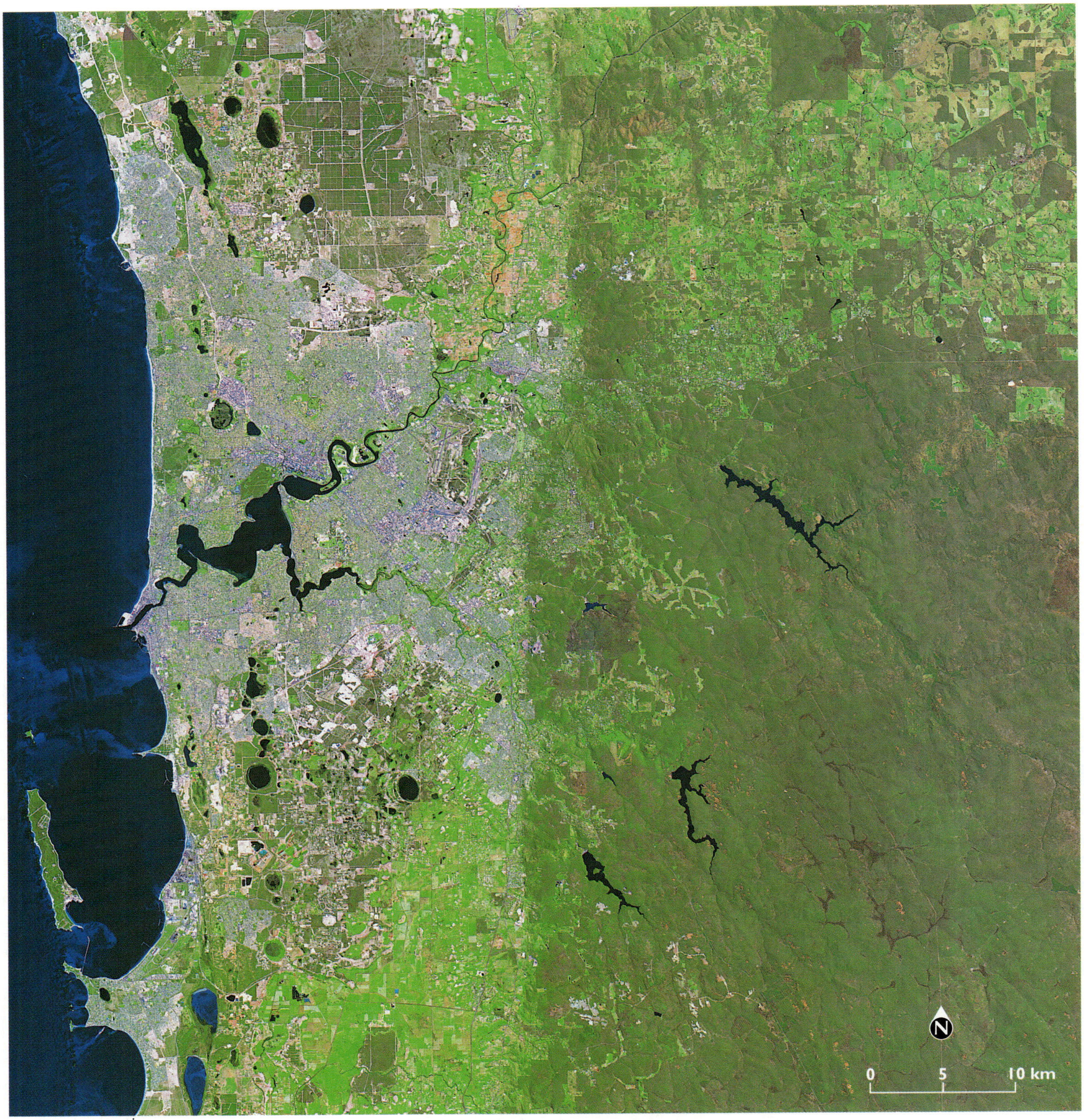

The urban area of Perth (shown in purple-blue) is centred around the Swan River, and spreads north and south along the coastline and east along the Swan and Canning Rivers. Dark green areas represent forest, and light green areas show farmland. Areas of lush growth, such as irrigated fields, show as pink-brown on this satellite image.

Social and economic patterns

POPULATION CHANGE

Between 1986 and 1991 there was significant growth in the outer areas, particularly to the north-west and south-west of the Central Business District.

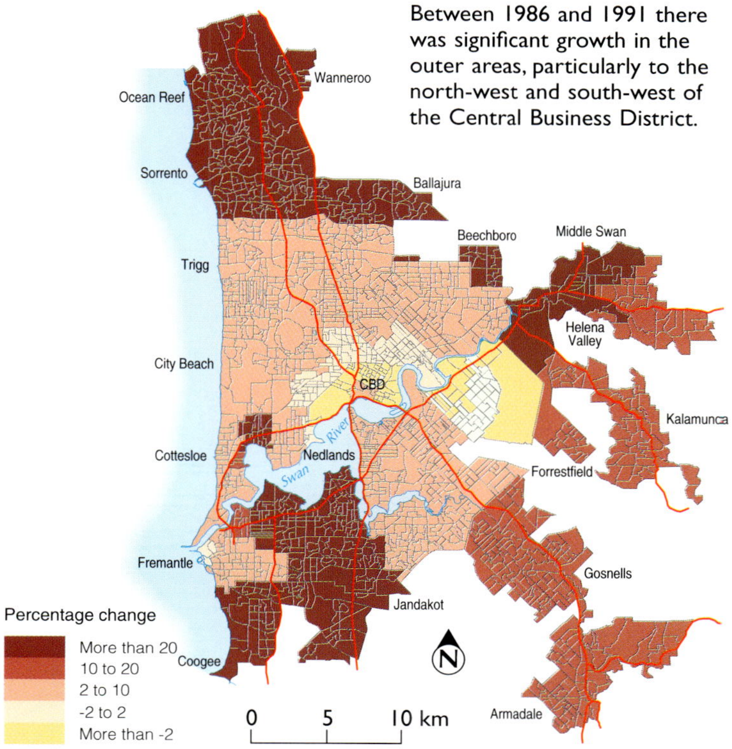

PEOPLE BORN OVERSEAS

In 1991, 34% of Perth's population was born overseas. New arrivals have traditionally settled around the city centre and in the northern city fringe areas.

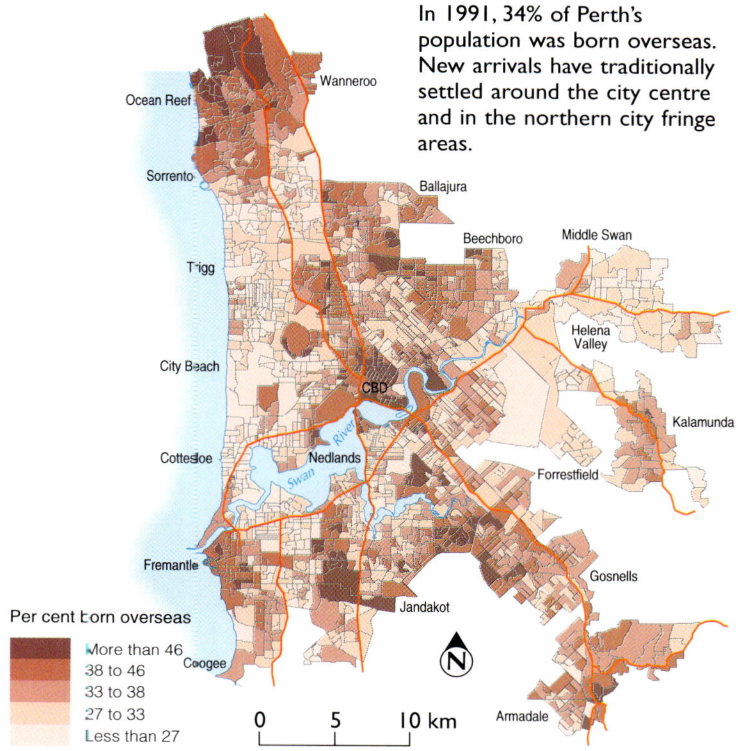

PEOPLE WITH UNIVERSITY QUALIFICATIONS

Many residents of the high-income suburbs around Nedlands hold university qualifications. The University of Western Australia is also located in this area.

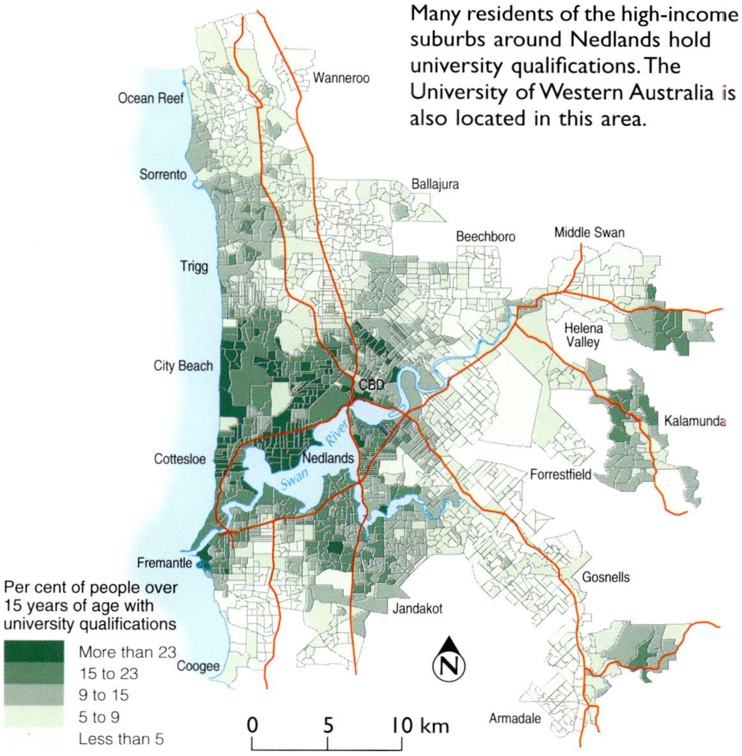

PEOPLE WITH INCOMES OVER $50 000

Many high-income earners live near the Swan River and the coast in Perth's western suburbs.

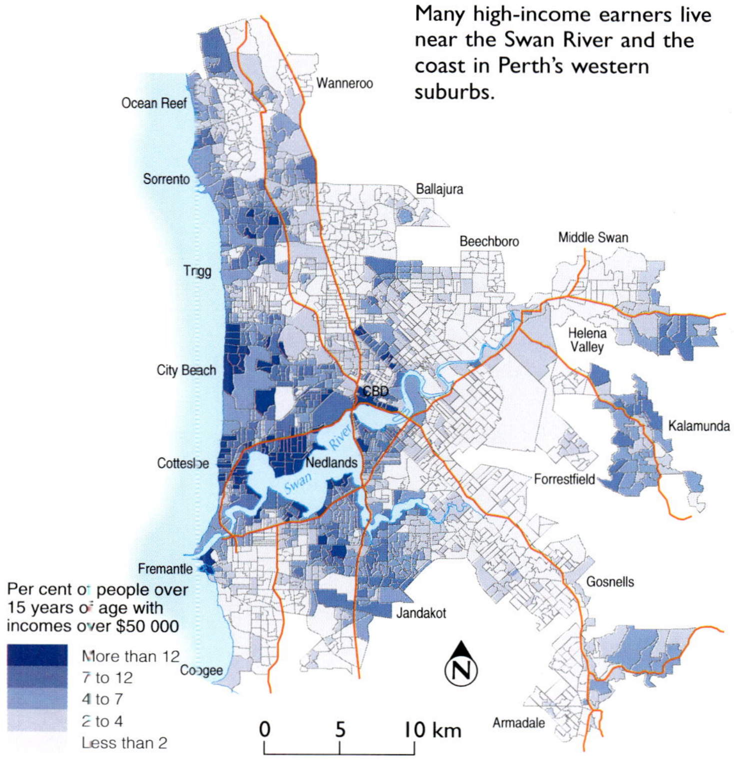

CANBERRA

National capital

Choosing a site

Canberra has three histories. Prior to the 1820s, it was home to Aboriginal peoples. For the next 100 years it was a quiet pastoral district and small village. In 1927, it became the national capital.

Debate about the choice of a site for the national capital went on for ten years before Canberra was finally chosen. Many sites were suggested. Victorian politicians suggested Melbourne, Sydney politicians suggested Sydney, and Tasmanians suggested Hobart.

Finally it was agreed that it should be in New South Wales, but inland, to protect it from possible naval and military attack. To appease the rivalry between Melbourne and Sydney, a site equidistant from both cities would be necessary. Canberra was not even suggested at first. It was a compromise put forward very late in the debate after politicians could not agree on Tumut or Albury.

Once the site was chosen, events moved quickly and 2590 square kilometres were acquired to form the Australian Capital Territory. An international competition for the capital's design was won by Walter Burley Griffin in 1912 and, in 1913, construction of the first public buildings began.

Today, many Australians make the journey to the national capital to visit buildings such as the War Memorial, Mint, National Gallery and the new Parliament House.

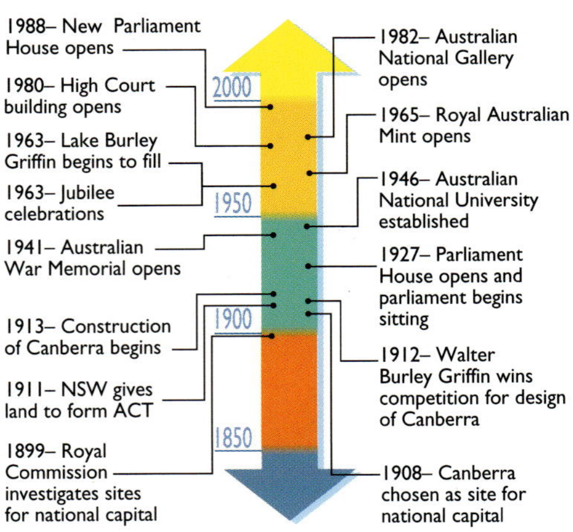

Walter Burley Griffin's amended 1913 plan for Canberra

SUBURBS OF CANBERRA

CENTRAL CANBERRA

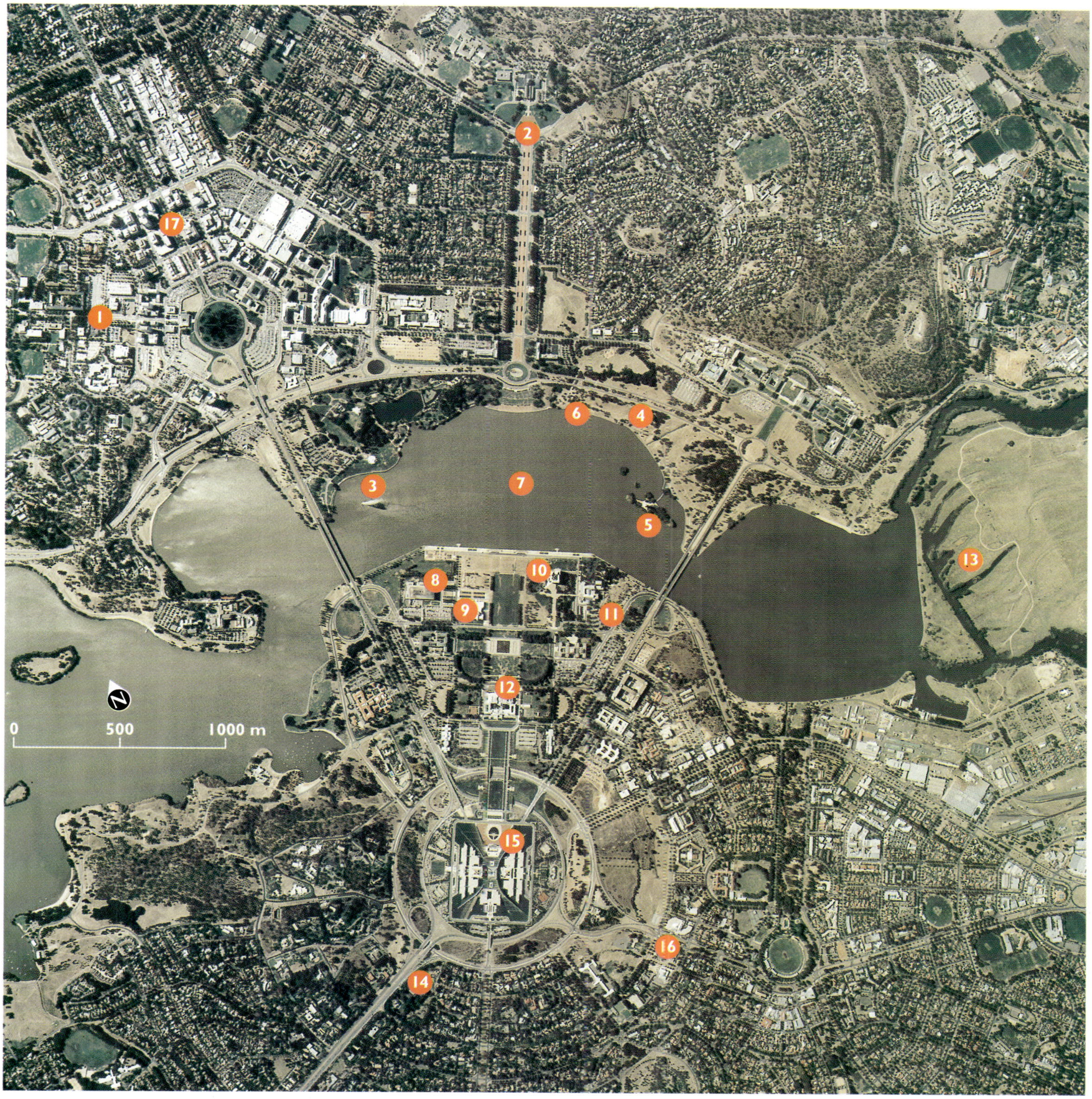

1. Australian National University
2. Australian War Memorial
3. Captain Cook Memorial Water Jet
4. Commonwealth Park
5. Carillon (belltower)
6. Blundell's Farmhouse
7. Lake Burley Griffin
8. National Library
9. National Science and Technology Centre
10. High Court
11. Australian National Gallery
12. Old Parliament House
13. Canberra Nature Park
14. Prime Minister's Lodge
15. Parliament House
16. Jewish Memorial Centre
17. Central Business District

NORTHERN TERRITORY

Settlement

Aboriginal language groups

Amarak — Aboriginal language group

Aboriginal language group boundary

Regional boundary

No published information available

Key

Built environment

State/territory border

Queensland — State/territory name

Major road

Minor road

Railway

no example — More than 1 000 000 people

no example — 500 000 to 1 000 000 people

no example — 100 000 to 500 000 people

• Darwin — 50 000 to 100 000 people

• Alice Springs — 10 000 to 50 000 people

• Elliott — Less than 10 000 people

▫ 'Elsey' — Homestead/mission

Scale

1 : 8 000 000

0 100 200 300 km

1 centimetre on the map represents 80 kilometres on the ground.

Natural environment

River

Lake

Groote Eylandt — Island

▲ Mt Leibig 1524 m — Mountain and height in metres

Land over 4000 metres

Land from 2000 to 4000 metres

Land from 1000 to 2000 metres

Land from 500 to 1000 metres

Land from 200 to 500 metres

Land from sea level to 200 metres

Land below sea level

Sea from sea level to 200 metres

Sea from 200 to 2000 metres

Sea from 2000 to 4000 metres

Sea from 4000 to 8000 metres

Sea over 8000 metres

David Gulpilil

David Gulpilil is a member of the Yolngu people of the Arnhem region and is a dancer, painter and musician. He is also widely recognised by Australian filmgoers as an actor. At 16, he was chosen for the leading role in the film *Walkabout*. Other roles in films such as *Storm Boy* and *Mad Dog Morgan* followed. In 1988 he was awarded the Order of Australia.

Currently, he performs the dances of his people at arts festivals in Australia and internationally. When not travelling to performances, he lives in the Northern Territory where he is planning to develop a cultural centre.

DARWIN

Change over time

Isolation on the northern coast

Darwin is located on a peninsula surrounding a natural harbour. The original settlement in 1863 was called Palmerston but was renamed Darwin in 1911 when the Federal Government took over control. Four earlier settlements on the north coast had failed between 1824 and 1864: Melville Island, Raffles Bay, Port Essington and Escape Cliffs.

A harsh tropical environment and isolation from the rest of Australia have always affected Darwin. Even though communications and transport networks improved in the late 1800s, Darwin grew only slowly. Discovery of gold at Pine Creek in 1872 brought a temporary influx of people. Growth was further slowed in the twentieth century by Japanese air-attacks during World War II. In 1974, the city was almost completely destroyed by Cyclone Tracy after which all but 12 000 of Darwin's inhabitants were evacuated south.

Today a re-built Darwin has recovered and prospered. Harbourside land is being developed for residential and tourist accommodation. The new Parliament House and the law courts are among Australia's finest. Road and air links have improved and many tourists use Darwin as a base for exploring the "Top End".

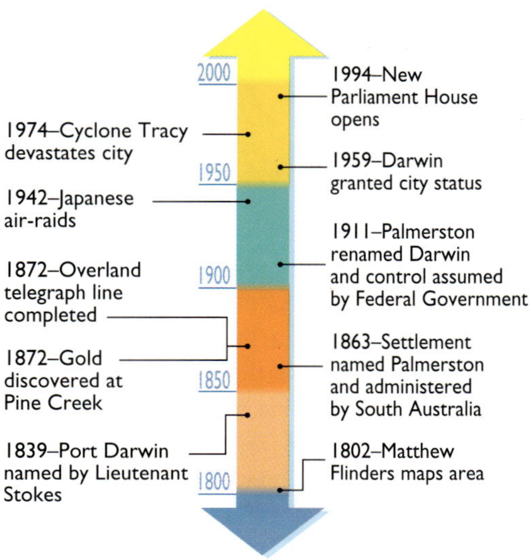

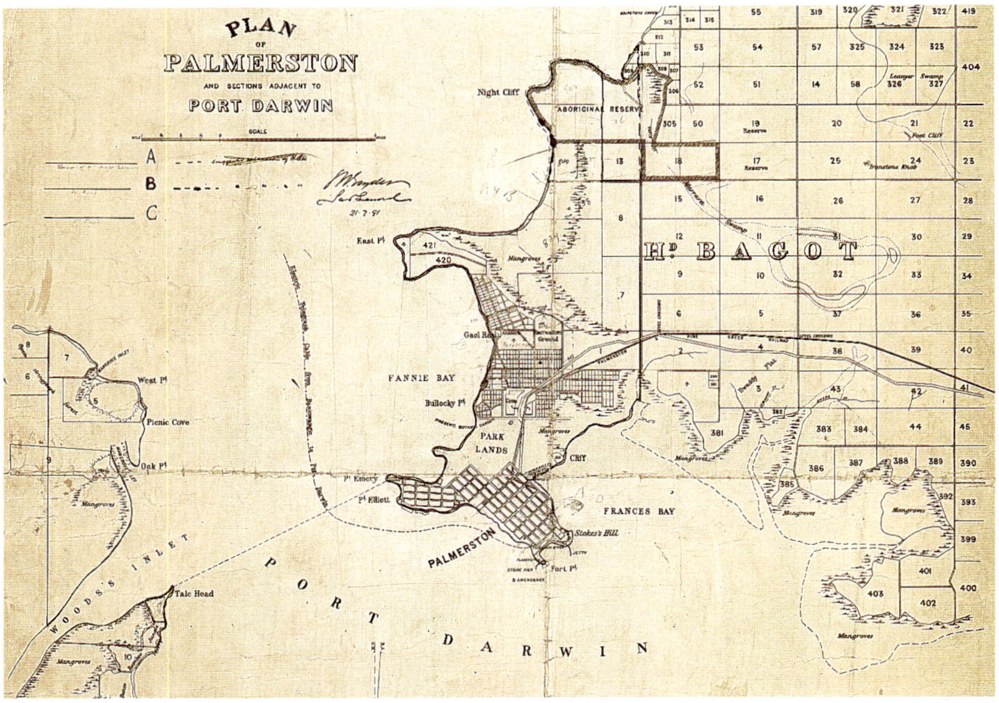

Plan of Palmerston and section adjacent to Port Darwin, 1869

Darwin today. Why are some areas close to the city still undeveloped?

The WORLD

PACIFIC

Countries and dependencies

A *country* is an independent, self-governing area of land (e.g. France).

A *dependency* is an area of land which is under the control of another (e.g. New Caledonia is under the control of France).

Fijian and Indian children play together. Current population figures show that Fijians now make up 49.7 per cent of Fiji's population and Indians 45.3 per cent.

Key

Built environment

(red line)	Country border
FIJI	Country name
Suva	Country capital underlined
Alaska	State/territory name
Guam (U.S.A.)	Dependency
● SAN DIEGO	More than 1 000 000 people
● Sendai	500 000 to 1 000 000 people
• Iligan	100 000 to 500 000 people
· Noumea	Less than 100 000 people

Natural environment

	River
	Lake
Jarvis	Island
▲ Mt Wilhelm 4509 m	Mountain and height in metres

Land over 4000 metres
Land from 2000 to 4000 metres
Land from 1000 to 2000 metres
Land from 500 to 1000 metres
Land from 200 to 500 metres
Land from sea level to 200 metres
Land below sea level
Sea from sea level to 200 metres
Sea from 200 to 2000 metres
Sea from 2000 to 4000 metres
Sea from 4000 to 8000 metres
Sea over 8000 metres

Scale

1 : 55 000 000

0 600 1200 1800 2400 km

1 centimetre on the map represents 550 kilometres on the ground.

PACIFIC
Earthquakes and volcanoes

DEATHS CAUSED BY VOLCANOES AND EARTHQUAKES

Tangshan (1976)	earthquake	650 000
Tokyo (1923)	earthquake	140 000
Chimbote (1970)	earthquake	67 000
Tambora (1815)	volcano	56 000
Krakatoa (1883)	volcano	36 500
Nevado del Ruiz (1985)	volcano	22 000
Mexico City (1985)	earthquake	10 000
Concepcion (1960)	earthquake	5000
Kobe (1995)	earthquake	5000
Lamington (1951)	volcano	3000
Paricutín (1943)	volcano	2800
Taiwan (1999)	earthquake	2400
Taal (1911)	volcano	1335
El Chichón (1982)	volcano	1000
Pinatubo (1991)	volcano	700
San Francisco (1906)	earthquake	452
San Francisco (1989)	earthquake	62
Mt St Helens (1980)	volcano	61
Los Angeles (1994)	earthquake	57
Newcastle (1989)	earthquake	13
Tavurvur, Vulcan (1994)	volcano	2

How do earthquakes occur?

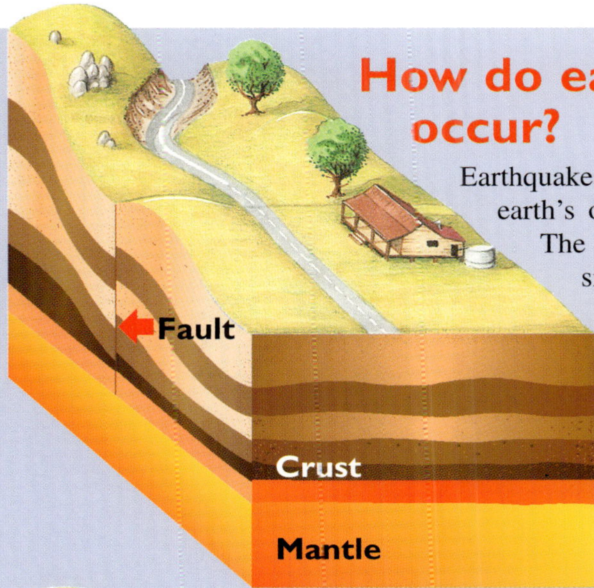

Earthquakes are caused by cracks in the earth's outer crust known as faults. The rough crust edges on either side of the fault grind past each other and become jammed. Eventually the tension becomes too much, the rocks snap and the crust edges jerk free causing an earthquake.

The place at which this happens is known as the focus of the earthquake. The point on the earth's surface directly above the focus is called the epicentre. The ground shakes as shock waves move through it away from the focus. Earthquakes can topple houses, trees and telephone lines and tear apart roads, railway lines and pipelines.

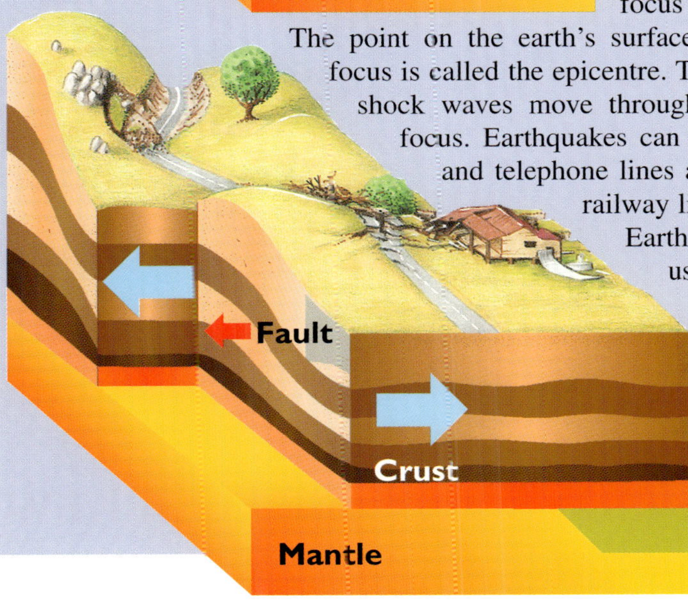

Earthquakes are measured using the Richter scale. A figure of 6.4 on this scale shows a moderate earthquake, while a figure in excess of 7 represents a major earthquake.

CASE STUDY
Kobe earthquake

On 17 January 1995 at 5.46 a.m., an earthquake of 20 seconds duration struck Kobe, Japan's second largest port. The earthquake measured 7.2 on the Richter scale and casualties included 5000 dead and 25 000 injured.

Buildings collapsed, trains were derailed, freeways split and fires broke out. Casualties would have been much greater had the quake struck during the morning commuter rush period.

The Kobe earthquake was the biggest to hit a major Japanese centre since the Tokyo earthquake of 1923 which killed 140 000.

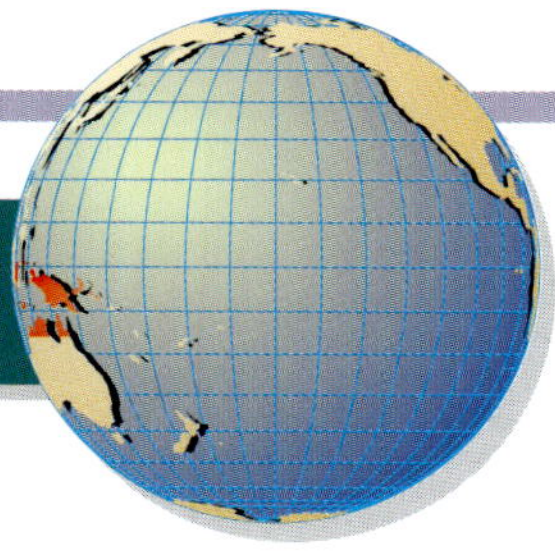

NEW GUINEA

Settlement, land and resources

MINERAL DEVELOPMENT IN PAPUA NEW GUINEA

Papua New Guinea's economy relies heavily on mining and agriculture. Its main exports include gold, copper, coffee, tea, palm oil, copra and timber.

Gross National Product (GNP), 1992
(GNP is often used to show a country's wealth.)

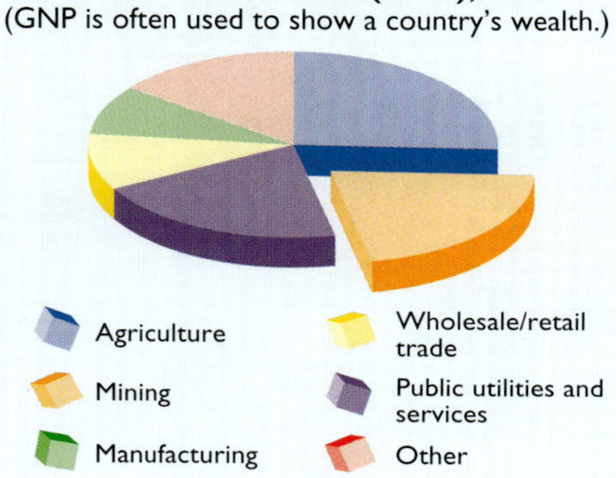

The Ok Tedi copper and gold mine established in 1981 makes a major contribution to the economy of Papua New Guinea.

Key

Built environment

(red line)	Country border
PAPUA NEW GUINEA	Country name
Port Moresby	Country capital underlined
Papua	State/province name
• Ambon	100 000 to 500 000 people
• Daru	Less than 100 000 people

Scale

1 : 10 000 000

0 100 200 300 400 km

1 centimetre on the map represents 100 kilometres on the ground.

Natural environment

	River
	Lake
Nila	Island
▲ Mt Murray 2243 m	Mountain and height in metres
	Land over 4000 metres
	Land from 2000 to 4000 metres
	Land from 1000 to 2000 metres
	Land from 500 to 1000 metres
	Land from 200 to 500 metres
	Land from sea level to 200 metres
	Land below sea level
	Sea from sea level to 200 metres
	Sea from 200 to 2000 metres
	Sea from 2000 to 4000 metres
	Sea from 4000 to 8000 metres
	Sea over 8000 metres

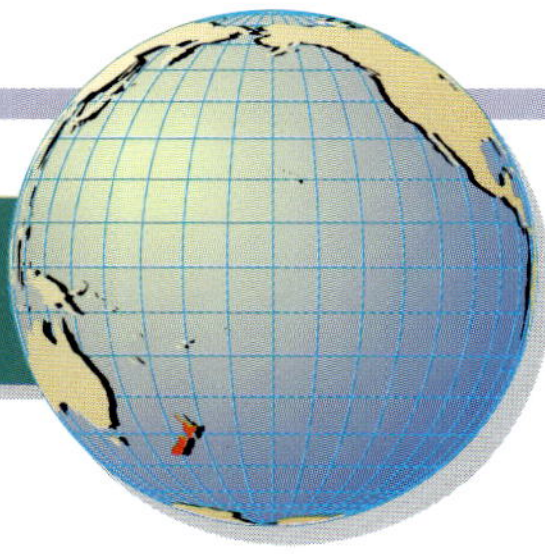

NEW ZEALAND

Settlement and landforms

CASE STUDY
Maori population

Less than 200 years ago New Zealand's population and culture was wholly Polynesian, but now the population is dominated by people of European origin. Today, 75 per cent of New Zealanders are of European origin, while only 10 per cent are Maori.

The indigenous Maori population at the time of early European contact in the 1770s was between 120 000 and 150 000 people. By 1840 the Maori population had decreased to approximately 100 000. This resulted mainly from the introduction of diseases such as tuberculosis. War with the Europeans and between tribes contributed to the further decline of the population: 60 000 in 1858 and 42 000 in 1896.

High birth rates throughout the twentieth century restored the Maori population to more than 400 000 in 1991. The current population is faced with problems of high unemployment, illiteracy, poor housing and ill-health. Maoris have been active in demanding political equality and land rights. They are challenging the Treaty of Waitangi, signed in 1840, which gave all the power and land to the British and led to the Maori Wars between 1860 and 1872.

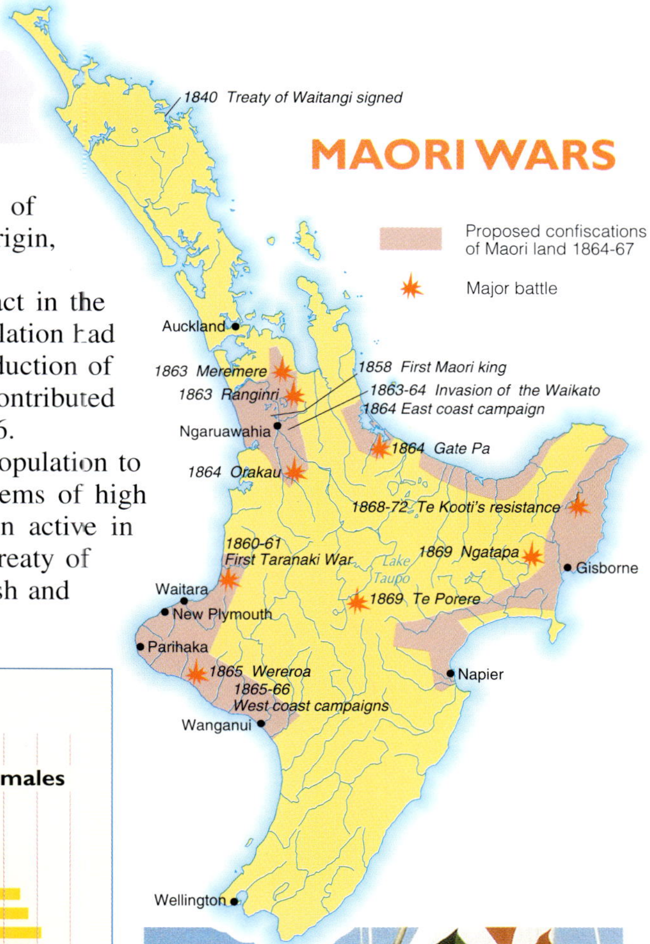

Key

Built environment

NEW ZEALAND	Country name
Wellington	Country capital underlined
• Auckland	500 000 to 1 000 000 people
• Christchurch	100 000 to 500 000 people
• Napier	Less than 100 000 people

Natural environment

	River
	Lake
White Island	Island
Mt Cook 3764 m	Mountain and height in metres
	Land over 4000 metres
	Land from 2000 to 4000 metres
	Land from 1000 to 2000 metres
	Land from 500 to 1000 metres
	Land from 200 to 500 metres
	Land from sea level to 200 metres
	Land below sea level
	Sea from sea level to 200 metres
	Sea from 200 to 2000 metres
	Sea from 2000 to 4000 metres
	Sea from 4000 to 8000 metres
	Sea over 8000 metres

Scale

1 : 6 700 000

0 80 160 240 km

1 centimetre on the map represents 67 kilometres on the ground.

Total population

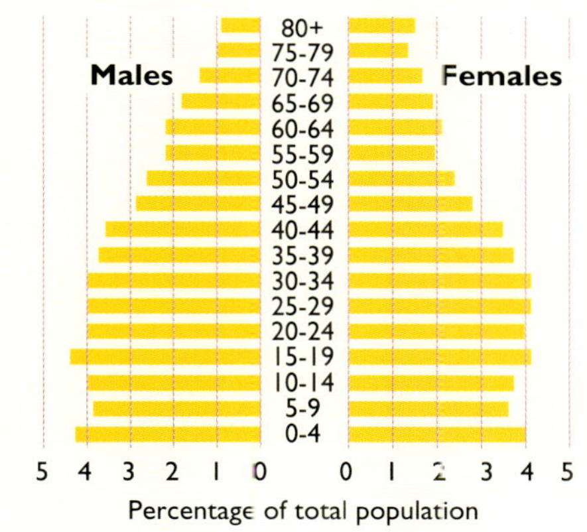

NZ Maori

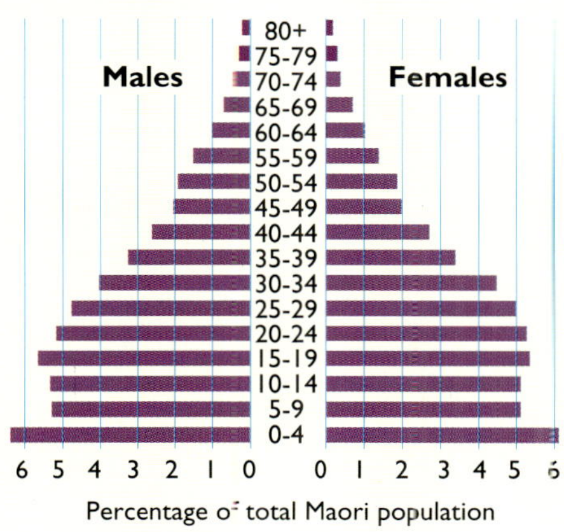

Maoris protesting on Waitangi Day, 1995

ANTARCTICA

Research

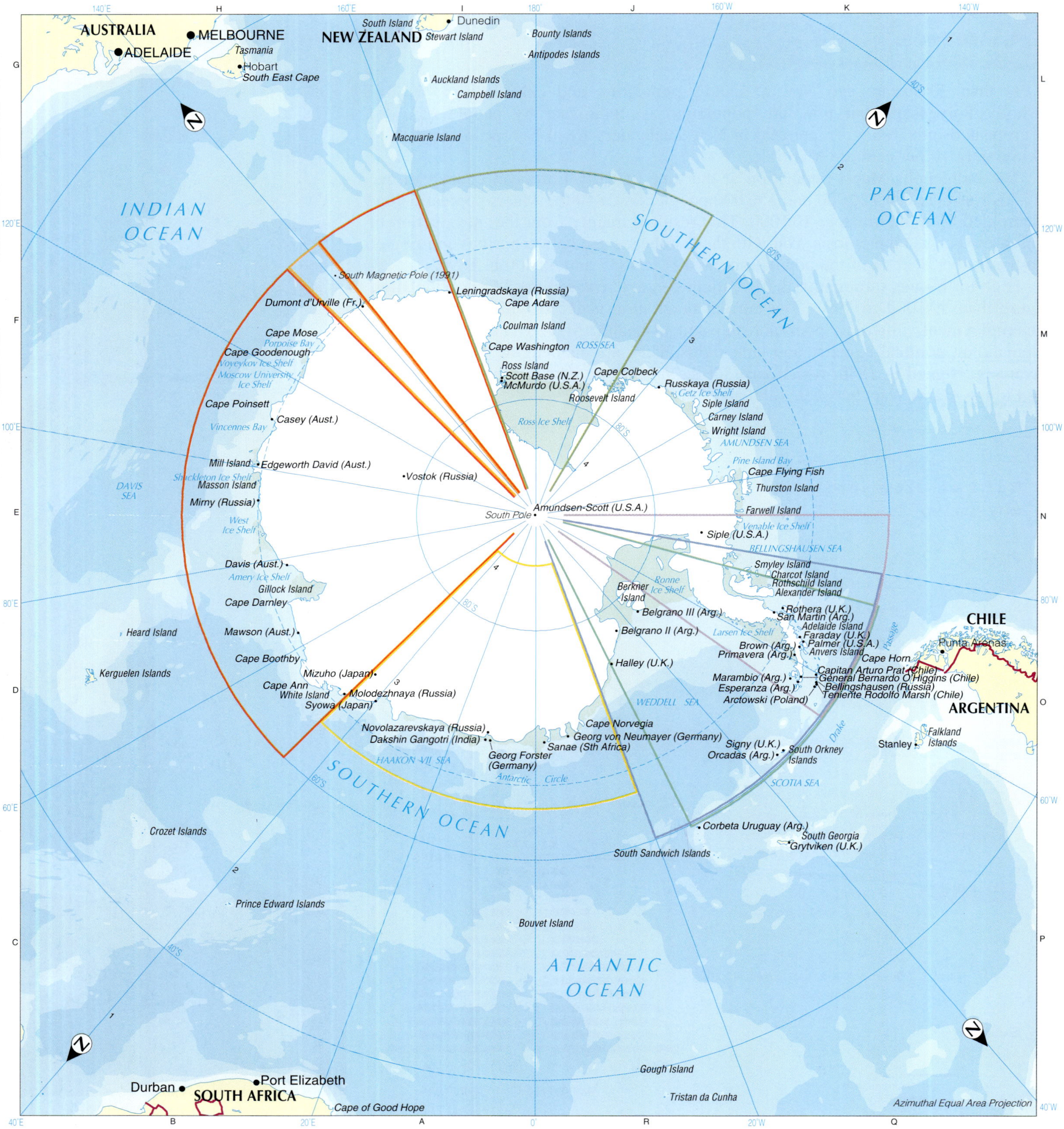

Human endeavour

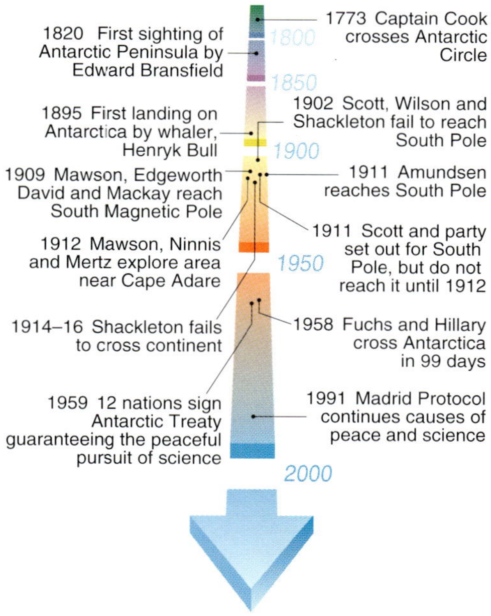

Key

Built environment

(line)	Country border
CHILE	Country name

Countries with territorial claims

(line)	Argentina
(line)	Australia
(line)	Chile
(line)	France
(line)	New Zealand
(line)	Norway
(line)	United Kingdom
• ADELAIDE	More than 1 000 000 people
• Durban	500 000 to 1 000 000 people
• Punta Arenas	100 000 to 500 000 people
• Stanley	Less than 100 000 people
• *Davis (Aust.)*	Scientific station (operating country)

Natural environment

(symbol)	Lake
South Georgia	Island
(symbol)	Ice shelf

Scale

1 : 48 000 000

0 500 1000 1500 2000 km

1 centimetre on the map represents 480 kilometres on the ground.

Working at Edgeworth David Station

Established in 1986, Edgeworth David Station is Australia's Antarctic summer research station in the Burger Hills. Unlike our winter bases, Casey, Mawson and Davis, this base is set up temporarily each summer using air-lifted igloos made from fibreglass. Scientists study the area's geological history, land forms and changes in the ice, unhampered by blizzard conditions. Because there is no need for large scale construction, this base does not create the environmental problems associated with permanent bases.

The research station quickly takes shape as Australian scientists prepare for their summer research projects.

EXPLORERS' ROUTES

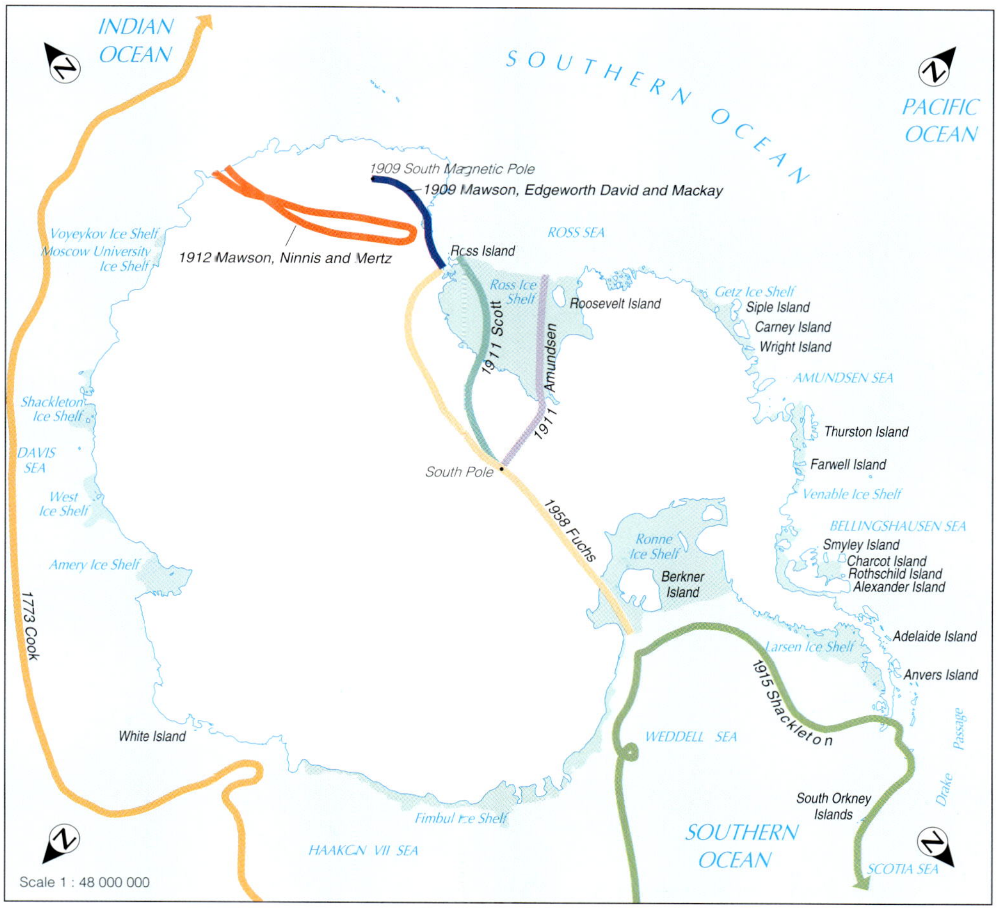

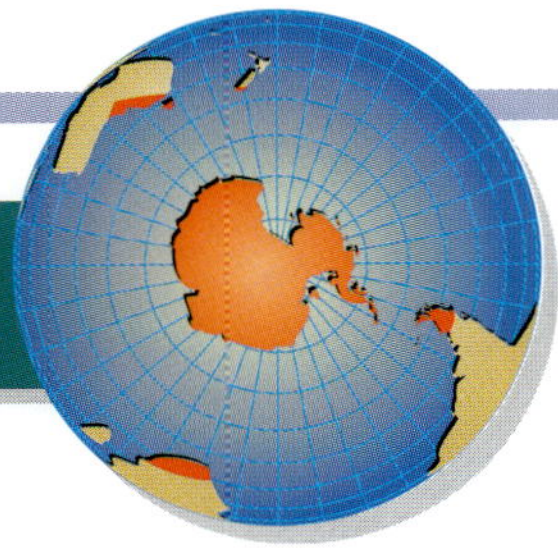

ANTARCTICA
Natural environment

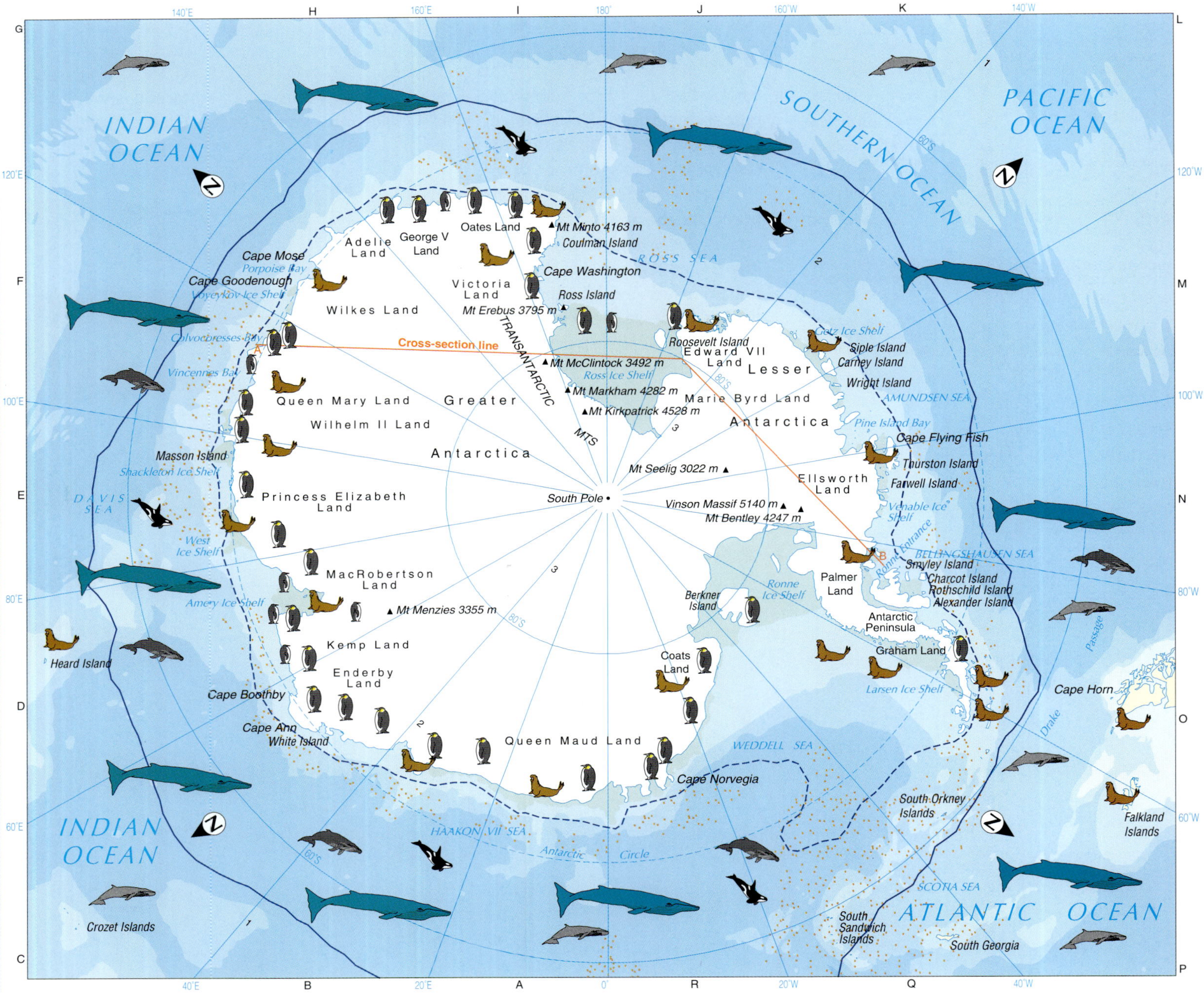

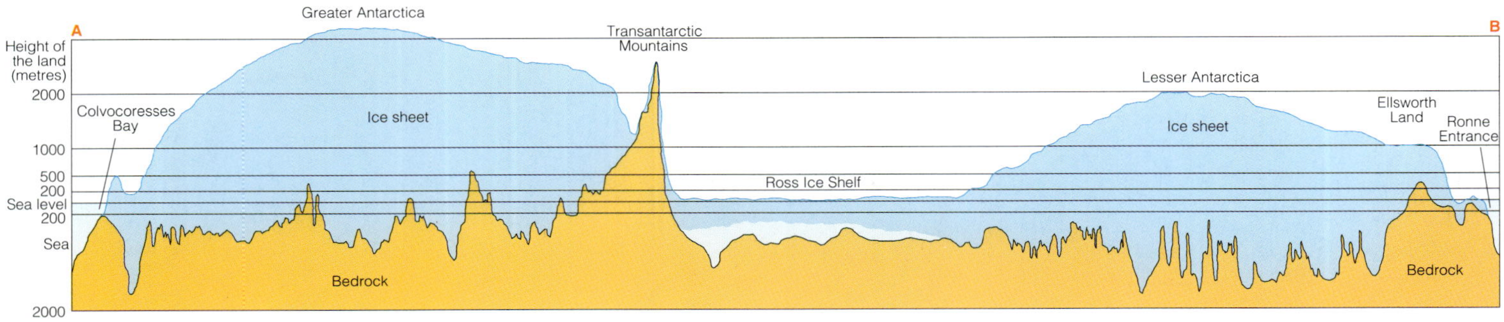

Cross-section from Colvocoresses Bay to Ronne Entrance (A — B)

CASE STUDY • • • • • • • • • • • • • • • • Antarctic food web

The food web in Antarctica begins with a mass of microscopic plant organisms called phytoplankton. This plankton provides the nourishment for krill, the most important organism in the Antarctic food web. These tiny, shrimp-like crustaceans live in large concentrations in the seasonal pack ice zone. It is estimated that the oceans around Antarctica contain up to 650 million tonnes of krill, providing the staple diet of seals, birds, whales and fish. As a predator of seals and other whales, even the killer whale at the top of the food web relies directly on krill. Krill harvesting by nations like Japan has led to concern about eventual serious consequences for the food web.

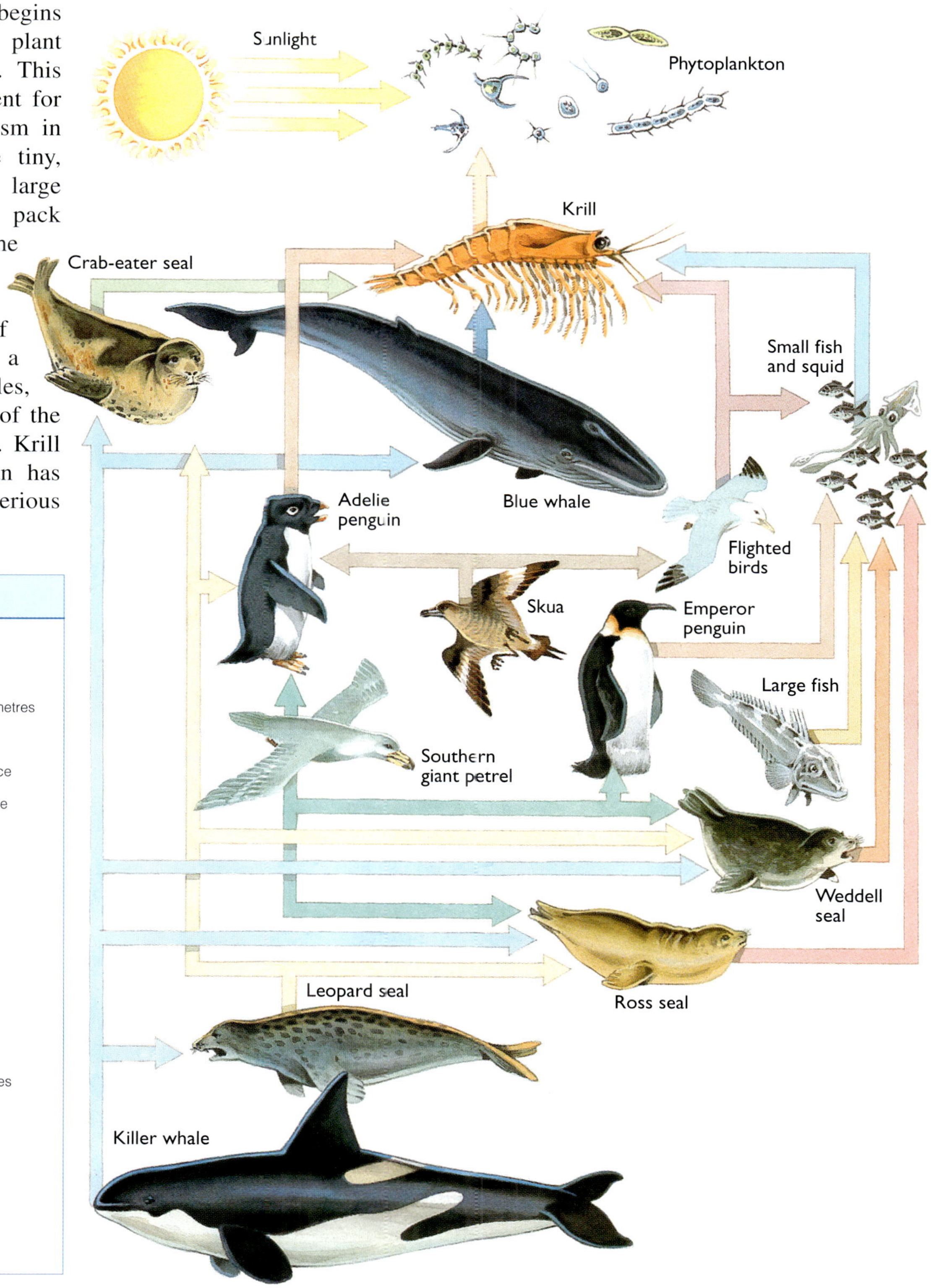

Key

Natural environment

South Georgia	Island
▲ Mt Menzies 3355 m	Mountain with height in metres
	Krill concentration
	Maximum extent of sea ice
	Minimum extent of sea ice
	Ice shelf
	Blue whales
	Humpback whales
	Sei whales
	Killer whales
	Seal breeding area
	Emperor penguin rookeries
	Adelie penguin rookeries

Scale

1 : 40 000 000

0 400 800 1200 1600 km

1 centimetre on the map represents 400 kilometres on the ground.

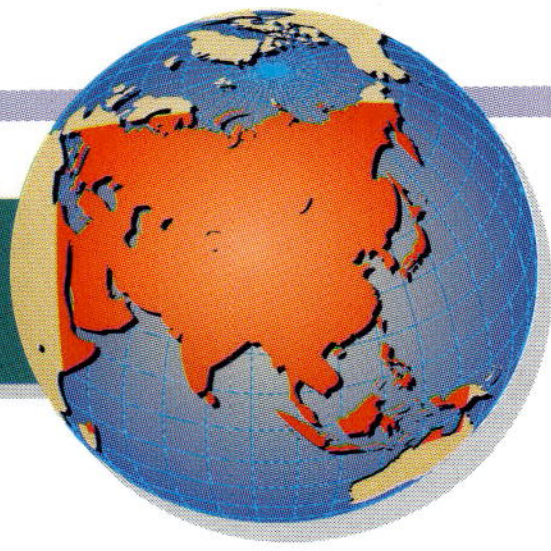

ASIA
Settlement

Key

Built environment

(line)	Country border
JAPAN	Country name
TOKYO	Country capital underlined
Guam (U.S.A.)	Dependency

● PUNE	More than 1 000 000 people
● Tabriz	500 000 to 1 000 000 people
• Chita	100 000 to 500 000 people
• Jayapura	Less than 100 000 people

Natural environment

(symbol)	River
(symbol)	Lake
Yap	Island

Scale

1 : 57 000 000

0 600 1200 1800 km

1 centimetre on the map represents 570 kilometres on the ground.

Geo Facts

Largest country (by area)	China ▶ 9 597 000 sq km
Largest country (by pop.)	*China ▶ 1 166 140 000
Largest city (by pop.)	*Tokyo ▶ 28 447 000
Highest GNP per head	Japan ▶ $28 217
Lowest GNP per head	Cambodia ▶ $103

*World record

Ancient cultures

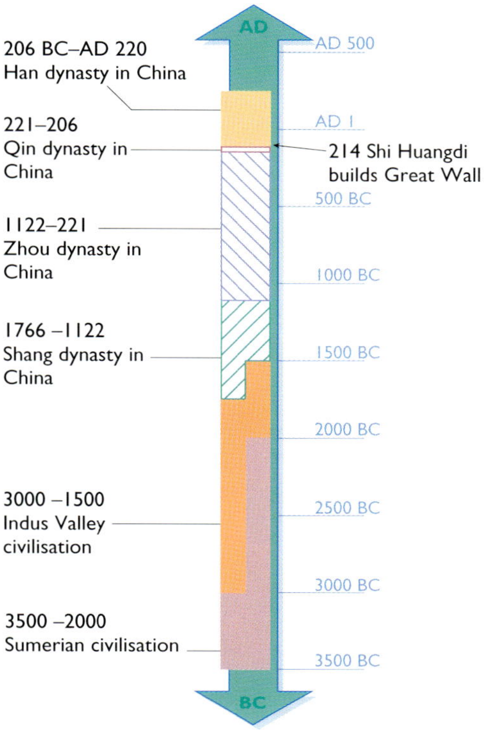

In 214 BC, Shi Huangdi, the first Emperor of China, began what could be called the greatest building project in human history. This project was the joining together of many small walls into one single wall to protect China from nomadic invasions from the north. When it was finished, the Great Wall ran for over 2500 kilometres and was 9 metres high and 4.5 metres wide. Soldiers patrolled the wall from over 25 000 watchtowers built along its length. More than 300 000 labourers — farmers, convicts and slaves — worked for many years using earth, stone and bricks to construct the wall. The wall was rebuilt during the Ming dynasty in the fifteenth century and the rebuilt sections have World Heritage listing.

A great myth of our time is that the Great Wall is visible from space. However, no astronaut has reported sighting it.

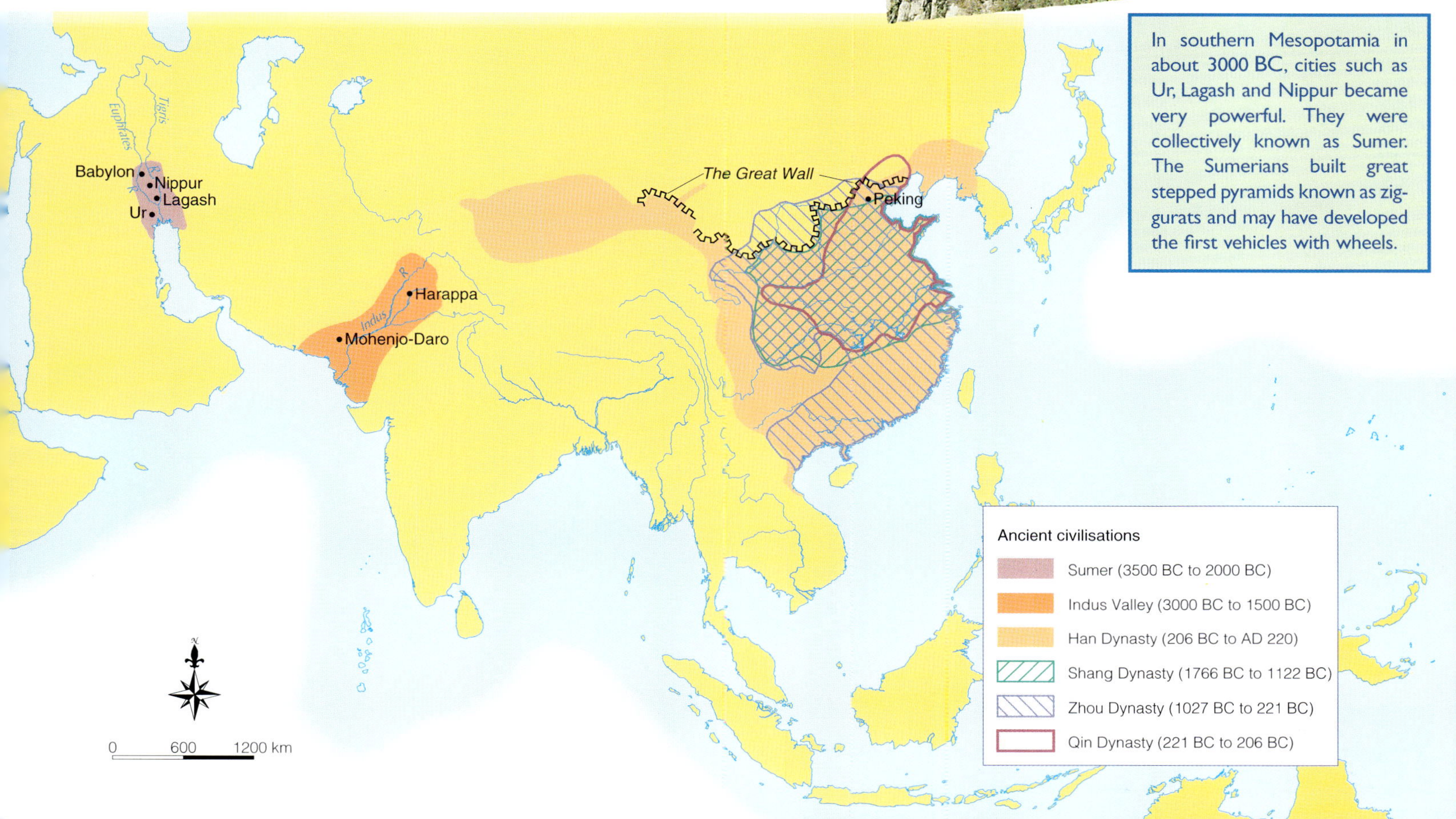

In southern Mesopotamia in about 3000 BC, cities such as Ur, Lagash and Nippur became very powerful. They were collectively known as Sumer. The Sumerians built great stepped pyramids known as ziggurats and may have developed the first vehicles with wheels.

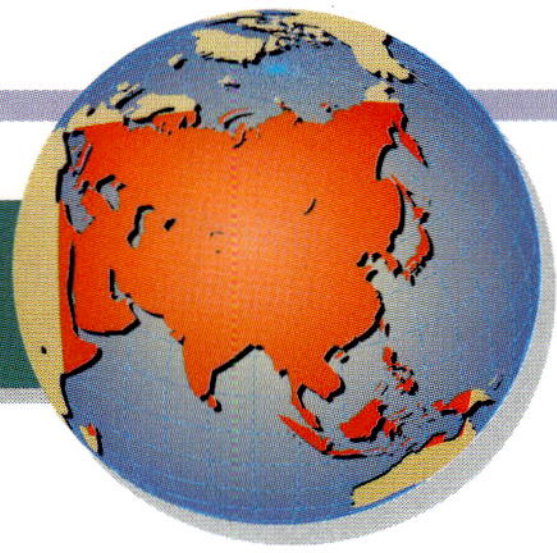

ASIA
Landforms and land use

Geo Facts

Highest mountain	*Everest, Nepal ▶ 8848 m	**Highest average annual rainfall**	*Cherrapunji, India ▶ 11 430 mm
Largest desert	Gobi, Mongolia ▶ 1 294 994 sq km	**Lowest average annual rainfall**	Aden, Yemen ▶ 46 mm
Longest river	Yangtze, China ▶ 6380 km	*World record	

TRANSECT

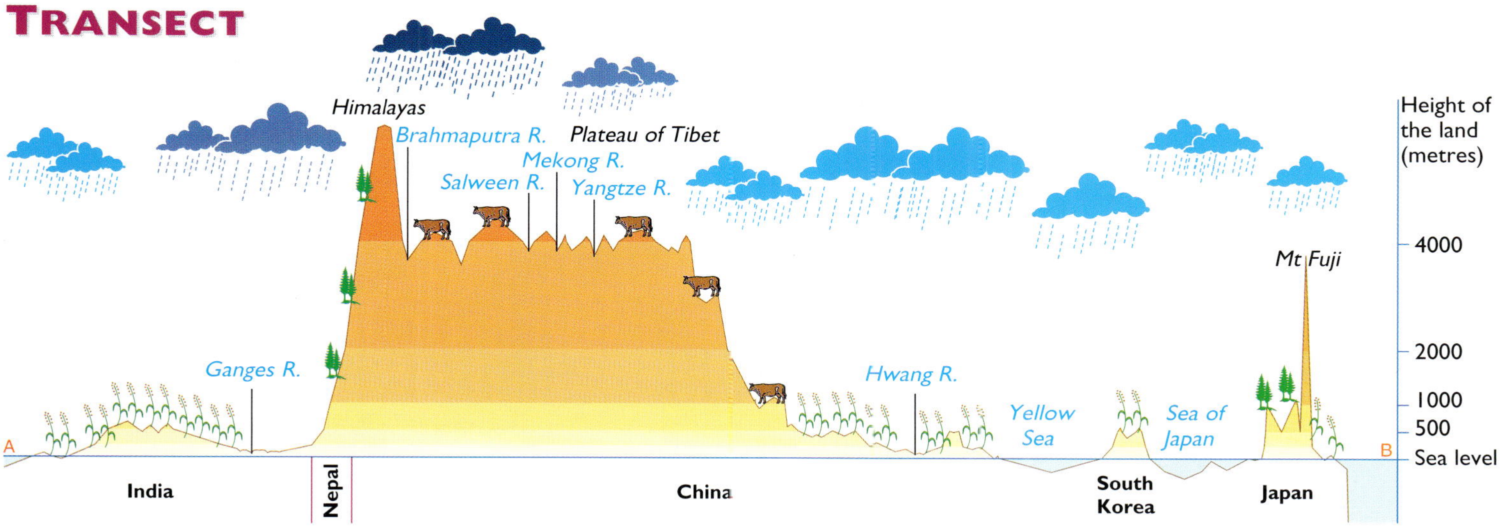

Key

Transect

A——B Ahmadabad to Tokyo

Annual rainfall *Land use*

Over 4000 mm — Grazing livestock

2000–4000 mm — Crops, i.e. rice, millet

1000–2000 mm — Forest

Built environment

Country border

Natural environment

River

Lake

Yap Island

▲*Mt Fuji 3776 m* Mountain and height in metres

Land over 4000 metres
Land from 2000 to 4000 metres
Land from 1000 to 2000 metres
Land from 500 to 1000 metres
Land from 200 to 500 metres
Land from sea level to 200 metres
Land below sea level
Sea from sea level to 200 metres
Sea from 200 to 2000 metres
Sea from 2000 to 4000 metres
Sea from 4000 to 8000 metres
Sea over 8000 metres

Scale

1 : 57 000 000

0 600 1200 1800 2400 km

1 centimetre on the map represents 570 kilometres on the ground.

CASE STUDY
Himalayas

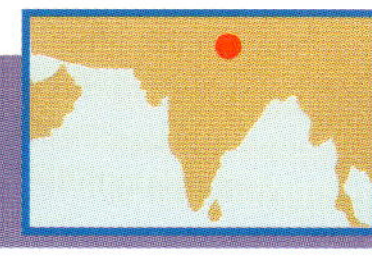

The Himalayas are the highest mountain range in the world. The white areas in the satellite image show snow and ice on the peaks and the southern edge of the image shows the vast floodplain of the Ganges River.

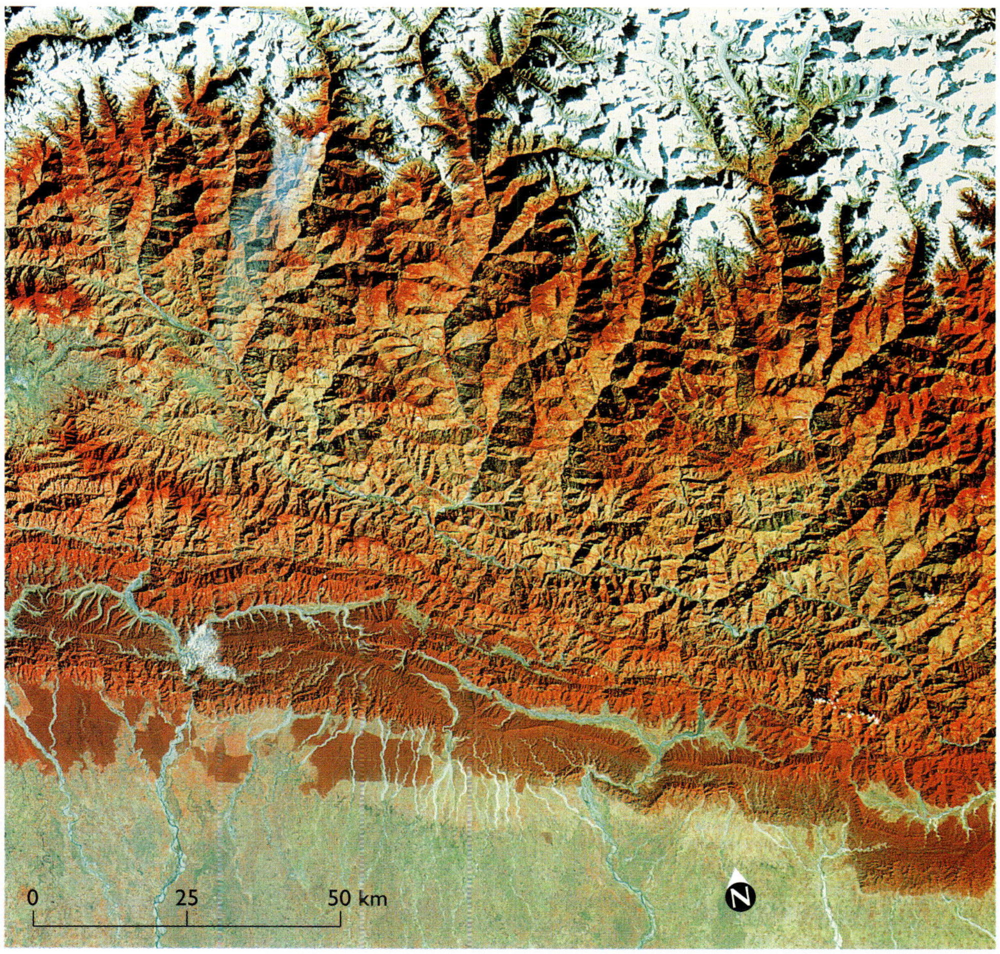

SOUTH-EAST ASIA

Settlement and landforms

Cambodian schoolgirls holding Cambodian (*left*) and Vietnamese (*right*) flags. These two countries share a common past in their struggle for independence and the resulting civil wars.

Key

Built environment

(line)	Country border
LAOS	Country name
Vientiane	Country capital underlined
(line)	State/territory border
Sabah	State/territory name
Guam (U.S.A.)	Dependency
● MEDAN	More than 1 000 000 people
● Malang	500 000 to 1 000 000 people
• Ambon	100 000 to 500 000 people
• Koror	Less than 100 000 people

Natural environment

(line)	River
(shape)	Lake
Koror	Island
▲Jaya Peak 5040 m	Mountain and height in metres
	Land over 4000 metres
	Land from 2000 to 4000 metres
	Land from 1000 to 2000 metres
	Land from 500 to 1000 metres
	Land from 200 to 500 metres
	Land from sea level to 200 metres
	Land below sea level
	Sea from sea level to 200 metres
	Sea from 200 to 2000 metres
	Sea from 2000 to 4000 metres
	Sea from 4000 to 8000 metres
	Sea over 8000 metres

Scale

1 : 18 500 000

0 200 400 600 800 km

1 centimetre on the map represents 185 kilometres on the ground.

EAST ASIA

Settlement and landforms

Religious ceremonies continue to play an important role in the cultural life of many Japanese.

Key

Built environment

(line)	Country border
TAIWAN	Country name
TAIPEI	Country capital underlined
● OSAKA	More than 1 000 000 people
● Ningbo	500 000 to 1 000 000 people
• Kashi	100 000 to 500 000 people
• Kuqa	Less than 100 000 people

Natural environment

	River
	Lake
Sado	Island
Mt Fuji 3776 m	Mountain and height in metres

Land over 4000 metres
Land from 2000 to 4000 metres
Land from 1000 to 2000 metres
Land from 500 to 1000 metres
Land from 200 to 500 metres
Land from sea level to 200 metres
Land below sea level
Sea from sea level to 200 metres
Sea from 200 to 2000 metres
Sea from 2000 to 4000 metres
Sea from 4000 to 8000 metres
Sea over 8000 metres

Scale

1 : 19 000 000

0 200 400 600 800 km

1 centimetre on the map represents 190 kilometres on the ground.

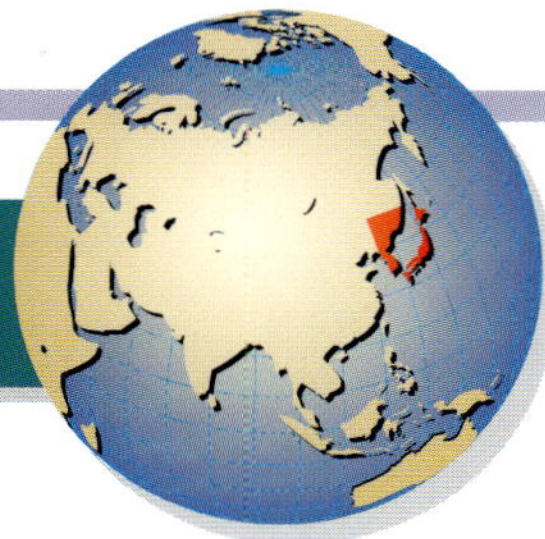

KOREA AND JAPAN
Settlement and landforms

CASE STUDY
Economic growth of South Korea

South Korea's economy was traditionally based on agriculture. Since the early 1960s the country has undergone rapid growth. Now more than a quarter of South Korea's workforce is employed in manufacturing, notably cars, electrical goods, machinery and ships.

South Korea's largest business is the electrical manufacturer, Samsung, which is ranked the eighteenth largest business in the world. Other internationally known companies include the Daewoo and Hyundai car manufacturing companies.

South Korea does not have great reserves of mineral resources and must import these raw materials from other countries such as Australia.

The fast growing economies are sometimes referred to as "tigers". The bar graph opposite shows the dominance of Asian countries in the world's fastest growing economies.

Fastest-growing world economies, 1985–92

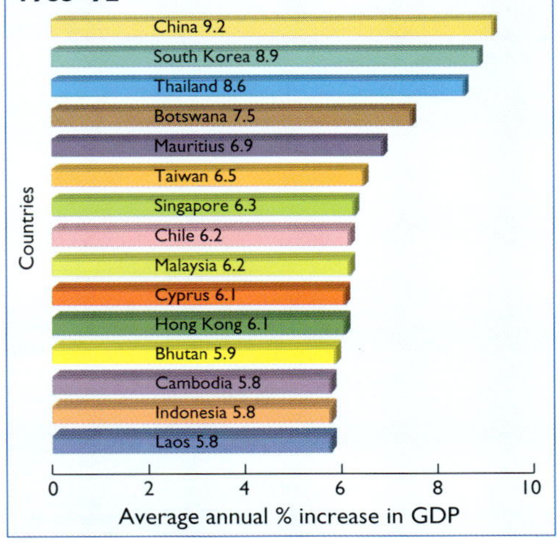

Key

Built environment

—	Country border
JAPAN	Country name
TOKYO	Country capital underlined
● NAGOYA	More than 1 000 000 people
• Okayama	500 000 to 1 000 000 people
• Chungju	100 000 to 500 000 people
· Kamaishi	Less than 100 000 people

Natural environment

	River
	Lake
Dogo	Island
▲ Mt Fuji 3776 m	Mountain and height in metres

Land over 4000 metres
Land from 2000 to 4000 metres
Land from 1000 to 2000 metres
Land from 500 to 1000 metres
Land from 200 to 500 metres
Land from sea level to 200 metres
Land below sea level
Sea from sea level to 200 metres
Sea from 200 to 2000 metres
Sea from 2000 to 4000 metres
Sea from 4000 to 8000 metres
Sea over 8000 metres

Scale

1 : 9 500 000

0 100 200 300 400 km

1 centimetre on the map represents 95 kilometres on the ground.

Workers at an electrical goods company

South Korea: employment as percentage of total workforce

Sector	1982	1992
Agriculture	32	16
Retail/wholesale	22	22
Manufacture	21	25
Public utilities and services	11	16
Construction	6	9
Unemployed	4	2
Other	4	10

South Korea – exports, 1992

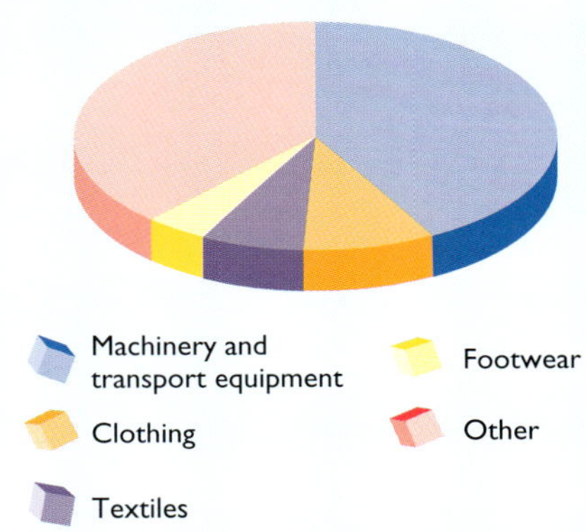

Total exports, 1992 US$ 77.6 billion
Main export destinations U.S.A. (23.6%), Japan (15.1%), Hong Kong (7.7%), Singapore (4.2%), Germany (3.8%)

South Korea – imports, 1992

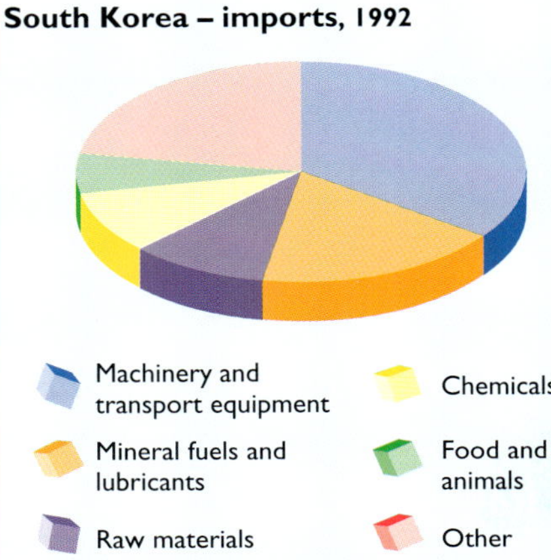

Total imports, 1992 US$ 81.8 billion
Main origins of imports Japan (23.8%), U.S.A. (22.4%), Germany (4.6%), Saudi Arabia (4.6%), Australia (3.8%)

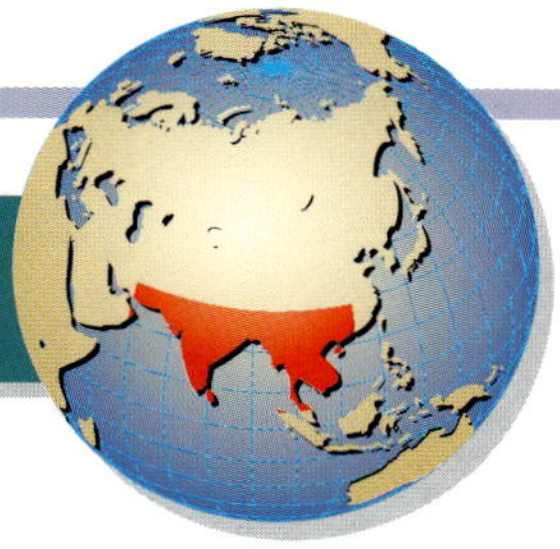

SOUTH ASIA
Settlement and landforms

In India where the adult literacy rate is 49 per cent, these Tamil schoolgirls understand the importance of education.

Key

Built environment

(line)	Country border
NEPAL	Country name
Kathmandu	Country capital underlined
● KARACHI	More than 1 000 000 people
• Patna	500 000 to 1 000 000 people
• Hue	100 000 to 500 000 people
· Galle	Less than 100 000 people

Natural environment

(line)	River
(outline)	Lake
Masirah	Island
▲ Mt Everest 8848 m	Mountain and height in metres

Land over 4000 metres
Land from 2000 to 4000 metres
Land from 1000 to 2000 metres
Land from 500 to 1000 metres
Land from 200 to 500 metres
Land from sea level to 200 metres
Land below sea level
Sea from sea level to 200 metres
Sea from 200 to 2000 metres
Sea from 2000 to 4000 metres
Sea from 4000 to 8000 metres
Sea over 8000 metres

Scale

1 : 17 000 000

0 200 400 600 km

1 centimetre on the map represents 170 kilometres on the ground.

MIDDLE EAST

Settlement and landforms

CASE STUDY
Ancient Sumer

Since 5000 BC, people have continuously occupied the region between the Tigris and Euphrates rivers. Ancient civilisations like Sumer rose and declined, dependent largely on the rivers' gift of water.

The Tigris and Euphrates changed their courses over time, and maintaining water supplies to Sumerian cities like Ur, Nippur and Lagash became impossible. During local civil wars between 1750 BC and 1460 BC, water was even used as a weapon when armies changed the water flow to starve out opponents.

Today the people of Iraq still depend on the rivers' life-giving waters in much the same way as the ancient Sumerians.

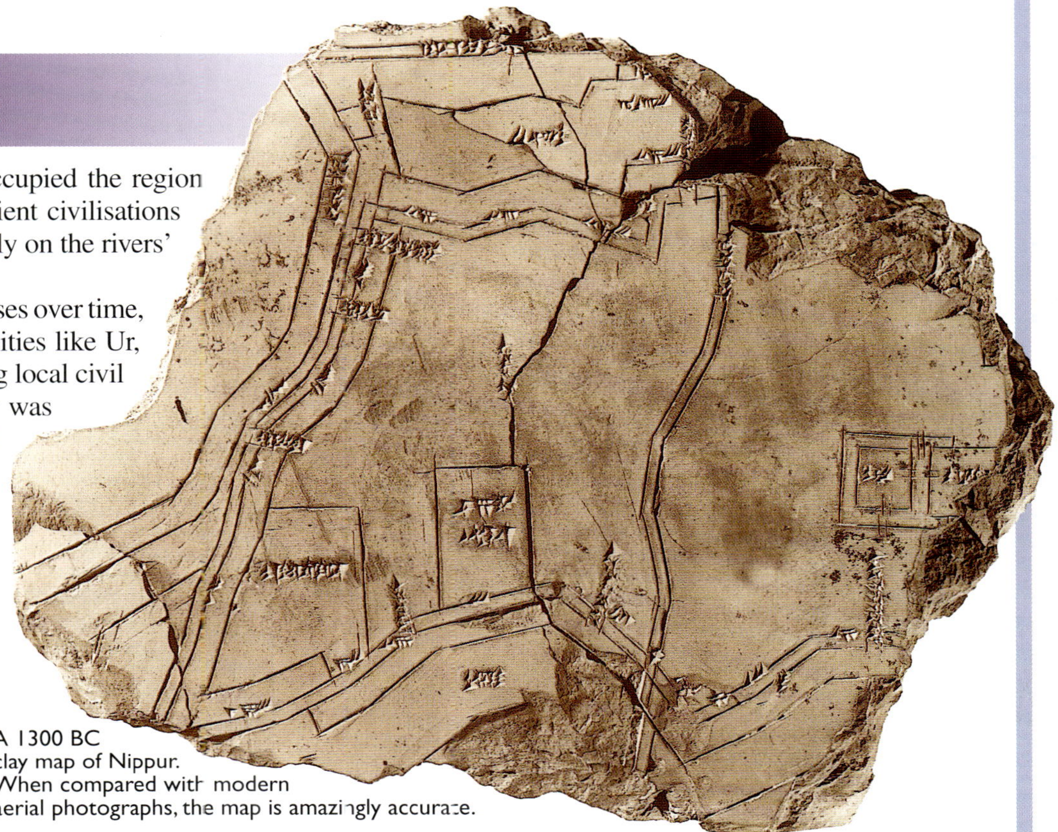

A 1300 BC clay map of Nippur. When compared with modern aerial photographs, the map is amazingly accurate.

Key

Built environment

(line)	Country border
QATAR	Country name
Doha	Country capital underlined
● IZMIR	More than 1 000 000 people
• Konya	500 000 to 1 000 000 people
• Yazd	100 000 to 500 000 people
· Dilam	Less than 100 000 people

Natural environment

(line)	River
(outline)	Lake
Masirah	Island
▲ Mt Sinai 2285 m	Mountain and height in metres
	Land over 4000 metres
	Land from 2000 to 4000 metres
	Land from 1000 to 2000 metres
	Land from 500 to 1000 metres
	Land from 200 to 500 metres
	Land from sea level to 200 metres
	Land below sea level
	Sea from sea level to 200 metres
	Sea from 200 to 2000 metres
	Sea from 2000 to 4000 metres
	Sea from 4000 to 8000 metres
	Sea over 8000 metres

Scale

1 : 16 000 000

0 150 300 450 600 km

1 centimetre on the map represents 160 kilometres on the ground.

ANCIENT SUMER

EUROPE

Settlement

Key

Built environment

(line)	Country border	● NAPLES	More than 1 000 000 people
SPAIN	Country name	● Penza	500 000 to 1 000 000 people
MADRID	Country capital underlined	• Hamar	100 000 to 500 000 people
Faroe Islands (Denmark)	Dependency	· Ayr	Less than 100 000 people

Natural environment

(line)	River
(shape)	Lake
Naxos	Island

Scale

1 : 28 000 000

0 300 600 900 km

1 centimetre on the map represents 280 kilometres on the ground.

Geo Facts

Largest country (by area) Russia ▶ 17 078 005 sq km

Largest country (by pop.) Russia ▶ 137 551 000

Largest city (by pop.) Moscow, Russia ▶ 10 800 000

Highest GNP per head *Switzerland ▶ $36 231

Lowest GNP per head Turkey ▶ $1280

*World record

Ancient cultures

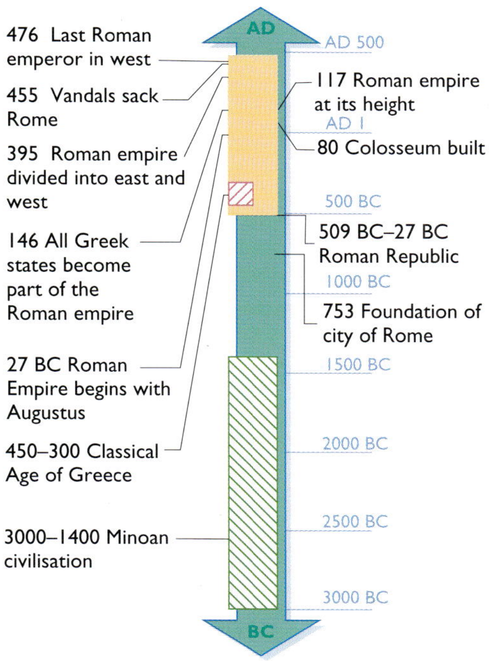

In AD 80, ancient Roman architects created the Colosseum, a huge amphitheatre where spectacular shows were performed for the entertainment of Roman citizens. Using a design based on the arch, the building was constructed of rock and marble, with concrete roofs, floors and pillars. It was 188 metres long, 156 metres across, 57 metres high and could seat 60 000 people. The 80 stairways and exits enabled the arena to be emptied in three minutes.

The gladiatorial contests were the main attraction. Specially trained gladiators locked in combat with each other or with wild animals. The Colosseum could be flooded with water for the staging of mock sea-battles or set up with exotic plants and wild animals to simulate a hunt. The ancient centre of Rome where the Colosseum still stands, is part of the World Heritage site that also includes the Forum, the Pantheon and the Roman baths.

The Colosseum was a gift to the people of Rome from the emperor Vespasian who did not live to see its opening. It was opened by his son Titus, in honour of his father.

Olympia was the site of Ancient Greece's Olympic Games. The games were held there every four years from 776 BC until they were abolished in AD 393. Lasting five days, the Olympic Games were part of a religious festival in honour of Zeus, the father of the Greek gods.

The Minoan civilisation was centred on Crete from 3000–1400 BC. Legends tell of the Minotaur, a half-man, half-bull creature kept in the palace of King Minos. The bull was a sacred symbol of the Minoans and paintings of acrobats leaping over bulls can be seen on the walls of the Palace of Knossos.

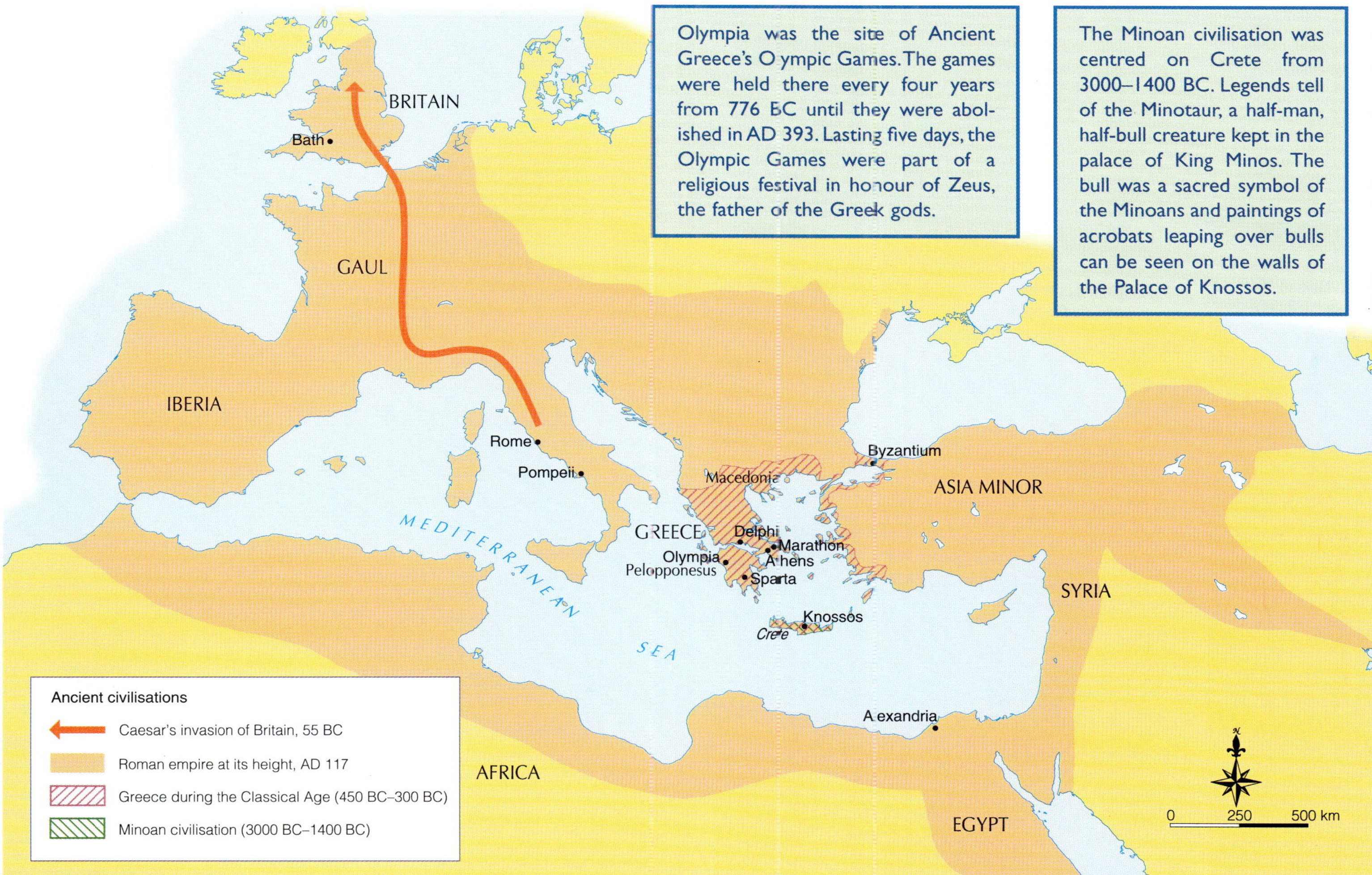

EUROPE

Landforms and land use

Geo Facts

Highest mountain Elbrus, Russia ▶ 5642 m
Longest river Volga, Russia ▶ 3531 km
Highest average annual rainfall Crkvice, Serbia and Montenegro ▶ 4648 mm
Lowest average annual rainfall Astrakhan, Russia ▶ 163 mm

TRANSECT

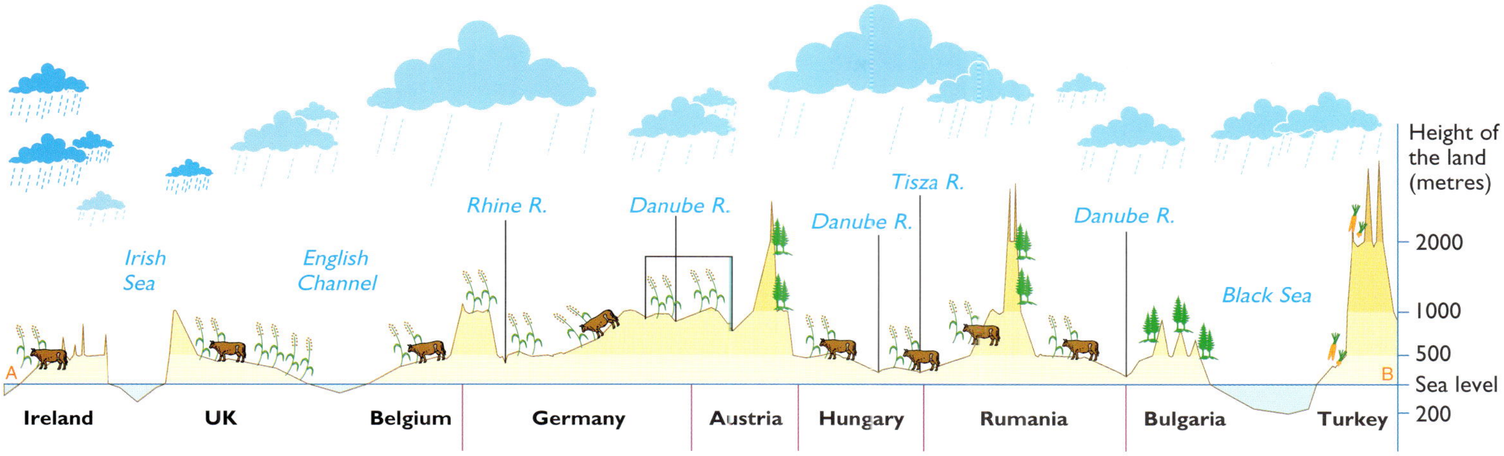

Key

Transect

A——B Galway to Ankara

Annual rainfall

1000–2000 mm

500–1000 mm

Land use

Crops and livestock

Crops, i.e. barley, rye

Fruit and vegetables

Forest

Built environment

Country border

Natural environment

River

Lake

Elba Island

Mt Elbrus 5642 m Mountain and height in metres

Land over 4000 metres
Land from 2000 to 4000 metres
Land from 1000 to 2000 metres
Land from 500 to 1000 metres
Land from 200 to 500 metres
Land from sea level to 200 metres
Land below sea level
Sea from sea level to 200 metres
Sea from 200 to 2000 metres
Sea from 2000 to 4000 metres
Sea from 4000 to 8000 metres
Sea over 8000 metres

Scale

1 : 28 000 000

0 300 600 900 1200 km

1 centimetre on the map represents 280 kilometres on the ground.

CASE STUDY
Mt Vesuvius

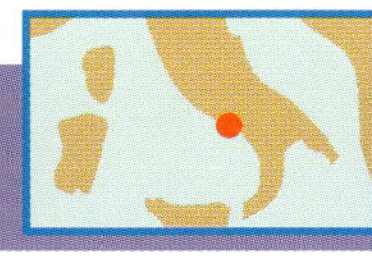

Mt Vesuvius is a 1277-metre-high active volcano near Naples, in Italy. It appears in the centre of the image. The volcano erupted in AD 79 burying the ancient cities of Pompeii, Stabiae and Herculaneum. The pale blue area around the crater represents volcanic ash, and the red patch to the south is lava.

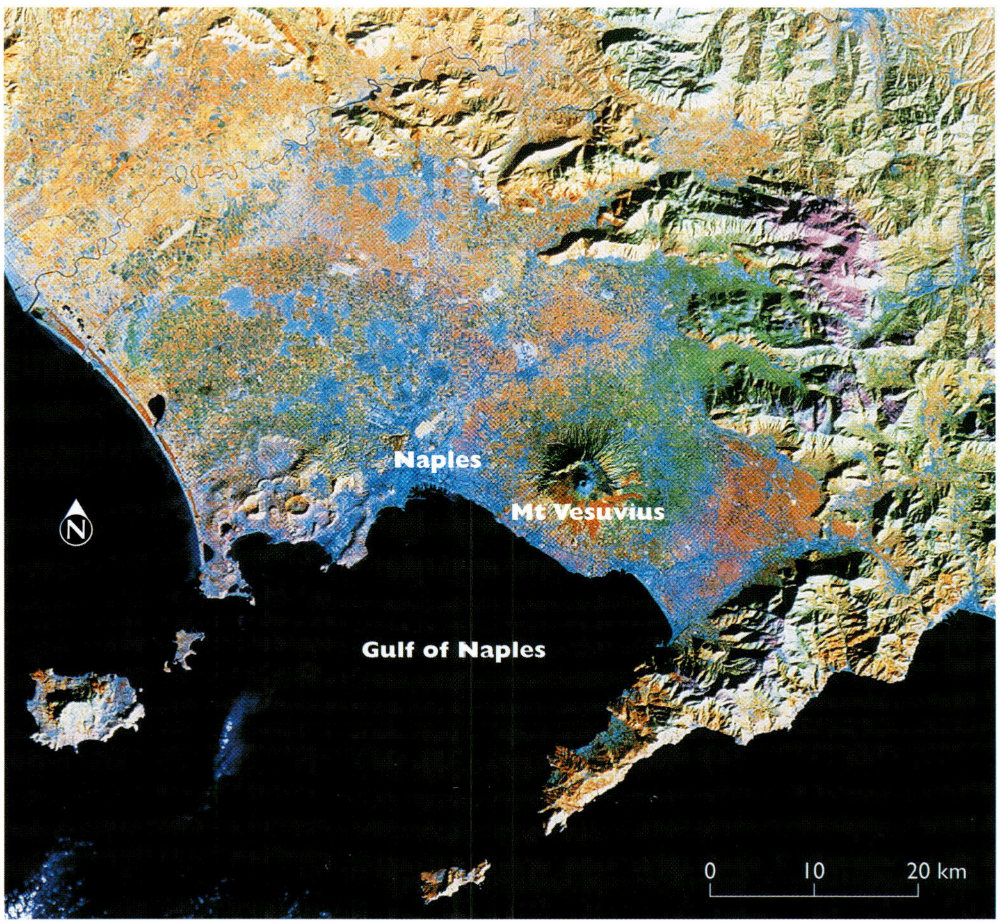

NORTH WEST EUROPE

Settlement and landforms

This Lapp woman in Finland cooks in the traditional way inside a tent made of reindeer hide.

Key

Built environment

(line)	Country border
LATVIA	Country name
Riga	Country capital underlined
Channel Islands (U.K.)	Dependency
● KHARKOV	More than 1 000 000 people
• Poznan	500 000 to 1 000 000 people
• Dijon	100 000 to 500 000 people
· Thurso	Less than 100 000 people

Natural environment

	River
	Lake
Gotland	Island
Glittertind 2472 m	Mountain and height in metres

Land over 4000 metres
Land from 2000 to 4000 metres
Land from 1000 to 2000 metres
Land from 500 to 1000 metres
Land from 200 to 500 metres
Land from sea level to 200 metres
Land below sea level
Sea from sea level to 200 metres
Sea from 200 to 2000 metres
Sea from 2000 to 4000 metres
Sea from 4000 to 8000 metres
Sea over 8000 metres

Scale

1 : 10 000 000

0 100 200 300 400 km

1 centimetre on the map represents 100 kilometres on the ground.

SOUTHERN EUROPE

Settlement and landforms

The people of Greece celebrate an ancient and rich culture through dance and song.

Key

Built environment

Country border

ITALY — Country name

ROME — Country capital underlined

Channel Islands (U.K.) — Dependency

● LYONS — More than 1 000 000 people

• Bursa — 500 000 to 1 000 000 people

• Florence — 100 000 to 500 000 people

• Gibraltar — Less than 100 000 people

Natural environment

River

Lake

Elba — Island

▲ Mt Olympus 2917 m — Mountain and height in metres

Land over 4000 metres
Land from 2000 to 4000 metres
Land from 1000 to 2000 metres
Land from 500 to 1000 metres
Land from 200 to 500 metres
Land from sea level to 200 metres
Land below sea level
Sea from sea level to 200 metres
Sea from 200 to 2000 metres
Sea from 2000 to 4000 metres
Sea from 4000 to 8000 metres
Sea over 8000 metres

Scale

1 : 10 000 000

0 100 200 300 400 km

1 centimetre on the map represents 100 kilometres on the ground.

COMMONWEALTH OF

Settlement and landforms

Boys in Tbilisi, Georgia, ride skateboards against a backdrop of devastation after Georgia proclaimed its independence from the U.S.S.R. in 1990.

Key

Built environment

(line)	Country border
CHINA	Country name
BEIJING	Country capital underlined
Svalbard (Norway)	Dependency
● DONETSK	More than 1 000 000 people
● Barnaul	500 000 to 1 000 000 people
• Grozny	100 000 to 500 000 people
• Mozyr	Less than 100 000 people

Natural environment

(river)	River
(lake)	Lake
Iturup	Island
▲Pobeda Peak 7439 m	Mountain and height in metres

Land over 4000 metres
Land from 2000 to 4000 metres
Land from 1000 to 2000 metres
Land from 500 to 1000 metres
Land from 200 to 500 metres
Land from sea level to 200 metres
Land below sea level
Sea from sea level to 200 metres
Sea from 200 to 2000 metres
Sea from 2000 to 4000 metres
Sea from 4000 to 8000 metres
Sea over 8000 metres

Scale

1 : 25 000 000

0 250 500 750 1000 km

1 centimetre on the map represents 250 kilometres on the ground.

AFRICA

Settlement

Geo Facts

Largest country (by area)	Sudan ▶ 2 505 813 sq km	Highest GNP per head	Libya ▶ $5141
Largest country (by pop.)	Nigeria ▶ 101 880 000	Lowest GNP per head	Sudan ▶ $55
Largest city (by pop.)	Cairo ▶ 11 155 000		

Ancient cultures

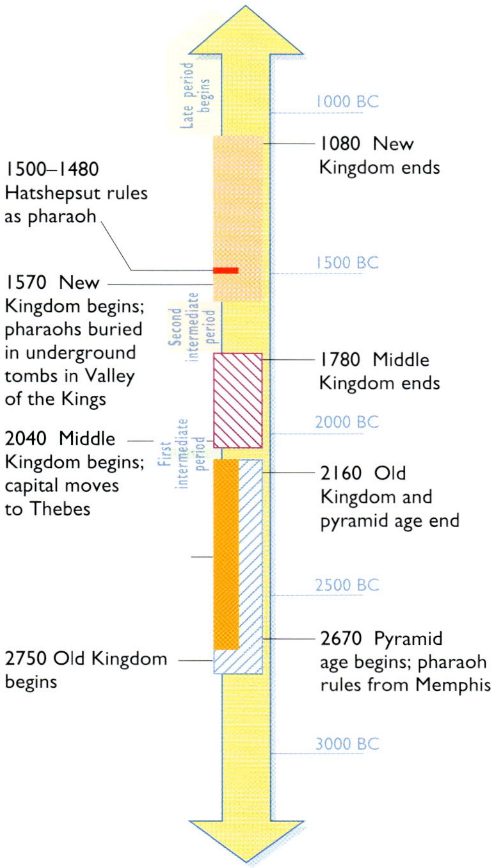

During the period of the Old Kingdom in Egypt, the pharaohs built mighty pyramids as tombs. In the New Kingdom, centred on Thebes, they dug their tombs deep into the hillsides on the western bank of the Nile in the Valley of the Kings. Here they hoped to rest undisturbed by robbers. However, by 1921, 60 tombs had been opened and looted. In 1922 a British archaeologist, Howard Carter found, undisturbed, the tomb of Tutankhamun and its treasures.

In 1995 another tomb complex was discovered which may prove to be the biggest ever found in Egypt. Archaeologists believe it was built for the sons of Rameses II, who was a great monument builder and reigned for 67 years. As part of ancient Thebes and its necropolis (cemetery), the Valley of the Kings is protected by World Heritage listing.

Hatshepsut's mortuary temple at Deir el-Bahari in the Valley of the Kings is a spectacular multi-terraced monument to Egypt's only female pharaoh.

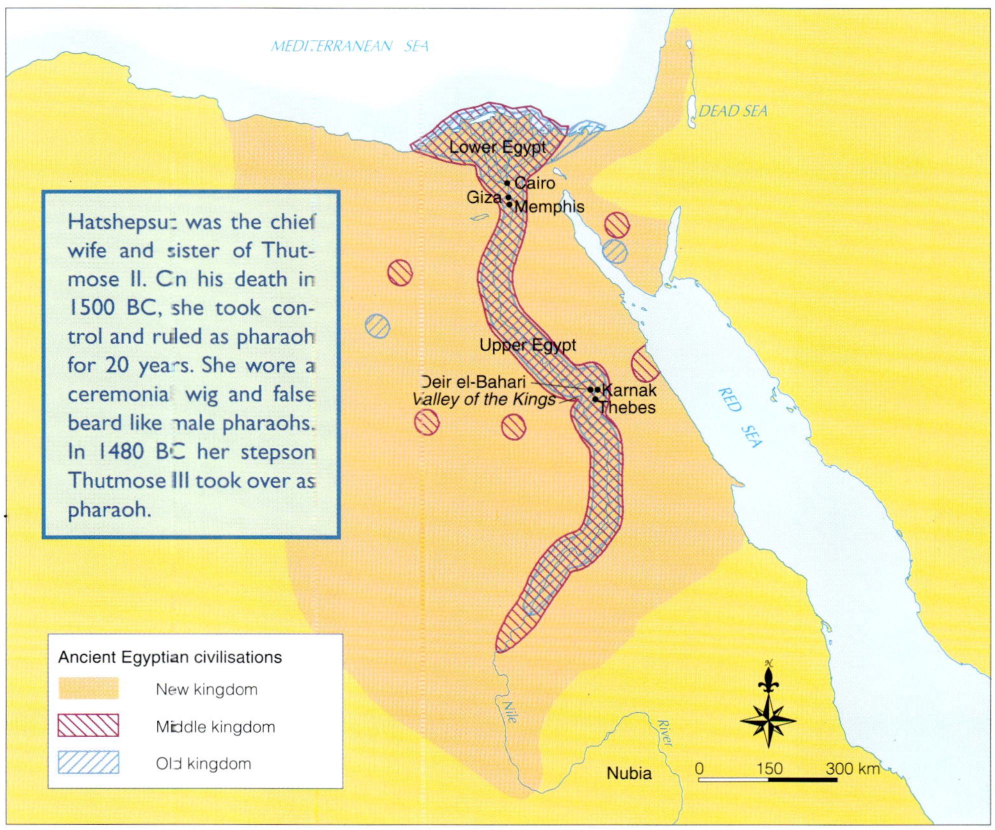

Hatshepsut was the chief wife and sister of Thutmose II. On his death in 1500 BC, she took control and ruled as pharaoh for 20 years. She wore a ceremonial wig and false beard like male pharaohs. In 1480 BC her stepson Thutmose III took over as pharaoh.

Key

Built environment

(line)	Country border
KENYA	Country name
NAIROBI	Country capital underlined
Reunion (France)	Dependency
● LAGOS	More than 1 000 000 people
• Oran	500 000 to 1 000 000 people
• Aswan	100 000 to 500 000 people
· Kindu	Less than 100 000 people

Natural environment

(line)	River
(outline)	Lake
Europa	Island

Scale

1 : 42 000 000

0 500 1000 1500 km

1 centimetre on the map represents 420 kilometres on the ground.

AFRICA

Landforms and land use

Geo Facts

Highest mountain	Kilimanjaro, Tanzania ▶ 5895 m	**Highest average annual rainfall**	Point Debundscha, Cameroon ▶ 10 277 mm
Largest desert	*Sahara ▶ 9 064 958 sq km	**Lowest average annual rainfall**	Wadi Halfa, Sudan ▶ 2.5 mm
Longest river	*Nile ▶ 6695 km		

*World record

TRANSECT

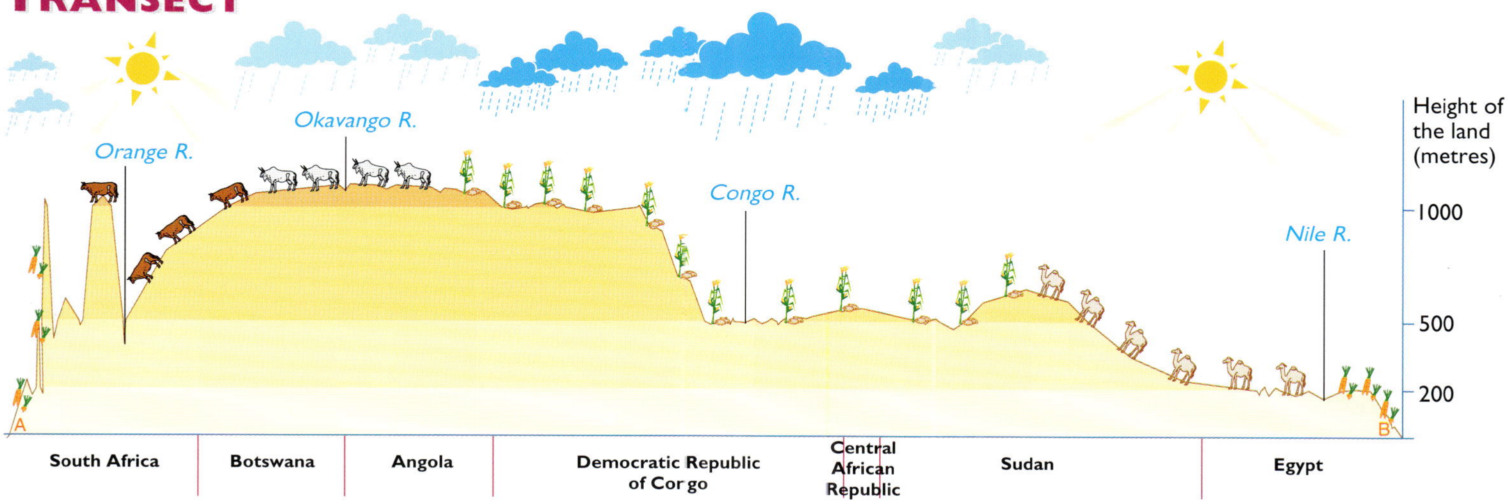

Key

Transect

A——B Cape Town to Nile Delta

Annual rainfall

1000–2000 mm

500–1000 mm

Under 500 mm

Land use

Grazing livestock

Nomadic herding – cattle

Nomadic herding – camels

Fruit and vegetables

Shifting cultivation

Built environment

Country border

Natural environment

River

Lake

Pemba Island

Mt Kenya 5199 m Mountain and height in metres

Land over 4000 metres

Land from 2000 to 4000 metres

Land from 1000 to 2000 metres

Land from 500 to 1000 metres

Land from 200 to 500 metres

Land from sea level to 200 metres

Land below sea level

Sea from sea level to 200 metres

Sea from 200 to 2000 metres

Sea from 2000 to 4000 metres

Sea from 4000 to 8000 metres

Sea over 8000 metres

Scale

1 : 42 000 000

0 500 1000 1500 km

1 centimetre on the map represents 420 kilometres on the ground.

CASE STUDY
Libyan Desert

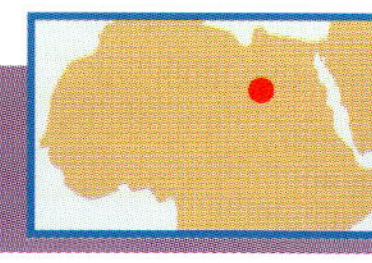

The Libyan Desert stretches from Libya into Egypt. The red circles in the centre of the image are circular irrigation fields in the desert. Water is pumped from the Al Kufrah oasis.

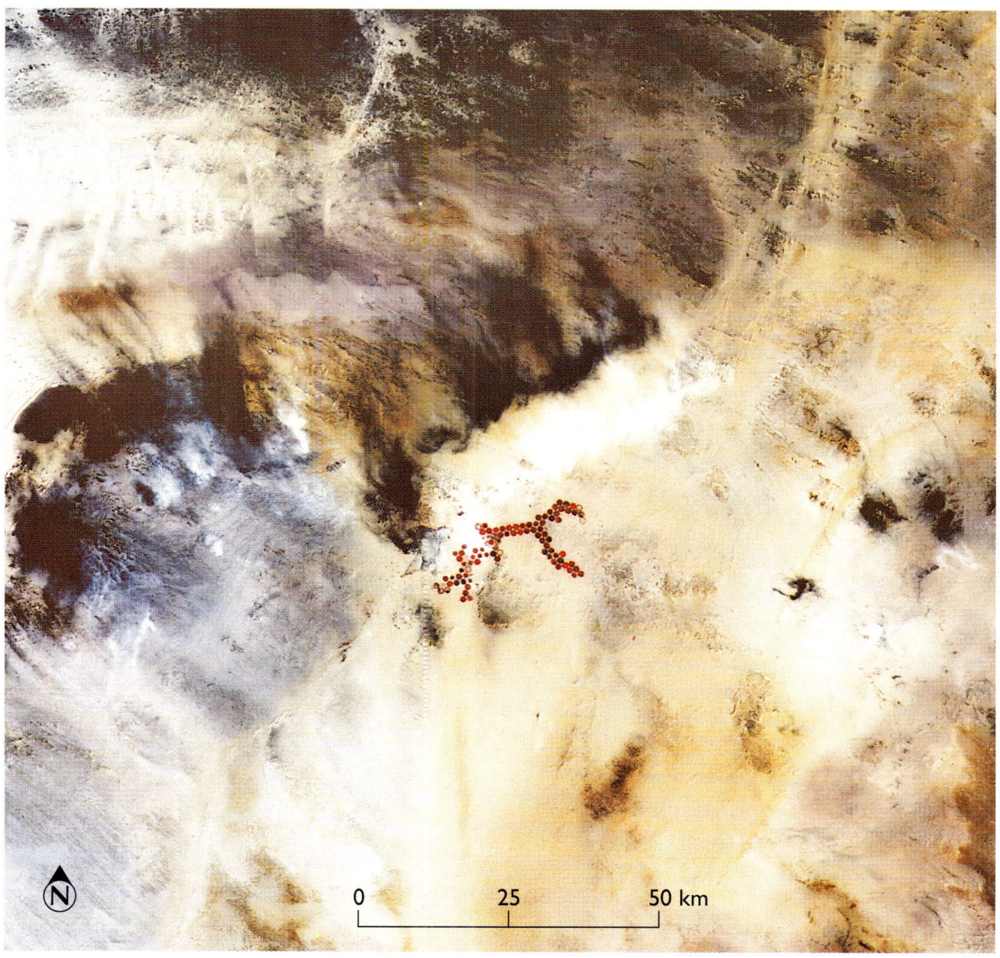

NORTHERN AFRICA
Settlement and landforms

Today, as in the past, 60 per cent of Somalis are nomadic. These women in north-west Somalia load tents onto their camels before moving south to seek grazing for their livestock.

Key

Built environment

(line)	Country border
EGYPT	Country name
CAIRO	Country capital underlined
Socotra (Yemen)	Dependency
● DOUALA	More than 1 000 000 people
• Fez	500 000 to 1 000 000 people
• Mombasa	100 000 to 500 000 people
· Timbuktu	Less than 100 000 people

Natural environment

(line)	River
(shape)	Lake
Bioko	Island
▲ *Mt Kenya 5199 m*	Mountain and height in metres

Land over 4000 metres
Land from 2000 to 4000 metres
Land from 1000 to 2000 metres
Land from 500 to 1000 metres
Land from 200 to 500 metres
Land from sea level to 200 metres
Land below sea level
Sea from sea level to 200 metres
Sea from 200 to 2000 metres
Sea from 2000 to 4000 metres
Sea from 4000 to 8000 metres
Sea over 8000 metres

Scale

1 : 21 000 000

0 200 400 600 800 km

1 centimetre on the map represents 210 kilometres on the ground.

SOUTHERN AFRICA

Settlement and landforms

CASE STUDY • • Rwandan refugee crisis

In April 1994, the presidents of Rwanda and Burundi were killed when a plane carrying them was shot down over Kigali, the capital of Rwanda. President Juvenal Habyarimana belonged to the Hutu, an ethnic group which accounted for 90 per cent of Rwanda's population. His death was blamed on the rival Tutsi ethnic group. Extremists in the Hutu government retaliated by slaughtering hundreds of thousands of Rwandans — mainly the Tutsis. The Tutsi-based Rwandan Patriotic Front hit back and forced government troops and other Hutu to flee into Zaire.

Over 500 000 people were killed in the civil war. More than 400 000 refugees left Rwanda for Zaire, Uganda and Burundi in four days in early July following the Tutsi reprisal. On 17 July 1994, the Rwandan Patriotic Front claimed victory in the civil war and, two days later, 500 000 Rwandan refugees fled into Zaire. Accurate numbers are difficult to obtain, but refugee estimates of between 40 and 50 per cent of Rwanda's total population of 8 000 000 (1994) reveal the extent of the problem.

RWANDAN REFUGEES

Key

Built environment

(line)	Country border
KENYA	Country name
NAIROBI	Country capital underlined
● DOUALA	More than 1 000 000 people
• Malange	500 000 to 1 000 000 people
• Blantyre	100 000 to 500 000 people
• Stanger	Less than 100 000 people

Natural environment

	River
	Lake
Pemba	Island
▲ Mt Mlanje 3002 m	Mountain and height in metres
	Land over 4000 metres
	Land from 2000 to 4000 metres
	Land from 1000 to 2000 metres
	Land from 500 to 1000 metres
	Land from 200 to 500 metres
	Land from sea level to 200 metres
	Land below sea level
	Sea from sea level to 200 metres
	Sea from 200 to 2000 metres
	Sea from 2000 to 4000 metres
	Sea from 4000 to 8000 metres
	Sea over 8000 metres

Scale

1 : 21 000 000

0 200 400 600 800 km

1 centimetre on the map represents 210 kilometres on the ground.

Mountain gorillas

Other innocent victims of the Rwandan crisis may be the endangered mountain gorillas. The 300 gorillas living in the high forests of the Virungu Mountains represent half of all remaining mountain gorillas.

Before the civil war, the gorillas were well protected against poachers and the destruction of their habitat in the Volcanoes National Park. By the end of 1995, two-thirds of the Rwandan park officers were dead and only 2 out of 50 vehicles survived the war. Only one gorilla was confirmed dead from the fighting, but there remains an on-going threat to their survival.

By mid 1995, 746 000 refugees (mainly Hutu) remained in makeshift tents on the Zairean side of the Virunga Mountains. More than 200 000 people lived in the largest tent city of Kibumba, with 16 000 of them going into the rainforest every day to gather wood for fuel. The foothills have been stripped of nearly every tree, with the mountain gorilla habitat only a few hours walk away.

NORTH AMERICA

Settlement

Geo Facts

Largest country (by area) Canada ▶ 9 976 000 sq km

Largest country (by pop.) USA ▶ 255 410 000

Largest city (by pop.) Mexico City ▶ 23 913 000

Highest GNP per head USA ▶ $23 119

Lowest GNP per head Haiti ▶ $360

Ancient cultures

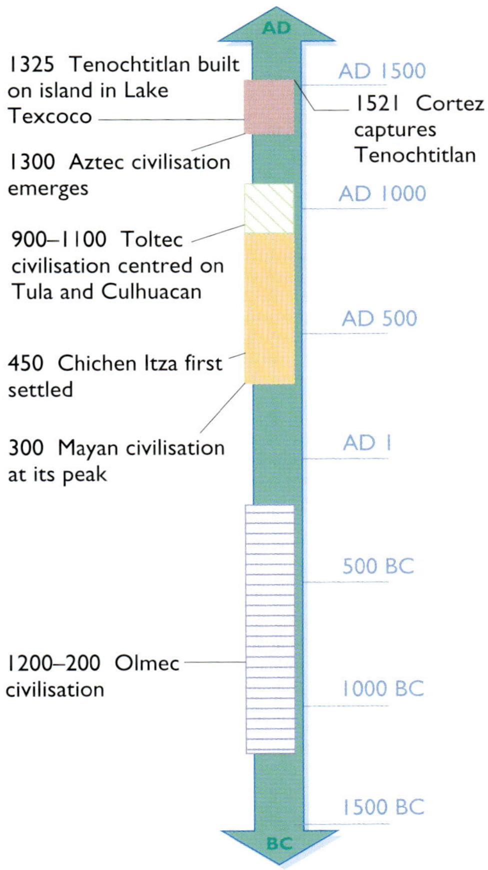

Mayan ruins at Chichen Itza include the House of Eagles and the Pyramid of Quetzalcoatl.

The great Mayan city of Chichen Itza was probably first settled around AD 450, with a later settlement period around AD 1000. Today it is a major archaeological site on the Yucatan Peninsula in Mexico. Buildings on the site include the Temple of Warriors, the House of Eagles, the Skull Wall and the Great Ball Court.

The Mayans were skilled astronomers and an observatory in the city enabled them to develop a highly accurate calendar. Their solar year had 18 months of 20 days each, making a total of 360 days. They added 5 "unlucky" days to make up 365 days. Mayan records were so accurate that they were able to predict eclipses of the sun.

The remains of the city have been listed as a World Heritage site for their historical significance. They represent a unique blend of Toltec and Mayan architecture.

Mystery surrounds the history of the Toltecs. They are referred to in many traditional legends. Their civilisation seems to have centred on Tula, but evidence of Toltec influence in architecture and religion suggests they migrated as far west as the Yucatan Peninsula. The Temple of Warriors at Chichen Itza looks as if it is a copy of a temple pyramid at Tula.

Key

Built environment

(line)	Country border
CUBA	Country name
HAVANA	Country capital underlined
Bermuda (U.K.)	Dependency
● SAN DIEGO	More than 1 000 000 people
• Calgary	500 000 to 1 000 000 people
• Acapulco	100 000 to 500 000 people
• Bismarck	Less than 100 000 people

Natural environment

(line)	River
(outline)	Lake
Guadalupe	Island

Scale

1 : 43 000 000

0 450 900 1350 1800 km

1 centimetre on the map represents 430 kilometres on the ground.

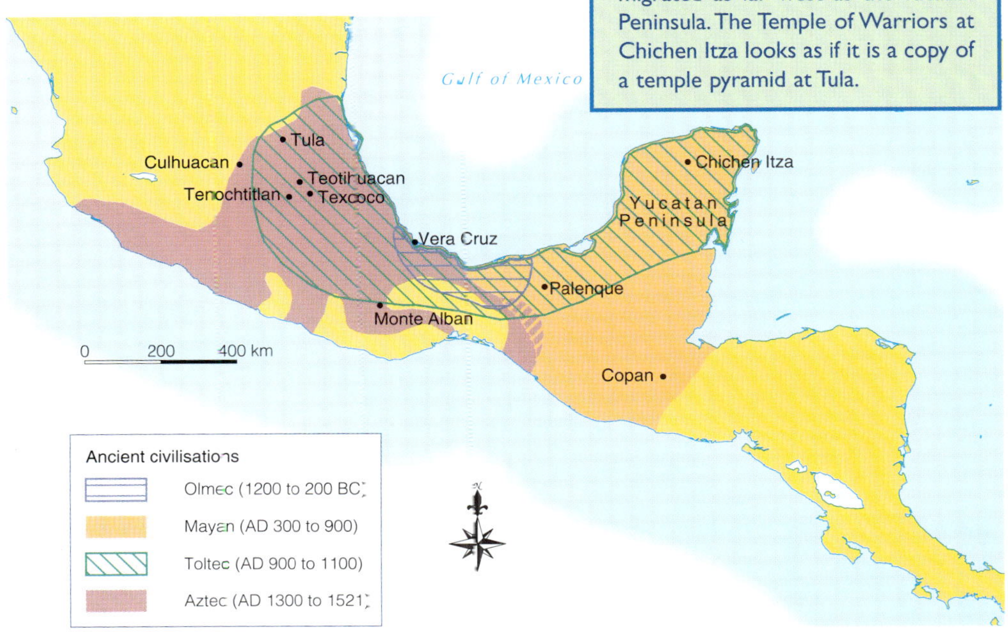

NORTH AMERICA
Landforms and land use

Geo Facts

Highest mountain McKinley, U.S.A. ▶ 6194 m
Largest desert Mojave, U.S.A. ▶ 65 000 sq km
Longest river Mississippi–Missouri, U.S.A. ▶ 6020 km
Highest average annual rainfall Henderson Lake, Canada ▶ 6655 mm
Lowest average annual rainfall Bateques, Mexico ▶ 30.5 mm

Transect

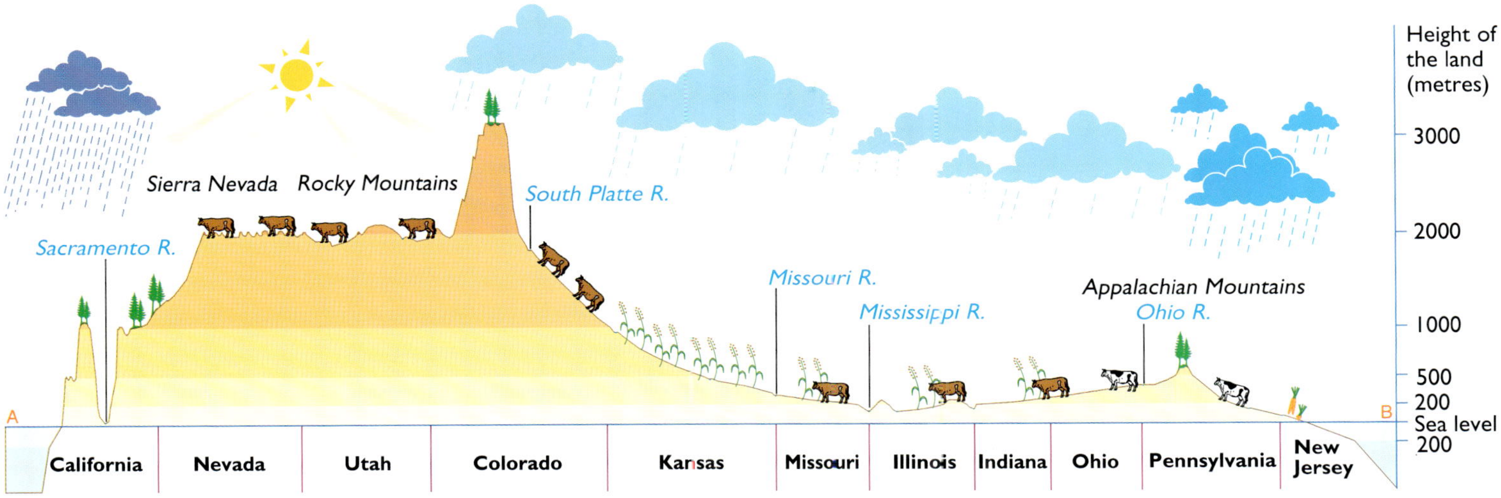

Key

Transect

A——B	Along 40° north latitude
Annual rainfall	*Land use*
Over 2000 mm	Grazing livestock
1000–2000 mm	Dairying
500–1000 mm	Crops, i.e. wheat, barley
Under 500 mm	Crops and livestock
	Fruit and vegetables
	Forest

Built environment

Country border

Natural environment

River

Lake

Andros Island

Mt Mitchell 2037 m Mountain and height in metres

Land over 4000 metres
Land from 2000 to 4000 metres
Land from 1000 to 2000 metres
Land from 500 to 1000 metres
Land from 200 to 500 metres
Land from sea level to 200 metres
Land below sea level
Sea from sea level to 200 metres
Sea from 200 to 2000 metres
Sea from 2000 to 4000 metres
Sea from 4000 to 8000 metres
Sea over 8000 metres

Scale

1 : 43 000 000

0 450 900 1350 1800 km

1 centimetre on the map represents 430 kilometres on the ground.

CASE STUDY
San Francisco Bay

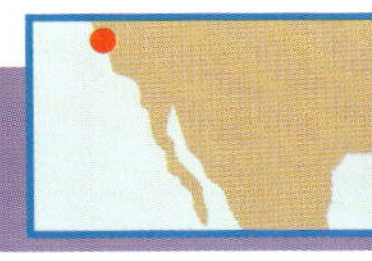

The city of San Francisco occupies the peninsula to the south of San Francisco Bay. Four bridges provide links across the bay, including the famous Golden Gate bridge across the entrance of San Francisco Bay.

CANADA

Settlement and landforms

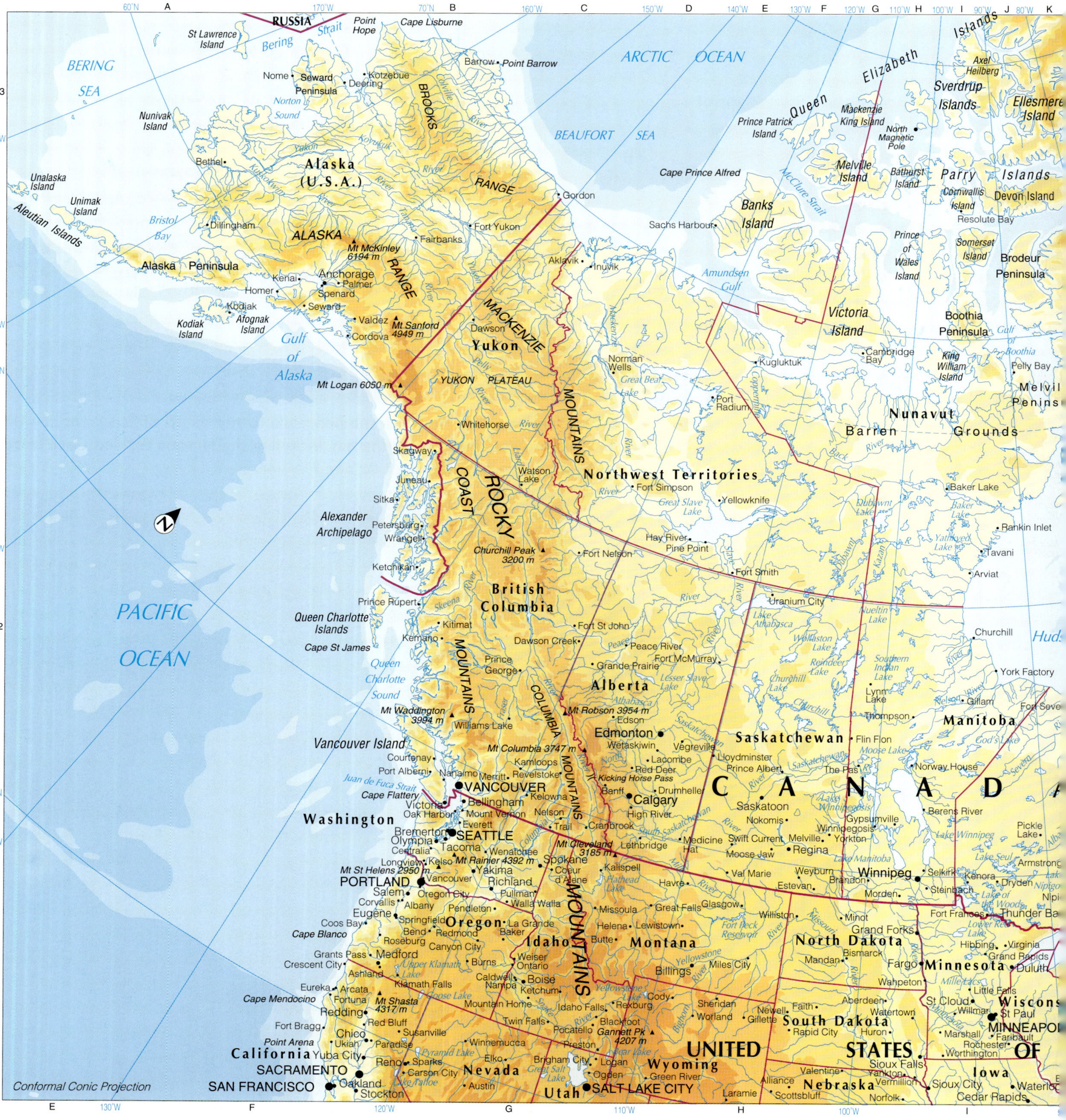

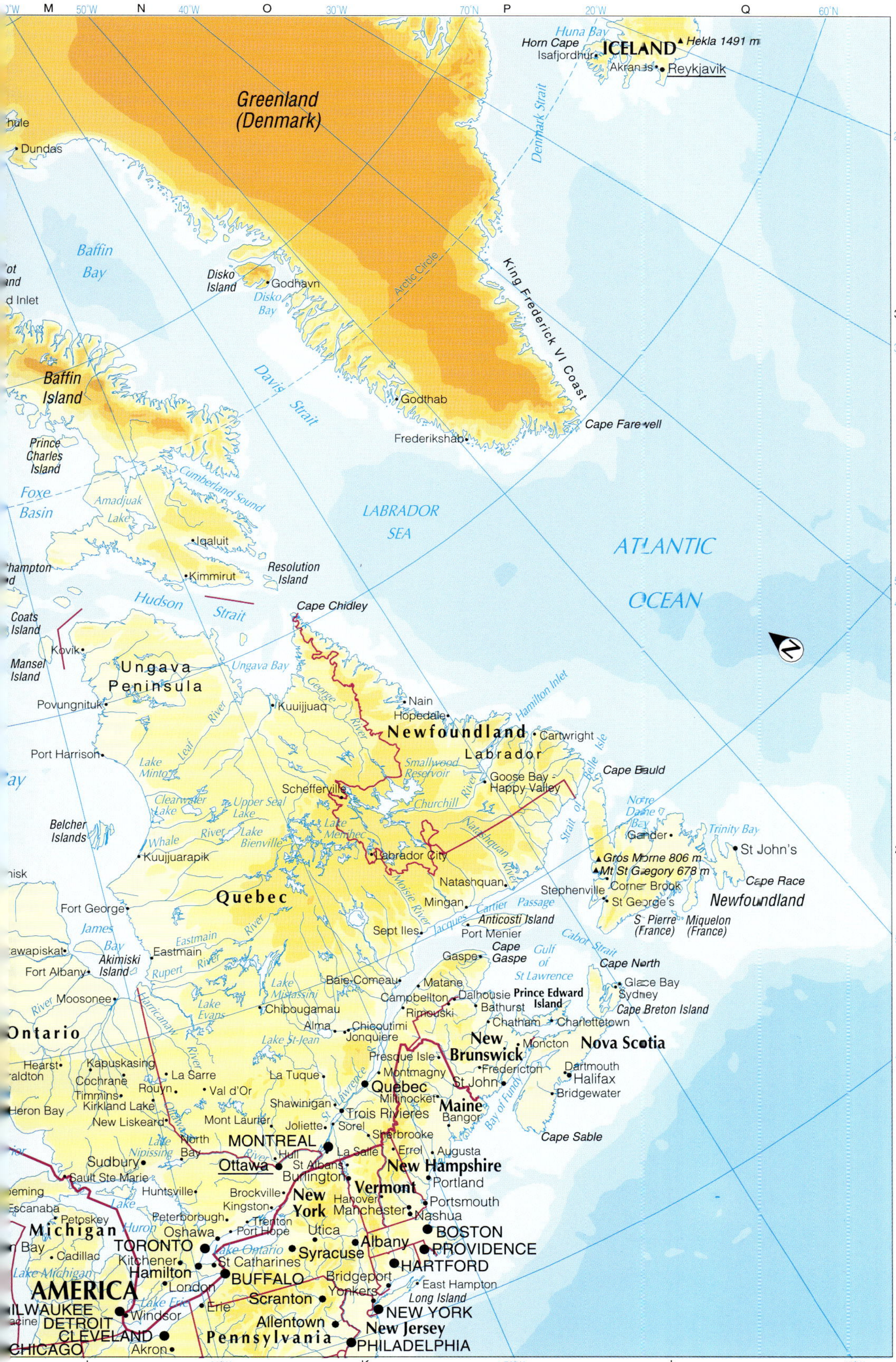

The Inuit people of Canada were once known as Eskimos. In 1999, the territory of Nunavut, an Inuit word meaning "our land", will be established for the Inuit.

Key

Built environment

(line)	Country border
CANADA	Country name
Ottawa	Country capital underlined
(line)	State/territory border
Quebec	State/territory name
Miquelon (France)	Dependency
● TORONTO	More than 1 000 000 people
● Calgary	500 000 to 1 000 000 people
• Regina	100 000 to 500 000 people
• Gaspe	Less than 100 000 people

Natural environment

	River
	Lake
Coats Island	Island
▲Mt Logan 6050 m	Mountain and height in metres

Land over 4000 metres
Land from 2000 to 4000 metres
Land from 1000 to 2000 metres
Land from 500 to 1000 metres
Land from 200 to 500 metres
Land from sea level to 200 metres
Land below sea level
Sea from sea level to 200 metres
Sea from 200 to 2000 metres
Sea from 2000 to 4000 metres
Sea from 4000 to 8000 metres
Sea over 8000 metres

Scale

1 : 20 000 000

0 200 400 600 800 km

1 centimetre on the map represents 200 kilometres on the ground.

UNITED STATES OF

Settlement and landforms

The Cheyenne is one of 266 American Indian tribes negotiating a better future for indigenous people in the U.S.A.

Key

Built environment

(line)	Country border
CUBA	Country name
HAVANA	Country capital underlined
(line)	State/territory border
Maine	State/territory name
Caicos Islands (U.K.)	Dependency
● SAN DIEGO	More than 1 000 000 people
• Omaha	500 000 to 1 000 000 people
• Abilene	100 000 to 500 000 people
· Missoula	Less than 100 000 people

Natural environment

(line)	River
(outline)	Lake
Andros	Island
▲Mt Mitchell 2037 m	Mountain and height in metres

Land over 4000 metres
Land from 2000 to 4000 metres
Land from 1000 to 2000 metres
Land from 500 to 1000 metres
Land from 200 to 500 metres
Land from sea level to 200 metres
Land below sea level
Sea from sea level to 200 metres
Sea from 200 to 2000 metres
Sea from 2000 to 4000 metres
Sea from 4000 to 8000 metres
Sea over 8000 metres

Scale

1 : 15 000 000

0 150 300 450 600 km

1 centimetre on the map represents 150 kilometres on the ground.

CENTRAL AMERICA

Settlement and landforms

Guatemala struggles to provide educational facilities for its children. This girl studies in a jungle school.

Key

Built environment

(line)	Country border
CUBA	Country name
HAVANA	Country capital underlined
(line)	State/territory border
Texas	State/territory name
Guadeloupe (France)	Dependency
● HOUSTON	More than 1 000 000 people
● Puebla	500 000 to 1 000 000 people
• Maturin	100 000 to 500 000 people
· Moron	Less than 100 000 people

Natural environment

(symbol)	River
(symbol)	Lake
Barbuda	Island
▲ *Tolima 5616 m*	Mountain and height in metres

Land over 4000 metres
Land from 2000 to 4000 metres
Land from 1000 to 2000 metres
Land from 500 to 1000 metres
Land from 200 to 500 metres
Land from sea level to 200 metres
Land below sea level
Sea from sea level to 200 metres
Sea from 200 to 2000 metres
Sea from 2000 to 4000 metres
Sea from 4000 to 8000 metres
Sea over 8000 metres

Scale

1 : 15 500 000

0 150 300 450 600 km

1 centimetre on the map represents 155 kilometres on the ground.

SOUTH AMERICA
Settlement

Geo Facts

Largest country (by area)	Brazil ▶ 8 511 965 sq km
Largest country (by pop.)	Brazil ▶ 153 850 000
Largest city (by pop.)	Sao Paulo ▶ 21 539 000
Highest GNP per head	Argentina ▶ $6051
Lowest GNP per head	Guyana ▶ $410

Ancient cultures

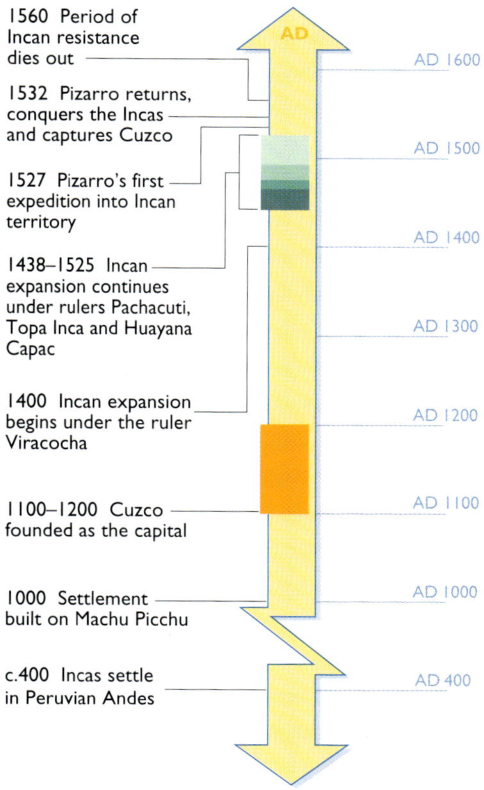

Incas first settled in the Peruvian Andes around AD 400. Gradually, over 1000 years, they expanded their empire to include much of the modern countries of Ecuador, Peru, Chile, Bolivia and parts of Colombia and Argentina. They were conquered by the Spanish in the early sixteenth century. The expansion of the Incan empire was helped by the excellent roads and suspension bridges built by the Incas. These engineering feats are remarkable because the Incas had no metal tools capable of cutting stone, nor the wheel to transport building materials. The skill of Incan architects in cutting stones to fit together with such precision that they needed no mortar can be seen today in the great stone buildings of Cuzco and the ruins of Machu Picchu. Both Cuzco and Machu Picchu have been listed as World Heritage sites for their cultural and historical significance.

Machu Picchu, high in the Andes, may have been used as a fortress by the Incas.

Key

Built environment

Country border

PERU — Country name

LIMA — Country capital underlined

South Georgia (U.K.) — Dependency

● CALI — More than 1 000 000 people

• La Plata — 500 000 to 1 000 000 people

• Caxias — 100 000 to 500 000 people

• Cayenne — Less than 100 000 people

Natural environment

River

Lake

Aruba — Island

Scale

1 : 40 000 000

0 400 800 1200 1600 km

1 centimetre on the map represents 400 kilometres on the ground.

The network of roads built by the Incas was used by the ruler himself, by his armies, by messengers and for the movement of supplies. One road system crossed the mountains and another wound along the coast. Shelters every 35 kilometres were provided for travellers. The total length of the road system was longer than even the Romans' network.

SOUTH AMERICA

Landforms and land use

Geo Facts

Highest mountain Aconcagua, Argentina ▶ 6960 m

Largest desert Atacama, Chile ▶ 180 000 sq km

Longest river Amazon ▶ 6437 km

Highest average annual rainfall Tutunendo, Colombia ▶ 11 770 mm

Lowest average annual rainfall *Arica, Chile ▶ .80 mm

*World record

TRANSECT

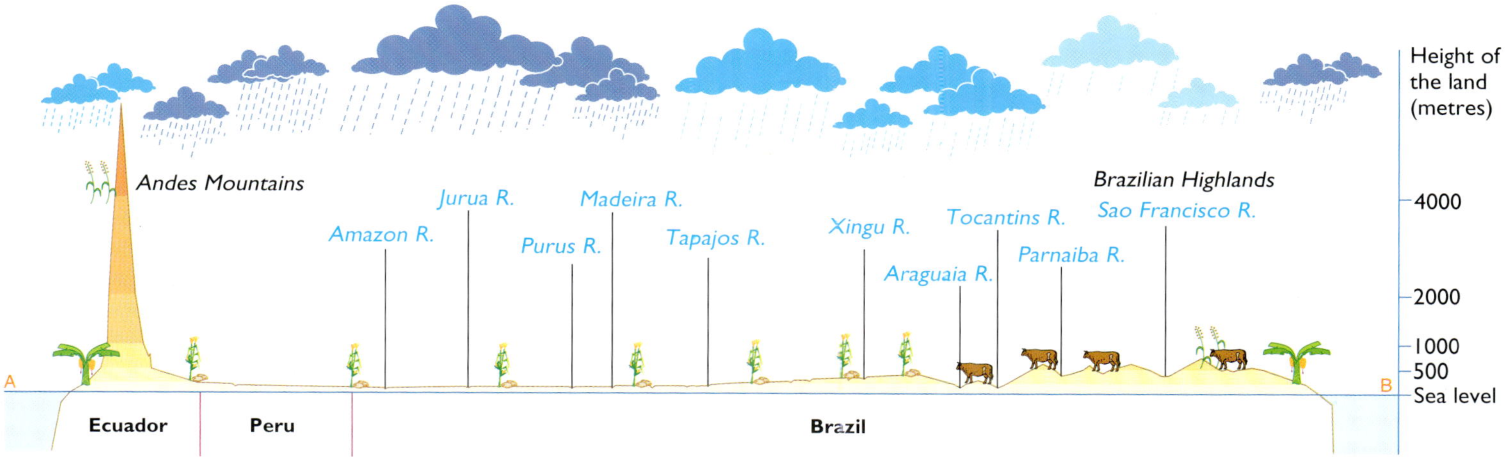

Key

Transect

A——B Quito to Sao Francisco River

Annual rainfall

Over 2000 mm

1000–2000 mm

500–1000 mm

Land use

Grazing livestock

Crops, i.e. coffee, corn

Crops and livestock

Shifting cultivation

Plantation

Built environment

Country border

Natural environment

River

Lake

Isabella Island

▲*Mt Chato 2420 m* Mountain and height in metres

Land over 4000 metres

Land from 2000 to 4000 metres

Land from 1000 to 2000 metres

Land from 500 to 1000 metres

Land from 200 to 500 metres

Land from sea level to 200 metres

Land below sea level

Sea from sea level to 200 metres

Sea from 200 to 2000 metres

Sea from 2000 to 4000 metres

Sea from 4000 to 8000 metres

Sea over 8000 metres

Scale

1 : 40 000 000

0 400 800 1200 1600 km

1 centimetre on the map represents 400 kilometres on the ground.

CASE STUDY

Amazon Basin

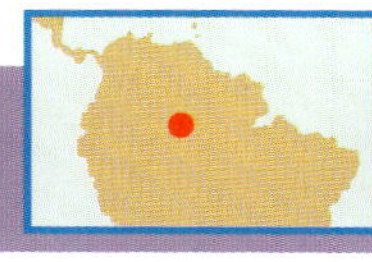

The Amazon Basin lies between the Guyana and Brazilian highlands. The red area in the satellite image shows dense rainforest. Sediment carried by the Amazon makes it appear blue. The city of Manaus is located on the northern bank of the Negro River, just before the river joins the Amazon.

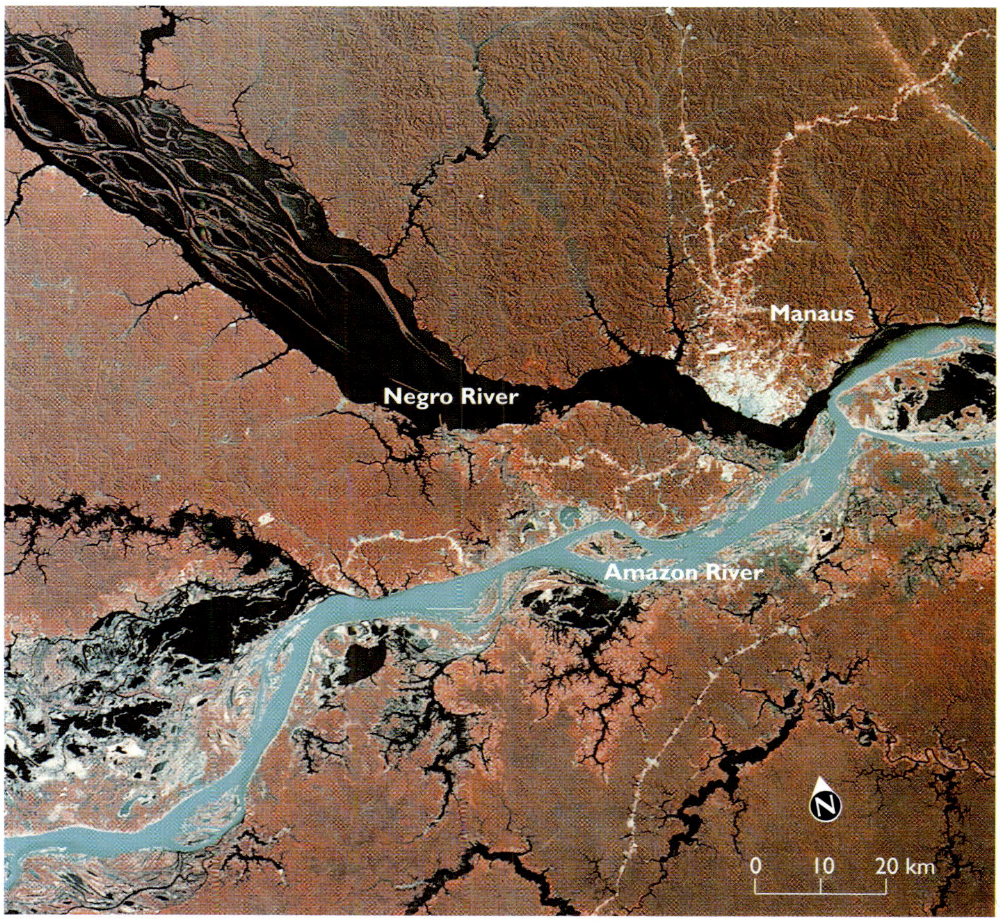

NORTHERN SOUTH
Settlement and landforms

AMERICA

Indians in the Amazon Basin have become very involved in the protection of their rainforest environments. Here a Kayaidos Indian uses a two-way radio to build support for the cause.

Key

Built environment

	Country border
ECUADOR	Country name
QUITO	Country capital underlined
French Guiana	State/territory name
Malpelo Island (Colombia)	Dependency
● CALI	More than 1 000 000 people
● Callao	500 000 to 1 000 000 people
• Vitoria	100 000 to 500 000 people
· Saul	Less than 100 000 people

Natural environment

	River
	Lake
Tobago	Island
▲ *Bandeira Peak 2894 m*	Mountain and height in metres

Land over 4000 metres
Land from 2000 to 4000 metres
Land from 1000 to 2000 metres
Land from 500 to 1000 metres
Land from 200 to 500 metres
Land from sea level to 200 metres
Land below sea level
Sea from sea level to 200 metres
Sea from 200 to 2000 metres
Sea from 2000 to 4000 metres
Sea from 4000 to 8000 metres
Sea over 8000 metres

Scale

1 : 18 000 000

0 200 400 600 800 km

1 centimetre on the map represents 180 kilometres on the ground.

SOUTHERN SOUTH AMERICA

Key

Built environment

(line)	Country border
CHILE	Country name
SANTIAGO	Country capital underlined
Falkland Islands (U.K.)	Dependency
● SAO PAULO	More than 1 000 000 people
● La Plata	500 000 to 1 000 000 people
• San Juan	100 000 to 500 000 people
• Stanley	Less than 100 000 people

Natural environment

(symbol)	River
(symbol)	Lake
Chiloe Island	Island
▲ *Aconcagua 6960 m*	Mountain and height in metres

- Land over 4000 metres
- Land from 2000 to 4000 metres
- Land from 1000 to 2000 metres
- Land from 500 to 1000 metres
- Land from 200 to 500 metres
- Land from sea level to 200 metres
- Land below sea level
- Sea from sea level to 200 metres
- Sea from 200 to 2000 metres
- Sea from 2000 to 4000 metres
- Sea from 4000 to 8000 metres
- Sea over 8000 metres

Scale

Scale 1 : 18 000 000

0 150 300 450 600 km

1 centimetre on the map represents 180 kilometres on the ground.

REFERENCE

THE POLITICAL WORLD

WORLD COUNTRIES AND CAPITAL CITIES

OCEAN
NORTH PACIFIC OCEAN
NORTH ATLANTIC OCEAN
SOUTH PACIFIC OCEAN
SOUTH ATLANTIC OCEAN
Greenland (Den.)
ICELAND
Alaska (U.S.A.)
CANADA
Ottawa
St Pierre & Miquelon (Fr.)
Azores (Port.)
Washington D.C.
UNITED STATES OF AMERICA
Bermuda (U.K.)
MEXICO
Mexico City
Hawaii (U.S.A.)
BAHAMAS
Nassau
Havana
CUBA
HAITI
DOMINICAN REPUBLIC
Port-au-Prince
Puerto Rico (U.S.A.)
ANTIGUA AND BARBUDA
BELIZE
Belmopan
JAMAICA
ST KITTS-NEVIS
GUATEMALA
Guatemala City
HONDURAS
Tegucigalpa
DOMINICA
ST LUCIA
ST VINCENT
GRENADA
BARBADOS
San Salvador
EL SALVADOR
NICARAGUA
Managua
TRINIDAD AND TOBAGO
Port of Spain
COSTA RICA
San Jose
Panama City
Caracas
PANAMA
VENEZUELA
Georgetown
Paramaribo
GUYANA
French Guiana (Fr.)
SURINAM
Bogota
COLOMBIA
Galapagos Is (Ecu.)
Quito
ECUADOR
PERU
Lima
BRAZIL
Brasilia
La Paz
BOLIVIA
Sucre
PARAGUAY
Asuncion
San Felix (Ch.)
Easter I (Ch.)
CHILE
Juan Fernandez Is (Ch.)
Santiago
Buenos Aires
URUGUAY
Montevideo
ARGENTINA
Falkland Is (U.K.)
South Georgia (U.K.)
MARSHALL ISLANDS
Majuro
Palikir
NAURU
Yaren
Bairiki
KIRIBATI
SOLOMON ISLANDS
Honiara
TUVALU
Funafuti
SAMOA
Apia
American Samoa (U.S.A.)
VANUATU
Port Vila
FIJI
Suva
French Polynesia (Fr.)
NIUE
Alofi
COOK ISLANDS
Avarua
New Caledonia (Fr.)
TONGA
Nukualofa
Norfolk I (Aust.)
Lord Howe I (Aust.)
International Date Line
NEW ZEALAND
Wellington
Chatham Is (N.Z.)
Auckland Is (N.Z.)
Macquarie I (Aust.)
H
I
J
K
L
M
N
180°
150°W
120°W
90°W
60°W
30°W
0°
30°E
60°N
30°N
0°
30°S
1
2
3
4
5

148 THE CHANGING EARTH

Continental plates

The earth's shell or crust is split into nine large crustal plates and about a dozen smaller ones. These plates fit together like a huge jigsaw puzzle. Some of these plates contain the continents (continental plates). The plates float on the semi-molten rocks of the earth's *mantle*.

Convection currents in the earth's **mantle** work just like those in any liquid. Heating from the core causes the semi-molten material in the mantle to churn in currents, carrying the crustal plates slowly along, up to as much as 15 centimetres a year. This movement has caused **continental drift**.

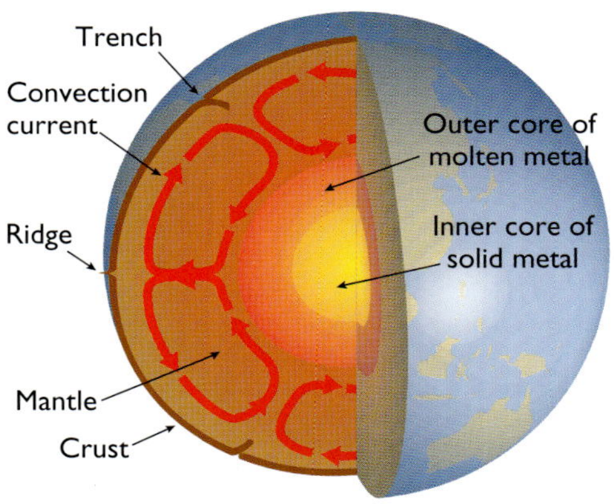

Sometimes the earth's plates collide, pushing up mountain ranges. The contact between these **converging plates** can also cause earthquakes (page 86) and volcanoes (page 150).

WORLD LANDFORMS AND TECTONIC PLATES

Mountain building

Most of the world's great mountain regions are formed when crustal rocks are buckled as the plates converge. This explains why mountain ranges such as the Rocky Mountains in North America and the Himalayas in Asia are located along converging plate boundaries. Which other mountain ranges can be found along the edges of converging plates?

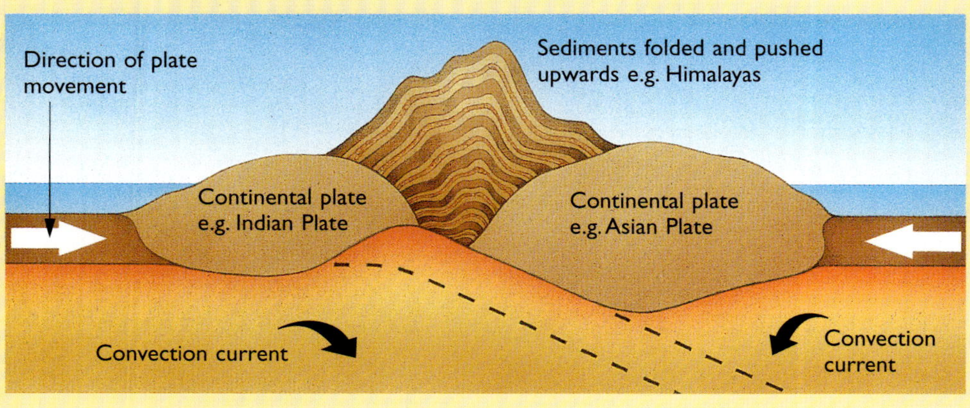

Key

Tectonic plates

- Converging plate boundary
- Diverging plate boundary
- Uncertain plate boundary
- Movement of plate
- Volcano
- Earthquake zone
- **Pacific Plate** — Plate name

Natural environment

- River
- Lake
- *Java* — Island
- Mt Cook 3764 m — Mountain and height in metres
- Land over 4000 metres
- Land from 2000 to 4000 metres
- Land from 1000 to 2000 metres
- Land from 500 to 1000 metres
- Land from 200 to 500 metres
- Land from sea level to 200 metres
- Land below sea level
- Sea from sea level to 200 metres
- Sea from 200 to 2000 metres
- Sea from 2000 to 4000 metres
- Sea from 4000 to 8000 metres
- Sea over 8000 metres

Scale

1 : 100 000 000

0 1000 2000 3000 4000 km

1 centimetre on the map represents 1000 kilometres on the ground.

Continental drift

More than 200 million years ago the huge landmass of Pangaea broke apart to form the continents we recognise today. Convection currents beneath the earth's surface have moved the world's continents to different positions on the earth's surface. By using computers to plot the present movements of the plates, scientists are predicting that in 250 million years a new large landmass may develop in the northern hemisphere.

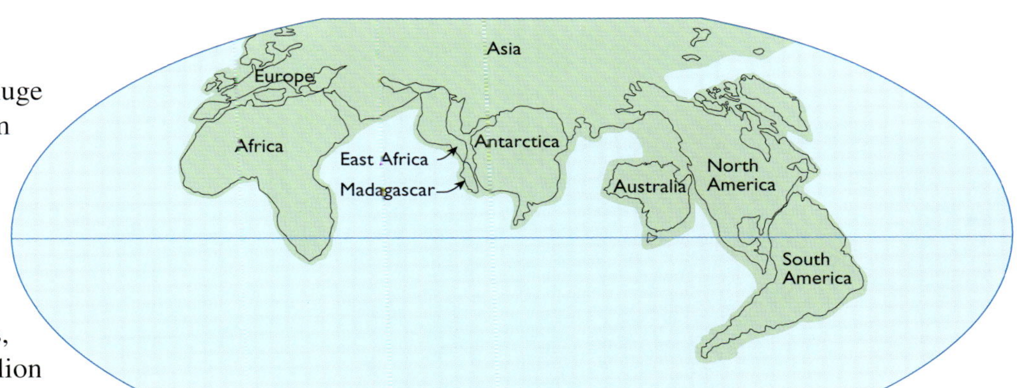

The world 250 million years from now

150 VOLCANOES

A volcano forms when **magma**, hot molten rock from beneath the crust, emerges on the earth's surface. When it does, this molten rock is called **lava**. The way it emerges and the lava characteristics determine the type of volcano.

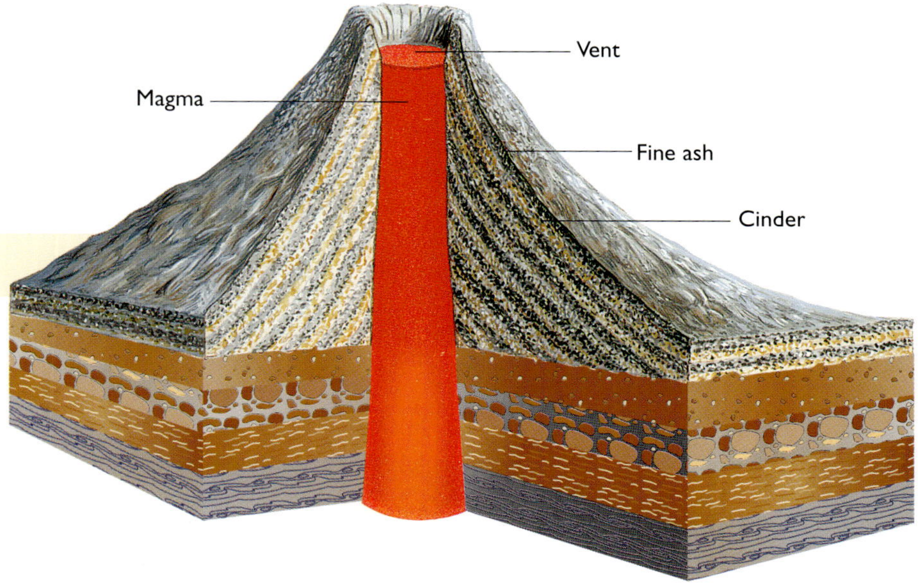

Ash or cinder cone

A **volcanic cone** forms when high gas concentrations cause explosive eruptions of volcanic ash and cinders. These fall back to form a steep concave-sided volcanic cone, for example Mt Paricutin in Mexico.

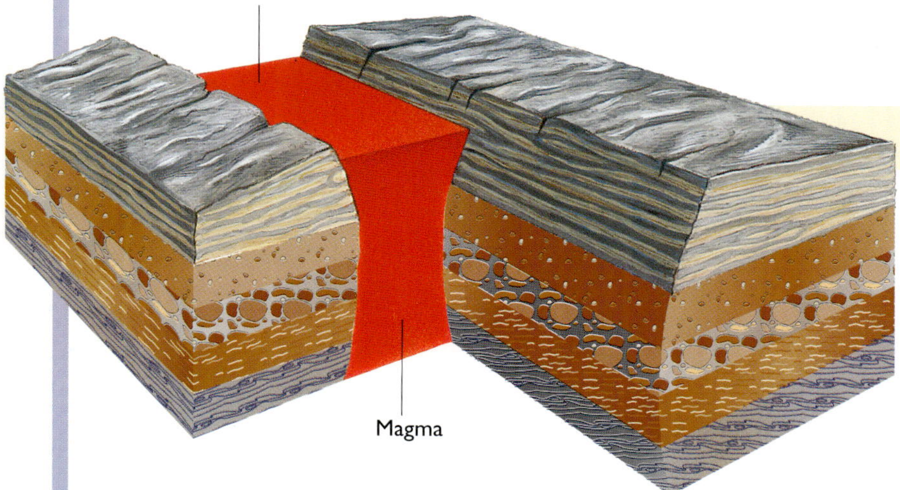

Fissure volcano

A **fissure** is a long crack in the earth's surface that can be caused by diverging plates. When lava escapes through one, a fissure volcano develops.

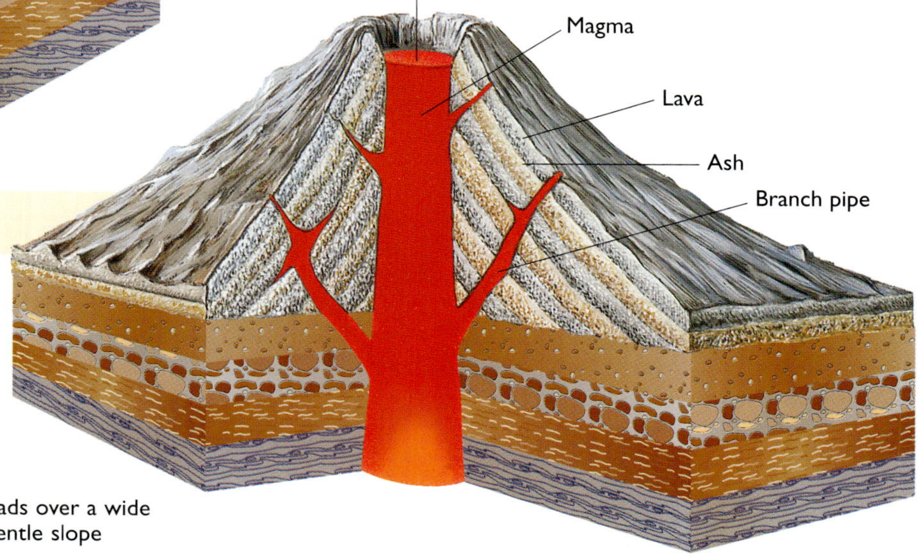

Composite cone

Composite cone volcanoes are made up of alternate layers of ash and lava because the eruptions alternate between explosive (ash) and quiet (lava). Mt Fuji in Japan and Mt Etna in Italy are examples of composite cones.

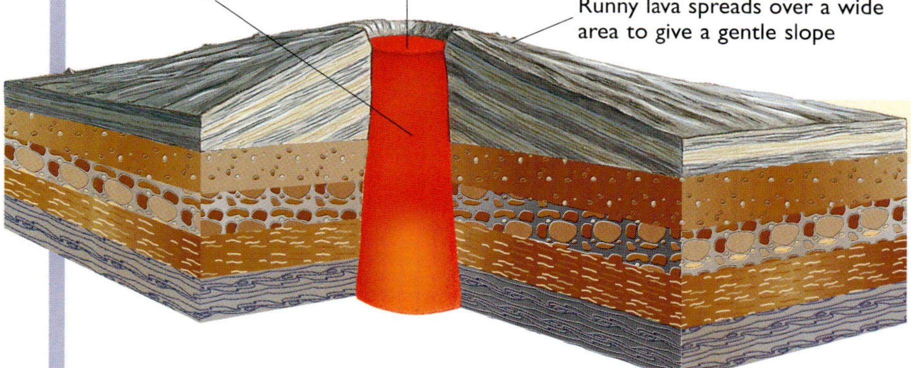

Shield volcano

This is a broad, low volcanic cone caused by very fluid (runny) lava. This type of volcano is non-explosive because little gas is present, for example Mt Warning, New South Wales. If the lava is very thick and sticky, a steep-sided dome volcano is formed because the lava will not spread out, for example Hanging Rock, Victoria.

Volcanic eruption

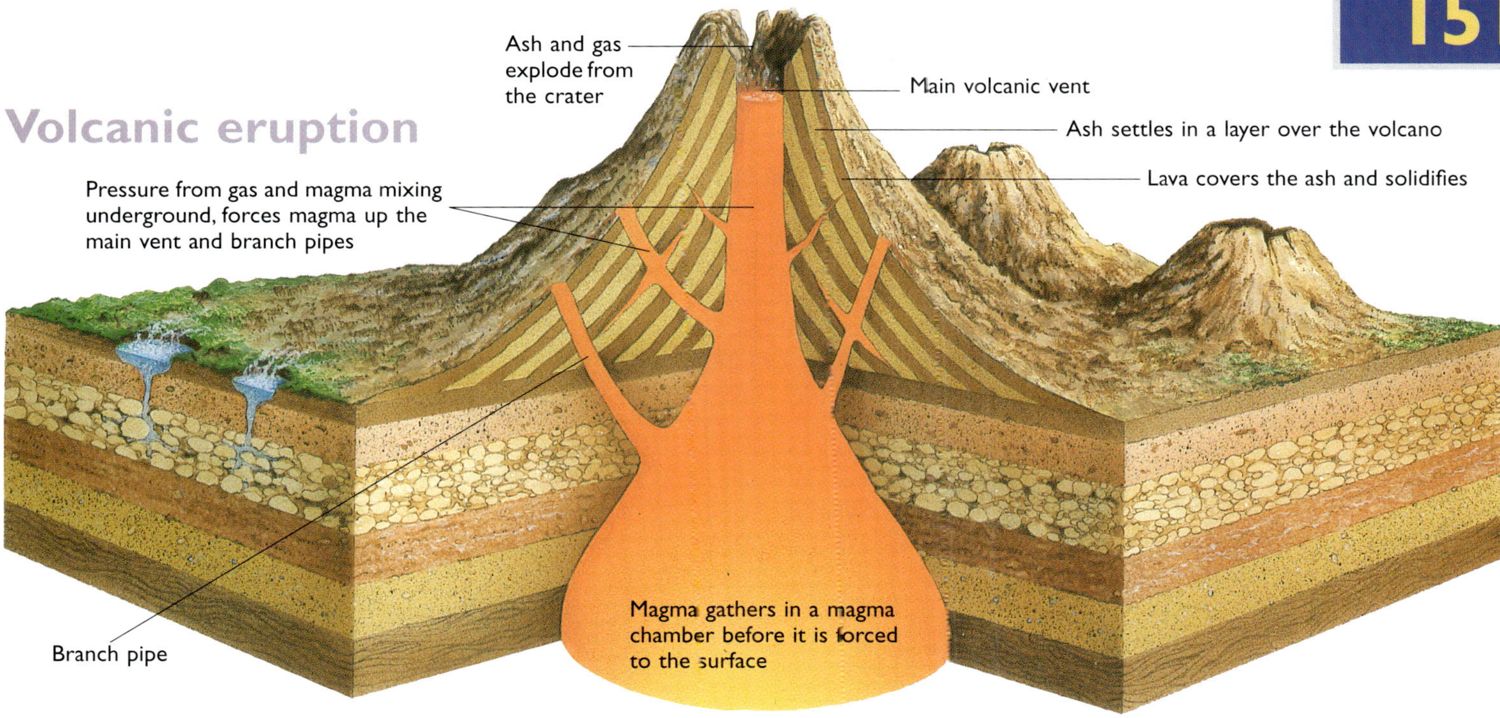

AUSTRALIAN VOLCANIC ACTIVITY

1. *Western Victoria*

The western district of Victoria has been widely affected by volcanic activity and the landscape is dotted with numerous cones. None rises more than 150–180 metres and some are as low as 30 metres, most cones having very gentle slopes. This area is one of the most recently affected by volcanic activity in Australia and contains shield volcanoes (for example Mt Hamilton), composite cones (for example Mt Napier) and calderas (for example Tower Hill).

A **caldera** is an enlarged crater caused by the collapse or explosion of a volcano. This example is Tower Hill in Victoria.

2. *Glass House Mountains, Queensland*

The Glass House Mountains, 70 kilometres north of Brisbane, are about 20 million years old. They are excellent examples of volcanic plugs. These are formed when the central vent areas remain after the cones that formerly surrounded them are eroded away. The highest is Mt Beerwah which rises 550 metres above the surrounding plain.

The Glass House Mountains in Queensland are made up of volcanic plugs.

RIVER SYSTEMS

Rivers are part of a system that begins with rain falling or ice melting (inputs) in mountains or hills. As the river flows downhill it wears away the land to form valleys. The eroded stones, sand and soil, known as **sediment**, are washed down to the mouth of the river. The erosion, transportation and deposition of material are examples of the **processes** of a river system.

The life of a river

Mountains
Rivers begin in mountains. The force of the water rushing down the steep slope erodes vertically and carries the material away. Mountain river valleys are steep sided and "V"-shaped.

Valley
A river is in the valley section when it reaches lower land downstream from the mountains. The valleys are deeper but wider as the river erodes both the floor and sides of the valley.

Flood plain
On the **flood plain**, the stream erodes horizontally causing a river valley that is broad and flat. The river can no longer hold all the sediment it carries and deposits material to form flood plains and **deltas**.

Waterfalls

Water in the **headwaters**, or start of a river, erodes the river bed. As the water travels downstream, it may meet harder rock which does not erode as easily.

A waterfall forms as the water spills over the harder rock. The falling water erodes the soft rock below to form a **plunge pool**.

The hard rock is eventually undermined by the water in the pool and collapses. A new waterfall forms further upstream. As this process continues, the areas formed by the collapsed waterfalls combine to form a very steep-sided valley known as a **gorge**.

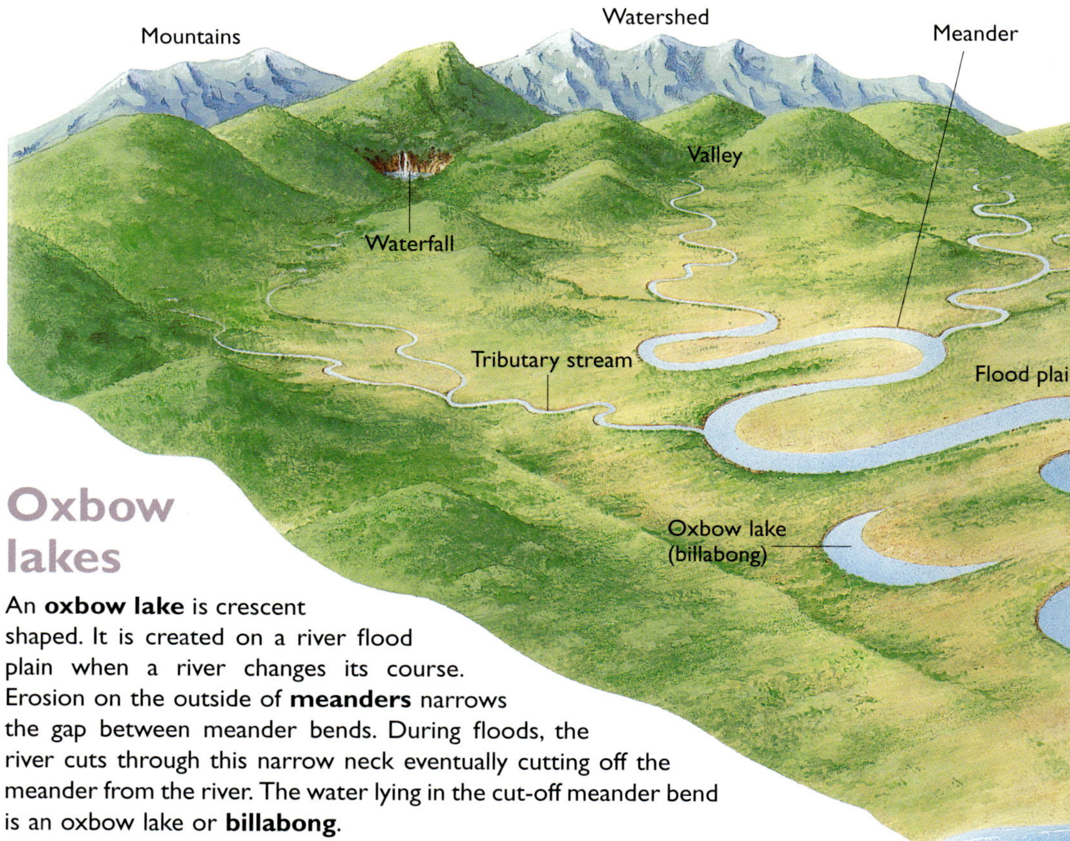

Oxbow lakes

An **oxbow lake** is crescent shaped. It is created on a river flood plain when a river changes its course. Erosion on the outside of **meanders** narrows the gap between meander bends. During floods, the river cuts through this narrow neck eventually cutting off the meander from the river. The water lying in the cut-off meander bend is an oxbow lake or **billabong**.

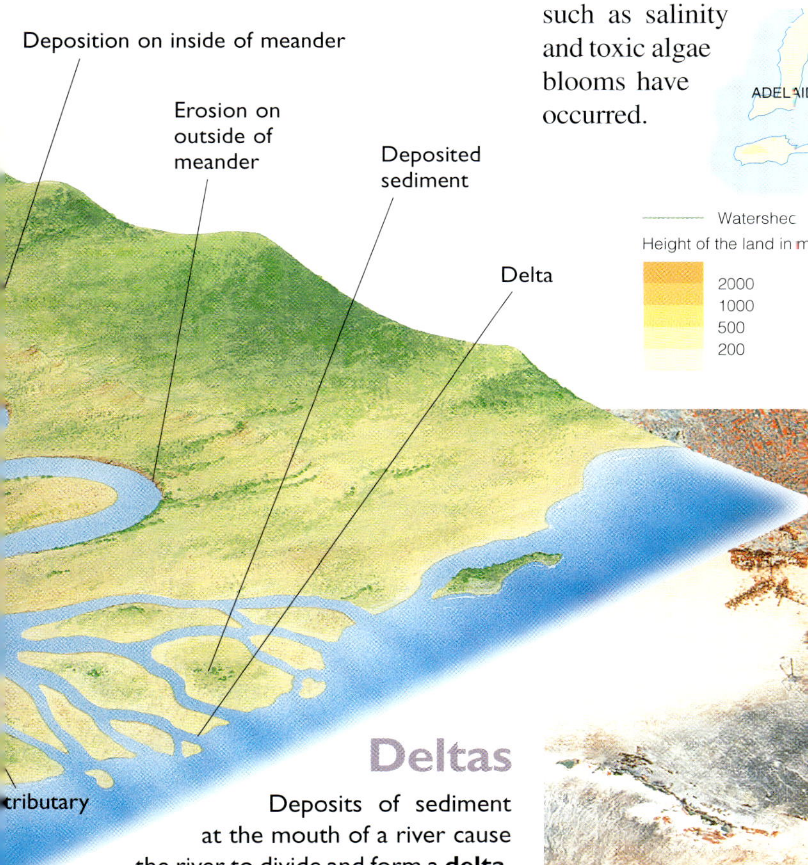

Murray–Darling Basin

A **drainage basin** is the area of land which feeds a river with water. The Murray–Darling Basin covers over one million square kilometres, about 14 per cent of Australia. It is drained by more than 20 major rivers which begin in mountain areas. The watershed shown on the map marks the peaks which divide the Murray–Darling Basin from other drainage basins. Since Europeans have begun to use the drainage basin for farms and industry, problems such as salinity and toxic algae blooms have occurred.

Deltas

Deposits of sediment at the mouth of a river cause the river to divide and form a **delta**. The river often splits into many smaller branches called **distributaries**.

The satellite image shows the delta of the Nile River. The river and its distributaries are shown in black. The red areas are irrigated cotton, corn and clover fields. They contrast with the surrounding desert area. Cairo, Africa's largest city, is seen as the large light blue shape at the head of the delta.

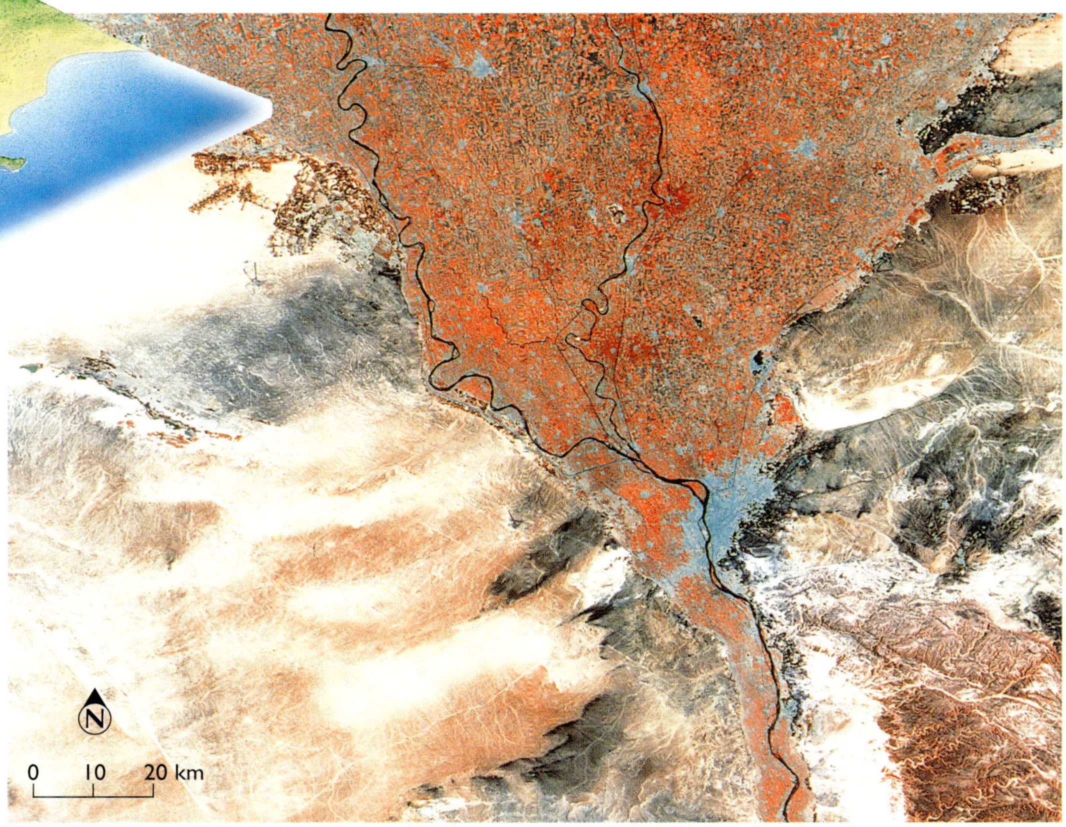

DESERT SYSTEMS

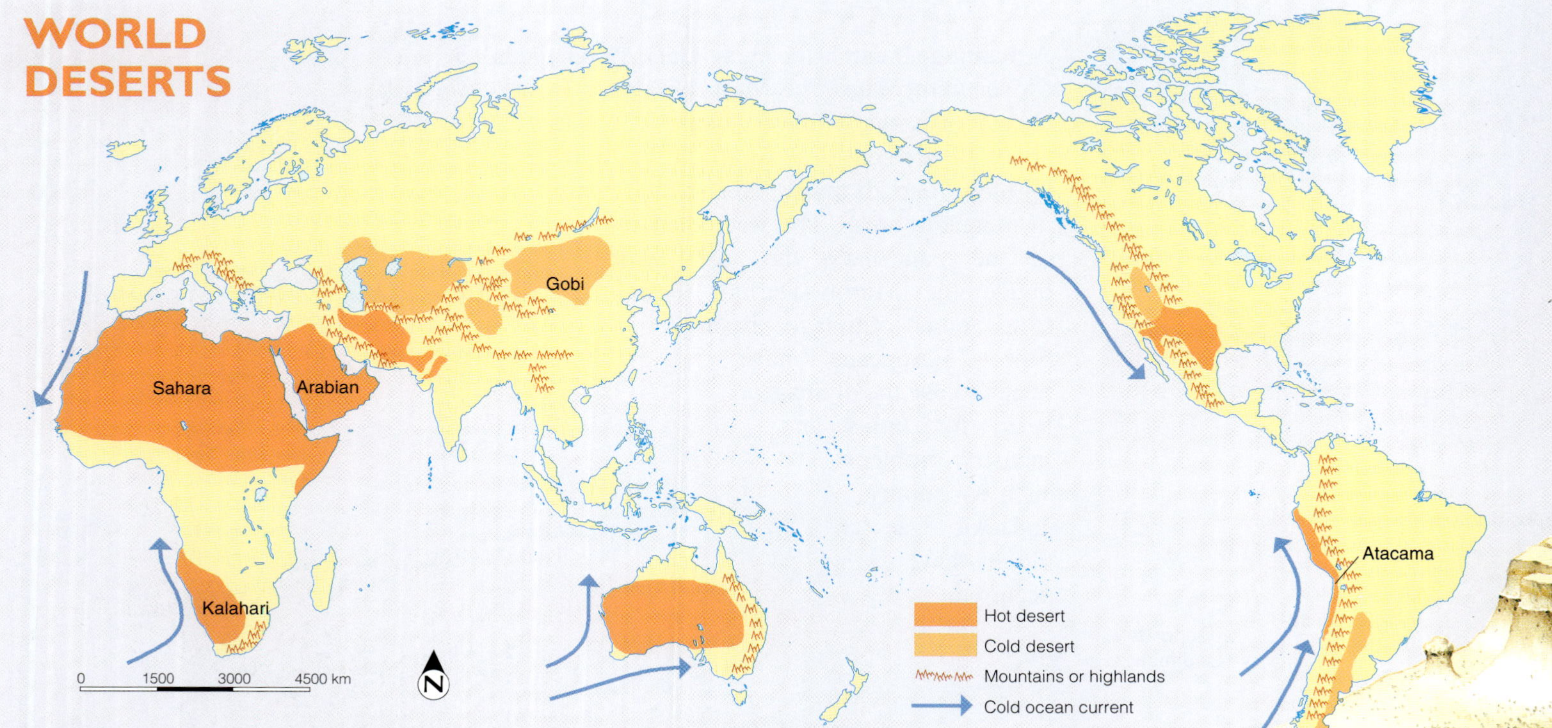

Deserts are areas of the earth's surface receiving less than 250 millimetres of rain a year. The world's deserts are located in areas where the air is dry. Rainfall occurs with rising air, and air over deserts is generally descending. When moist air is forced to rise over mountains where it produces rain, only dry air is left to pass over desert regions. This is the situation with Australia's Great Dividing Range and some of Australia's deserts. Cold ocean currents can also lower rainfall and extend deserts to the coast.

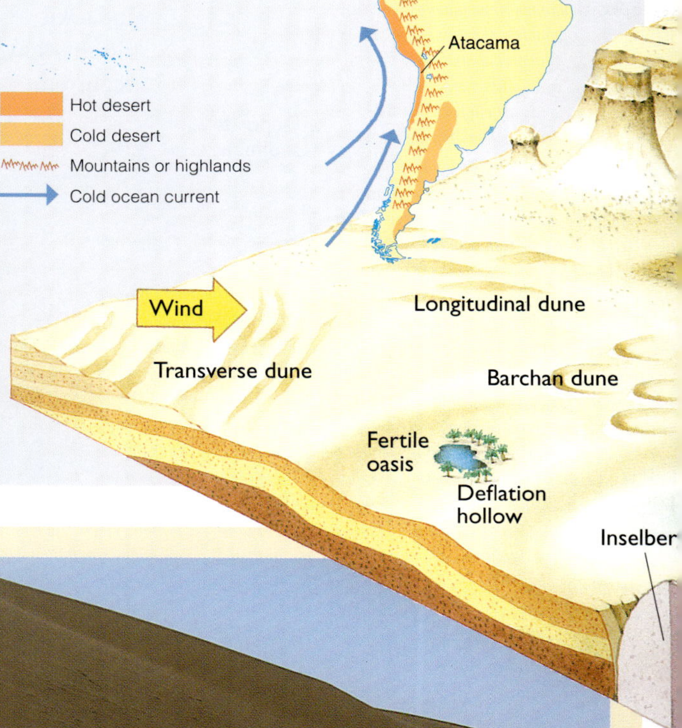

Wind processes in deserts

Sand dunes are one of the most obvious features of desert areas. When the wind blows, the sand builds up and dunes are formed.

The blowing away of fine surface sand and dust is known as **deflation**. The shapes of the dunes vary according to the strength and direction of the wind, the amount of sand available and the vegetation patterns. Dunes are named according to their shape and direction, for example **longitudinal dunes, star dunes**, **transverse** and **crescent dunes**. This barchan dune (pictured) in the Namib Desert in Africa is a type of crescent dune.

Water processes in deserts

Mount Connor is a mesa in the Northern Territory.

Water is the most important factor in the development of desert landforms. This is because it commonly comes in torrential thunderstorms and the rapid surface run-off is channelled into dry river beds (**wadis**), producing flash floods. Erosion by rivers can cut through plateaus to form hills called **mesas** and **buttes**. The eroded materials flow on to the lowlands to form **alluvial fans**, the fine sands being carried into the desert stream basins (**bolsons**) to be shaped later into dunes by the wind. Where water flows into a desert depression (**deflation hollow**), **playa lakes** form which produce salt or clay pans as they evaporate.

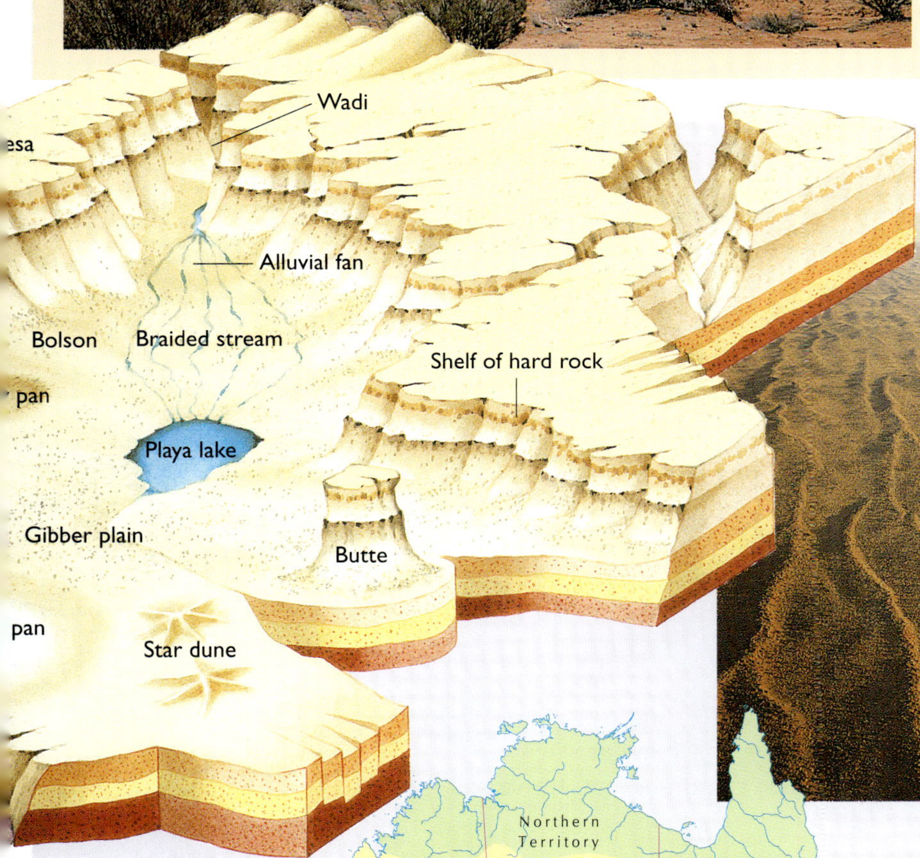

The Simpson Desert

Longitudinal dunes in the Simpson Desert

The Simpson Desert is located in the corner where the boundaries of the Northern Territory, South Australia and Queensland meet. It covers an area greater than 70 000 square kilometres and the characteristic landforms of the area are longitudinal dunes. These dunes are long ridges of sand which are usually formed and shaped by winds blowing from two directions. The main wind blows parallel to the dunes, forming a channel between them. A weaker crosswind blows the sand up onto the dunes.

Longitudinal dunes in the Simpson Desert run south-east to north-west. They are between 6 and 30 metres high, 300 to 500 metres apart and continue over great distances.

AUSTRALIAN DESERTS

COASTAL SYSTEMS

Headland system

Headlands are the high pieces of land which jut out into the sea at either end of a bay. The headland is changed into a variety of landforms by the energy of the waves and the wind which causes weathering and erosion.

Sand and sediment eroded from the headland are transported to beaches and dunes by wind and waves.

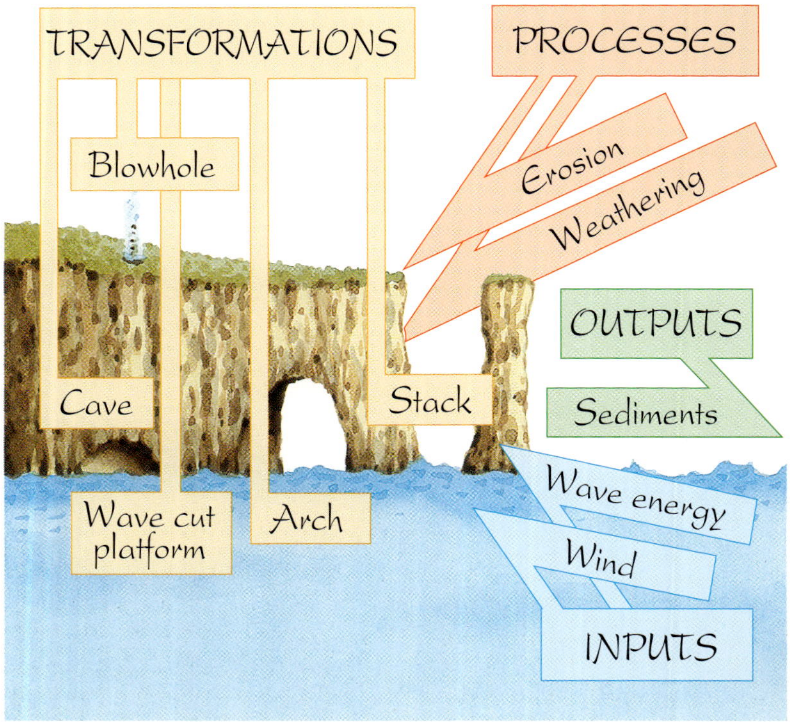

The impact and effect of peopl

Dune blowouts on Fraser Island, Queensland

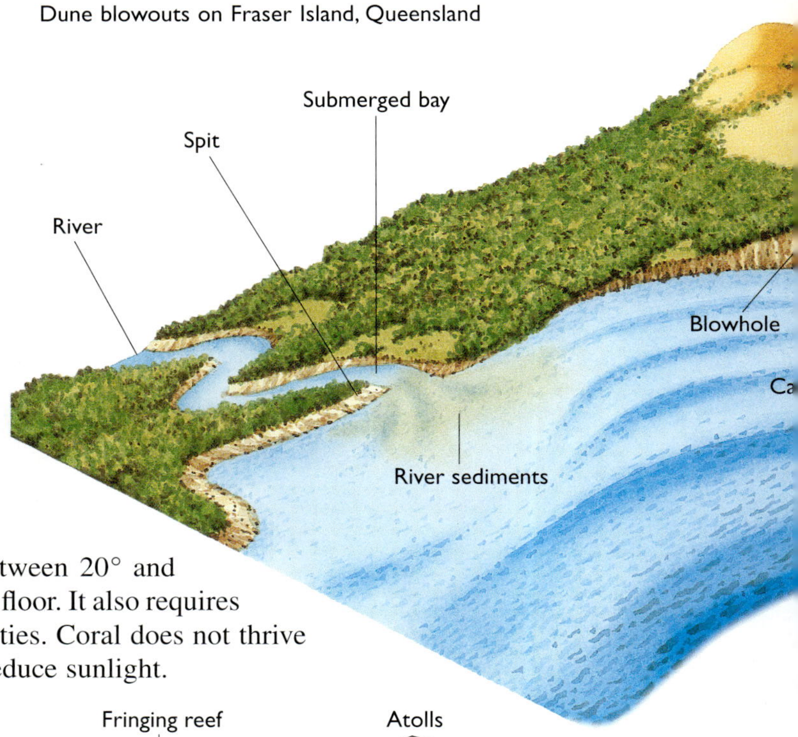

Coral reef system

The development of coral requires salt water, sea temperatures between 20° and 30°C and water shallow enough for sunlight to penetrate to the ocean floor. It also requires high oxygen input and low levels of silt, sediment and other impurities. Coral does not thrive near river mouths because the silt and lower concentration of salt reduce sunlight.

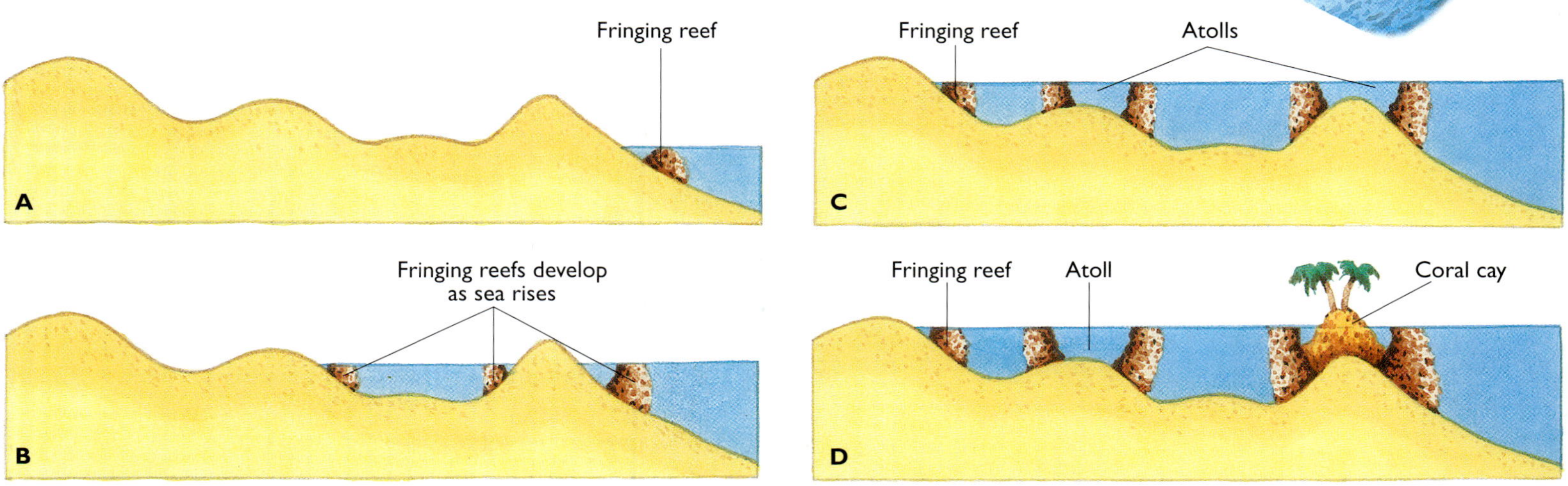

n coastal systems

Natural systems can be greatly influenced and disrupted by people. In coastal areas, the removal of vegetation leads to erosion of beach dune systems. Without he roots of plants to bind the sand ogether, the wind blows the dune sands nland. In many coastal areas vegetation as been replaced, and fences built to eep people out.

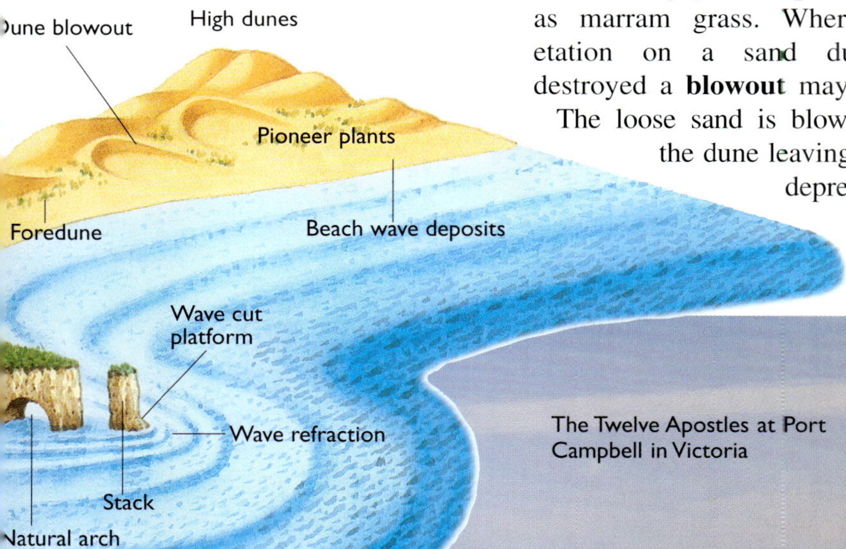

Beach–dune system

This system combines two main inputs — wave energy and wind energy. Waves transport and deposit sand along beaches through a process called **longshore drift**, while wind transports this sand inland to create dunes. The dunes at the back of the beach (**foredunes**) are stabilised by **pioneer plants** such as marram grass. Where vegetation on a sand dune is destroyed a **blowout** may occur. The loose sand is blown from the dune leaving a bare depression.

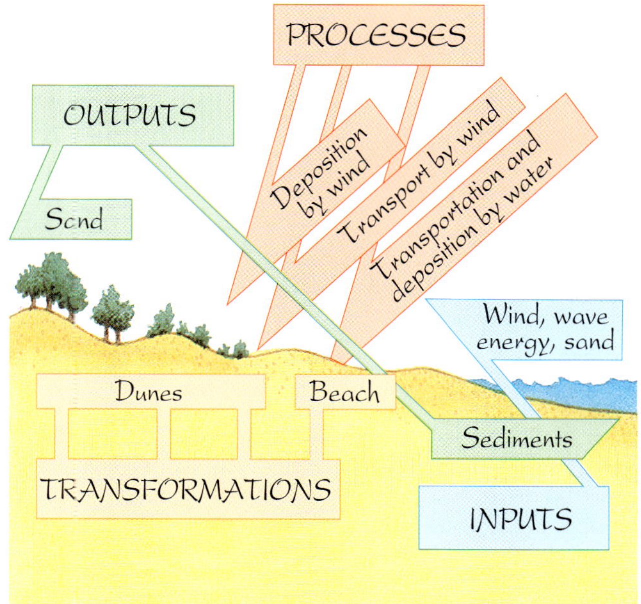

The Twelve Apostles at Port Campbell in Victoria

Port Campbell

The Port Campbell area of Western Victoria contains many spectacular landforms created by wave erosion. The **stacks** of the Twelve Apostles and natural **arches** of London Bridge (part of which remains) and Island Arch are well-known landforms. These have been formed by the waves pounding on the soft sandstone and eroding the cliffs. The more resistant parts remain at present, but will eventually be removed by erosion.

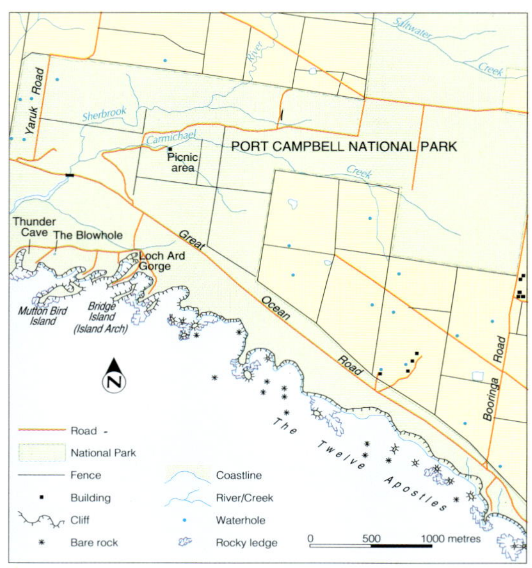

WORLD CLIMATE

World records

Antarctica is the world climate record holder. It is the coldest continent with an average annual temperature of –49°C.

The world's lowest temperature of –86.6°C was recorded at Russia's Vostok Station. Commonwealth Bay on the George V coast of Antarctica is the world's windiest place. In 1972 a wind speed of 320 km per hour was recorded. The driest continent of all, Antarctica records no rainfall. The South Pole on Antarctica receives the least sunshine on earth. There is no sunshine for 182 days every year. Ozone levels reached a world record low of 91 Dobson units over Antarctica in October 1993. Adequate protection from solar radiation is 300 Dobson units.

Highest average annual rainfall

Lowest average annual rainfall

Highest recorded temperature

Lowest recorded temperature

Most sunny days

Most rainy days

WORLD AVERAGE ANNUAL RAINFALL

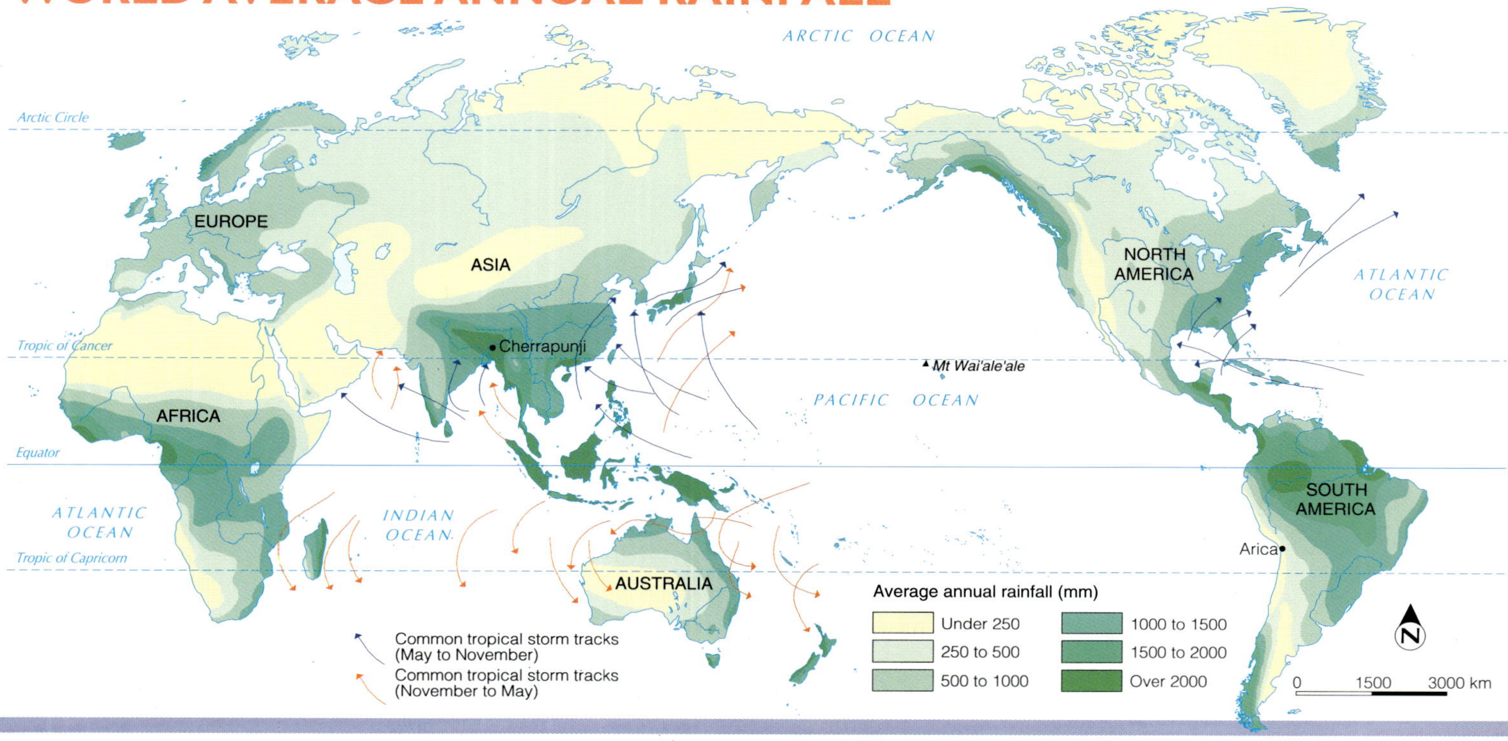

WORLD AVERAGE TEMPERATURES — JANUARY

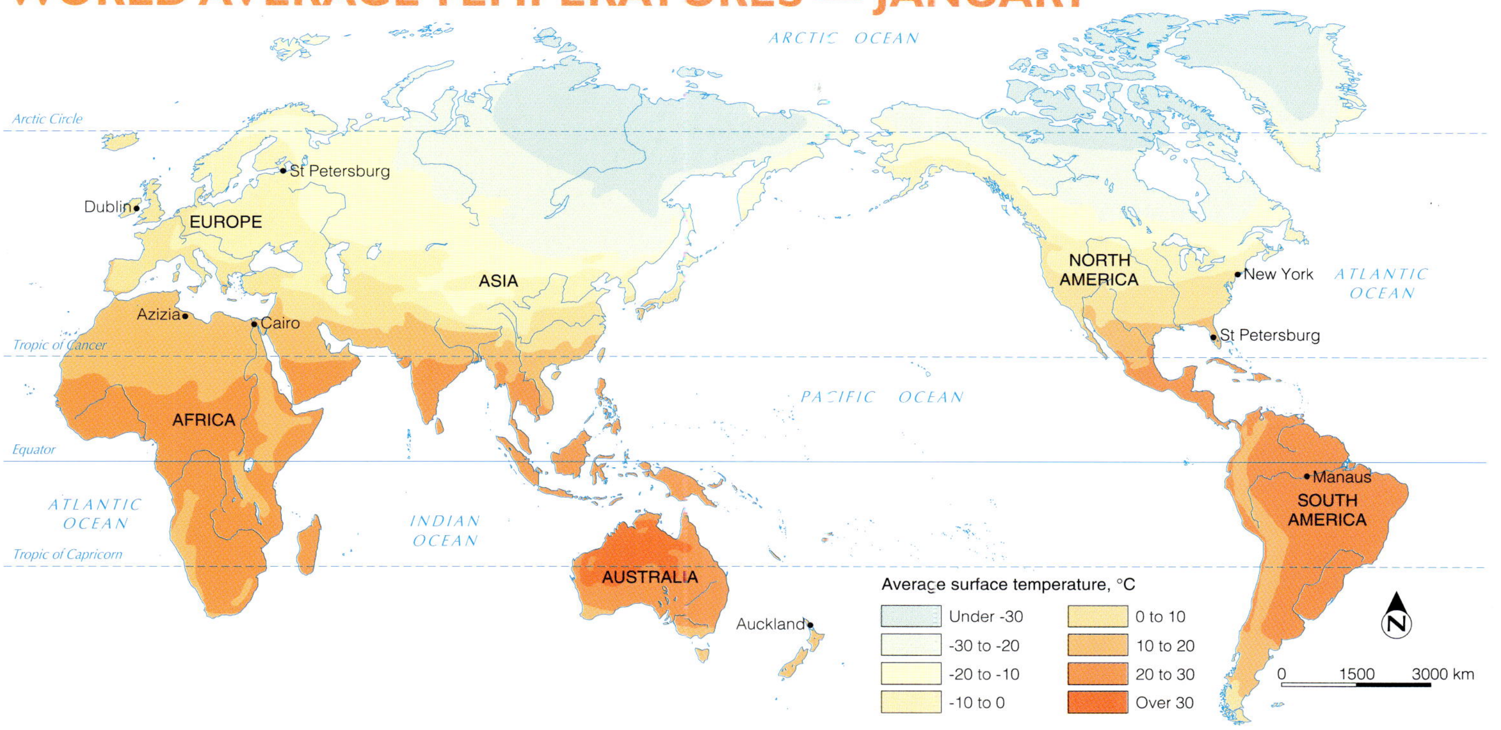

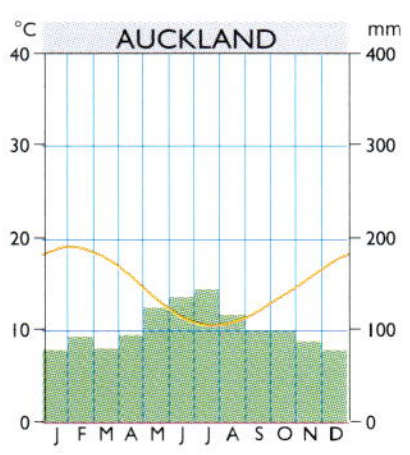

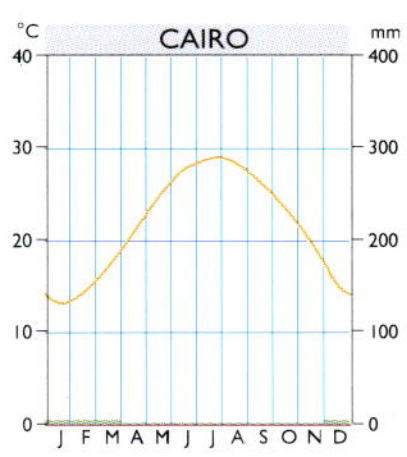

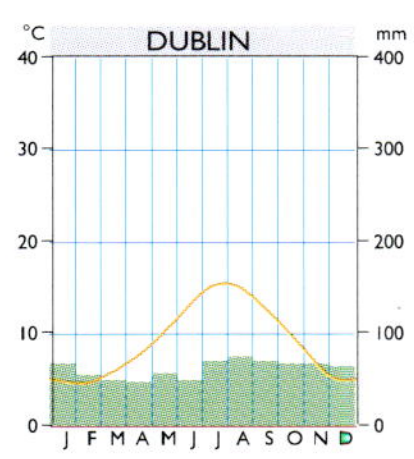

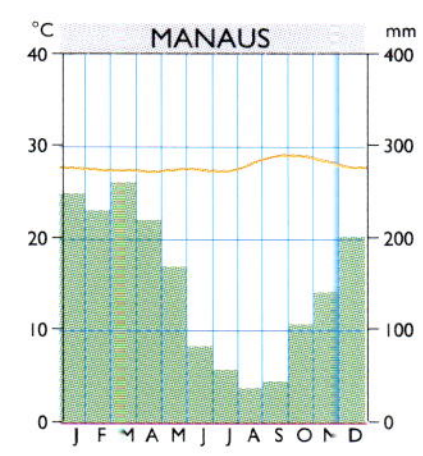

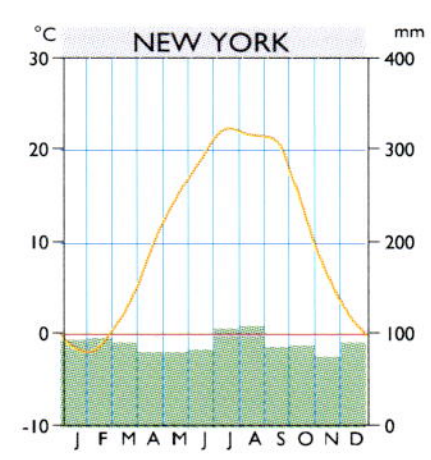

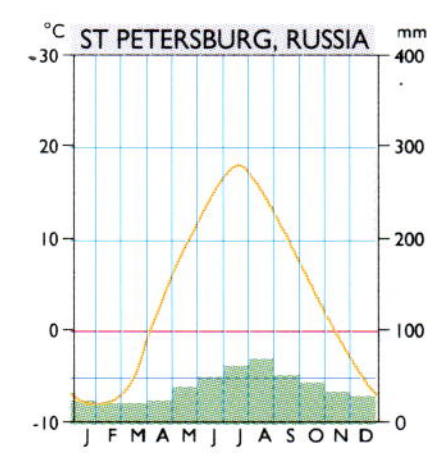

WORLD AVERAGE TEMPERATURES — JULY

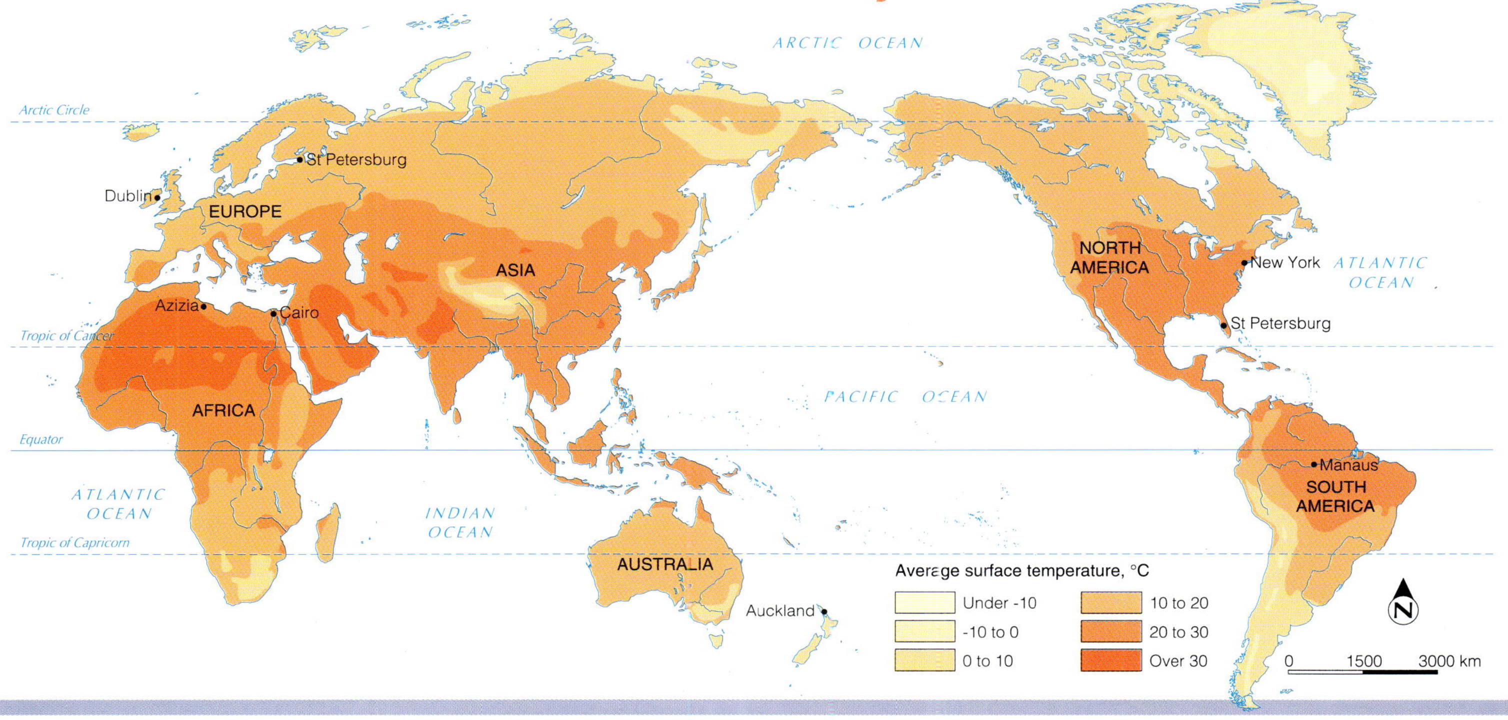

AUSTRALIA'S CLIMATE

El Nino

El Nino is an irregular climatic event which occurs in the Pacific Ocean about every three to eight years. In Australia, El Nino years are often years of drought. In these years the Pacific Ocean is warmer near South America, so there is not a large temperature difference between Australian and South American waters. If there is very little difference in temperature, there is little or no difference in air pressure. This causes weaker trade winds which, in turn, cause weaker ocean currents and cooler waters near Australia. As a result of this, summer rainfall in the coastal areas of eastern Australia is well below normal. Eastern Australia has experienced severe drought conditions from 1990 to 1994 and this has been attributed to El Nino conditions.

However, the 1995–96 summer in eastern Australia saw significant rainfall and floods in areas like western Queensland. Meteorologists believe that the El Nino effect has given way to its opposite, La Nina. The conditions created by La Nina are a reversal of those of El Nino, and promise a return to more normal summer rainfall patterns in eastern Australia.

The Warrego Highway near Miles in Queensland is cut by floods in January 1996.

CLIMATE ZONES

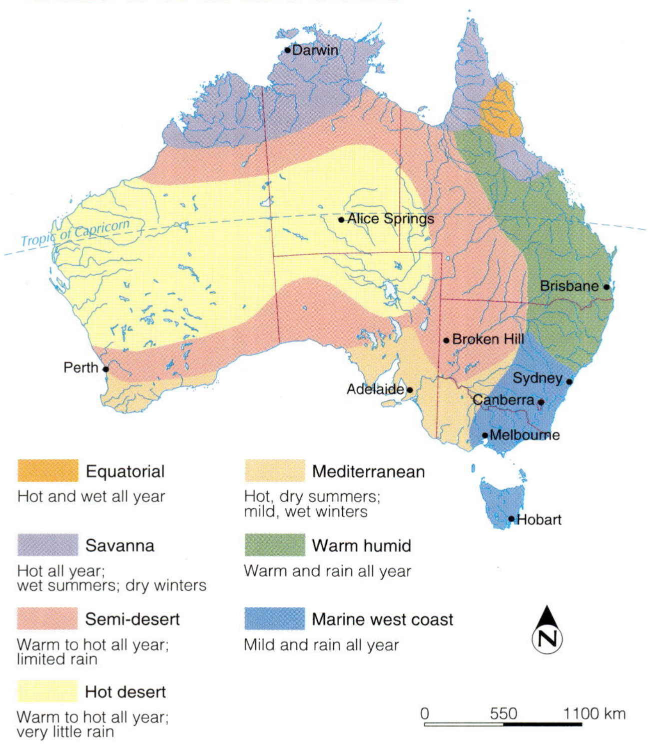

Climographs for selected cities

Lines show average monthly temperature
Bars show average monthly rainfall

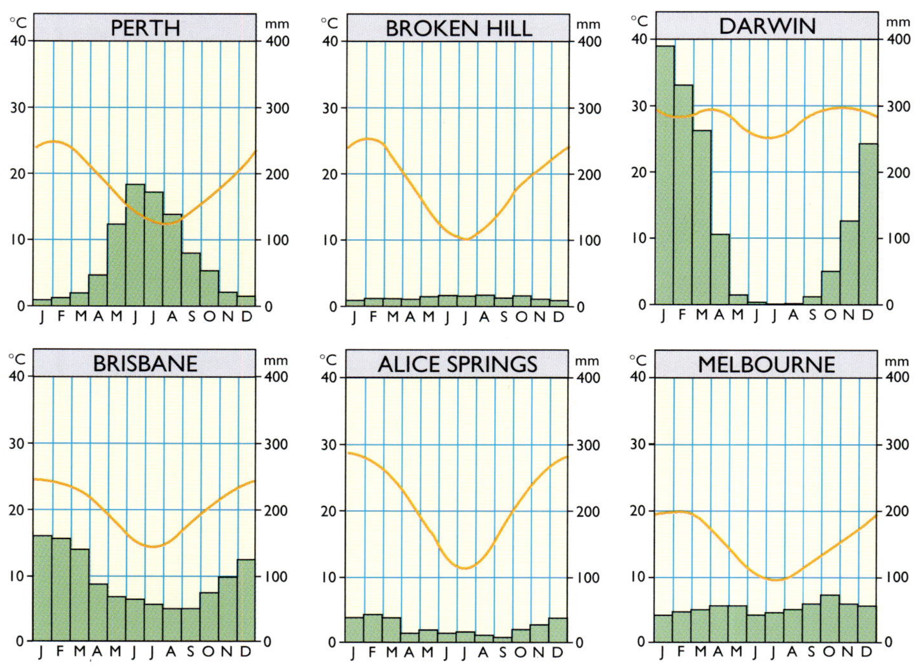

SEASONAL CLIMATIC PATTERNS

Air masses

Summer

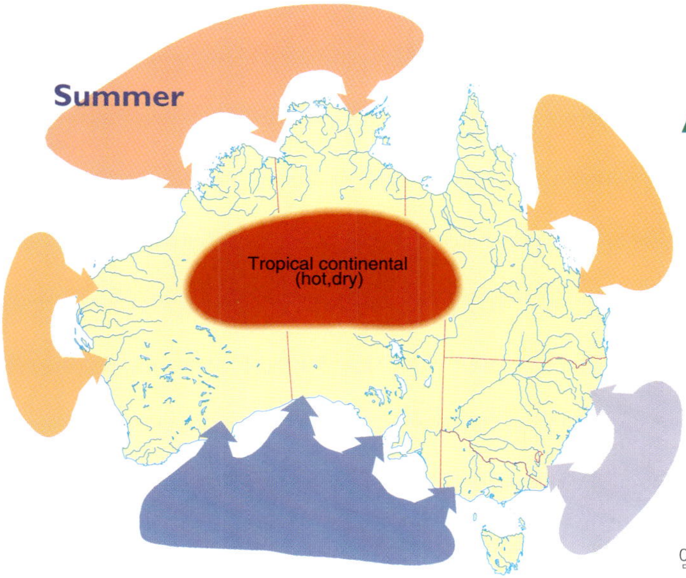

Winter

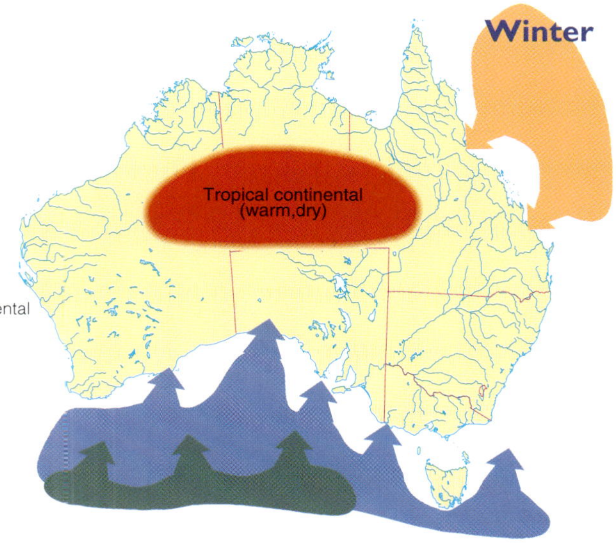

- Equatorial maritime (hot, humid)
- Tropical maritime (warm, humid)
- Southern maritime (mild, moist)
- Southern maritime (cool, moist)
- Polar maritime/continental (cold, dry)

N

0 600 1200 km

Rainfall

January

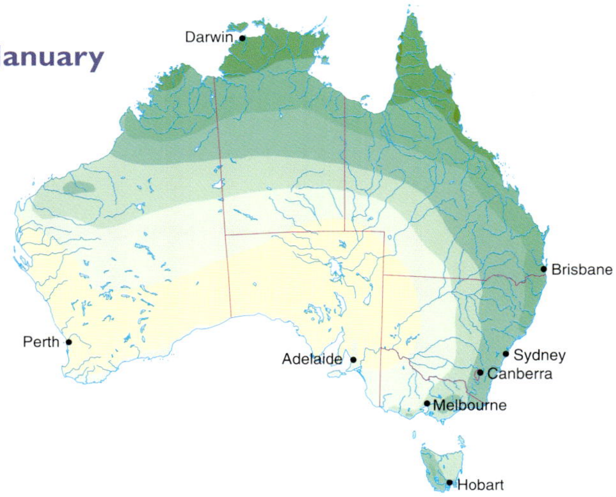

July

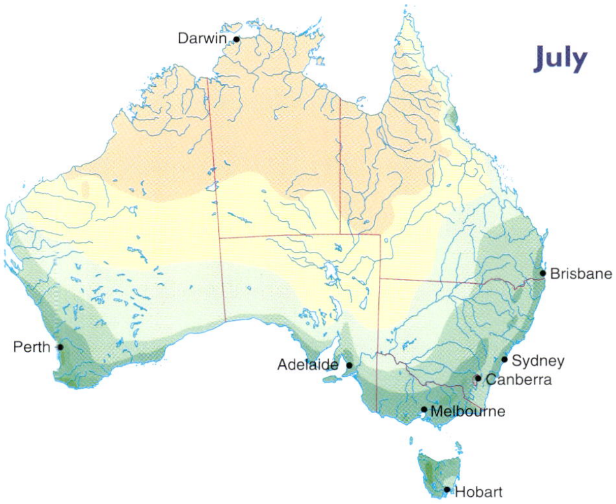

Average monthly rainfall

millimetres
400
200
100
50
25
10
0
no rain

Temperature

January

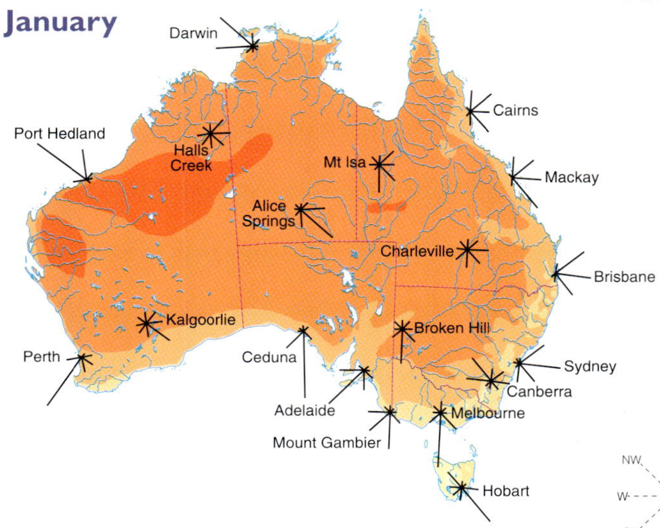

July

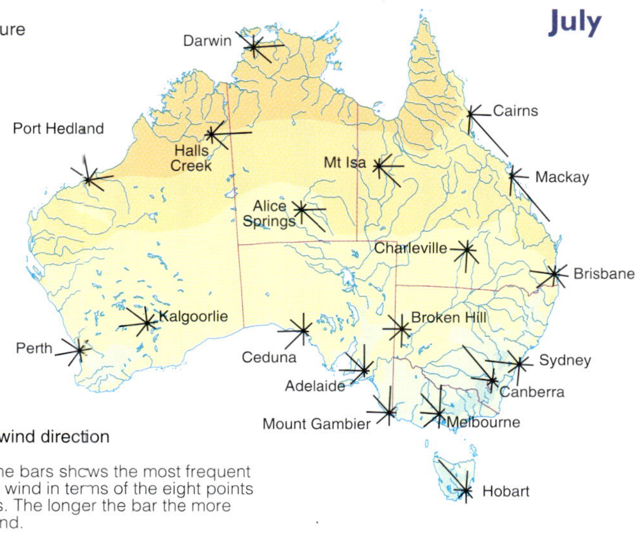

Average daily temperature

°C Highest
39
33
27
21
15
9
3 Lowest

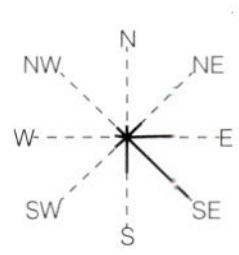

Frequency of wind direction

The length of the bars shows the most frequent direction of the wind in terms of the eight points of the compass. The longer the bar the more frequent the wind.

Weather Systems

Weather refers to the atmospheric conditions at a particular time or place. The day-to-day characteristics of rain, wind, cloud, heat, sunlight, humidity and atmospheric pressure describe the particular weather experienced that day (or week). Weather maps or **synoptic charts** illustrate such features by using symbols to represent conditions.

The main feature of the weather map is a series of lines which link places of equal air pressure. The lines are known as **isobars**. High pressure zones form where colder air is sinking and under higher pressure than the surrounding air. Lighter, warm rising air forms low pressure zones. We can use the pressure zones to read and predict the direction and strength of winds.

In the southern hemisphere, air moves in an anticlockwise direction in a high pressure zone. The moving air, known as wind, always moves out of a high and into a low pressure zone. Air circulates in a clockwise direction in a low pressure zone.

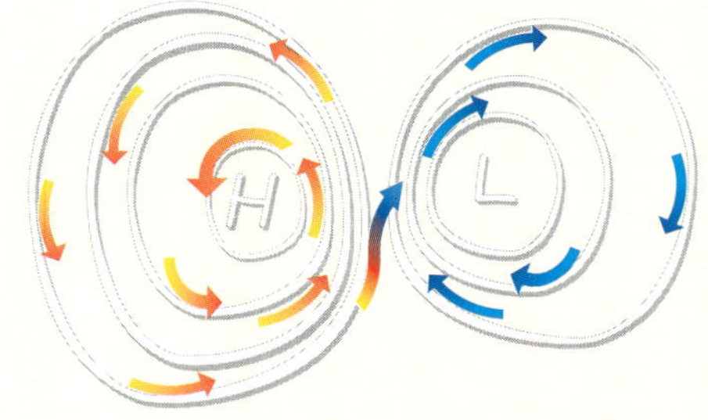

Air moves from high pressure zones to low pressure zones. Where isobars are closer together, the wind is stronger.

Many people confuse pressure zones with air masses. An **air mass** is a large section of the atmosphere containing air with a similar temperature and moisture content. On a synoptic chart a line with triangles marks the **front** where two air masses meet.

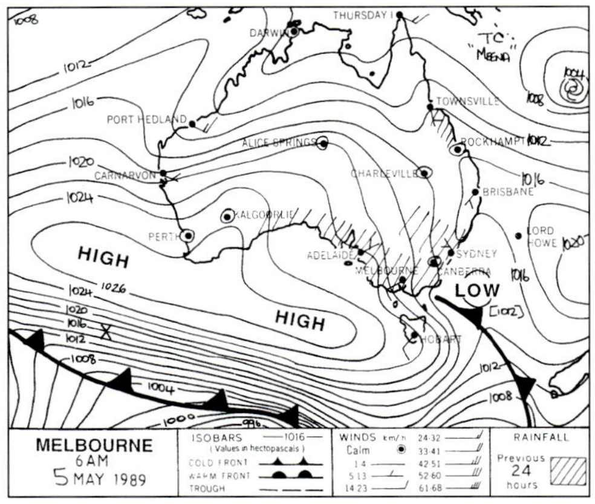

Cold fronts

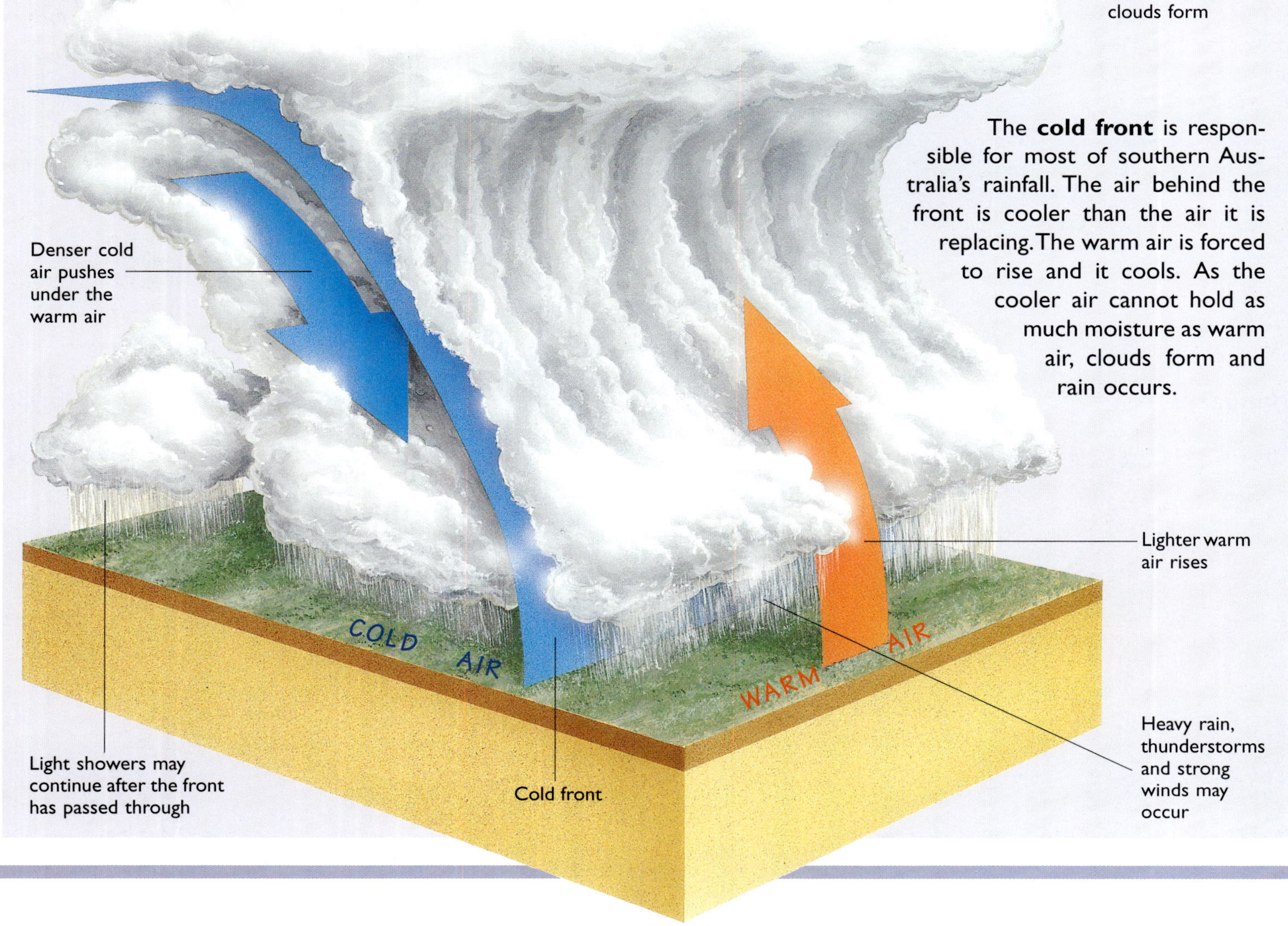

The **cold front** is responsible for most of southern Australia's rainfall. The air behind the front is cooler than the air it is replacing. The warm air is forced to rise and it cools. As the cooler air cannot hold as much moisture as warm air, clouds form and rain occurs.

Cyclones

A tropical **cyclone** is probably the most powerful and dangerous of all weather events. These whirlpools of wind are called hurricanes in the Americas and typhoons in Asia. The furious winds of up to 250 kilometres per hour and torrential rain can cause massive destruction.

The energy that powers a cyclone is warm moist rising air. Tropical cyclones only form over oceans with temperatures greater than 27°C. This means they only form in the latitudes less than 20° north and south of the Equator.

One of the odd things about a tropical cyclone is the "eye". This is an area up to 30 kilometres across in the centre of the cyclone which has light winds, no rain and little cloud.

Every year an average of about nine tropical cyclones form in the shallow tropical oceans around northern Australia, with January to March being the peak months. They normally last about six days from the time they are detected to when they weaken.

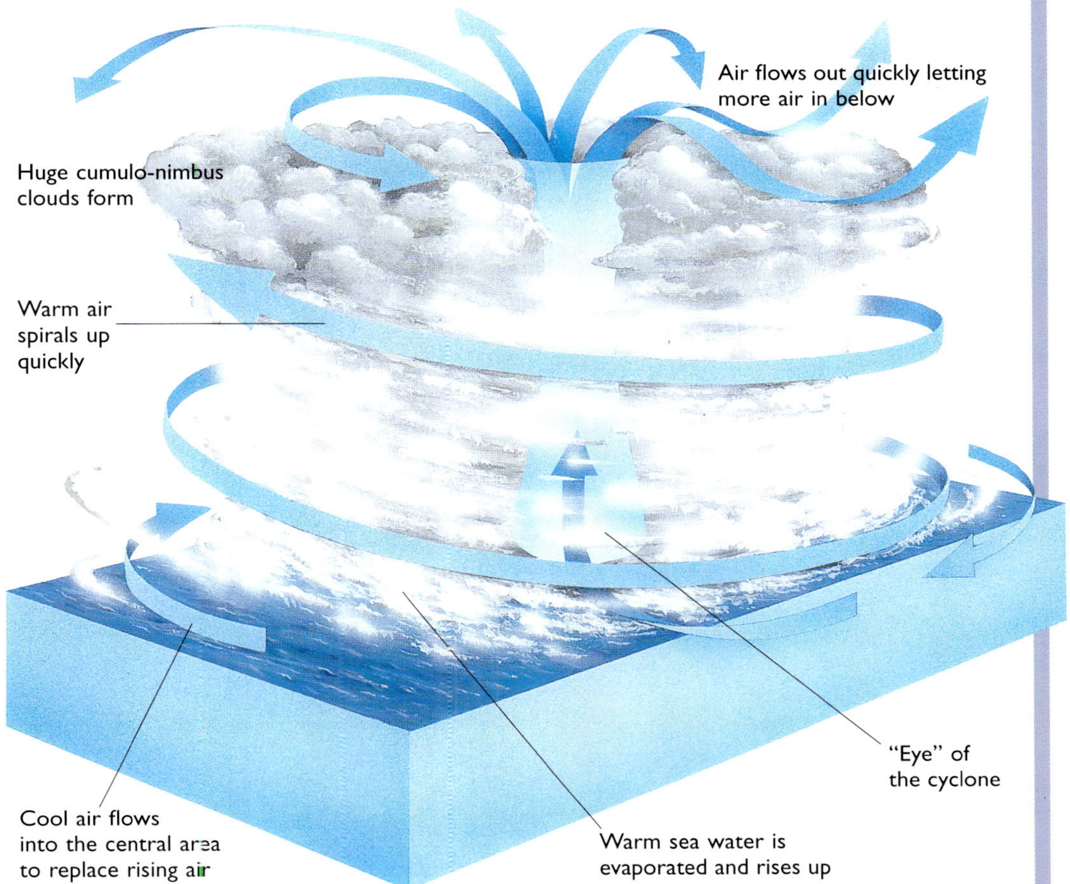

Satellite image of a cyclone off the coast of Queensland

MAJOR AUSTRALIAN CYCLONES SINCE 1970

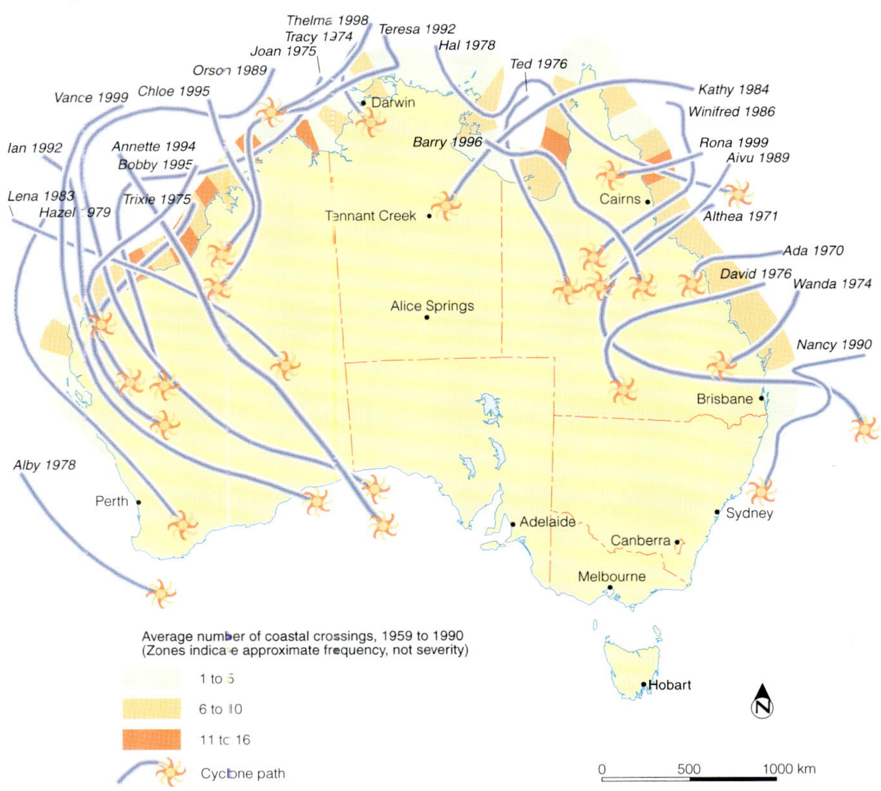

THE NATURAL WORLD

Biomes

A biome is a large collection of plants and animals adapted to survive within a particular geographical zone. The soil and animal life of an area of the earth are closely related to its vegetation which is strongly affected by climate. Such a system which includes soil, climate, vegetation and animals on a global scale is called a biome.

Orang-utans live in the **rainforests** of Asia. Rainforest trees are mostly evergreen, tall and smooth-barked. Their branches and leaves are close together, forming a canopy.

The Arctic fox lives in the **tundra** of Europe, Asia and North America. Small shrubs, grasses, lichens and mosses grow there. Mountainous areas have some trees and shrubs.

WORLD BIOMES

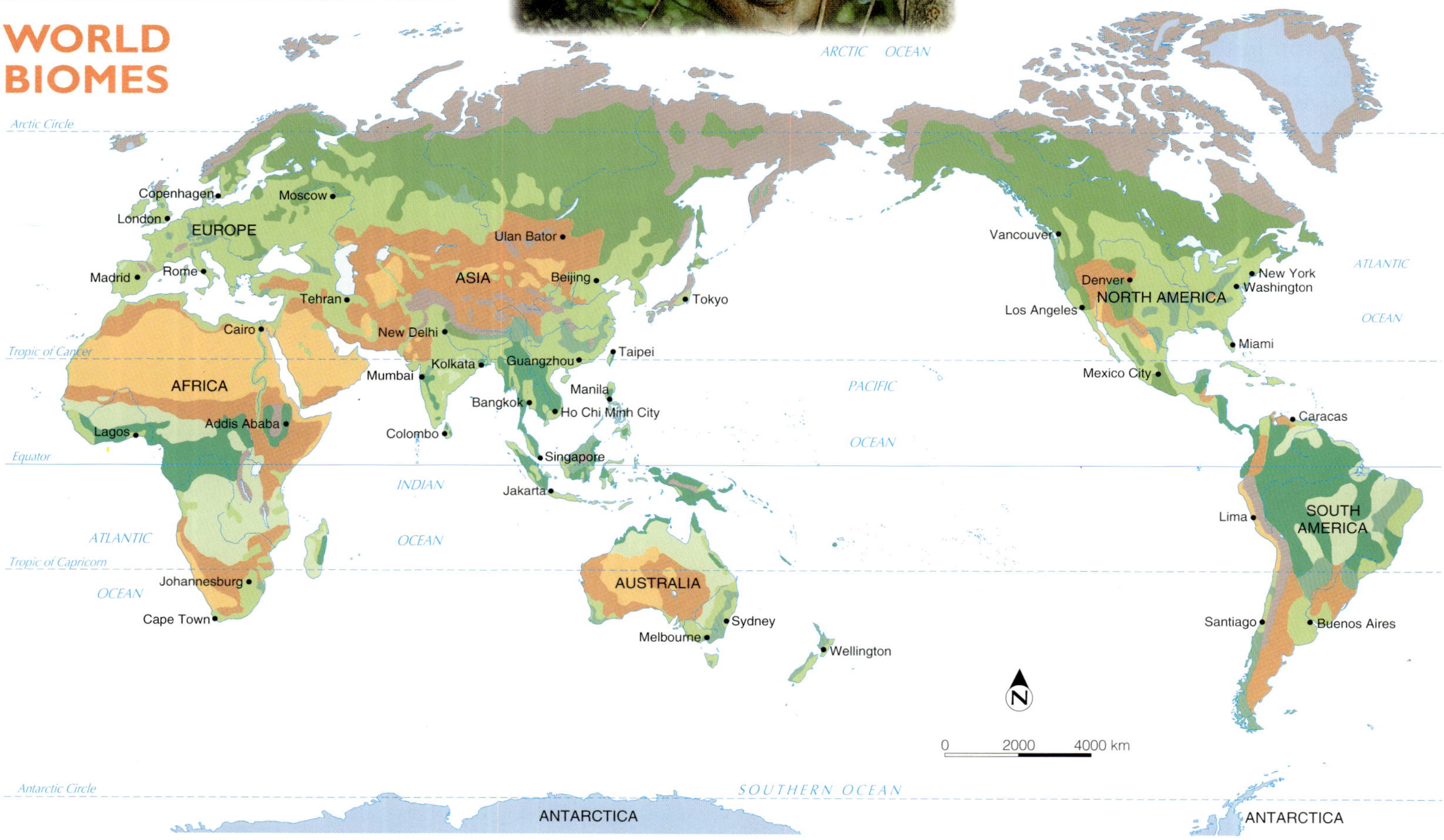

Giraffes live in the **savanna** areas of Africa. The area is mainly grassland with scattered, short trees which are mainly deciduous.

Emperor penguins live on the Antarctic **ice cap**, where there is no vegetation and a permanent ice cover.

Kangaroos are found in a wide range of habitats, including Australia's **open-forest** areas. These forest areas have tall straight evergreen trees, sometimes with a dense undergrowth of ferns and shrubs.

Reindeer live in the **coniferous** forests of Europe and Asia. The trees are evergreen, tall and straight. Very few plants grow under the trees.

Llamas live in **grassland** areas in South America. Short or tall grasses grow in these treeless areas.

Frill-neck lizards live in the dry woodlands and **deserts** of Australia. In deserts much of the ground is bare. Drought-resistant plants grow in some areas.

Galahs are often found in areas of **cultivated land** in Australia. Natural vegetation has been cleared for growing crops and grazing animals.

Rainforest ecosystem

1. Cluster fig
2. Leaf-tailed gecko
3. Double-eyed fig parrot
4. Cicada
5. King parrot
6. Bobuck possum
7. Bird's nest fern
8. Common ringtail possum
9. Scrub turkey
10. Eastern yellow robin
11. Southern angle-headed dragon
12. Bracken fern
13. Purple-winged stick insect
14. Cunjevoi
15. Lawyer cane
16. Strangler fig
17. Green catbird
18. Red-necked pademelon
19. Red-eyed tree frog
20. Giant barred frog
21. Noisy pitta
22. Regent bowerbird
23. Rufous fantail
24. Basket fern
25. Crimson rosella
26. Red-crowned pigeon
27. Tree fern

ENVIRONMENTS IN DANGER

The world's natural environments are under threat. Humans continue to consume natural resources or misuse the environments they have created.

Tropical rainforests cover seven per cent of the world's land surface but contain half of its animal species. Logging for timber, firewood collection and clearance for farms have reduced tropical rainforest coverage to less than half of its original extent. It is estimated that nearly all tropical rainforest will disappear by the year 2040.

The world's wetlands are under threat from chemical pollution and from housing, industry, and farm development. Coastal wetlands are particularly important as they:

- provide habitats for birds and animals
- are the spawning grounds for most of the world's commercial fisheries
- reduce coastal erosion
- filter pollutants from sea water
- purify fresh groundwater.

Half of the Florida Everglades wetlands has been cleared for housing, leaving native grasses, trees, wading birds and the American crocodile in danger of extinction.

Useable land is being lost or poisoned by **land degradation**. This is a broad term used to describe the loss of soil productivity through erosion, salinity, compaction, or desertification. In Australia, six tonnes of topsoil are lost for every tonne of produce grown.

Clearance of trees and removal of vegetation by grazing animals leaves the soil able to be blown or washed away. In arid areas the loss of vegetation can lead to **desertification**, where the desert takes over the degraded land. Tree clearance also allows groundwater to rise, bringing salts from the subsoil closer to the surface. Salt kills most plants, so the land degradation cycle begins again.

Major causes of world soil degradation by region, 1990s

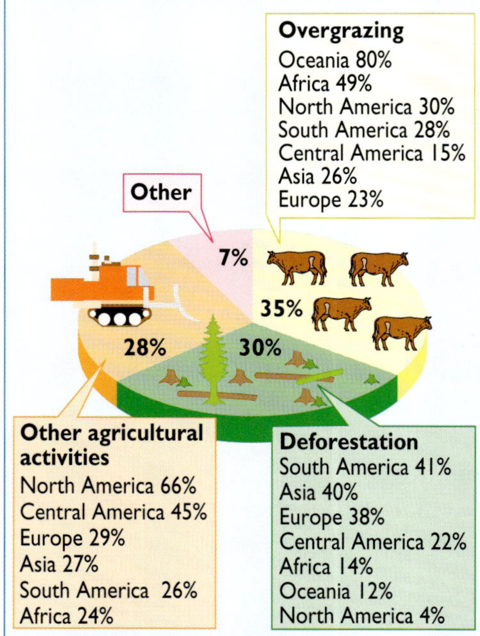

Source: World Resources, 1992–93

WORLD LAND DEGRADATION

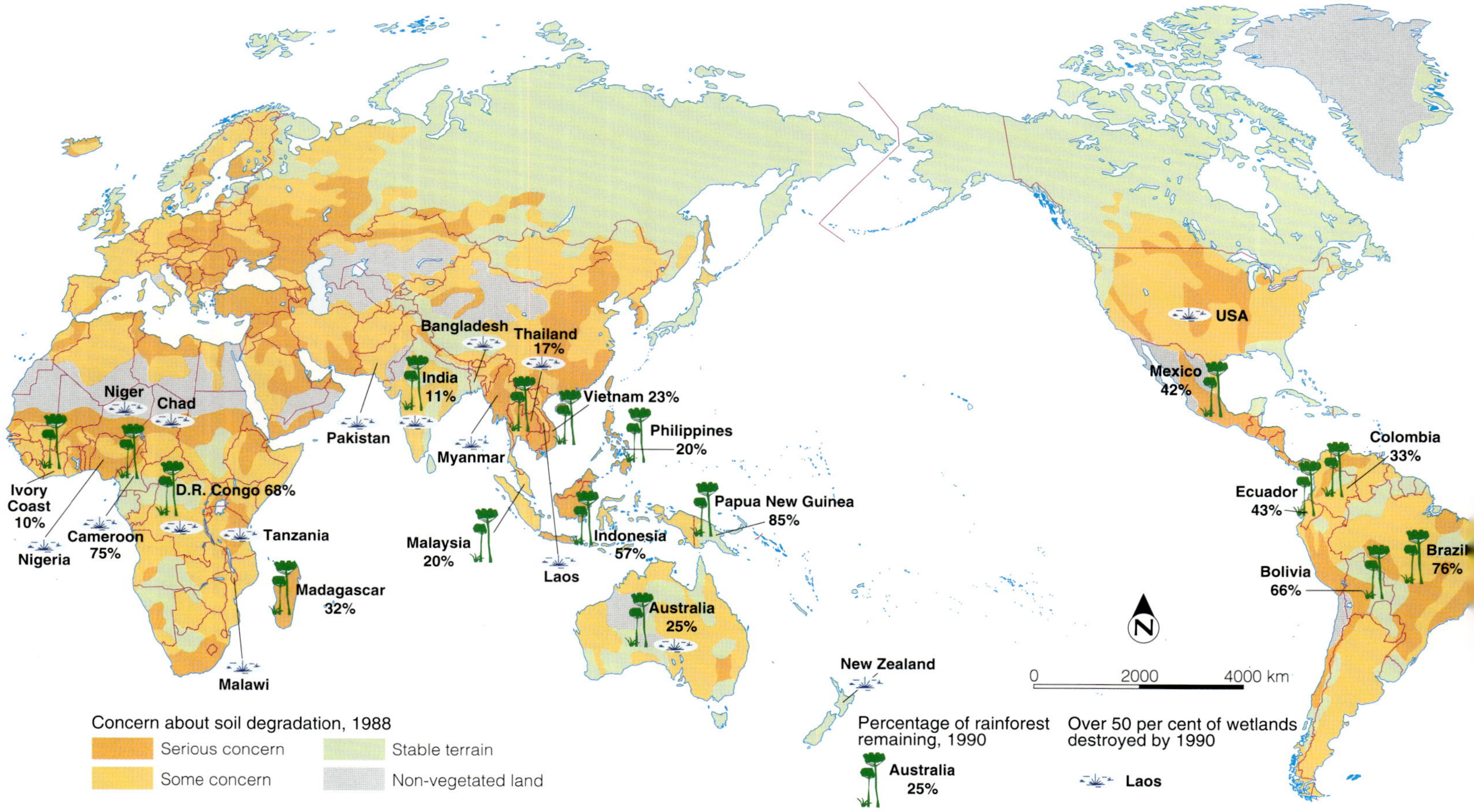

Endangered species

The loss of habitat has forced many animal and plant species to the brink of extinction. Many others are threatened by illegal hunting or poaching. In 1990, over 4000 animal species were threatened with extinction; in the year 2000 the figure will be closer to 40 000. All species of tiger are extinct or in danger of becoming extinct. There are fewer than 8000 tigers surviving in the wild. Tigers have suffered from the clearance of their forest habitat and demand for their skins.

Most of the endangered animals are found in countries where the greatest deforestation is taking place. Brazil, for example clears 100 square kilometres of rainforest every day.

Countries with greatest number of endangered mammals, 1994

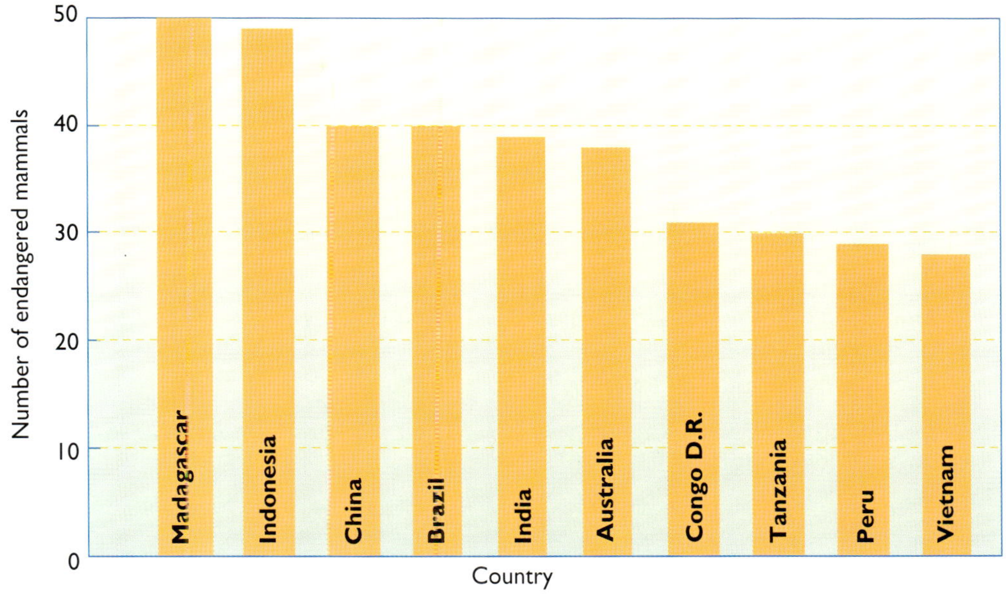

Endangered animals of Shark Bay, Western Australia

The habitats of many of Australia's small native animals have been replaced by farms. The native animals are forced to compete for food and shelter with introduced species such as rabbits, foxes, rats and cats. Seventeen small Australian mammals are now extinct and a further 29 are endangered. They have retreated to isolated areas of Australia such as Shark Bay on the Western Australian coast.

Shark Bay is listed as a World Heritage site for its outstanding natural features, vegetation, wildlife and wildlife habitats. It is the last retreat for some of Australia's endangered small mammals.

Shark Bay, Western Australia

Banded hare-wallaby
Once common over southern Western Australia, the banded hare-wallaby is now restricted to scrub on Bernier and Dorre Islands in Shark Bay. Its decline was mainly due to the spread of farming. The last mainland sighting was in 1906.

Burrowing bettong
The burrowing bettong was once found over most of mainland Australia. The introduction of foxes, rabbits and the use of poison grain have all contributed to its decline. The burrowing bettong is now found only on Bernier and Dorre Islands in Shark Bay and on Barrow and Boodie Islands.

Western barred bandicoot
Once abundant in the shrubs and sand ridges of arid Australia, the western barred bandicoot is now found only in the sandhills on Bernier and Dorre Islands. It is thought the introduction of feral goats and pigs, rabbits, foxes and cats was responsible for its decline.

REVEALING THE PAST

The skill of the archaeologist

Archaeologists study pottery, wood, seeds, plants, carvings, paintings, monuments, documents and even remains of human bodies preserved by ash, ice or other means, in their search for clues about the past.

The methods used to uncover the past include searching an ancient rubbish tip, picking up **sherds** (bits of broken pottery) or excavating buried buildings, temples, houses and even whole towns. Using special tools and recording what they find, archaeologists can build up a picture of what life was like for ancient peoples and cultures.

Scientific methods are more complex. **Radiocarbon dating**, for example, measures the amount of radiocarbon (or C-14) left in living things when they die. These bits of radiocarbon decay at a set rate so an old bone necklace, shell brooch, house post, preserved plant or animal can be accurately identified as being a hundred, a thousand or even several thousand years old. Other scientific methods can date tree-rings, pottery and archaeological material in the layers between rocks.

Cave paintings at Lascaux in France show how early peoples recorded their lives.

EARLY HUMANS

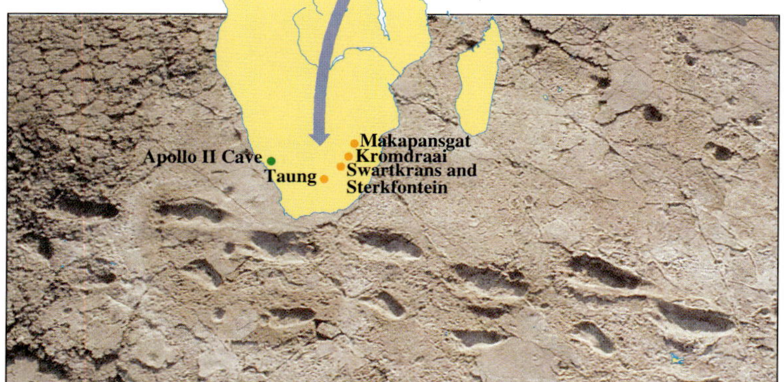

Footprints of an adult and child, preserved at Laetoli in the African Rift Valley

Key

Movement of early humans

40 000 B.P. = 40 000 years before present day

Evidence of early humans

(Note: Some of the sites show evidence of many stages of human evolution. In most cases, only the earliest findings at each site are shown here.)

- *Australopithecus africanus* – 4 million to 1.5 million years ago. The first hominid to walk upright.
- *Homo habilis "Tool making man"* – 3 million to 2 million years ago. The first evidence of tool making.
- *Homo erectus "Standing man"* – 1.5 million to 500 000 years ago. The first to use fire for cooking and warmth.
- *Homo sapiens neanderthalensis "Wise man"* – 100 000 to 35 000 years ago. In Europe and Asia Neanderthals used tools and fire, made clothes and held ceremonies. Spreading from Africa, this species showed their great intelligence through artwork and sculpture.
- *Homo sapiens sapiens "Wise wise man"* – 35 000 years ago to present day.

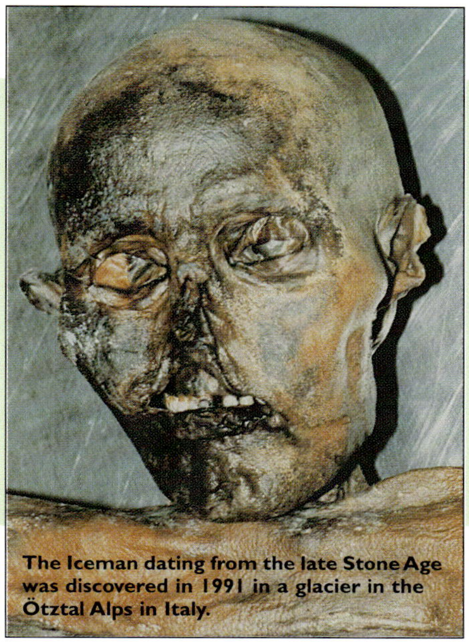

The Iceman dating from the late Stone Age was discovered in 1991 in a glacier in the Ötztal Alps in Italy.

Early humans

The earliest humans (or **hominids**) were walking on Earth at least 3.7 million years ago. By the last great Ice Age, 35 000 years ago, there were men, women and children much like us today, living around the world. The earlier hominid forms had disappeared, leaving a modern species *Homo sapiens sapiens* ("the wise, wise man"), to migrate and establish itself as the only human species.

The skeletons, skulls and other remains of these prehistoric and Ice Age hominids have been found everywhere: "Peking Man" at Zhoukoudian in China; "Java Man" at Trinil in Indonesia; "Lucy" at Hadar in Ethiopia; the "Laetoli footprints" in the Rift Valley in Africa; "Iceman" in the Ötztal Alps in Italy; and burial and cremation sites at Lake Mungo in Australia. The cremation of "Mungo Woman" at Lake Mungo occurred 26 000 years ago and is the earliest evidence of cremation in the world.

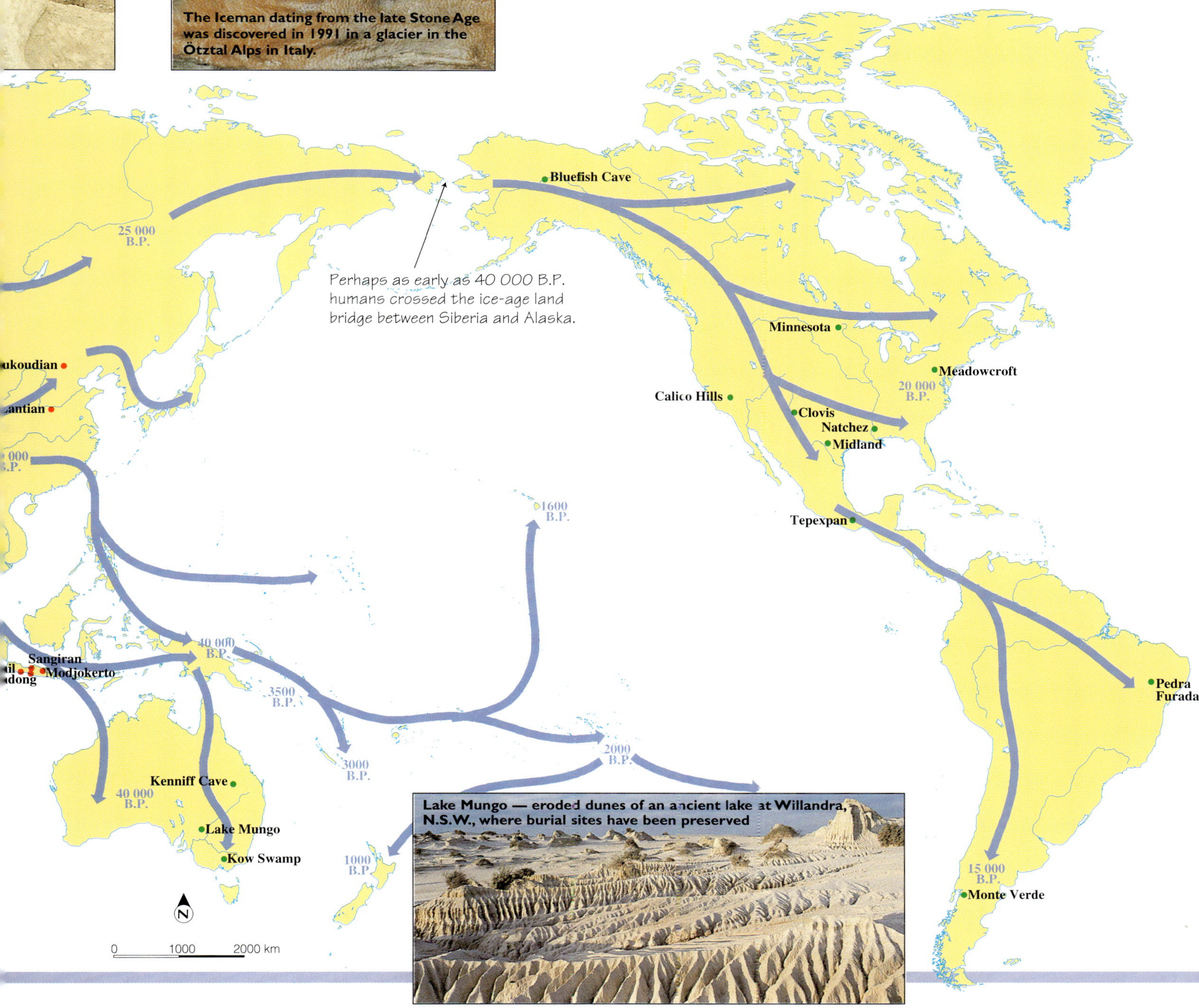

Lake Mungo — eroded dunes of an ancient lake at Willandra, N.S.W., where burial sites have been preserved

WORLD HERITAGE

What is worth keeping?

The World Heritage Commission lists 440 sites around the world. Why should we protect and conserve these natural and cultural places and things?

We protect "Head-Smashed-in Buffalo Jump" in western Canada because it is a sacred site for Native Americans. Long ago they gathered food, bone and skin supplies by herding buffalo along a narrow ridge and over a steep cliff. We protect Sagarmartha National Park (including Mt Everest) in Nepal for its natural and scenic beauty. Kakadu and Uluru/Kata Tjuta are protected in Australia for their importance as geological sites and because they contain evidence of the earliest human occupation and lifestyles.

To be included in the "Natural" **(N)** category of the World Heritage List, a site must be an outstanding example of:

- the earth's natural processes
- plant and animal diversity
- exceptional phenomena or natural beauty
- natural habitat.

To be included in the "Cultural" **(C)** category, a site must be either:

- a monument (exceptional historical, art or scientific values)
- a group of buildings and sites (showing the interaction of humans and environment)
- an archaeological "dig" and research area.

Australia's World Heritage sites

(C) Interior of the Great Mosque at Cordoba, Spain
Cordoba has been invaded and rebuilt by the Romans, Visigoths (AD 500–719), Arabs and modern Spaniards. The present mosque was built and extended in 786, 961 and 987, using pillars of earlier Roman and Visigoth churches. The interior is adorned with calligraphy from the Koran. The mosque remains important for both its architectural and religious significance.

WORLD HERITAGE SITES

(C) Nubian monuments, Egypt
These monuments, along with other famous sites including Memphis and the pyramids, are relics of a spectacular empire whose achievements in medicine, science, architecture, art, government and technology were outstanding. The Nubian monuments are to be found from Abu Simbel to Philae along the Nile River. These Egyptian sites are perhaps the most well known (but endangered) archaeological sites in the world today.

(C) Lalibela rock-hewn churches, Ethiopia
Eleven of these churches have been carved out of solid rock with "the help of God's angels" (according to legend). Built during the 13th century by the Christian King Lalibela, they represent his plan for a "New Jerusalem" in Ethiopia. Each church is meant to represent a stage in the life of Christ.

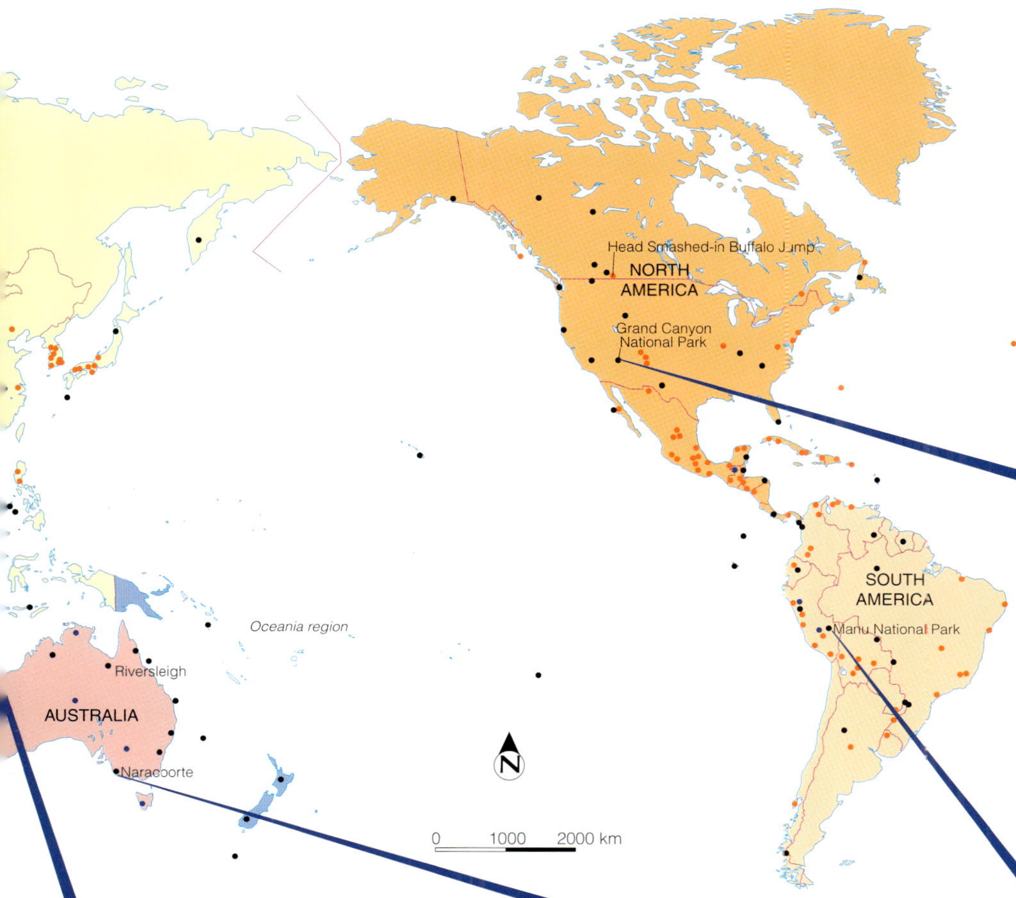

(N) Grand Canyon National Park, United States of America
The Colorado River carved a huge, deep and twisting canyon over centuries as the earth rose and erosion occurred. There are several important Native American sites along the river. It has been the inspiration for a symphony (Grofe's *Grand Canyon Suite*). It is also one of the world's most popular canoeing, hiking, riding and tourist destinations.

(C) Angkor, Cambodia
This remarkable spiritual, architectural and cultural site demonstrates the achievements of early Asian empires. The site has been severely damaged by civil wars, pillaging by later settlers and migrants and neglect by governments. Now, only a shadow of the greatness that once was Angkor Wat remains.

(N) Naracoorte (South Australia) mammal site
This site and Riversleigh (Queensland) make up the World Heritage listed Australian Fossil Mammal Sites. These two sites contain outstanding examples of evolution and ecological and biological change going back twenty million years. Some extinct mammal remains are perhaps 25 million years old. The sites contain evidence of changes from the Ice Age to rainforest and semi-arid grasslands, sea-level changes and how mammals adapted to these changing climates and vegetation.

(N) Manu National Park, Peru
This is the world's largest park. It is considered to be of international importance to the biosphere. Its forests are the world's lungs. As well as containing outstanding flora, fauna and natural features, it includes the headwaters of the great Amazon River basin and several small communities of indigenous peoples.

INDIGENOUS PEOPLE OF THE

Indigenous peoples make up one tenth of the world's population. These 350 million people, in seventy modern nations, are descendants of the original inhabitants of these lands.

There are perhaps 5000 different cultural groups, some of which are shown on the adjacent map. War, natural and human disasters and harsh political conditions have caused much hardship for indigenous groups. As a result many leave their traditional lands and move to other areas. Those that remain are often deprived of their land and have no say in the government of their country. Education and health services are often very poor and there are very few opportunities for people to improve their situation.

Indigenous people live in the modern world and should have the opportunities offered to others. It is widely agreed that this should happen, but the claims of indigenous people are often ignored.

Tibetans

The Tibetan people live in a landlocked mountainous country bordered by five other nations. Since China's occupation of Tibet in 1950, the life of the indigenous Tibetan people has changed dramatically. Traditionally, their lives revolved around the Buddhist religion and farming. Today, 80 per cent of Tibetans still live in rural areas, but Chinese migrants now outnumber Tibetans. Many Tibetans have fled as refugees to seek political or religious freedom, as well as economic opportunities denied them in their homeland.

Tuareg

The Tuareg live in Africa's great Sahara Desert. About 900 000 Tuareg live a nomadic way of life moving with their camels, goats and household goods in search of water and pasture. This traditional nomadic existence has limited the pressure on the environment where fertile land and water are scarce. Today, government programs in countries like Niger, Mali and Burkina Faso are taking over much of the Tuareg land, threatening their livelihood.

Inuit

The Inuit are the indigenous people of the Arctic region. Numbering about 100 000, they live in Alaska, Canada, Greenland and Russia.

Traditionally, the Inuit lived in harmony with their harsh Arctic environment, hunting, fishing and trapping. Today, regions like Alaska and Canada have oil and mining developments and less than ten per cent of the Inuit earn a living in traditional ways.

Torres Strait islanders

The 6000 inhabitants of the Torres Strait islands live in a narrow strait between Cape York Peninsula and Papua New Guinea. They rely on the ocean for their livelihood.

Since the 1970s, the islanders, shown here in traditional costume, have been linked to Aboriginal people in Australia, while still recognised as a separate community. In 1992, the Mabo case upheld the islanders' claim as original landholders. Currently the Torres Strait islanders are seeking a form of self-government within Australia.

Kayapo

The Kayapo Indians live in the Amazon rainforest in Brazil and rely on farming, hunting and gathering for their livelihood. The forest is both their spiritual and physical home. The area is under increasing threat from deforestation, dam construction and pollution. With other Amazon Indians, the Kayapo recently asked for world support to halt the construction of a dam on the Xingu River.

EXPLORING, DISCOVERING,

For thousands of years indigenous people travelled back and forth to neighbouring regions. They traded with neighbours for goods which were carried by camels, canoes and human porters from faraway places they had heard of but never seen. They knew the world was bigger than just their own village, city, valley or plain.

People other than their neighbours were considered strangers. These strangers seemed to come from out of the sky, from underground, or from the spirit world. The worst strangers were those that conquered, or who raided for products, slaves or wives or demanded payment just because they were powerful. The most worrying strangers were those with a different skin colour and a different technology.

Towards the end of the 15th century, Europeans began to travel to other parts of the world. When Columbus discovered the New World in 1492, he began a long line of "first contact" descriptions.

Cave paintings, songs and dance were used by indigenous peoples to record the arrival of new strangers.

WORLD EXPLORATION

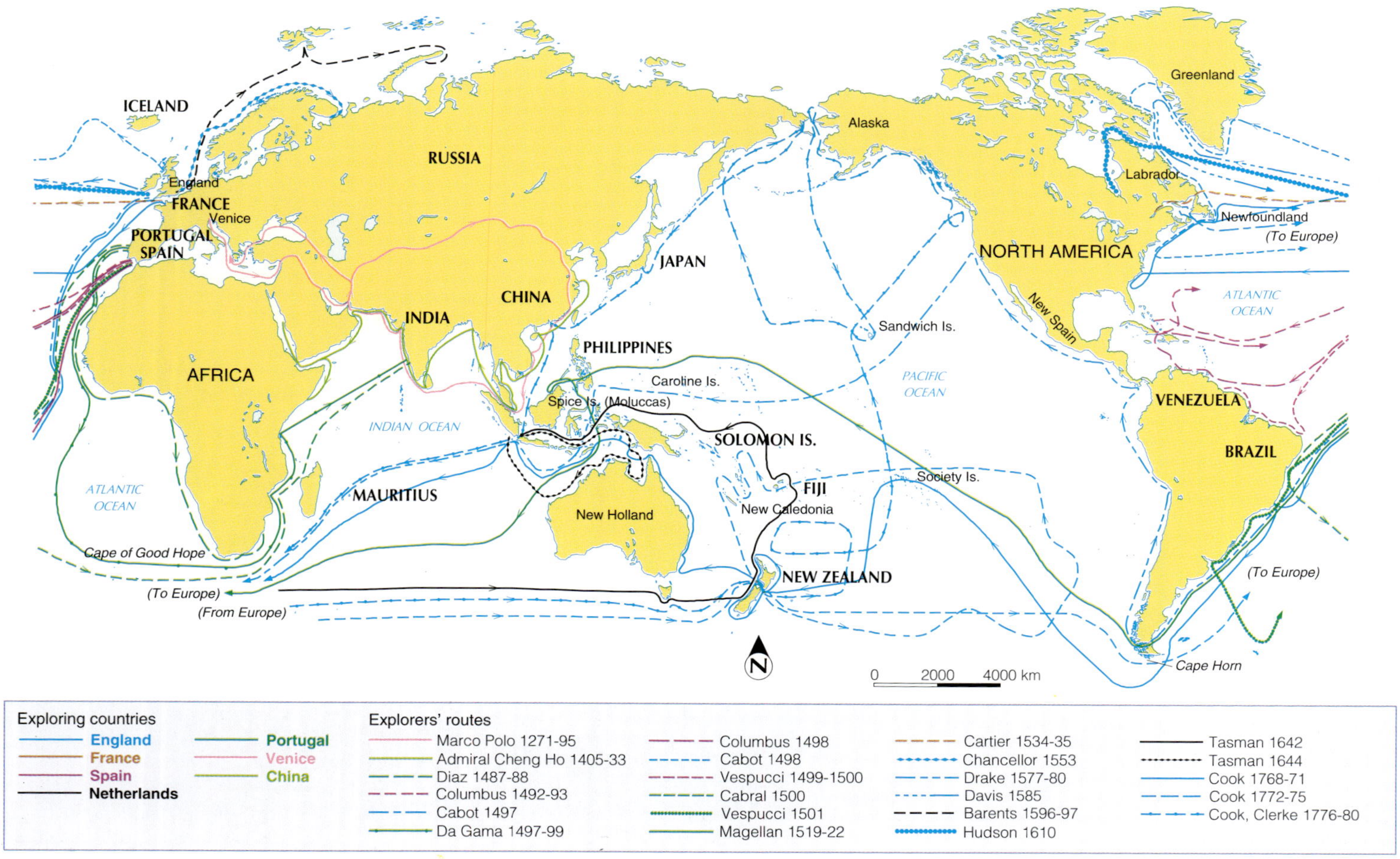

First contact in the Pacific

First contacts between Europeans and Pacific islanders were sometimes violent, but mostly people watched in amazement, danced, traded, made speeches and enjoyed the presence of strangers from distant unknown places.

The word for stranger in the Pacific varies from island to island. **I-Matang** in Kiribati means man from the west; **Tupua** in Maori means goblin or strange being; **Te Aoe** in the Marquesas Islands means outsider or stranger. The first white-skinned men in the Marquesas were called **atua** (men gods from beyond the sky). Later, every plant, animal, item and custom brought by Europeans was called atua by the Marquesans.

These so-called "first contacts" have been described by Europeans in books, diaries, drawings and paintings. But the songs, chants, stories, carvings, tattoos and oral history of indigenous peoples provide the other side of the story.

A visitor's song

This song was composed by a man called Tioi (*c*1780) and later performed by Mangaia Island people to record Cook's visit, the first European to visit their island. (The song was written down much later by missionaries who settled on Mangaia.)

Visitor's song (extract)
Chorus (using falsetto):
They say they're Britons
"We're from Britain."
Yes they're Britons, they are Britons,
Beretane is their land.
And here is Mai.
They talk through Mai: "O Mai! O Mai!"

Chorus
We come in hundreds,
Warriors! Hundreds!
With our spears put up to fight
We lusty people of Mangaia
Will destroy this boastful ship.
Kua ta Mangaia,
Kua ta te pa'i!

Solo
Tangaroa has sent a ship,
It has come bursting through the sky.

Solo
And here's a stranger call him Mai!

Chorus
It is Cook who's come to visit
Call him Tute bringing Mai.

Solo
Who has come, then?

Chorus
What a boatload?
Here them jabber "Jabber-jabber-jabber!"
"Ouaraurau, Ouaraurau.
"Ouaraurau, raurauae!"

Solo
No good counting them, that boatload,
They are foreigners — so's Mai!

Solo
Lord of the winds

Chorus
Send down some seabird
On their shoulders hold them nigh!

NOTES
Mai — A Tahitian taken by James Cook from Tahiti to England.
Tangaroa — supreme God in Polynesian mythology.
Tute — Cook.

A goblin comes to Whitianga

The following story was told by Horeta Te Taniwha, who was a small child when Cook's *Endeavour* sailed into Whitianga (Mercury Bay), New Zealand, in November 1769.

We lived at Whitianga, and a vessel came there, and when our old men saw the ship they said it was an atua, a god, and the people on board were tupua, strange beings or 'goblins'. The ship came to anchor, and the boats pulled on shore. As our old men looked at the manner in which they came on shore, the rowers pulling with their backs to the bows of the boat, the old people said, 'Yes, it is so: these people are goblins; their eyes are at the back of their heads; they pull on shore with their backs to the land to which they are going.' When these goblins came on shore we (the children and women) took notice of them, but we ran away from them into the forest, and the warriors alone stayed in the presence of those goblins; but, as the goblins stayed some time, and did not do any evil to our braves, we came back one by one, and gazed at them, and we stroked their garments with our hands, and we were pleased with the whiteness of their skins and the blue of the eyes of some of them.

...There was one supreme man in that ship. We knew that he was the lord of the whole by his perfect gentlemanly and noble demeanour. He seldom spoke, but some of the goblins spoke much. But this man did not utter many words: all that he did was to handle our mats and hold our mere, spears, and waha-ika, and touch the hair of our heads. He was a very good man, and came to us — the children — and patted our cheeks, and gently touched our heads. His language was a hissing sound, and the words he spoke were not understood by us in the least.

A. Salmon, *Two Worlds* Penguin Books (NZ) Ltd pages 88–89

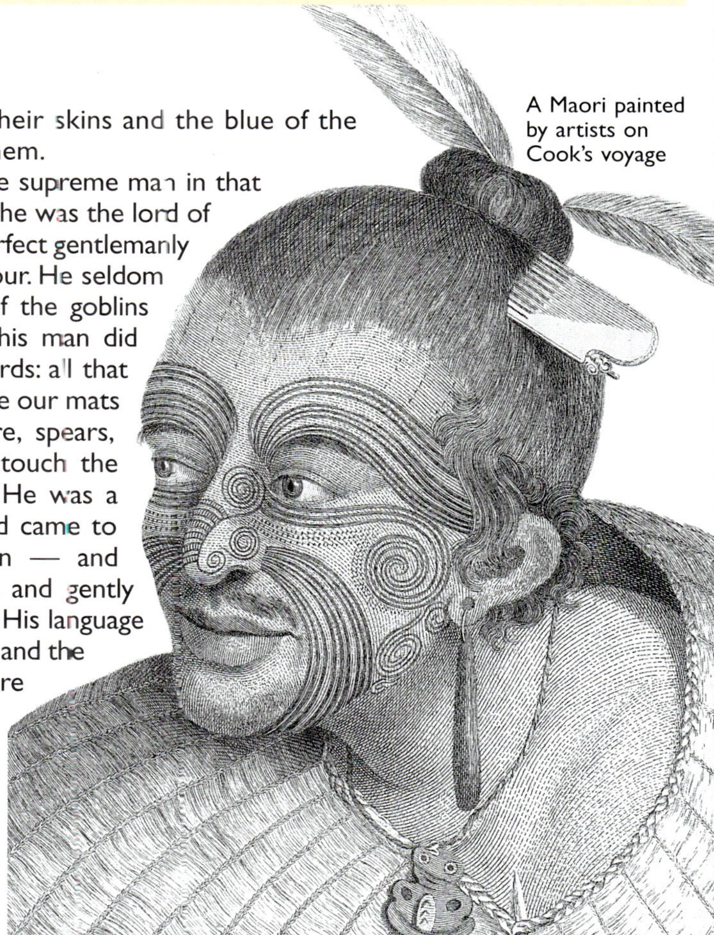

A Maori painted by artists on Cook's voyage

CULTURAL GROUPS

Australia

Australia is one of the most multicultural countries in the world. Our population of 18 million consists of people from over 140 different ethnic (cultural) backgrounds, who speak almost 100 different languages and follow 80 different religions. Since 1970, many Asian migrants and refugees have arrived and together with earlier European migrants, have created a truly multicultural country. Official sources estimate that, by the year 2001, 1 in 17 people in Australia will have an Asian background. Two million people over the age of five speak a language other than English at home, with Italian, Greek and Vietnamese accounting for almost 1 million of these.

Growth in religions in Australia, 1981–91

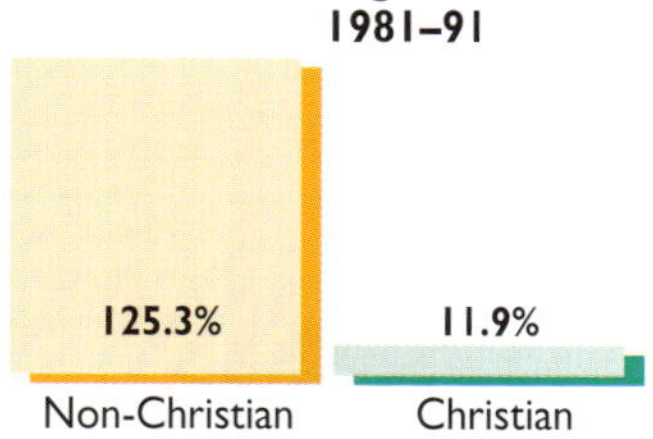

Source: Australian Bureau of Statistics

English speaking ability of people born overseas (who also speak another language)

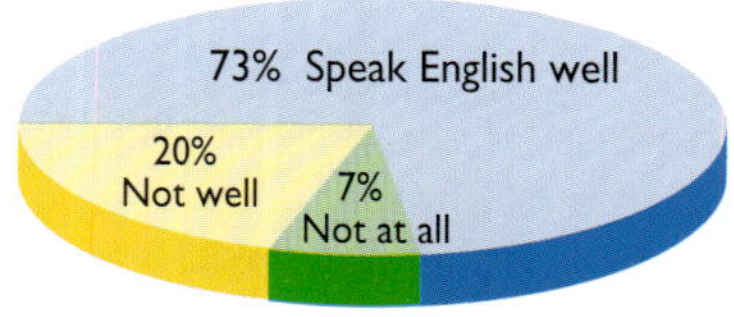

Source: Australian Bureau of Statistics

Australia's top 10 sources of migrants, 1994

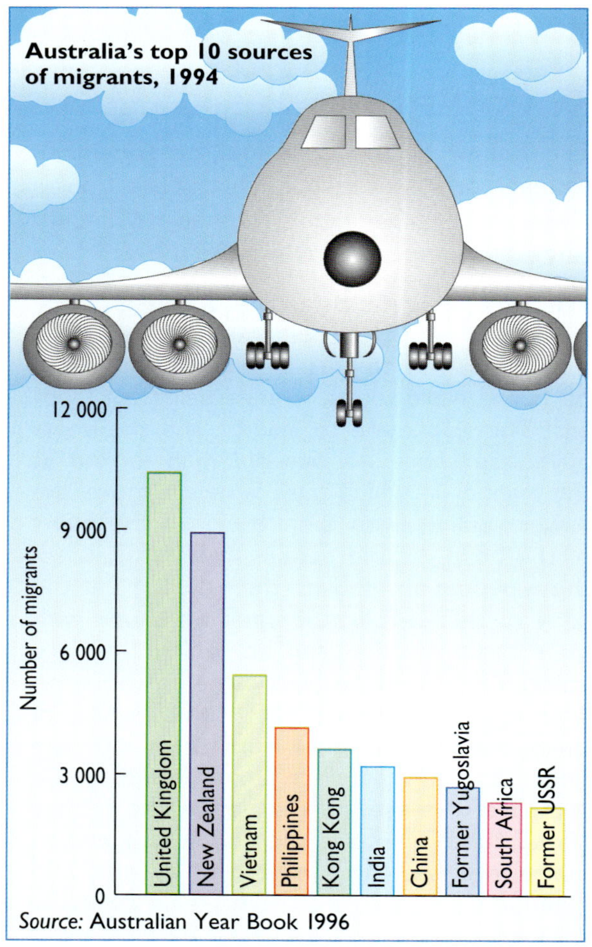

Source: Australian Year Book 1996

WORLD LANGUAGES

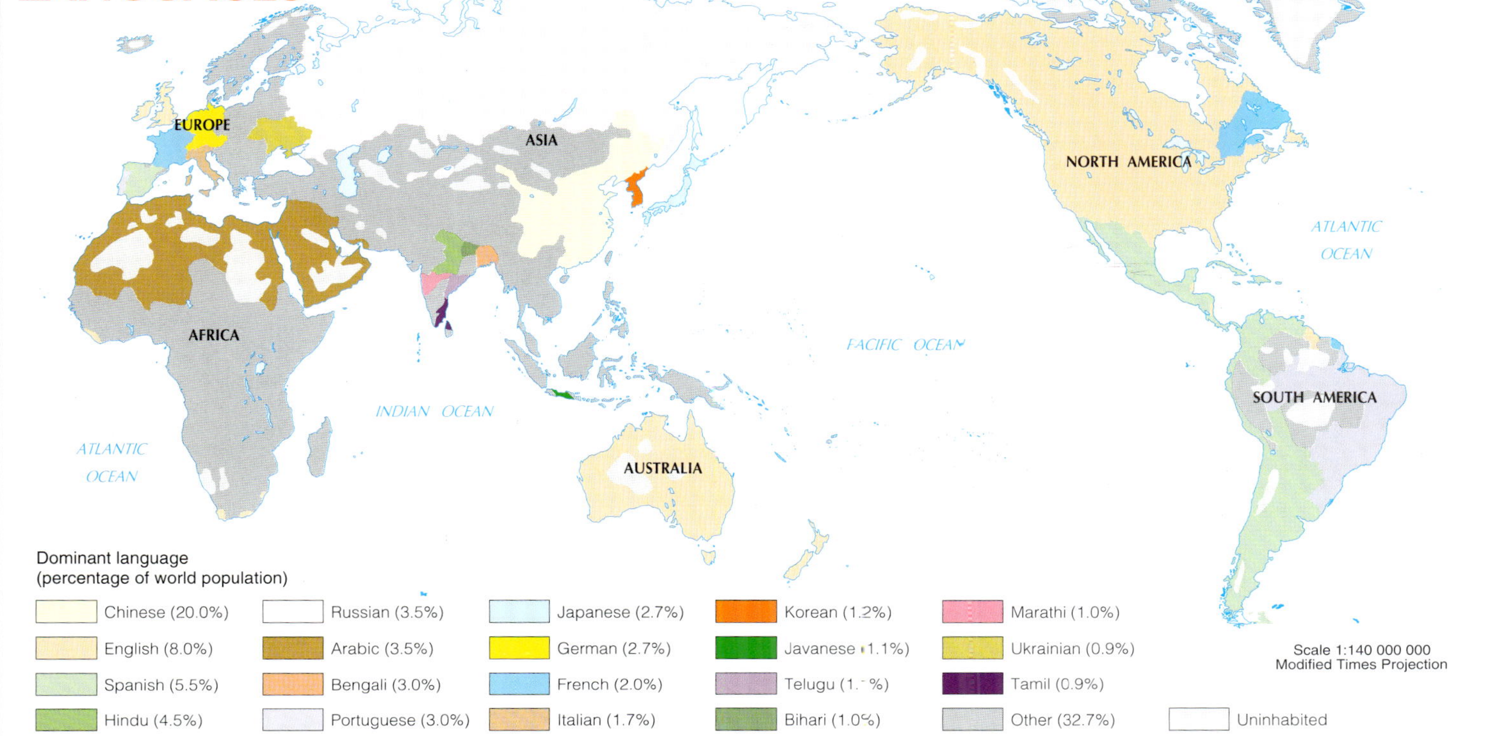

WORLD RELIGIONS

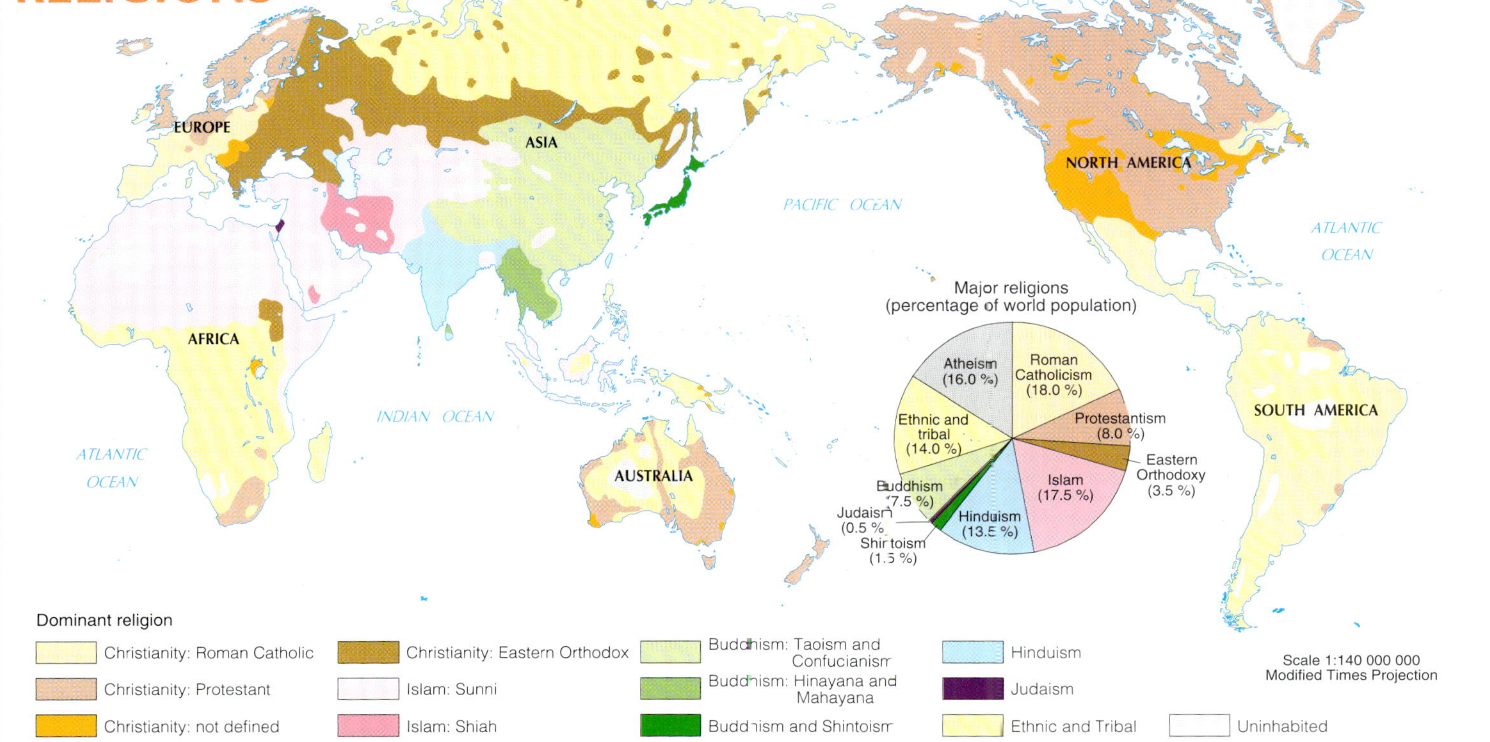

THE WORLD AT WAR

WORLD WAR I

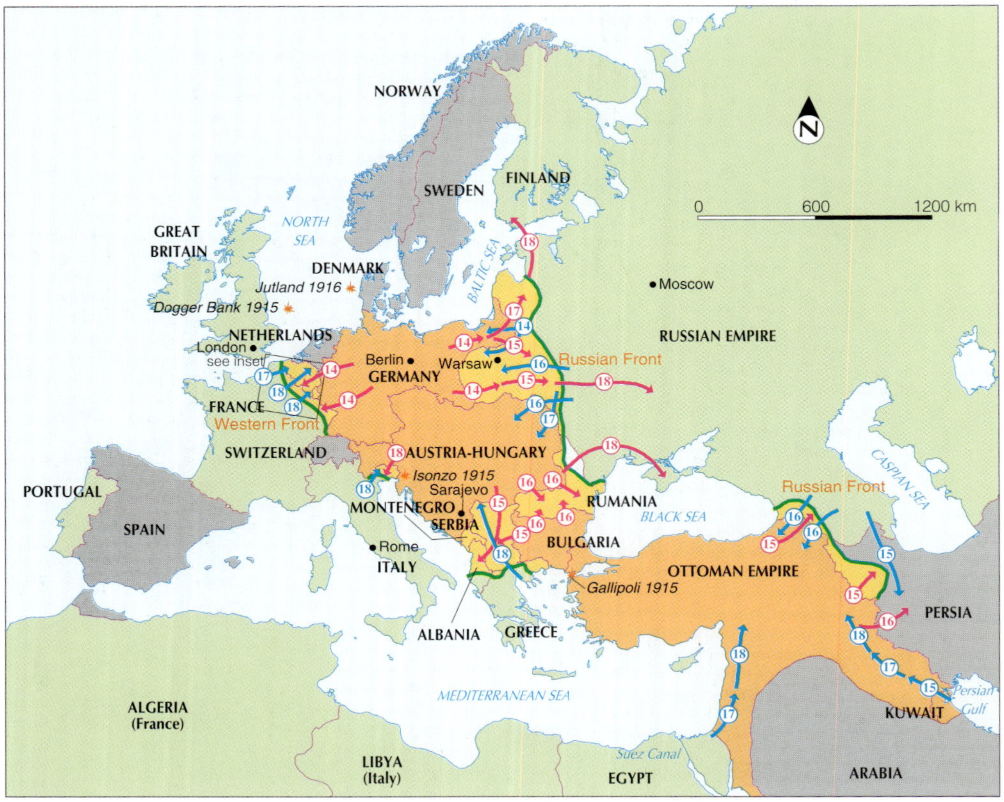

- Allies
- Neutral States
- Central Powers
- Occupied by the Central Powers
- Battle front
- Armistice line November 1918
- Major battle, year
- Allied offensives, year
- Central Powers offensives, year

THE WESTERN FRONT

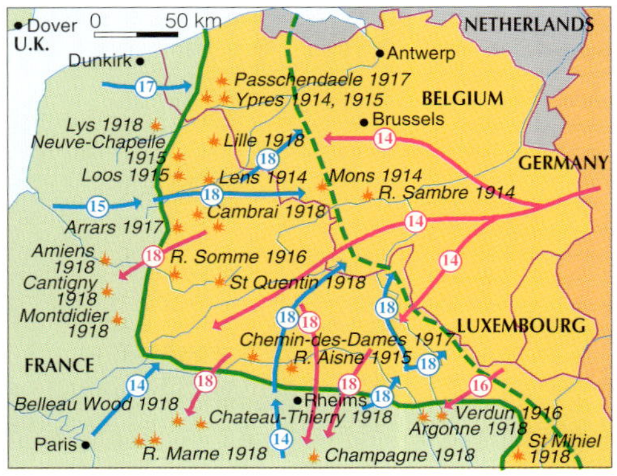

WORLD WAR II

- Allies
- Occupied by the Allies
- Neutral States
- Axis Powers
- Occupied by the Axis Powers
- 8/44 Major battle, month/year
- Allied offensives, year
- Axis movements
- (1) Germans invade Poland 1939
- (2) Germans invade Denmark and Norway 1940
- (3) Germans invade Netherlands, Belgium and Luxembourg 1940
- (4) Germans invade France 1940
- (5) German air attack on Britain 1940
- (6) Germans invade Yugoslavia and Greece 1941
- (7) Germans invade Crete 1941
- (8) Germans invade the U.S.S.R. 1941
- (9) Southern France occupied 1942
- (10) German counter attack in Belgium "The Bulge" 1944

Australians at war

When Japanese soldiers were on the ridges looking down on Port Moresby in Papua and a huge invasion fleet was heading for land at Bootless Bay near Port Moresby aerodrome, it seemed Australia was next. Then two decisive victories saw the Japanese advance falter — the battle of the Coral Sea, in which ships never saw each other and all fighting was by aircraft, and along the Kokoda Trail.

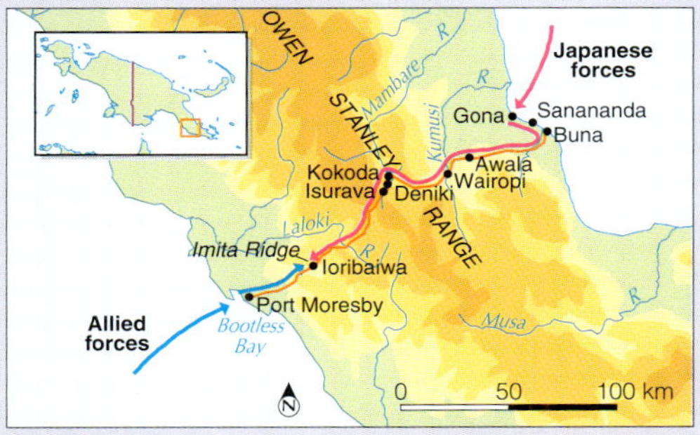

The Japanese were also defeated on land at Milne Bay, Buna, Gona and Sananda and by US and allied troops on Guadalcanal in the Solomons. This meant Japan was pushed back and the immediate danger to Australia was over. That was in 1942. The war in the Pacific lasted three more years.

However, the war in Papua New Guinea ruined the prosperity, health and property of many Papua New Guineans. Some were conscripted as labour while others volunteered. Papua New Guineans worked and fought for both sides, as control of areas changed.

Papua New Guineans also celebrate on Australia's Anzac Day when they acknowledge those who helped restore peace to their country. Australians remember, with gratitude, the help given by the "Fuzzy wuzzy angels" during the war.

(Australian War Memorial Negative Number 014028)

Soldiers on the Kokoda Trail in 1942 (Australian War Memorial Negative Number 027025)

PACIFIC CAMPAIGNS IN WORLD WAR II

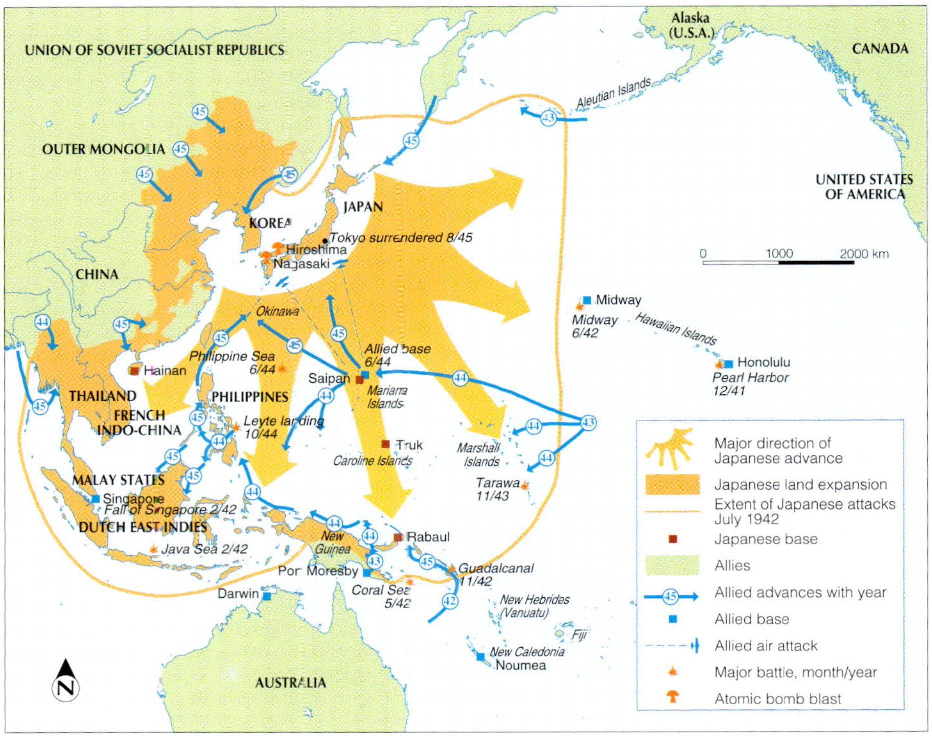

AGRICULTURAL LAND USE IN AUSTRALIA

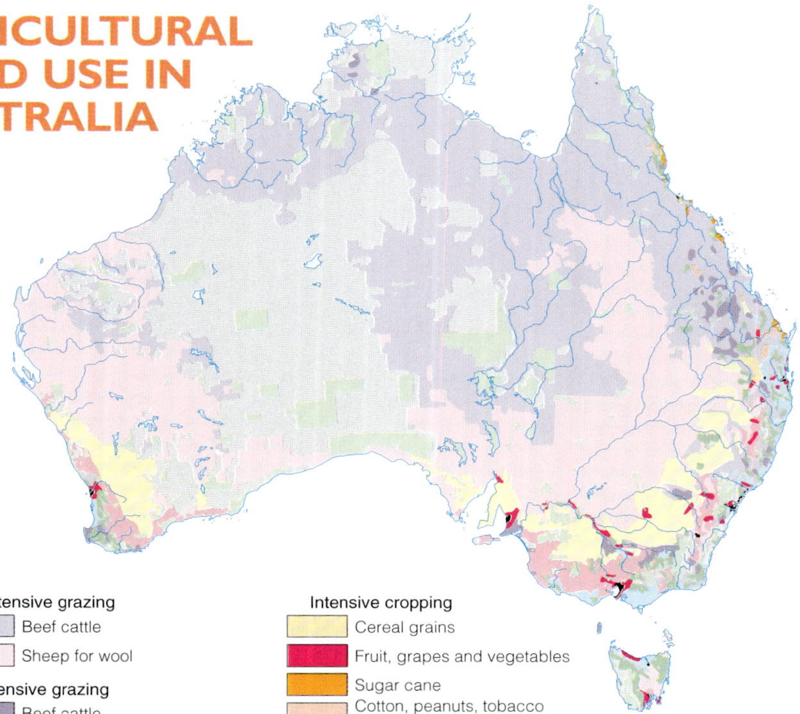

Resources and manufacturing

One of the biggest changes in the twentieth century has been the move to the **mass production** of manufactured goods. Throughout history the entire process of manufacturing a good was usually the responsibility of a single worker. Today, individual **components** are made in great numbers by the machines of specialist manufacturers. Components produced in a range of factories or **imported** from overseas are then put together on an **assembly line**.

To manufacture products, profitably, there needs to be a **market** for the goods and a ready supply of a number of different resources.

MINERAL MINING AND MANUFACTURING SITES IN AUSTRALIA

Transport and location factors

Manufacturing industries must take into account a number of factors when deciding on a location. They must try to be as close as possible to the suppliers of components and to road, rail and port facilities. Large manufacturing industries also need vast areas of flat land for factory and storage areas, a power supply, work force and, of course, a market in which to sell the goods.

Holden's car plant at Elizabeth

Raw materials

Primary industries produce **raw materials** and food directly from nature. Mining is a primary industry which extracts mineral or fuel resources from the ground. In the case of minerals, the rocks (ore) are transported to a smelter, where the ore is **refined** into metal (**processing industry**).

Manufacturing industries

The products of processing industries or other manufacturing industries are used to create a new product. For example, a motor vehicle factory requires processed resources such as steel and other metals, plastics, rubber, paint, tyres, electrical equipment, fabric, glass and many other components to assemble a car. Processing and manufacturing are examples of **secondary industries**.

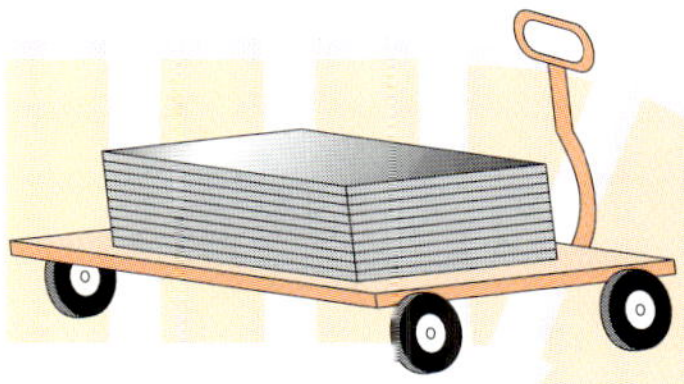

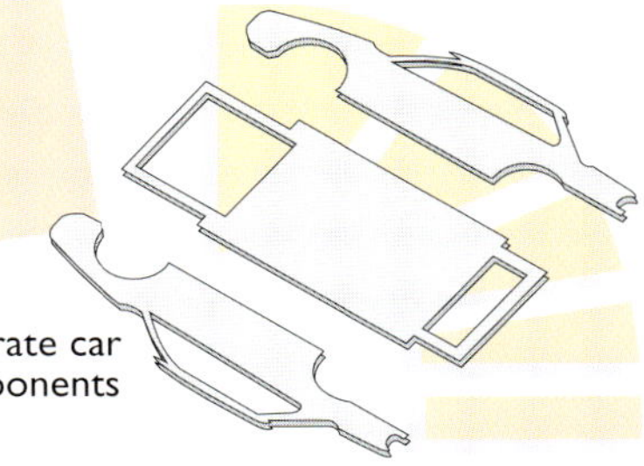

Separate car components

A steel smelter

Robot

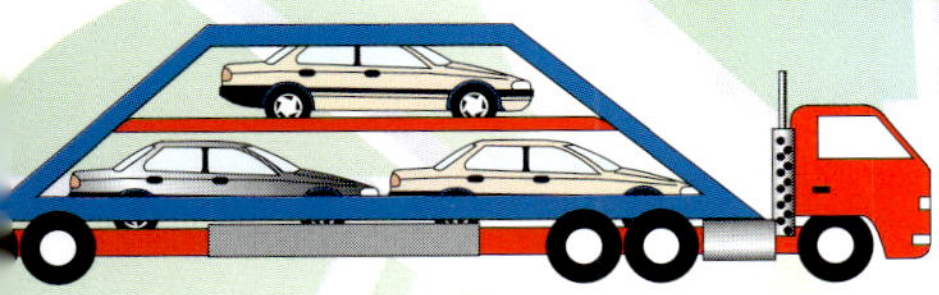

Completed car loaded onto trailer

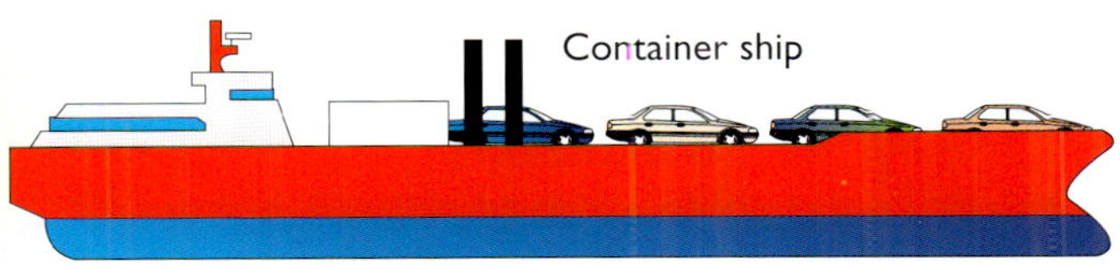

Container ship

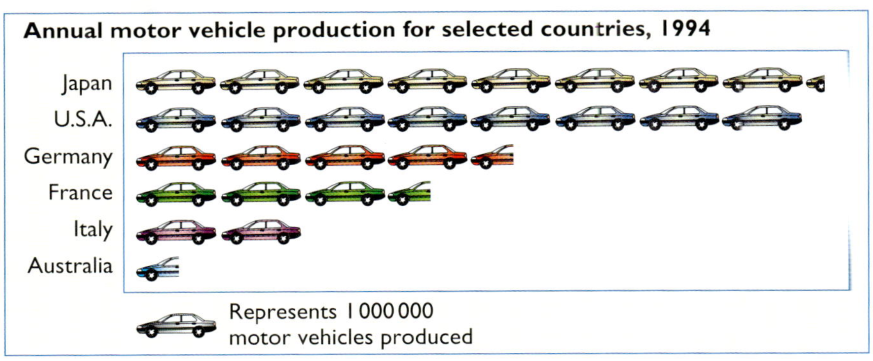

Labour resources

Many companies employ a large number of workers to provide **labour**. On most car assembly lines, including Holden's, robots and other automated systems perform tasks such as welding, which streamlines production and creates a safer working environment. A wide range of tradespeople, along with designers, engineers, marketers and sales people are involved in the motor industry. Workers who provide services rather than goods are employed in **tertiary** or **service** industries.

AN UNEQUAL WORLD

Feast and famine

There is enough food produced to feed the world's population. Hunger is not due to scarcity: it is due to unequal distribution of wealth. The wealthiest 20 per cent of the population receive 85 per cent of the world's income. The poorest 20 per cent of people in the world receive just 1.4 per cent of world wealth.

Who has enough to eat?

100% represents the daily supply of calories required for healthy living.

Most countries have a surplus of food but the poor people cannot afford to buy it. India has the largest number of hungry people in the world, and a very high percentage (65%) of malnourished children. Its food surplus is sold on the world market. By using only 10 per cent of this surplus, India could feed all the hungry in its own country. The photograph shows an Indian woman preparing breakfast for her family.

Wealth and health

Poverty is the world's deadliest disease. Life expectancies can be nearly twice as high in wealthy countries as they are in poor ones. For every 1000 births in Australia, seven children die before they reach the age of one year. In Afghanistan, 162 children under the age of one die per 1000 births.

THE HAVES AND HAVE NO

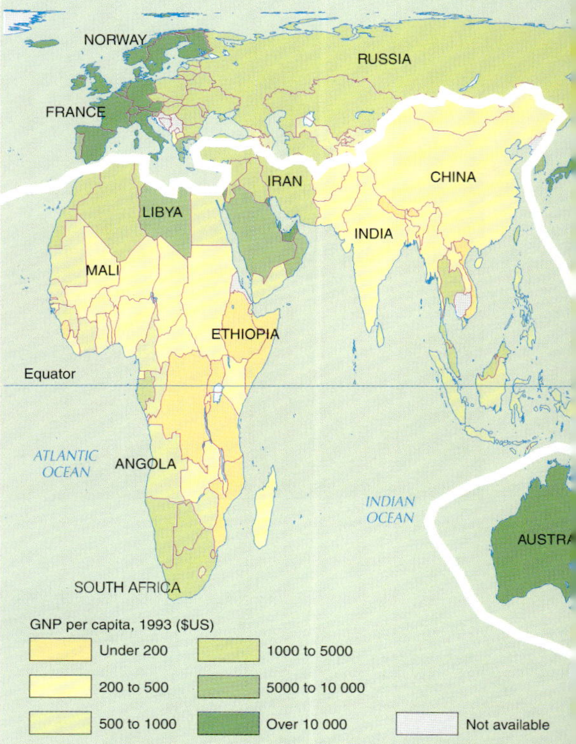

Life expectancy and wealth

A country's wealth is often measured by its Gross National Product (GNP) per capita. For example, Australia produces $17 000 of goods and services in a year for every person in the country. In contrast, Uganda has a GNP of only $120 per capita.

Countries **north** of this line are seen as the wealthy nations of the world, with high standards of living.

Countries **south** of this line generally have a lower standard of living. There are obvious exceptions such as Singapore and Saudi Arabia.

ıth Africa (LE 63) ıP $2670 | Bangladesh (LE 53) GNP $220 | Uganda (LE 43) GNP $170

Education and literacy

Twenty-five per cent of the world's adults cannot read and write and most children of secondary school age never see the inside of a classroom. In Mali, Rwanda and Tanzania only three per cent of children attend secondary school.

Who can read and write?

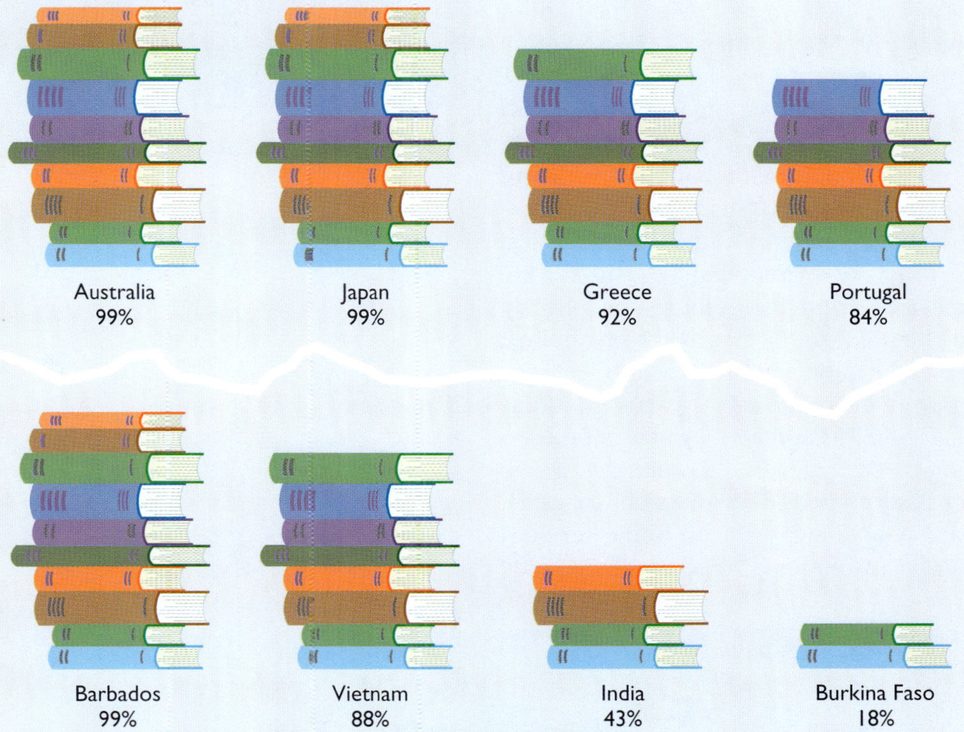

There are many countries in the world where lack of wealth has not stood in the way of education. Vietnam produced only $130 of goods and services for every person in the country in 1991. This was the seventh lowest figure of all countries in the world. Vietnam, however, had a literacy rate of 88 per cent in the same year. This is one of the highest figures in the developing world. The photograph shows a school in rural Vietnam.

CITIES OF THE WORLD

On every continent cities and towns are growing rapidly. In some parts of the world this is because of **rural-urban migration**. People move from rural areas to the cities in search of employment and the services cities offer. City growth is also boosted by natural population increase.

Africa and Asia presently have the least urbanised populations of the world. However, it is in Asia that the largest number of "super cities" — those with a population over 3 million — are located. By 2000, five of the biggest ten cities will be Asian: Tokyo–Yokohama (Japan), Kolkata (India), Mumbai (India), Seoul (South Korea) and Shanghai (China).

Rural immigrants to the world's fastest growing city, Sao Paulo, Brazil, are "pushed" from their homes by poverty and "pulled" to the city by the prospect of jobs, education and better health services.

Even in Mexico City with problems of flooding, landslides, water shortages, major earthquakes, pollution, traffic congestion and overstretched services, rural people are attracted to the services and opportunities offered by a large city.

City of Sao Paulo

World's ten largest cities

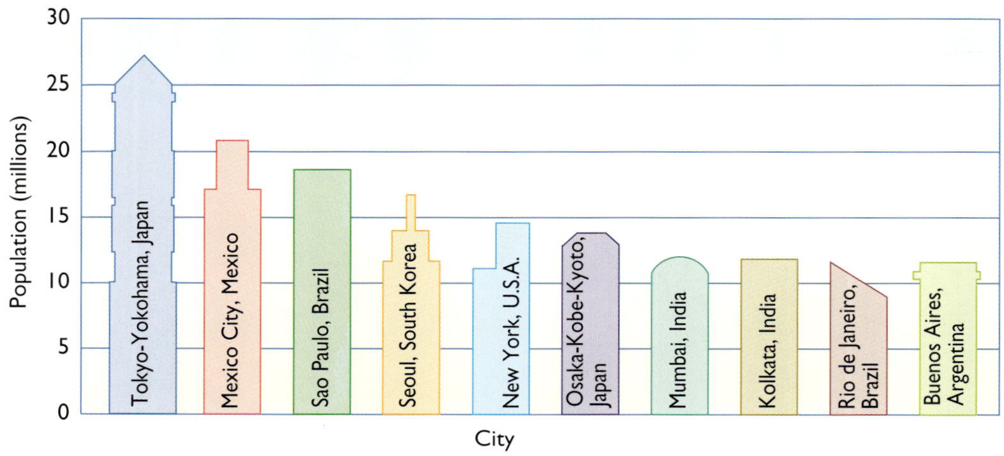

WORLD'S LARGEST CITIES

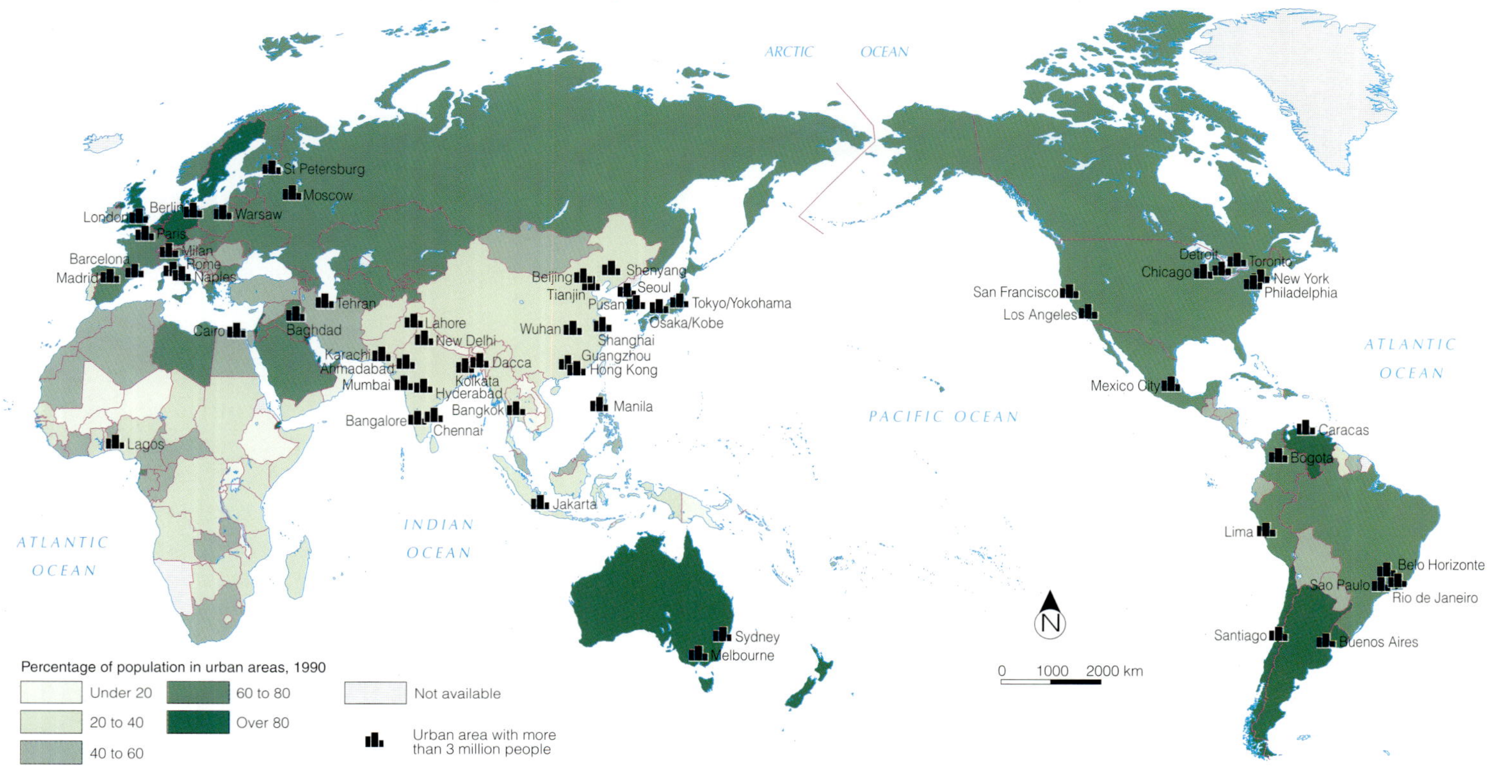

The location of towns and cities

Towns and cities are usually closely tied to places that can fulfil the basic human needs for shelter from the elements, good supplies of fresh water and food. Resources to provide a livelihood, a pleasant environment, good transport links and defence of territory also play a part in decisions on where to locate settlements.

Because of change over time and improved technology in transport, engineering and environmental management, it may be difficult to find the original factors that influenced where towns and cities were established. However, it is possible to identify some places that are commonly used as sites for settlements.

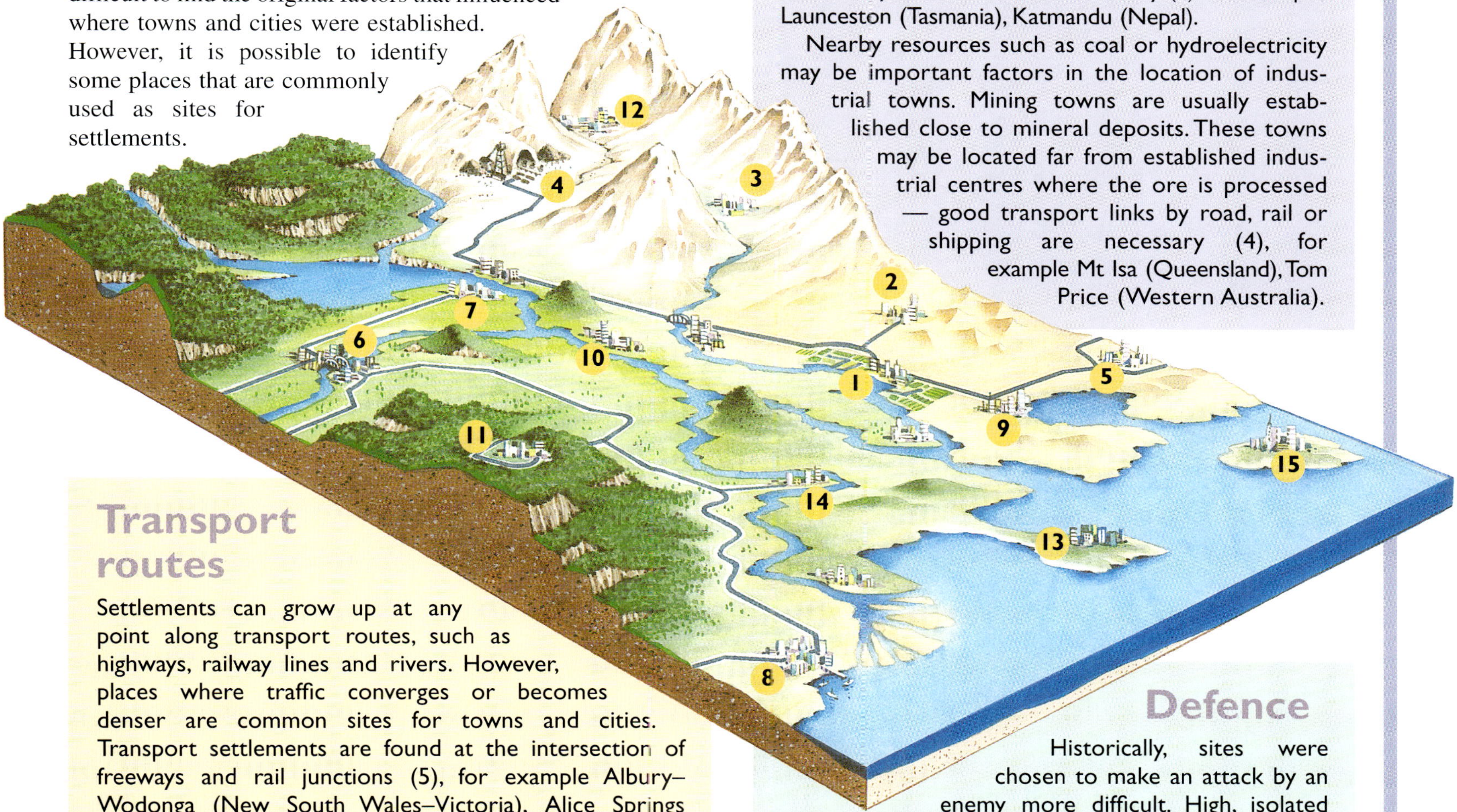

Resources

Abundant fresh water for drinking, washing, irrigation, watering animals or industry means that settlements may grow along rivers, especially at points where the river is deep and the ground well drained (1), for example Mildura (Victoria), Frankfurt (Germany). In deserts, towns grow up around oases (2), for example Al Faiyum (Egypt), Palm Springs (USA). In a mountainous area, a valley may be the only level site for a town or city (3), for example, Launceston (Tasmania), Katmandu (Nepal).

Nearby resources such as coal or hydroelectricity may be important factors in the location of industrial towns. Mining towns are usually established close to mineral deposits. These towns may be located far from established industrial centres where the ore is processed — good transport links by road, rail or shipping are necessary (4), for example Mt Isa (Queensland), Tom Price (Western Australia).

Transport routes

Settlements can grow up at any point along transport routes, such as highways, railway lines and rivers. However, places where traffic converges or becomes denser are common sites for towns and cities. Transport settlements are found at the intersection of freeways and rail junctions (5), for example Albury–Wodonga (New South Wales–Victoria), Alice Springs (Northern Territory), river crossings (6), for example Perth (Western Australia), London (UK) or where tributaries meet a river or lake (7), for example Kuala Lumpur (Malaysia), Manaus (Brazil).

Places where goods are transferred from one form of transport to another are points where towns and cities grow — deep water harbours (8), for example Sydney (New South Wales), Cape Town (South Africa), bays, inlets and estuaries (9), for example Melbourne (Victoria), Oslo (Norway) and the highest point of navigation along a stream (10), for example Brisbane (Queensland), Minneapolis (USA).

Defence

Historically, sites were chosen to make an attack by an enemy more difficult. High, isolated hilltops (11), for example Athens (Greece), Jerusalem (Israel) and sites at mountain passes (12), for example Kabul (Afghanistan) are easy to defend. Sites surrounded by water including peninsulas (13), for example, Gallipoli (Turkey), Bombay (India), meanders in a river (14), for example Dhaka (Bangladesh), New Orleans (USA) or islands (15), for example Hong Kong, Copenhagen (Denmark) provide protection from enemies who are slowed by having to cross the water. Given the technological advances in weaponry, many of the reasons for the choice of these sites are no longer relevant.

Slavery

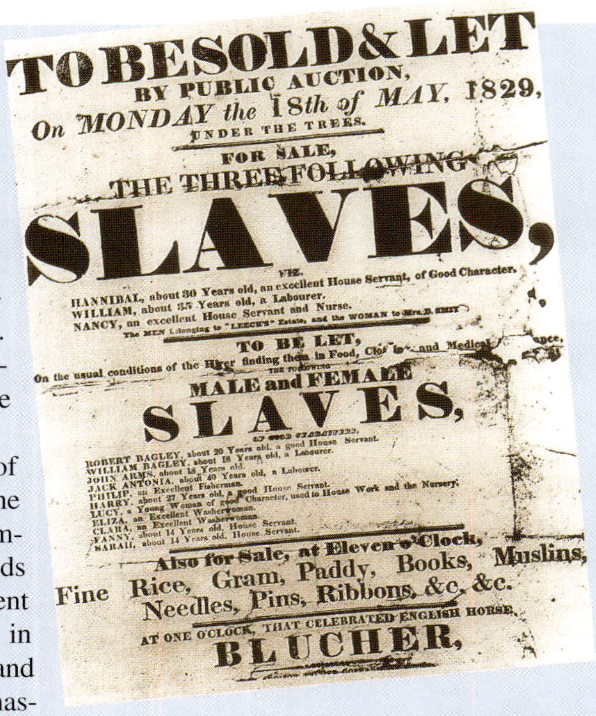

TO BE SOLD & LET
BY PUBLIC AUCTION,
On MONDAY the 18th of MAY, 1829,
UNDER THE TREES.
FOR SALE,
THE THREE FOLLOWING
SLAVES,
VIZ.
HANNIBAL, about 30 Years old, an excellent House Servant, of Good Character.
WILLIAM, about 35 Years old, a Labourer.
NANCY, an excellent House Servant and Nurse.
The MEN belonging to "LEECH'S" Estate, and the WOMAN to
TO BE LET,
On the usual conditions of the Hirer finding them in Food, Clo and Medical
THE FOLLOWING
MALE and FEMALE
SLAVES,
ROBERT BAGLEY, about 20 Years old, a good House Servant.
WILLIAM BAGLEY, about 18 Years old, a Labourer.
JOHN ARMS, about 18 Years old.
JACK ANTONIA, about 40 Years old, a Labourer.
PHILIP, an Excellent Fisherman.
HARRY, about 27 Years old, a good House Servant.
LUCY, a Young Woman of good Character, used to House Work and the Nursery.
ELIZA, an Excellent Washerwoman.
CLARA, an Excellent Washerwoman.
FANNY, about 14 Years old, House Servant.
SARAH, about 14 Years old, House Servant.
Also for Sale, at Eleven o'Clock,
Fine Rice, Gram, Paddy, Books, Muslins,
Needles, Pins, Ribbons, &c. &c.
AT ONE O'CLOCK, THAT CELEBRATED ENGLISH HORSE,
BLUCHER,

Throughout history there have been many instances of the use of slavery. Slavery involves forcing someone to work for little or no return. Slaves were obtained through raids or conquests of other people's lands. They were regarded as the property of their owner and could be bought, sold or traded.

Slavery was an accepted part of life in ancient civilisations. The ancient Egyptians used large numbers of slaves to build pyramids and other monuments. In Ancient Greece slaves were employed in domestic service, as labourers and rowers. The Ancient Roman masters had the lawful right to decide on the life or death of their slaves. Tens of thousands of slaves were captured through wars. Aztec, Incan and Mayan cultures used slavery for agriculture and warfare.

The first African slaves were brought to America in 1619. With the development of large plantations in the southern colonies of America the number of slaves grew to 3 953 760 by 1860. The American Civil War (1861–65) brought freedom to nearly four million blacks, but the pro-slavery attitude that had operated in the South for hundreds of years did not end with the war. Tensions and problems between African Americans and the white population still exist today.

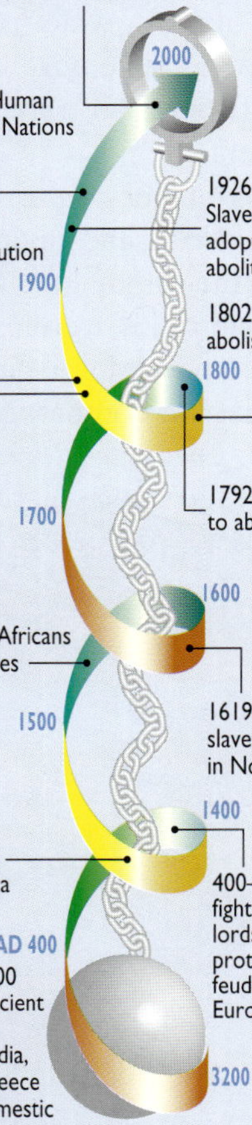

Informal work

In developing countries such as the Philippines, people unable to find work create their own ways of earning money. This informal sector of the work force includes occupations such as selling souvenirs, begging, shining shoes, washing car windows, street entertainment, carpentry and sorting garbage. More than a billion people in the world feed, clothe and shelter their families from their own resources.

Every city in the developing world has its garbage pickers. They collect glass, aluminium, paper and plastic products and sell them to recycling plants. Other items are restored or made into new products for sale.

Women and work

Women make up more than one-third of the world's paid labour force. But they generally remain in a limited range of occupations such as nursing, child care, teaching, sales, and assembly line work. As countries become industrialised, more women obtain jobs in more occupations. However, recognition by society and governments of women's contribution to paid and unpaid work is not high.

Over the last 20 years, Swedish women's average hourly earnings have risen from 66 to 90 per cent of men's earnings. The Swedish government made major changes to female education, child-care, and marriage and divorce laws, to enable equal work opportunities. Japan has retained its traditional attitudes toward working women, who are expected to retire when they have children.

Most women work in the fields and marketplaces of developing nations in much of Africa, Asia, the Middle East and Latin America. Some progress is being made in widening women's work opportunities in developing countries. However, the status of women is commonly low due to cultural and religious traditions. Many women still do not have equal access to education and training programs.

In the former USSR the ideal of equality under communism did not happen in reality. Even educated women generally remained poorly paid with the worst jobs. In the developed world today, the so-called "glass ceiling" prevents women from reaching the top positions in industry and government.

Percentage of men's wages earned by women in non-agricultural work (late 1980s)

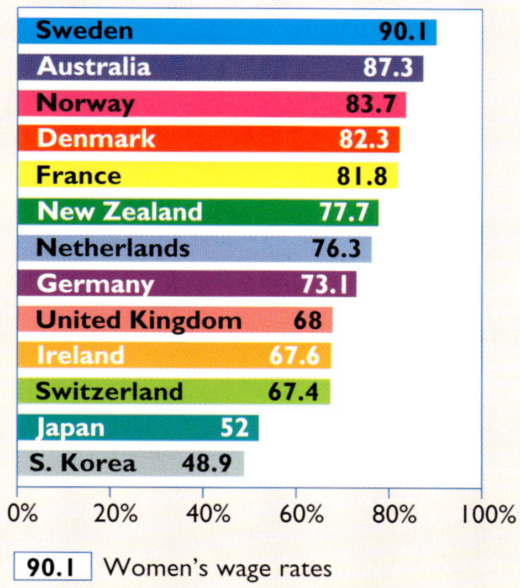

Source: ILO Year Book of Labour Statistics, 1991 50th Issue. Copyright © 1991, International Labour Organisation, Geneva

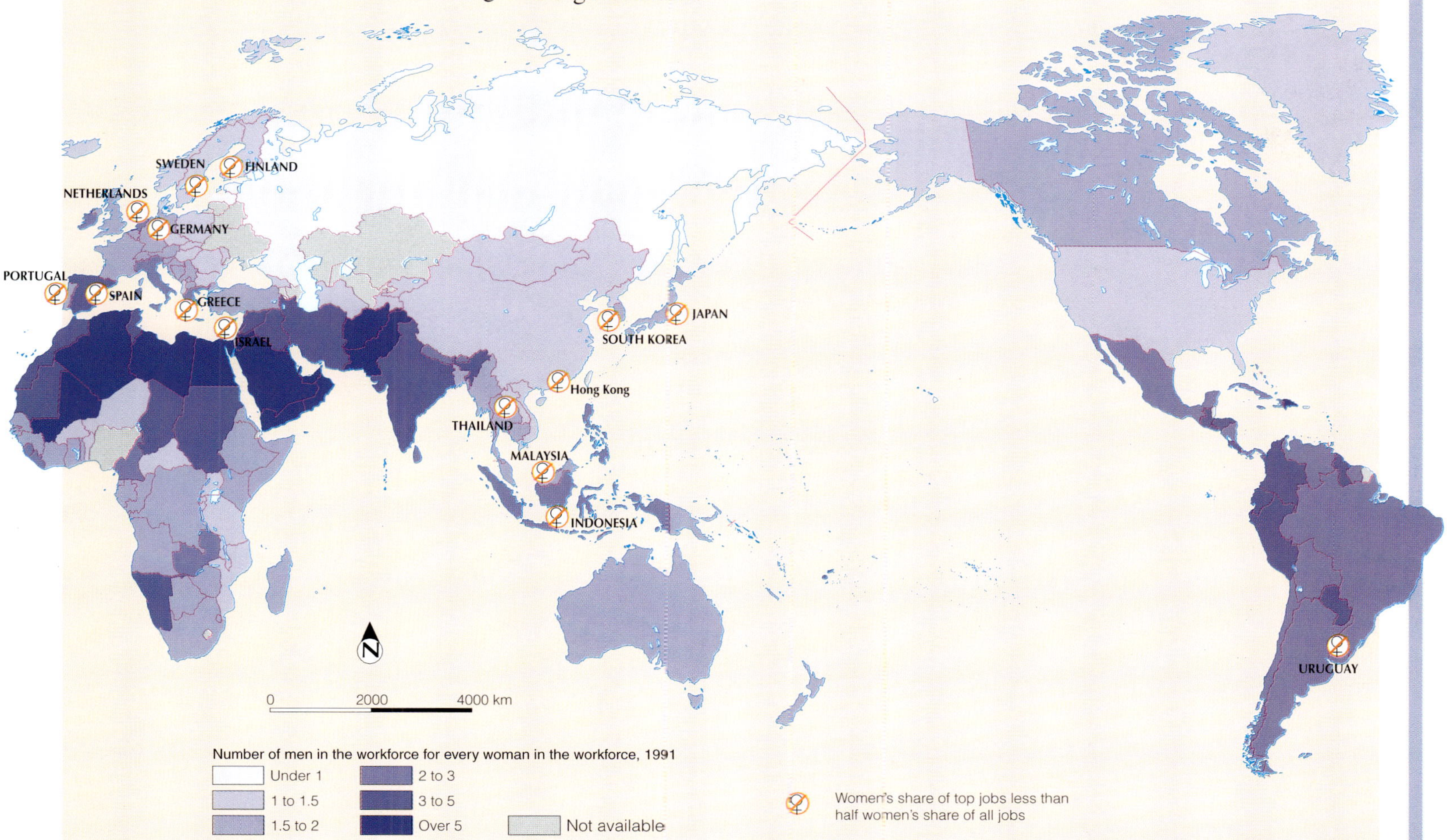

STATISTICS

Country	Capital city	Area (sq km)	Time at 12 noon (Australian EST)	Population ('000s) 1995	Male/ female life expectancy (years) 1995	Crude birth rate (per 1000 people) 1995	Infant mortality rate (per 1000 births) 1995	Persons per doct 1995
Afghanistan	Kabul	647 497	6.30 a.m.	16 494	45/44	44	159	5 578
Albania	Tirane	28 748	3.00 a.m.	3 334	70/76	23	32	599
Algeria	Algiers	2 381 741	2.00 a.m.	27 256	66/68	30	54	1 157
Angola	Luanda	1 246 700	3.00 a.m.	9 545	43/47	46	149	14 418
Antigua & Barbuda	St Johns	442	10.00 p.m.*	64	71/75	18	19	1 333
Argentina	Buenos Aires	2 766 889	11.00 p.m.*	33 533	68/75	20	30	349
Armenia	Yerevan	29 800	5.00 a.m.	3 481	68/75	26	28	237
Australia	Canberra	7 686 848	12.00 noon	17 827	74/81	14	7	487
Austria	Vienna	83 849	3.00 a.m.	7 915	73/80	12	7	329
Azerbaijan	Baku	86 600	5.00 a.m.	7 573	67/75	24	36	270
Bahamas	Nassau	13 935	9.00 p.m.*	269	68/76	19	32	745
Bahrain	Manama	622	4.00 a.m.	568	71/76	27	20	1 062
Bangladesh	Dacca	143 998	8.00 a.m.	122 255	55/54	35	109	5 821
Barbados	Bridgetown	431	10.00 p.m.*	255	71/76	16	21	1 049
Belarus	Minsk	207 600	5.00 a.m.	10 370	66/76	13	19	249
Belgium	Brussels	30 513	3.00 a.m.	10 041	73/80	12	7	297
Belize	Belmopan	22 965	8.00 p.m.*	204	66/70	36	37	2 125
Benin	Porto Novo	112 622	3.00 a.m.	5 167	50/53	48	113	14 234
Bhutan	Thimphu	47 000	7.30 a.m.	700	51/50	40	123	4 459
Bolivia	Sucre and La Paz	1 098 581	10.00 p.m.*	7 544	60/65	33	77	2 377
Bosnia & Herzegovina	Sarajevo	51 130	3.00 a.m.	4 619	72/78	14	13	667
Botswana	Gaborone	600 372	4.00 a.m.	1 326	60/66	33	41	5 525
Brazil	Brasilia	8 511 965	9.00–11.00 p.m.*	156 664	58/67	22	62	924
Brunei	Bandar Seri Begawan	5 765	10.00 a.m.	277	69/73	27	26	1 620
Bulgaria	Sofia	110 912	4.00 a.m.	8 831	70/76	12	13	326
Burkina Faso	Ouagadougou	274 200	2.00 a.m.	9 853	47/48	49	120	35 189
Burundi	Bujumbura	27 834	4.00 a.m.	5 985	39/43	45	116	35 625
Cambodia	Phnom Penh	181 035	9.00 a.m.	9 899	48/51	46	112	23 292
Cameroon	Yaoundé	475 442	3.00 a.m.	15 756	55/59	41	79	18 915
Canada	Ottawa	9 976 139	10.30–5.00 p.m.*	27 770	75/82	14	7	475
Cape Verde Islands	Praia	4 033	1.00 a.m.	411	60/64	47	60	5 338
Central African Republic	Bangui	622 984	3.00 a.m.	3 074	41/44	43	139	19 961
Chad	N'Djamena	1 284 000	3.00 a.m.	5 351	39/42	42	134	56 926
Chile	Santiago	756 945	10.00 p.m.*	13 740	71/77	21	16	937
China	Beijing	9 596 961	10.00 a.m.	1 177 585	67/69	18	52	651
Colombia	Bogota	1 138 914	9.00 p.m.*	34 943	69/75	23	30	1 190
Comoros	Moroni	2 171	5.00 a.m.	512	55/60	47	82	12 190
Congo	Brazzaville	342 000	3.00 a.m.	2 389	46/50	41	113	4 213
Congo, Democratic Rep. of	Kinshasa	2 345 409	3.00–4.00 a.m.	41 346	45/49	48	113	31 370
Costa Rica	San Jose	50 700	8.00 p.m.*	3 265	76/80	26	12	1 045
Croatia	Zagreb	56 537	3.00 a.m.	4 694	70/77	11	9	524
Cuba	Havana	110 861	11.00 p.m.*	10 957	75/79	17	11	315
Cyprus	Nicosia	9 251	4.00 a.m.	723	74/78	17	9	599
Czech Republic	Prague	78 860	3.00 a.m.	10 389	69/77	13	10	N/A
Denmark	Copenhagen	43 069	3.00 a.m.	5 176	73/79	13	7	363
Djibouti	Djibouti	22 000	5.00 a.m.	402	47/51	43	113	4 144

* previous day

Number of AIDS cases eported (per 0 000 people) 1993	Secondary students per teacher 1994	Education funding (percentage of GNP) 1994	Literacy rate (per cent) 1995	Total exports (millions of $US) 1993	Total imports (millions of $US) 1993	Electricity consumption (millions of kWh) 1992	Protected area (percentage of total area) 1990	Number of threatened mammals (species) 1990	CO_2 emission (thousands of metric tonnes) 1993
0.00	44.60	2.00	29	252	624	1 015	0.10	13	6 273
0.00	21.20	N/A	72	N/A	N/A	3 155	0.80	2	9 732
0.34	17.10	9.10	57	12 964	8 777	16 584	2.00	12	46 492
5.39	30.10	7.30	42	N/A	N/A	1 340	0.90	14	4 965
7.81	12.50	2.50	89	33	326	96	N/A	N/A	N/A
5.43	7.10	1.50	95	12 078	7 387	54 923	1.90	23	118 157
N/A	N/A	N/A	100	N/A	N/A	12 076	N/A	N/A	N/A
20.28	14.20	5.10	100	41 930	38 499	156 883	2.40	38	257 480
10.42	8.00	5.40	99	40 136	52 186	52 249	7.70	2	51 699
N/A	N/A	N/A	100	N/A	N/A	N/A	N/A	N/A	N/A
347.21	15.20	4.40	90	308	1 229	965	N/A	N/A	N/A
0.53	15.00	5.40	77	3 469	3 662	3 495	0.00	1	12 161
0.00	33.00	2.20	35	1 672	3 259	8 930	0.30	15	14 114
123.53	15.00	8.00	99	147	599	527	0.00	1	971
0.05	N/A	N/A	100	N/A	N/A	51 744	N/A	N/A	N/A
12.19	7.20	5.20	99	108 762	108 132	70 098	1.00	2	98 104
25.98	17.10	N/A	91	124	189	105	1.30	8	180
4.78	24.90	5.10	23	379	511	203	3.00	11	667
0.00	22.00	3.70	18	73	105	183	8.00	15	33
0.65	17.80	2.40	78	760	804	2 148	2.50	21	5 064
N/A	19.10	N/A	86	N/A	N/A	15 201	N/A	N/A	N/A
26.62	16.70	8.10	72	1 753	1 606	N/A	6.80	9	1 700
20.02	13.20	3.90	81	34 375	18 263	261 161	1.00	40	206 957
0.72	10.50	4.90	77	N/A	N/A	1 030	N/A	N/A	N/A
0.18	14.10	5.80	93	N/A	N/A	40 990	0.50	3	106 989
12.82	44.60	2.30	18	304	593	157	1.10	10	520
84.41	2.10	3.50	50	73	189	159	0.60	4	176
0.00	19.20	N/A	35	N/A	N/A	70	0.00	21	451
8.93	25.90	3.30	54	1 841	1 221	2 712	1.80	27	5 774
24.81	13.40	7.40	99	127 459	121 529	489 485	2.00	5	455 530
12.65	26.80	2.90	66	11	110	36	0.00	0	77
60.31	55.60	2.80	27	127	198	96	2.60	12	264
7.14	37.60	1.80	30	155	240	85	0.00	18	202
4.17	N/A	2.90	93	8 929	7 354	19 961	7.30	9	31 833
0.00	2.20	2.40	73	58 919	50 176	680 400	0.90	30	2 388 613
8.46	19.90	2.90	87	7 080	5 108	36 661	3.30	25	53 831
0.59	24.80	6.50	48	18	45	16	0.00	3	51
145.75	27.70	5.50	57	1 029	458	547	1.60	12	1 773
43.98	21.70	0.90	72	2 138	1 539	5 972	1.50	31	3 822
12.83	16.20	4.60	93	1 366	1 833	3 892	4.80	10	2 557
N/A	15.40	N/A	97	N/A	N/A	11 902	N/A	N/A	N/A
1.25	10.80	6.70	94	N/A	N/A	16 255	2.60	11	36 292
3.32	11.90	3.60	94	847	2 305	2 077	0.00	1	4 192
N/A	15.40	4.70	99	6 600	5 700	N/A	N/A	N/A	110 147
20.71	29.70	7.40	99	36 841	31 982	34 331	4.00	1	47 009
65.92	27.90	2.50	48	N/A	N/A	178	0.20	6	326

Country	Capital city	Area (sq km)	Time at 12 noon (Australian EST)	Population ('000s) 1995	Male/ female life expectancy (years) 1995	Crude birth rate (per 1000 people) 1995	Infant mortality rate (per 1000 births) 1995	Persons per doct 1995
Dominica	Roseau	751	10.00 p.m.*	87	74/80	21	11	2 351
Dominican Republic	Santo Domingo	48 734	10.00 p.m.*	7 684	66/70	26	54	1 048
Ecuador	Quito	283 561	9.00 p.m.*	10 461	67/72	27	41	1 069
Egypt	Cairo	1 001 449	4.00 a.m.	59 586	59/62	33	78	1 903
El Salvador	San Salvador	21 041	8.00 p.m.*	5 637	64/69	33	43	1 382
Equatorial Guinea	Malabo	28 051	3.00 a.m.	399	50/54	41	105	4 433
Eritrea	Asmara	117 598	5.00 a.m.	3 467	N/A	N/A	135	48 153
Estonia	Tallinn	45 100	5.00 a.m.	1 608	65/75	14	20	268
Ethiopia	Addis Ababa	1 104 301	5.00 a.m.	53 278	51/54	45	109	42 932
Fiji	Suva	18 274	2.00 p.m.	757	63/67	25	18	2 523
Finland	Helsinki	337 032	4.00 a.m.	5 051	72/80	13	5	409
France	Paris	547 026	3.00 a.m.	57 566	74/82	13	7	389
Gabon	Libreville	267 667	3.00 a.m.	1 223	51/57	29	97	2 165
Gambia	Banjul	11 295	2.00 a.m.	930	47/52	47	126	15 246
Georgia	Tbilisi	69 700	5.00 a.m.	5 634	69/76	16	24	176
Germany	Berlin	356 755	3.00 a.m.	80 768	73/79	11	7	321
Ghana	Accra	238 537	2.00 a.m.	16 699	53/57	45	85	26 591
Greece	Athens	131 944	4.00 a.m.	10 470	75/80	10	9	316
Grenada	St Georges	344	10.00 p.m.*	94	68/73	31	13	1 679
Guatemala	Guatemala	108 889	8.00 p.m.*	10 466	61/67	36	56	2 924
Guinea	Conakry	245 857	2.00 a.m.	6 237	41/46	45	142	9 281
Guinea-Bissau	Bissau	36 125	1.00 a.m.	1 072	45/49	41	122	8 787
Guyana	Georgetown	214 969	11.00 p.m.*	734	61/68	21	49	6 613
Haiti	Port-au-Prince	27 750	9.00 p.m.*	6 385	43/47	41	110	6 764
Honduras	Tegucigalpa	112 088	8.00 p.m.*	5 170	65/70	36	47	1 783
Hungary	Budapest	93 030	3.00 a.m.	10 324	67/75	12	13	294
Iceland	Reykjavik	103 000	2.00 a.m.	261	76/81	17	4	365
India	New Delhi	3 287 590	7.30 a.m.	903 159	58/59	29	81	2 474
Indonesia	Jakarta	1 904 569	9.00–11.00 a.m.	197 232	58/62	25	70	7 659
Iran	Tehran	1 648 000	5.30 a.m.	63 370	64/66	43	62	2 880
Iraq	Baghdad	434 924	5.00 a.m.	19 162	64/66	45	72	2 046
Ireland	Dublin	70 283	2.00 a.m.	3 530	73/78	14	8	681
Israel	Jerusalem	20 770	4.00 a.m.	4 919	76/80	21	9	414
Italy	Rome	302 225	3.00 a.m.	58 019	74/81	11	8	388
Ivory Coast	Yamoussoukro	322 463	2.00 a.m.	13 808	47/51	47	97	27 506
Jamaica	Kingston	10 991	9.00 p.m.*	2 530	72/76	22	18	2 255
Japan	Tokyo	372 313	11.00 a.m.	124 712	76/82	10	4	589
Jordan	Amman	97 740	4.00 a.m.	3 824	70/74	39	33	658
Kazakhstan	Astana	2 715 100	7.00–8.00 a.m.	17 156	63/73	20	42	254
Kenya	Nairobi	582 646	5.00 a.m.	27 372	52/56	43	75	8 381
Kirgizia	Bishkek	198 500	8.00 a.m.	4 626	63/72	27	48	282
Kuwait	Kuwait	17 818	5.00 a.m.	1 698	72/77	30	13	576
Laos	Vientiane	236 800	9.00 a.m.	4 569	50/53	44	104	3 895
Latvia	Riga	63 700	5.00 a.m.	2 736	64/75	14	22	256
Lebanon	Beirut	10 400	4.00 a.m.	3 552	67/72	28	41	1 012
Lesotho	Maseru	30 355	4.00 a.m.	1 896	60/64	35	72	18 408

* previous day

Number of AIDS cases eported (per 0 000 people) 1993	Secondary students per teacher 1994	Education funding (percentage of GNP) 1994	Literacy rate (per cent) 1995	Total exports (millions of $US) 1993	Total imports (millions of $US) 1993	Electricity consumption (millions of kWh) 1992	Protected area (percentage of total area) 1990	Number of threatened mammals (species) 1990	CO_2 emission (thousands of metric tonnes) 1993
13.79	29.40	5.80	94	60	104	31	N/A	N/A	N/A
23.54	42.80	1.50	83	735	1 793	5 330	0.00	1	6 745
2.14	14.30	2.70	86	2 851	2 207	6 952	15.30	21	15 316
0.10	18.10	6.70	48	3 604	10 303	40 460	0.30	9	79 483
6.78	20.90	1.80	73	558	1 220	2 302	0.50	6	2 352
3.26	28.90	1.70	50	33	44	18	0.00	15	106
N/A	N/A	N/A	20	N/A	N/A	N/A	N/A	N/A	N/A
N/A	N/A	N/A	100	N/A	N/A	9 856	N/A	N/A	N/A
7.47	40.10	4.80	62	298	915	973	2.20	25	2 565
0.53	18.30	5.00	86	494	591	472	0.10	1	678
2.22	19.00	5.80	100	26 089	25 322	65 427	1.00	3	51 300
37.33	13.10	5.50	99	207 084	217 223	401 472	3.50	6	357 163
17.58	24.30	5.70	61	2 471	760	914	2.70	17	7 826
19.35	27.00	5.20	27	111	141	68	0.40	7	183
N/A	N/A	N/A	100	N/A	N/A	N/A	N/A	N/A	N/A
11.01	14.90	4.10	99	378 030	354 970	575 052	5.60	2	964 028
21.03	17.20	3.40	60	891	1 199	5 867	1.80	13	3 521
6.58	16.50	2.80	93	6 365	16 543	36 457	0.30	4	70 920
34.04	20.60	4.60	98	28	92	52	N/A	N/A	N/A
2.61	14.70	1.80	55	1 126	1 484	2 330	0.30	10	4 071
5.42	14.40	1.40	24	N/A	N/A	521	0.20	17	1 000
17.63	6.80	2.80	36	N/A	N/A	41	0.00	5	147
45.37	28.70	8.80	95	214	209	225	0.00	12	660
48.32	19.50	1.80	53	163	300	475	0.10	1	725
38.22	19.90	4.60	73	967	964	1 273	2.60	7	1 979
1.02	16.70	6.10	99	9 688	9 330	37 401	2.20	2	64 076
8.43	N/A	5.40	100	1 589	1 509	4 494	3.60	1	1 942
0.03	22.00	3.20	48	16 144	22 254	310 330	1.70	39	651 936
0.01	11.70	0.90	77	29 120	24 453	44 660	3.80	49	137 726
0.09	23.50	4.10	54	10 709	10 608	56 900	1.90	15	166 074
0.04	23.00	5.10	60	10 838	5 867	20 810	0.00	9	68 898
8.33	18.20	6.60	98	23 357	19 387	15 147	0.20	0	29 352
3.90	11.50	6.00	92	12 180	16 476	20 960	4.40	8	32 903
25.48	20.50	5.00	97	169 940	169 216	257 123	1.70	3	389 747
78.16	26.10	7.00	54	3 120	1 701	2 376	2.60	18	7 595
14.27	18.00	5.90	98	1 157	1 625	2 735	0.00	5	4 899
0.41	18.50	4.70	99	306 580	203 490	888 085	2.60	5	1 040 554
0.63	30.20	4.40	80	1 109	1 883	3 723	0.40	5	9 416
N/A	N/A	N/A	100	N/A	N/A	118 805	N/A	N/A	N/A
113.93	22.30	6.40	69	1 011	2 009	3 361	2.30	17	5 192
N/A	N/A	N/A	100	N/A	N/A	N/A	N/A	N/A	N/A
0.41	12.30	5.00	73	11 396	5 525	9 100	0.00	5	31 181
0.02	11.80	1.00	84	N/A	N/A	411	0.00	23	227
0.07	N/A	N/A	100	N/A	N/A	10 230	N/A	N/A	N/A
0.99	12.50	N/A	80	N/A	N/A	4 750	0.10	4	8 720
3.38	18.50	4.00	59	57	776	N/A	0.10	2	0

Country	Capital city	Area (sq km)	Time at 12 noon (Australian EST)	Population ('000s) 1995	Male/ female life expectancy (years) 1995	Crude birth rate (per 1000 people) 1995	Infant mortality rate (per 1000 births) 1995	Persons per docto 1995
Liberia	Monrovia	111 369	2.00 a.m.	2 875	55/60	44	116	32 303
Libya	Tripoli	1 759 540	4.00 a.m.	4 873	61/66	46	66	924
Liechtenstein	Vaduz	157	3.00 a.m.	30	74/81	13	5	1 034
Lithuania	Vilnius	65 200	5.00 a.m.	3 820	66/76	15	17	278
Luxembourg	Luxembourg	2 586	3.00 a.m.	398	73/80	13	7	510
Macedonia (FYROM)	Skopje	25 416	3.00 a.m.	2 194	71/75	16	30	507
Madagascar	Antananarivo	587 041	5.00 a.m.	13 006	52/55	46	91	10 939
Malawi	Lilongwe	118 484	4.00 a.m.	9 832	40/41	51	142	37 527
Malaysia	Kuala Lumpur	329 749	10.00 a.m.	18 845	66/72	29	27	2 865
Maldives	Malé	298	7.00 a.m.	243	63/65	44	58	6 075
Mali	Bamako	1 240 000	2.00 a.m.	8 869	44/47	52	108	25 413
Malta	Valletta	316	3.00 a.m.	364	74/79	14	8	452
Marshall Islands	Majuro	179	2.00 p.m.	52	61/64	47	51	2 737
Mauritania	Nouakchott	1 030 700	2.00 a.m.	2 125	45/50	48	87	11 364
Mauritius	Port Louis	2 045	6.00 a.m.	1 107	66/74	20	19	1 016
Mexico	Mexico City	1 927 547	6.00–8.00 p.m.*	90 420	69/76	28	29	696
Micronesia	Palikir	627	12 noon	118	65/69	28	38	3 688
Moldova	Kishinev	33 700	5.00 a.m.	4 456	64/72	16	31	261
Mongolia	Ulan Bator	1 565 000	10.00 a.m.	2 367	64/68	33	45	375
Morocco	Rabat	446 550	2.00 a.m.	27 955	66/69	29	53	4 935
Mozambique	Maputo	801 590	4.00 a.m.	16 342	46/50	45	131	47 784
Myanmar (Burma)	Yangon	676 552	8.30 a.m.	43 456	58/62	29	66	3 549
Namibia	Windhoek	824 292	4.00 a.m.	1 541	59/64	44	64	4 756
Nepal	Kathmandu	140 797	8.00 a.m.	20 535	52/52	38	86	17 170
Netherlands	Amsterdam	40 844	3.00 a.m.	15 275	74/81	13	6	408
New Zealand	Wellington	268 676	2.00 p.m.	3 369	72/80	16	9	326
Nicaragua	Managua	130 000	8.00 p.m.*	3 987	61/66	36	55	1 876
Niger	Niamey	1 267 000	3.00 a.m.	8 337	43/46	57	113	52 106
Nigeria	Abuja	923 768	3.00 a.m.	95 060	54/56	44	77	5 940
North Korea	Pyongyang	220 277	11.00 a.m.	22 646	66/73	24	29	393
Norway	Oslo	324 219	3.00 a.m.	4 297	74/81	14	6	305
Oman	Muscat	212 459	6.00 a.m.	1 645	65/69	41	38	1 042
Pakistan	Islamabad	803 943	7.00 a.m.	125 214	57/58	43	104	2 413
Panama	Panama City	77 082	9.00 p.m.*	2 579	72/77	25	17	911
Papua New Guinea	Port Moresby	461 691	12 noon	4 101	55/57	34	65	11 360
Paraguay	Asuncion	406 752	10.00 p.m.*	5 071	71/75	33	26	1 695
Peru	Lima	1 285 216	9.00 p.m.*	23 210	63/67	26	56	1 062
Philippines	Manila	300 000	10.00 a.m.	68 464	63/68	28	52	1 195
Poland	Warsaw	312 677	3.00 a.m.	38 519	68/77	14	14	465
Portugal	Lisbon	92 082	3.00 a.m.	10 486	71/79	12	10	380
Qatar	Doha	11 000	5.00 a.m.	499	70/75	20	23	743
Romania	Bucharest	237 500	4.00 a.m.	23 172	68/74	14	21	554
Russia	Moscow	17 075 380	5.00 a.m.–2.00 p.m.	149 300	64/74	13	28	227
Rwanda	Kigali	26 338	4.00 a.m.	8 139	40/42	50	119	45 725
Samoa	Apia	2 842	3.00 p.m.*	200	65/70	33	39	4 545
San Marino	San Marino	61	3.00 a.m.	24	77/85	11	6	400

* previous day

Number of AIDS cases reported (per 100 000 people) 1993	Secondary students per teacher 1994	Education funding (percentage of GNP) 1994	Literacy rate (per cent) 1995	Total exports (millions of $US) 1993	Total imports (millions of $US) 1993	Electricity consumption (millions of kWh) 1992	Protected area (percentage of total area) 1990	Number of threatened mammals (species) 1990	CO_2 emission (thousands of metric tonnes) 1993
0.97	38.30	5.70	40	375	312	450	0.50	18	773
0.14	13.70	9.60	64	11 362	7 582	19 500	0.00	12	37 842
N/A	8.00	N/A	100	N/A	N/A	N/A	N/A	N/A	N/A
0.05	9.10	N/A	100	N/A	N/A	16 614	N/A	N/A	N/A
13.82	9.90	4.40	100	N/A	N/A	5 135	0.00	1	9 266
N/A	17.80	N/A	89	N/A	N/A	5 888	N/A	N/A	N/A
0.02	21.80	1.90	80	313	314	569	0.70	50	901
226.81	24.50	3.50	22	297	253	750	3.60	10	634
0.24	18.70	5.50	78	33 714	33 977	28 339	1.40	23	49 061
0.00	11.80	6.90	92	58	117	28	N/A	N/A	N/A
12.53	19.80	3.30	32	345	432	276	0.30	16	425
6.87	14.30	3.80	84	866	1 328	1 [illegible]10	N/A	N/A	1 674
3.85	19.60	N/A	93	N/A	N/A	N/A	N/A	N/A	N/A
1.69	17.20	4.90	34	448	349	143	0.70	14	3 023
0.99	21.00	3.50	61	1 207	1 477	813	0.90	3	1 000
12.20	16.00	4.10	87	27 121	38 184	124 250	1.90	25	319 702
1.69	17.10	N/A	90	N/A	N/A	N/A	N/A	N/A	N/A
N/A	N/A	N/A	100	N/A	N/A	20 161	N/A	N/A	N/A
0.04	21.10	7.00	90	N/A	N/A	3 650	1.50	9	10 303
0.43	17.00	7.40	50	4 210	6 282	9 834	0.30	9	22 120
3.29	26.80	1.20	33	N/A	N/A	815	0.00	10	1 205
0.03	19.30	1.90	81	331	621	2 400	0.10	23	5 009
20.18	28.40	1.60	38	N/A	N/A	N/A	5.10	11	0
0.06	31.40	2.90	26	218	667	854	2.80	22	934
15.25	10.90	6.50	99	122 639	110 330	83 406	3.80	2	124 990
10.33	15.30	7.90	99	9 283	8 452	29 556	4.30	1	26 176
0.78	43.00	2.60	57	236	718	1 233	0.10	8	2 180
5.96	24.70	3.10	28	311	369	356	0.50	15	1 008
0.19	22.40	1.70	51	13 585	4 932	9 855	0.70	25	79 263
0.00	28.80	3.70	99	N/A	N/A	53 500	0.20	5	151 488
6.59	11.80	7.90	99	34 172	25 217	108 133	6.00	3	46 009
1.64	14.30	3.70	41	5 488	2 519	5 548	0.10	6	10 259
0.02	18.20	3.40	35	5 351	8 050	47 334	1.80	15	60 773
15.04	20.00	5.70	88	4 043	4 981	3 046	6.90	13	2 730
1.10	21.90	4.70	52	1 319	1 341	1 790	0.00	5	2 250
1.01	17.50	1.00	90	1 392	1 354	2 980	1.20	14	1 722
2.65	21.30	3.50	85	3 276	2 885	14 828	1.70	29	21 174
0.12	31.70	3.00	90	8 186	12 206	22 484	0.80	12	40 960
0.31	15.80	4.90	98	15 837	12 248	132 078	2.90	4	440 929
9.60	39.20	4.90	85	16 251	24 003	29 963	2.00	6	40 912
6.21	9.40	3.40	76	N/A	N/A	4 716	N/A	N/A	13 308
8.95	19.60	3.10	98	4 125	5 345	63 959	1.00	2	212 193
0.06	N/A	7.90	100	29 500	25 800	1 068 200	N/A	N/A	N/A
104.23	16.20	4.20	50	103	228	186	5.00	11	381
N/A	22.00	5.90	97	9	73	47	N/A	N/A	N/A
4.17	9.00	N/A	96	N/A	N/A	N/A	N/A	N/A	N/A

Country	Capital city	Area (sq km)	Time at 12 noon (Australian EST)	Population ('000s) 1995	Male/ female life expectancy (years) 1995	Crude birth rate (per 1000 people) 1995	Infant mortality rate (per 1000 births) 1995	Persons per docto 1995
Sao Tome & Principe	Sao Tome	964	2.00 a.m.	133	61/65	35	65	3 325
Saudi Arabia	Riyadh	2 149 690	5.00 a.m.	17 615	66/69	39	55	776
Senegal	Dakar	196 192	2.00 a.m.	8 463	55/57	43	78	20 794
Serbia and Montenegro	Belgrade	153 597	3.00 a.m.	24 175	70/76	14	21	1 156
Seychelles	Victoria	280	6.00 a.m.	71	66/73	22	12	1 164
Sierra Leone	Freetown	71 740	2.00 a.m.	4 511	43/49	45	145	15 037
Singapore	Singapore City	581	10.00 a.m.	2 826	73/79	17	6	713
Slovakia	Bratislava	49 010	3.00 a.m.	5 376	68/77	15	11	334
Slovenia	Ljubljana	20 251	3.00 a.m.	1 968	70/78	12	8	483
Solomon Islands	Honiara	28 446	1.00 p.m.	373	68/73	39	29	12 032
Somalia	Mogadishu	637 657	5.00 a.m.	6 515	33/33	42	163	14 478
South Africa	Pretoria (Administrative)	1 221 037	4.00 a.m.	42 793	62/68	34	48	1 739
South Korea	Seoul	98 484	11.00 a.m.	44 614	67/74	16	23	1 048
Spain	Madrid	504 782	3.00 a.m.	39 207	74/81	11	7	256
Sri Lanka	Colombo	65 610	7.30 a.m.	17 838	69/74	19	23	6 022
St Kitts Nevis	Basseterre	306	10.00 p.m.*	40	63/69	24	21	1 429
St Lucia	Castries	616	10.00 p.m.*	144	67/72	24	19	2 400
St Vincent	Kingstown	388	10.00 p.m.*	115	70/73	21	18	2 875
Sudan	Khartoum	2 505 813	4.00 a.m.	28 730	53/55	43	81	11 946
Surinam	Paramaribo	163 265	10.30 p.m.*	416	67/72	26	32	1 391
Swaziland	Mbabane	17 363	4.00 a.m.	907	52/60	43	96	11 338
Sweden	Stockholm	449 964	3.00 a.m.	8 730	75/81	14	6	400
Switzerland	Bern	41 288	3.00 a.m.	6 987	75/82	12	7	318
Syria	Damascus	185 180	4.00 a.m.	14 339	65/67	44	44	1 703
Tadzhikistan	Dushanbe	143 000	8.00 a.m.	5 836	66/71	36	64	414
Taiwan	Taipei	35 760	10.00 a.m.	21 092	72/78	16	6	844
Tanzania	Dar-es-Salaam	945 087	5.00 a.m.	27 286	42/46	46	110	25 621
Thailand	Bangkok	514 000	9.00 a.m.	58 722	65/72	20	39	4 690
Togo	Lomé	56 785	2.00 a.m.	4 105	54/59	48	91	14 766
Trinidad & Tobago	Port-of-Spain	5 130	10.00 p.m.*	1 314	68/73	20	17	1 638
Tunisia	Tunis	163 610	3.00 a.m.	8 571	71/75	24	36	1 987
Turkey	Ankara	780 576	4.00 a.m.	60 898	68/73	27	52	1 203
Turkmenistan	Ashkhabad	488 100	7.00 a.m.	3 915	61/69	31	71	284
Uganda	Kampala	236 036	5.00 a.m.	19 344	38/39	50	112	27 634
Ukraine	Kiev	603 700	5.00 a.m.	51 821	65/75	12	21	226
United Arab Emirates	Abu Dhabi	83 600	6.00 a.m.	2 657	72/70	28	23	890
United Kingdom	London	244 046	2.00 a.m.	57 970	74/79	14	7	629
United States of America	Washington DC	9 372 614	6.00–9.00 p.m.*	258 104	72/79	15	8	420
Uruguay	Montevideo	176 215	11.00 p.m.*	3 175	71/77	18	18	350
Uzbekistan	Tashkent	449 600	7.00 a.m.	22 128	65/72	31	54	295
Vanuatu	Port Vila	14 763	1.00 p.m.	166	57/61	33	70	8 300
Venezuela	Caracas	912 050	10.00 p.m.*	20 118	70/76	26	29	617
Vietnam	Hanoi	329 556	9.00 a.m.	71 788	63/67	28	46	3 081
Yemen	San'a and Aden	527 968	5.00 a.m.	10 742	50/52	51	116	3 967
Zambia	Lusaka	752 614	4.00 a.m.	8 926	45/46	47	84	11 185
Zimbabwe	Harare	390 580	4.00 a.m.	10 838	41/44	38	75	8 719

* previous day

Number of AIDS cases eported (per 0 000 people) 1993	Secondary students per teacher 1994	Education funding (percentage of GNP) 1994	Literacy rate (per cent) 1995	Total exports (millions of $US) 1993	Total imports (millions of $US) 1993	Electricity consumption (millions of kWh) 1992	Protected area (percentage of total area) 1990	Number of threatened mammals (species) 1990	CO_2 emission (thousands of metric tonnes) 1993
8.27	23.00	4.30	57	N/A	N/A	15	N/A	N/A	N/A
0.26	13.70	5.80	62	44 283	21 490	47 710	1.10	9	173 776
7.66	35.60	3.70	38	912	1 176	756	4.50	11	3 151
1.29	19.10	N/A	90	13 799	13 287	39 453	1.30	3	132 901
0.00	12.10	9.10	58	18	147	102	N/A	N/A	N/A
0.89	17.10	1.40	21	140	160	230	0.60	13	671
1.52	19.90	3.40	88	56 819	60 948	16 597	1.70	N/A	35 860
N/A	15.60	N/A	99	N/A	N/A	N/A	N/A	N/A	126 200
N/A	14.20	N/A	99	N/A	N/A	12 377	N/A	N/A	N/A
0.00	17.50	4.70	54	83	128	30	0.00	2	161
0.20	18.20	0.40	24	68	346	110	0.00	17	960
3.08	20.70	3.80	76	23 715	17 449	163 963	2.10	25	278 468
0.02	23.30	3.60	96	63 123	65 127	132 228	2.40	6	221 104
38.24	18.40	4.80	95	55 221	86 958	155 025	2.80	6	203 227
0.11	19.70	2.70	88	1 853	2 326	3 377	4.80	7	4 034
92.50	16.20	3.00	98	25	103	40	N/A	N/A	N/A
33.33	20.00	7.20	67	105	267	105	N/A	N/A	N/A
35.65	18.30	6.90	96	75	115	51	N/A	N/A	N/A
2.26	22.30	4.00	27	327	649	1 329	1.30	17	3 338
29.33	14.40	9.70	95	466	374	1 400	1.90	11	1 440
21.72	21.20	7.20	55	557	632	N/A	0.90	0	443
8.51	10.30	7.80	99	54 538	48 560	146 435	1.60	1	58 888
37.08	N/A	4.80	99	77 488	83 865	55 006	1.10	2	39 326
0.13	17.70	4.40	64	4 221	2 062	12 179	0.0	4	28 154
N/A	N/A	N/A	100	N/A	N/A	N/A	N/A	N/A	N/A
N/A	21.40	3.60	86	81 500	72 000	80 977	N/A	N/A	N/A
126.46	17.90	5.80	46	408	1 186	901	5.10	30	2 099
1.55	16.20	3.80	93	22 811	29 561	53 041	4.00	26	77 680
31.13	25.20	5.20	43	396	502	350	N/A	N/A	627
82.57	20.00	4.10	95	1 935	948	3 640	1.20	1	18 580
1.33	23.10	6.00	65	4 021	4 895	5 513	0.10	6	13 923
0.15	23.90	1.80	81	13 672	20 998	60 591	0.10	5	126 078
N/A	N/A	N/A	100	N/A	N/A	8 337	N/A	N/A	N/A
178.92	17.10	3.40	48	178	491	671	3.00	16	879
N/A	N/A	N/A	100	45 700	53 800	248 200	N/A	N/A	N/A
0.30	12.60	1.90	68	N/A	N/A	13 790	0.00	4	50 944
11.23	12.50	4.70	99	183 091	201 081	338 540	7.70	3	538 451
93.82	11.60	5.30	98	416 510	490 110	3 101 357	4.30	27	4 869 005
9.76	18.60	3.30	96	1 605	1 544	5 226	0.10	5	4 749
N/A	N/A	N/A	100	N/A	N/A	54 800	N/A	N/A	N/A
0.00	30.50	4.40	53	14	80	26	N/A	N/A	N/A
10.80	9.10	4.10	88	14 892	10 101	57 150	9.00	19	95 887
0.00	12.10	3.00	88	N/A	N/A	8 550	1.10	28	18 170
0.00	31.50	6.10	38	720	1 837	1 750	0.00	6	3 495
73.44	25.60	2.90	73	1 340	774	6 925	3.40	10	2 612
115.46	27.70	8.20	67	1 665	1 164	10 460	2.90	9	16 059

Index

Entry / ***Page*** / ***Ref.*** / *Lat.* / *Long.*

A

Aanekoski **115 J5** 62.38N 25.40E
Aba **124 D2** 5.06N 7.21E
Abadan **108 C4** 30.20N 48.15E
Abadla **124 C5** 31.01N 2.45W
Abaiang, island **84 M6**
Abakan **118 J4** 53.43N 91.25E
Abancay **142 C2** 13.37S 72.52W
Abarqu **108 D4** 31.09N 53.18E
Abashiri **104 L9** 44.02N 144.17E
Abau **89 E1** 10.11S 148.42E
Abaya, Lake **125 G2**
Abbeville **116 D5** 50.06N 1.51E
Abbieglassie, homestead **42 F2** 27.15S 147.35E
Abd al Kuri, island **108 D2**
Abdanan **108 C4** 32.57N 47.27E
Abeche **124 F3** 13.49N 20.49E
Abemama, island **84 M6**
Abeokuta **124 D2** 7.10N 3.26E
Aberdeen, U.K. **114 D4** 57.10N 2.04W
Aberdeen, U.S.A. **134 E3** 45.28N 98.30W
Abha **108 C2** 18.14N 42.31E
Abi Addi **125 G3** 13.39N 39.03E
Abidjan **124 C2** 5.19N 4.01W
Abilene **134 E2** 32.27N 99.45W
Abomey **124 D2** 7.14N 2.00E
Abrantes **116 B2** 39.28N 8.12W
Abri **125 G4** 20.46N 30.21E
Abu Dhabi **108 D3** 24.28N 54.25E
Abu Hamed **125 G3** 19.32N 33.20E
Abuja **124 D2** 9.07N 7.29E
Abu Kemal **108 C4** 34.29N 40.56E
Abu Tig **125 G4** 27.06N 31.17E
Acaponeta **136 B3** 22.30N 105.25W
Acapulco **136 C2** 16.51N 99.56W
Acarigua **142 D4** 9.35N 69.12W
Accra **124 C2** 5.33N 0.15W
Aceh, province **100 A3**
Achegour **124 E3** 19.10N 11.54E
Acklins Island **137 E3**
Aconcagua, mountain **144 C3** 32.40S 70.02W
Acqui **116 E3** 44.41N 8.28E
Acre **108 B4** 32.55N 35.04E
Acri **117 G2** 39.29N 16.23E
Ada **134 E2** 34.47N 96.41W
Adamawa Highlands **124 E2**
Adana **108 B4** 37.00N 35.19E
Adarama **125 G3** 17.10N 34.52E
Adare, Cape **92 I3** 71.30S 170.24E
Adaut **88 B2** 8.10S 131.09E
Addis Ababa **125 G2** 9.03N 38.42E
Adelaide **66 G3** 34.56S 138.36E
Adelaide Island **92 O3**
Adelaide River **80 C8**
Adelaide River **80 C8** 13.14S 131.08E
Adelbert Range **89 E3**
Adelie Land **94 G2**
Aden **108 C2** 12.47N 45.03E
Aden, Gulf of **108 C2**
Adi, island **88 B3**
Adieu, Cape **66 D4** 32.00S 132.09E
Adirondack Mountains **135 G3**
Adi Ugri **125 G3** 14.47N 38.50E
Adiyaman **108 B4** 37.46N 38.15E
Adjud **117 I4** 46.07N 27.10E
Admiralty Gulf **72 D6**
Admiralty Islands **89 E3**
Ado **124 D2** 6.38N 3.00E
Adoni **106 C2** 15.38N 77.16E
Adrar **124 C4** 27.51N 0.19W
Adriatic Sea **117 G3**
Aegean Sea **117 H2**
Afghanistan, republic **106 B4**
Aflou **124 D5** 34.08N 2.03E
Afognak Island **132 C3**
Afyon **108 B4** 38.46N 30.32E
Agadez **124 D3** 17.00N 7.56E

Entry / ***Page*** / ***Ref.*** / *Lat.* / *Long.*

Agadir **124 C5** 30.30N 9.40W
Agats **88 C2** 5.36S 138.05E
Agboville **124 C2** 5.55N 4.15W
Agdam **108 C4** 39.59N 48.57E
Agen **116 D3** 44.12N 0.38E
Aginskoye **119 L4** 51.10N 114.32E
Agordat **108 B2** 15.35N 37.55E
Agra **106 C3** 27.09N 78.00E
Agrihan, island **101 F4**
Agrinion **117 H2** 38.38N 21.25E
Agua Prieta **134 D2** 31.20N 109.32W
Aguascalientes **136 B3** 21.51N 102.18W
Aguilas **116 C2** 37.25N 1.35W
Agulhas, Cape **126 C1** 34.50S 20.00E
Agulhas Negras, mountain **143 F1** 22.20S 44.43W
Ahaggar, mountains **124 D4**
Ahmadabad **106 C3** 23.03N 72.40E
Ahmadnagar **106 C2** 19.08N 74.48E
Ahvaz **108 C4** 31.17N 48.43E
Aigion **117 H2** 38.15N 22.05E
Ailinglapalap, island **84 L6**
Ain Beida **116 E2** 35.44N 7.22E
Ain Dalla **125 F4** 27.20N 27.23E
Ain M'Lila **116 E2** 36.02N 6.30E
Ain Oussera **116 D2** 35.24N 2.54E
Ain Salah **124 D4** 27.12N 2.29E
Ain Temouchent **116 C2** 35.18N 1.09W
Air, mountains **124 D3**
Airbangis **100 A3** 0.09N 99.25E
Aisega **89 E2** 5.44S 148.22E
Aitape **88 D3** 3.10S 142.17E
Aitutaki, island **84 P4**
Aix-en-Provence **116 E3** 43.31N 5.27E
Aiyura **89 E2** 6.20S 145.55E
Aizuwakamatsu **104 I5** 37.30N 139.50E
Ajaccio **116 E3** 41.55N 8.43E
Ajedabia **124 F5** 30.48N 20.15E
Ajmer **106 C3** 26.29N 74.40E
Ajo **134 C2** 32.24N 112.51W
Aketi **126 C5** 2.42N 23.51E
Akhaltsikhe **108 C5** 41.37N 42.59E
Akhisar **108 A4** 38.54N 27.50E
Akhmim **108 B3** 26.35N 31.48E
Akimiski Island **133 J3**
Akita **104 J6** 39.44N 140.05E
Akjoujt **124 B3** 19.44N 14.26W
Aklavik **132 E4** 68.15N 135.02W
Akola **106 C3** 20.40N 77.05E
Akranes **133 P4** 64.19N 22.05W
Akron **133 J2** 41.04N 81.31W
Aksehir **108 B4** 38.22N 31.24E
Aksha **119 L4** 50.18N 113.15E
Aksu **102 C4** 41.10N 80.20E
Aktau **118 F3** 43.37N 51.11E
Aktyubinsk **118 F4** 50.16N 57.13E
Ala, Mount **108 B4** 37.42N 35.09E
Alabama River **135 F2**
Alabama, state **135 F2**
Alagoinhas **143 G2** 12.09S 38.21W
Al Ahmadi **108 C3** 29.05N 48.04E
Alakol, Lake **118 I3**
Alamagan, island **101 F4**
Alamogordo **134 D2** 32.55N 106.00W
Alamosa **134 D2** 37.25N 105.50W
Alaska, Gulf of **132 D3**
Alaska Peninsula **132 C3**
Alaska Range **132 C4**
Alaska, state **132 C4**
Alawoona **66 H3** 34.44S 140.31E
Albacete **116 C2** 39.00N 1.52W
Albacutya, Lake **54 A3**
Alba-Iulia **117 H4** 46.04N 23.33E
Albania, socialist people's republic **117 G3**
Albany, Australia **72 C2** 35.00S 117.52E
Albany, Georgia **135 F2** 31.37N 84.10W
Albany, New York **135 G3** 42.40N 73.49W
Albany, Oregon **134 B3** 44.38N 123.07W
Albany River **132 J3**

Entry / ***Page*** / ***Ref.*** / *Lat.* / *Long.*

Albatross Bay **42 C9**
Albatross Island **60 B4**
Albenga **116 E3** 44.03N 8.13E
Alberga **66 E7** 27.10S 135.22E
Alberga Creek **66 D7**
Alberrie Creek **66 F6** 29.38S 137.32E
Alberta, province **132 G3**
Albert Edward, Mount **89 E2** 8.23S 147.24E
Albert, Lake **126 D5**
Albi **116 D3** 43.56N 2.08E
Albina, Point **126 B3** 15.52S 11.44E
Alborg **114 F4** 57.03N 9.56E
Albuquerque **134 D2** 35.05N 106.38W
Albury **48 D2** 36.03S 146.53E
Alcala de Henares **116 C3** 40.28N 3.22W
Alcamo **116 F2** 37.58N 12.58E
Alcazar **116 C2** 39.24N 3.12W
Alcoy **116 C2** 38.42N 0.29W
Aldabra Islands **122 H4**
Aldan **119 M4** 58.44N 125.22E
Aldan River **119 M4**
Aldermen Islands, The **90 G6**
Aleksandrov **115 L4** 56.23N 38.45E
Aleksin **115 L3** 54.31N 37.07E
Alencon **116 D4** 48.25N 0.05E
Alenquer **143 E3** 1.58S 54.45W
Aleppo **108 B4** 36.14N 37.10E
Ales **116 D3** 44.08N 4.05E
Alessandria **116 E3** 44.55N 8.37E
Alesund **114 F5** 62.28N 6.11E
Aleutian Islands **132 B3**
Aleutian Trench **84 N11**
Alexander Archipelago **132 E3**
Alexander Island **92 O3**
Alexander Bay **126 B2** 28.40S 16.30E
Alexander City **135 F2** 32.57N 86.00W
Alexandra, Australia **54 C2** 37.12S 145.43E
Alexandra, New Zealand **90 C2** 45.15S 169.24E
Alexandria, Egypt **125 F5** 31.13N 29.55E
Alexandria, Rumania **117 I3** 43.59N 25.19E
Alexandria, U.S.A. **134 E2** 31.19N 92.29W
Alexandria, homestead **80 F5** 19.03S 136.42E
Alexandrina, Lake **66 G3**
Alexandroupolis **117 I3** 40.51N 25.53E
Aleysk **102 C5** 52.32N 82.17E
Al Faiyum **125 G4** 29.19N 30.50E
Al Fallujah **108 C4** 33.21N 43.46E
Al Fashir **125 F3** 13.37N 25.22E
Alga **118 F3** 49.49N 57.16E
Algeciras **116 B2** 36.08N 5.27W
Algeria, democratic republic **124 D4**
Alghero **116 E3** 40.34N 8.19E
Algiers **124 D5** 36.50N 3.00E
Al Hadhr **108 C4** 35.35N 42.44E
Al Hayy **108 C4** 32.11N 46.03E
Alicante **116 C2** 38.21N 0.29W
Alice Springs **80 D3** 23.42S 133.52E
Ali Curung **80 E4** 21.01S 134.25E
Aligarh **106 C3** 27.54N 78.04E
Alitus **115 I3** 54.24N 24.03E
Aliwal North **126 C1** 30.42S 26.43E
Al Jahrah **108 C3** 29.22N 47.40E
Al Khasab **108 D3** 26.14N 56.15E
Alkmaar **114 E3** 52.38N 4.44E
Al Kut **108 C4** 32.30N 45.51E
Allahabad **106 D3** 25.27N 81.50E
Allentown **135 G3** 40.37N 75.30W
Alleppey **106 C1** 9.30N 76.22E
Alliance **134 D3** 42.06N 102.52W
Alma **133 K2** 48.33N 71.39W
Alma-Ata **118 H3** 43.19N 76.55E
Almada **116 B2** 38.40N 9.09W
Almansa **116 C2** 38.52N 1.06W
Almeria **116 C2** 36.50N 2.26W
Al Minya **125 G4** 28.06N 30.45E
Almirante **136 D1** 9.20N 82.22W
Alofi **84 O4** 19.10S 169.55W

Entry / ***Page*** / ***Ref.*** / *Lat.* / *Long.*

Alonnah **60 E1** 43.20S 147.15E
Alor, island **100 D2**
Alora **116 C2** 36.50N 4.43W
Alor Setar **100 B3** 6.06N 100.23E
Alotau **89 F1** 10.18S 150.27E
Aloysius, Mount **72 E3** 26.02S 128.36E
Alpine **134 D2** 30.22N 103.40W
Alps, The, mountains **116 E4**
Al Qunfidha **108 C2** 19.09N 41.07E
Al Qurna **108 C4** 31.00N 47.26E
Alroy Downs, homestead **80 F5** 19.18S 136.04E
Alta **118 C6** 70.00N 23.15E
Altai Mountains **102 C4**
Altamira **143 E3** 3.13S 52.15W
Altamura **117 G3** 40.49N 16.34E
Altay, China **102 C4** 47.48N 88.07E
Altay, Mongolia **102 D4** 46.18N 96.16E
Alto Nevado, Mount **144 B2** 44.31S 72.15W
Altoona **135 G3** 40.32N 78.23W
Altus **134 E2** 34.39N 99.21W
Al Ubayyid **125 G3** 13.11N 30.13E
Aluksne **115 J4** 57.22N 27.00E
Alula **125 I3** 11.59N 50.48E
Alvsbyn **118 C5** 65.41N 21.00E
Alwar **106 C3** 27.32N 76.35E
Alyangula **80 F8** 13.51S 136.25E
Amadeus, Lake **80 C2**
Amadjuak Lake **133 K4**
Amagasaki **104 G4** 34.42N 135.23E
Amakusa Islands **104 D3**
Amalias **117 H2** 37.48N 21.21E
Amami Island **103 G2**
Amanab **88 D3** 3.35S 141.10E
Amapa, northern Brazil **143 E4** 2.00N 50.50W
Amapa, southern Brazil **144 D4** 29.30S 51.40W
Amara **108 C4** 31.51N 47.10E
Amarillo **134 D2** 35.14N 101.50W
Amata **66 C7** 26.09S 131.09E
Amazon, Mouths of the **143 F4**
Amazon River **142 D3**
Ambala **102 B3** 30.19N 76.49E
Ambato **142 C3** 1.18S 78.39W
Ambon **88 A3** 3.41S 128.10E
Ambovombe **126 E2** 25.10S 46.05E
Ambre, Cape **122 H3** 11.58S 49.14E
Ambriz **126 B4** 7.50S 13.09E
Ambunti **88 D3** 4.15S 142.50E
American Samoa, islands **84 O4**
Americus **135 F2** 32.04N 84.14W
Amery Ice Shelf **92 D3**
Amga River **119 M4**
Amguid **124 D4** 26.28N 5.21E
Amiens **116 D4** 49.54N 2.18E
Amistad Reservoir **136 B3**
Amman **108 B4** 31.57N 35.56E
Ammaroo, homestead **80 E4** 21.45S 135.14E
Amol **108 D4** 36.23N 52.20E
Amravati **106 C3** 20.58N 77.50E
Amritsar **106 C4** 31.35N 74.56E
Amroha **102 B2** 28.54N 78.29E
Amsterdam **114 E3** 52.21N 4.54E
Amu River **118 F3**
Amundsen Gulf **132 F5**
Amundsen Sea **92 M3**
Amundsen-Scott, antarctic base **92 N4** 90.00S
Amuntai **100 C2** 2.24S 115.14E
Amur River **119 M3**
Ana **108 C4** 34.29N 41.57E
Anaa, island **85 Q4**
Anaconda **134 C3** 46.07N 112.55W
Anacortes **134 B3** 48.30N 122.37W
Anadyr **119 R5** 64.41N 177.32E
Anadyr, Gulf of **119 S5**
Anadyr River **119 R5**

*Entry / **Page** / **Ref.** / Lat. / Long.*

Anaheim **134 C2** 33.50N 117.56W
Anaiza **108 C3** 26.06N 43.56E
Anambas Islands **100 B3**
Anantapur **106 C2** 14.42N 77.05E
Anapolis **143 F2** 16.19S 48.58W
Anarak **108 D4** 33.20N 53.44E
Anardara **106 B4** 32.45N 61.38E
Anchorage **132 C4** 61.10N 150.00W
Ancohuma, mountain **142 D2** 15.54S 68.30W
Ancona **116 F3** 43.37N 13.31E
Ancud **144 B2** 41.53S 73.50W
Anda **103 G4** 46.25N 125.20E
Andado, homestead **80 E2** 25.25S 135.17E
Andalusia **135 F2** 31.20N 86.30W
Andaman Islands **107 E2**
Andaman Sea **107 E2**
Andamooka **66 F5** 30.27S 137.10E
Anderson **135 F3** 40.05N 85.41W
Anderson Bay **60 E4**
Andes Mountains **140 D6**
Andizhan **118 H3** 40.48N 72.23E
Andong **104 D5** 36.37N 128.44E
Andorra, principality **116 D3**
Andorra-la-Vella **116 D3** 42.30N 1.30E
Andrews **134 D2** 32.19N 102.34W
Andreyevka **118 I3** 45.50N 80.34E
Andria **117 G3** 41.14N 16.18E
Andropov Reservoir **118 D4**
Andros, island, Bahamas **136 E3**
Andros, island, Greece **128 L3**
Aneto, Pico de, mountain **116 D3** 42.37N 0.40E
Angara River **118 J4**
Angarsk **118 K4** 52.31N 103.55E
Angaston **66 G3** 34.31S 139.03E
Angel Falls **142 D4** 5.56N 62.30W
Angelholm **114 G4** 56.15N 12.50E
Angemuk, mountain **88 C3** 3.29S 138.36E
Angepena, homestead **66 G5** 30.34S 138.51E
Angers **116 C4** 47.29N 0.32W
Angkor **100 B4** 13.26N 103.50E
Angoche **126 D3** 16.10S 39.57E
Angol **144 B3** 37.47S 72.45W
Angola, people's republic **126 B3**
Angoram **89 D3** 4.04S 144.04E
Angostura Reservoir **136 C2**
Angouleme **116 C4** 45.40N 0.10E
Angren **118 H3** 41.01N 70.10E
Anguilla, island **137 F2**
Angurugu **80 F8** 13.59S 136.27E
Anina **117 H4** 45.05N 21.51E
Aniva Bay **104 K10**
Aniva, Cape **104 K10** 46.03N 143.30E
Anju **104 B6** 39.36N 125.42E
Ankara **108 B4** 39.55N 32.50E
Ankavandra **126 E3** 18.43S 45.20E
Anklam **114 G3** 53.52N 13.42E
Ann, Cape **92 C3** 66.13S 51.10E
Annaba **124 D5** 36.55N 7.47E
Anna Creek **66 F6** 28.54S 136.10E
An Nafud, desert **108 C3**
An Najaf **108 C4** 31.59N 44.19E
Annapurna, mountain **106 D3** 28.34N 83.50E
An Nasiriya **108 C4** 31.04N 46.17E
An Nawfaliyah **124 E5** 30.47N 17.50E
Annecy **116 E4** 45.54N 6.07E
An Nhon **100 B4** 13.53N 109.07E
Anningie, homestead **80 D4** 21.51S 133.07E
Anniston **135 F2** 33.38N 85.50W
Annitowa, homestead **80 F4** 21.13S 136.28E
Anqing **103 F3** 30.46N 117.00E
Anser Group **54 D1**
Anshan **103 G4** 41.05N 122.58E
Anshun **102 E2** 26.15N 105.51E
Anson Bay **80 C8**
Antahan, island **101 F4**
Antakya **108 B4** 36.12N 36.10E
Antalya **108 B4** 36.53N 30.42E
Antalya, Gulf of **108 B4**
Antananarivo **120 H3** 18.52S 47.30E

*Entry / **Page** / **Ref.** / Lat. / Long.*

Antarctic Peninsula **94 O2**
Antarctica, continent **146 F1**
Antequera **116 C2** 37.01N 4.34W
Anthony Lagoon, homestead **80 E6** 17.59S 135.33E
Anti Atlas, mountains **124 C4**
Anticosti Island **133 L2**
Antigo **135 F3** 45.10N 89.10W
Antigua, island **137 F2**
Antigua and Barbuda, independent state **137 F2**
Antioquia **137 E1** 6.36N 75.53W
Antipodes Islands **92 J2**
Antofagasta **144 B4** 23.40S 70.23W
Antwerp **114 E3** 51.13N 4.25E
Anvers Island **92 O3**
Anxious Bay **66 E4**
Anyang **102 F3** 36.04N 114.20E
Anzio **116 F3** 41.27N 12.38E
Aola **89 H2** 9.31S 160.30E
Aomori **104 J7** 40.50N 140.43E
Aouk River **124 F2**
Aoulef **124 D4** 27.00N 1.04E
Apalachee Bay **135 F1**
Aparri **100 D4** 18.22N 121.40E
Apennines, mountains **116 F3**
Apia **84 N4** 13.49S 171.45W
Apio **89 H2** 9.38S 161.23E
Apo, Mount **100 D3** 6.58N 125.17E
Apollo Bay **54 B1** 38.46S 143.40E
Apolo **142 D2** 14.48S 68.31W
Appalachian Mountains **135 F2**
Appleton **135 F3** 44.17N 88.24W
April River **88 D3**
Apsley **60 E2** 42.26S 147.09E
Apure River **142 C4**
Aqaba **108 B3** 29.32N 35.00E
Aquidauana **143 E1** 20.28S 55.48W
Arabian Desert **125 G4**
Arabian Peninsula **108 C3**
Arabian Sea **106 B3**
Aracaju **143 G2** 10.54S 37.07W
Aracati **143 G3** 4.32S 37.45W
Aracatuba **143 E1** 21.12S 50.24W
Arad **117 H4** 46.10N 21.19E
Arafura Sea **88 B2**
Araguacema **143 F3** 8.48S 49.31W
Araguaia River **143 E2**
Araguari **143 F2** 18.38S 48.13W
Araguatins **143 F3** 5.38S 48.07W
Arak, Algeria **124 D4** 25.20N 3.46E
Arak, Iran **108 C4** 34.05N 49.42E
Arakan Range **107 E3**
Araks River **108 C4**
Aral Sea **118 F3**
Aralsk **118 G3** 46.56N 61.43E
Aramac **42 E4** 22.58S 145.14E
Aranda de Duero **116 C3** 41.40N 3.41W
Arandai **88 B3** 2.10S 133.00E
Aranos **126 B2** 24.08S 19.07E
Araouane **124 C3** 18.53N 3.31W
Arapiraca **143 G3** 9.45S 36.39W
Araraquara **143 F1** 21.46S 48.08W
Ararat **54 B2** 37.17S 142.56E
Ararat, Mount **108 C4** 39.44N 44.15E
Arauca **137 E1** 7.04N 70.41W
Arauco **144 B3** 37.15S 73.22W
Aravalli Range **106 C3**
Arawa **89 G2** 6.14S 155.34E
Arbaikhere **102 E4** 46.15N 102.46E
Arbroath **114 D4** 56.34N 2.35W
Arcachon **116 C3** 44.40N 1.11W
Arcadia **135 F1** 27.12N 81.52W
Arcata **134 B3** 40.54N 124.05W
Arc Dome, mountain **134 C2** 38.50N 117.20W
Archer Bay **42 C9**
Archer River **40 E4**
Arctic Ocean **118 F7**
Arctowski, antarctic base **92 P3** 62.09S 58.28W
Ardabil **108 C4** 38.15N 48.18E
Ardlethan **48 D2** 34.22S 146.54E
Arena, Point **134 B2** 38.58N 123.44W
Arendal **114 F4** 58.27N 8.49E

*Entry / **Page** / **Ref.** / Lat. / Long.*

Arequipa **142 C2** 16.25S 71.32W
Areyonga **80 D2** 24.05S 132.16E
Arfak, Mount **88 B3** 1.14S 134.01E
Argentan **114 E2** 48.45N 0.01W
Argentina, federal republic **144 C3**
Argentino, Lake **144 B1**
Argos **117 H2** 37.38N 22.43E
Argyle, Lake **72 E5**
Arhus **114 G4** 56.10N 10.13E
Arica **142 C2** 18.30S 70.20W
Aripuana River **142 E3**
Arizona, state **134 C2**
Arkadelphia **134 E2** 34.07N 93.06W
Arkansas River **134 D2**
Arkansas, state **134 E2**
Arkansas City **134 E2** 37.03N 97.02W
Arkhangelsk **118 E5** 64.32N 40.40E
Arkticheskiy, Cape **118 J7** 83.30N 95.59E
Arles **116 D3** 43.41N 4.38E
Arlon **114 F2** 49.41N 5.49E
Armagh **114 C3** 54 21N 6.39W
Armavir **118 E3** 44.59N 41.10E
Armenia **142 C4** 4.32N 75.40W
Armenia, republic **108 C5**
Armidale **48 F4** 30.32S 151.40E
Armstrong **132 J3** 50.20N 89.02W
Armu River **104 G9**
Arnhem, Cape **80 F8** 12.21S 136.59E
Arnhem Land **80 E8**
Arno Bay **66 F4** 33.55S 136.34E
Arorae, island **84 M5**
Arrah **102 C2** 25.34N 84.40E
Arrecife **124 B4** 28.57N 13.33W
Arta **117 H2** 39.09N 20.59E
Artesia **134 D2** 32.51N 104.24W
Arthur River **60 B3**
Arthurs Lake **60 E3**
Arthur's Pass **90 D3** 42.55S 171.34E
Arthur's Town **137 E3** 24.40N 75.41W
Artigas **144 D3** 30.25S 56.28W
Artvin **108 C5** 41.12N 41.48E
Aru Islands **88 B2**
Arua **126 D5** 3.02N 30.56E
Aruba, island **137 C5**
Arumpo, homestead **48 B3** 33.48S 142.53E
Arusha **126 D4** 3.21S 36.40E
Aruwimi River **126 C5**
Arviat **132 I4** 61.10N 94.15W
Arvika **114 G4** 59.41N 12.38E
Arzamas **118 E4** 55.24N 43.48E
Asahi Peak **104 K8** 43.42N 142.54E
Asahikawa **104 K8** 43.46N 142.23E
Asan Bay **104 C5**
Asansol **102 C2** 23.40N 86.59E
Asaralta, mountain **102 E4** 48.30N 107.29E
Ascension, island **122 B4**
Aschersleben **114 G3** 51.46N 11.28E
Aseb **125 H3** 13.01N 42.47E
Ashburton **90 D3** 43.54S 171.45E
Ashburton River **72 C4**
Asheville **135 F2** 35.35N 82.35W
Ashibetsu **104 K8** 43.29N 142.14E
Ashizuri, Cape **104 F3** 32.42N 133.00E
Ashkhabad **118 F2** 37.58N 58.24E
Ashland **134 B3** 42.12N 122.42W
Asilah **116 B2** 35.32N 6.04W
Asimiro **100 D3** 2.00N 127.46E
Asir, Cape **125 I3** 11.50N 51.16E
Asmara **125 G3** 15.20N 38.58E
Aspiring, Mount **90 C2** 44.23S 168.44E
As Salman **108 C4** 30.30N 44.32E
Assen **114 F3** 53.00N 6.34E
Assens **114 F4** 55.16N 9.54E
Astana **118 H4** 51.10N 71.28E
Asti **116 E3** 44.54N 8.13E
Astorga **116 B3** 42.27N 6.04W
Astrakhan **118 E3** 46.22N 48.04E
Astrolabe Bay **89 E2**
Asuncion **144 D4** 25.15S 57.40W
Asuncion, island **101 F4**
Aswan **125 G4** 24.05N 32.56E
Aswan High Dam **125 G4** 23.54N 32.52E
Asyut **125 G4** 27.14N 31.07E
Atacama Desert **144 C4**
Atakpame **124 D2** 7.34N 1.14E

*Entry / **Page** / **Ref.** / Lat. / Long.*

Atar **124 B4** 20.32N 13.08W
Atascadero **134 B2** 35.29N 120.40W
Atbara **125 G3** 17.42N 34.00E
Atbara River **125 G3**
Athabasca, Lake **132 H3**
Athabasca River **132 G3**
Athens, Greece **117 H2** 38.00N 23.44E
Athens, U.S.A. **135 F2** 35.27N 84.38W
Atherton **42 E7** 17.16S 145.29E
Ati **124 E3** 13.11N 18.20E
Atlanta **135 F2** 33.45N 84.23W
Atlantic Ocean **122 C3**
Atlas Mountains **124 C5**
Attawapiskat **133 J3** 53.00N 82.30W
Atuel River **144 C3**
Aua, island **88 D3**
Auburn **135 G3** 44.04N 70.26W
Auburn, homestead **42 H3** 25.57S 150.37E
Auburn Range **42 G2**
Aucanquilcha, mountain **142 D1** 21.13S 68.26W
Auch **116 D3** 43.40N 0.36E
Auckland **90 F6** 36.52S 174.45E
Auckland Islands **92 I2**
Aue **114 G3** 50.35N 12.42E
Augathella **42 F3** 25.48S 146.35E
Augsburg **114 G2** 48.21N 10.54E
Augusta, Australia **72 C2** 34.19S 115.09E
Augusta, Georgia **135 F2** 33.29N 82.00W
Augusta, Italy **117 G2** 37.14N 15.14E
Augusta, Maine **135 H3** 44.17N 69.48W
Augusta, Lake **60 D3**
Augustus Island **72 D5**
Augustus, Mount **72 C4** 24.20S 116.49E
Augustus Downs, homestead **42 B6** 18.33S 139.52E
Auld, Lake **72 D4**
Auponhia **100 D2** 1.56S 125.30E
Aurangabad **106 C2** 19.52N 75.22E
Aurillac **116 D3** 44.56N 2.26E
Aurukun, homestead **42 C9** 13.21S 141.43E
Austin, Minnesota **134 E3** 43.40N 92.58W
Austin, Nevada **134 C2** 39.30N 117.05W
Austin, Texas **134 E2** 30.18N 97.47W
Austin, Lake **72 C3**
Austral Islands **85 P3**
Austral Downs, homestead **80 F4** 20.30S 137.45E
Australia, commonwealth **38 D3**
Australian Capital Territory **48 E2**
Austria, republic **117 F4**
Autun **116 D4** 46.58N 4.18E
Auxerre **116 D4** 47.48N 3.35E
Aux Sources, Mount **122 F2** 28.46S 28.54E
Avarua **84 P3** 21.12S 159.46W
Aveiro **116 B3** 40.38N 8.40W
Avieme **89 D3** 4.55S 143.36E
Avignon **116 D3** 43.56N 4.48E
Avila **116 C3** 40.39N 4.42W
Aviles **116 B3** 43.33N 5.55W
Avoca **54 B2** 37.05S 143.29E
Avoca River **54 B2**
Avon Downs, homestead **80 F4** 20.02S 137.30E
Avranches **116 C4** 48.42N 1.21W
Avu Avu **89 H2** 9.51S 160.24E
Awash **125 H2** 9.01N 40.10E
Awash River **125 G2**
Awio **89 F2** 6.12S 150.07E
Awjilah **124 F4** 29.05N 21.12E
Axel Heilberg, island **132 I6**
Ayacucho **144 D3** 37.09S 58.30W
Ayaguz **118 I3** 47.59N 80.27E
Ayamonte **116 B2** 37.13N 7.24W
Ayan **119 N4** 56.29N 138.07E
Ayapel **142 C4** 8.16N 75.10W
Aydin **108 A4** 37.51N 27.51E
Ayers Rock see Uluru
Ayon Island **119 Q5**
Ayr, Australia **42 F6** 19.34S 147.24E
Ayr, U.K. **114 D4** 55.28N 4.38W
Aytos **117 I3** 42.41N 27.15E
Ayutthaya **100 B4** 14.20N 100.35E
Ayvalik **117 I2** 39.18N 26.42E
Azamgarh **106 D3** 26.03N 83.10E

*Entry / **Page** / **Ref.** / Lat. / Long.*

Azerbaijan, republic **108 C5**
Azores, islands **122 A8**
Azov, Sea of **115 L2**
Azrou **124 C5** 33.27N 5.14W
Azul **144 D3** 36.46S 59.50W
Az Zubayr **108 C4** 30.24N 47.45E

B

Baalbek **108 B4** 34.00N 36.12E
Babar, island **88 A2**
Babar Islands **88 A2**
Babelthuap, island **101 E3**
Babinda **42 E7** 17.21S 145.56E
Babinda Hill **48 D4** 31.57S 146.28E
Babo **88 B3** 2.33S 133.20E
Babol **108 D4** 36.32N 52.42E
Babuyan Islands **100 D4**
Bacabal **143 F3** 4.15S 44.45W
Bacan, island **88 A3**
Bacau **117 I4** 46.33N 26.58E
Bacchus Marsh **54 C2** 37.41S 144.26E
Back River **132 H4**
Bacolod **100 D4** 10.38N 122.58E
Badajoz **116 B2** 38.53N 6.58W
Badanah **108 C4** 31.00N 41.05E
Bad Ischl **114 G2** 47.43N 13.38E
Badu Island **88 D1**
Baena **116 C2** 37.37N 4.20W
Bafata **124 B3** 12.09N 14.38W
Baffin Bay **133 L5**
Baffin Island **133 K4**
Bafia **124 E2** 4.39N 11.14E
Bafoulabe **124 B3** 13.49N 10.50W
Bafq **108 D4** 31.35N 55.21E
Bafra **108 B5** 41.34N 35.56E
Bagansiapiapi **100 B3** 2.07N 100.53E
Bage **144 D3** 31.22S 54.06W
Baghdad **108 C4** 33.20N 44.26E
Baghlan **106 B4** 36.11N 68.44E
Bagoe River **124 C3**
Bagua **142 C3** 5.37S 78.36W
Baguio **100 D4** 16.25N 120.37E
Bahamas, commonwealth **137 E3**
Bahawalpur **106 C3** 29.24N 71.47E
Bahia Blanca **144 C3** 38.45S 62.15W
Bahia Blanca, bay **144 C3**
Bahia Grande, bay **144 C1**
Bahir Dar **125 G3** 11.33N 37.25E
Bahraich **106 D3** 27.35N 81.36E
Bahrain, Islamic republic **108 D3**
Baia-Mare **117 H4** 47.39N 23.36E
Baibung, Point **100 B3** 8.38N 104.42E
Baicheng **103 G4** 45.37N 122.48E
Baidoa **125 H2** 3.08N 43.34E
Baie-Comeau **133 L2** 49.12N 68.10W
Baimuru **89 D2** 7.34S 144.49E
Bairam Ali **108 E4** 37.44N 62.13E
Bairiki **84 M6** 1.29N 172.59E
Bairnsdale **54 D2** 37.50S 147.37E
Baja **115 H2** 46.11N 18.58E
Baja, Point **134 C1** 29.59N 115.50W
Bajimba, Mount **48 G5** 29.18S 152.06E
Bajone, Point **126 E3** 15.14S 40.41E
Bake **100 B2** 3.04S 100.17E
Baker **134 C3** 44.46N 117.50W
Baker, island **84 N6**
Baker Lake **132 I4**
Baker Lake **132 I4** 64.20N 96.10W
Bakersfield **134 C2** 35.25N 119.00W
Bakhta **118 I5** 62.30N 89.02E
Bakhtaran **108 C4** 34.19N 47.04E
Bako **125 G2** 5.51N 36.32E
Baku **108 C5** 40.22N 49.53E
Bakumpai **100 C2** 1.29S 113.05E
Balabac **100 C3** 8.01N 117.02E
Balabac, island **100 C3**
Balabac Strait **100 C3**
Balaghat **106 D3** 21.48N 80.16E
Balaklava **66 G3** 34.09S 138.25E
Balazard **118 F2** 30.55N 59.06E
Balcarce **144 D3** 37.51S 58.16W
Balchik **117 I3** 43.24N 28.10E
Balclutha **90 C1** 46.14S 169.44E
Baldy Peak **134 D2** 33.55N 109.35W

*Entry / **Page** / **Ref.** / Lat. / Long.*

Balearic Islands **116 D2**
Baler **100 D4** 15.46N 121.33E
Bali, island **100 C2**
Bali, province **100 C2**
Balikesir **108 A4** 39.37N 27.51E
Balikpapan **100 C2** 1.15S 116.50E
Balimo **88 D2** 8.02S 142.58E
Balingian **100 C3** 2.59N 112.34E
Balkan Mountains **117 H3**
Balkan Peninsula **117 H3**
Balkhash **118 H3** 46.50N 74.57E
Balkhash, Lake **118 H3**
Ballarat **54 B2** 37.34S 143.51E
Ballard, Lake **72 D3**
Ballina **48 G5** 28.52S 153.34E
Balonne River **42 F1**
Balranald **48 B2** 34.38S 143.34E
Balsas **143 F3** 7.30S 46.00W
Balsas River **136 B2**
Balsfjord **118 B5** 69.20N 19.20E
Balta **115 J2** 47.58N 29.39E
Baltic Sea **114 H4**
Baltimore **135 G2** 39.18N 76.38W
Baltiysk **114 H3** 54.41N 19.59E
Balurghat **106 D3** 25.12N 88.50E
Bam **108 D3** 29.07N 58.20E
Bamako **124 C3** 12.40N 7.59W
Bambari **124 F2** 5.40N 20.37E
Bamberg **114 G2** 49.54N 10.54E
Bampur **108 E3** 27.13N 60.28E
Banaba, island **84 L5**
Banam **100 B4** 11.20N 105.17E
Banas, Cape **125 G4** 23.54N 35.48E
Banda **102 C2** 25.28N 80.20E
Banda, island **88 B3**
Banda Sea **88 A2**
Banda Aceh **100 A3** 5.30N 95.20E
Banda Banda, Mount **48 G4** 31.10S 152.26E
Bandar Abbas **108 D3** 27.12N 56.15E
Bandar-e Lengeh **108 D3** 26.34N 54.52E
Bandar Seri Begawan **100 C3** 4.56N 114.58E
Bandeira Peak **143 F1** 20.25S 41.45W
Bandirma **108 A5** 40.21N 27.58E
Bandundu **126 B4** 3.20S 17.24E
Bandung **100 B2** 6.57S 107.34E
Banemo **101 D3** 0.21N 128.35E
Banff **132 G3** 51.10N 115.34W
Bangalore **106 C2** 12.58N 77.35E
Bangassou **124 F2** 4.41N 22.48E
Bangeta, Mount **89 E2** 6.17S 147.05E
Banggai **100 D2** 1.34S 123.33E
Banggai Islands **100 D2**
Bangka, island **100 B2**
Bangkal **100 C2** 2.44S 112.25E
Bangkok **100 B4** 13.44N 100.30E
Bangladesh, people's republic **107 D3**
Bangor, U.K. **114 D3** 53.13N 4.08W
Bangor, U.S.A. **135 H3** 44.49N 68.47W
Bangui **124 E2** 4.23N 18.37E
Bangweulu, Lake **126 C3**
Baniara **89 E2** 9.46S 149.51E
Banja Luka **117 G3** 44.47N 17.11E
Banjarmasin **100 C2** 3.22S 114.33E
Banjul **124 B3** 13.28N 16.39W
Banka Banka, homestead **80 E5** 18.48S 134.02E
Banks Island **132 F5**
Banks Peninsula **90 D3**
Banks Strait **60 E4**
Ban Me Thuot **100 B4** 12.41N 108.02E
Banska Bystrica **115 H2** 48.44N 19.10E
Bantaeng **100 C2** 5.32S 119.58E
Ban Takua Pa **100 A3** 8.55N 98.20E
Banyak Islands **100 A3**
Banyo **124 E2** 6.47N 11.50E
Baoding **102 F3** 38.54N 115.26E
Baoji **102 E3** 34.23N 107.16E
Baoshan **102 D2** 25.07N 99.08E
Baotou **102 F4** 40.38N 109.59E
Ba'quba **108 C4** 33.45N 44.40E
Bar **117 G3** 42.05N 19.06E
Baradine **48 E4** 30.57S 149.04E
Barahona **137 E2** 18.13N 71.07W

*Entry / **Page** / **Ref.** / Lat. / Long.*

Baralaba **42 G3** 24.11S 149.49E
Baranovichi **115 J3** 53.09N 26.00E
Barbacena **143 F1** 21.13S 43.47W
Barbados, independent state **137 G2**
Barbas, Cape **124 B4** 22.20N 16.42W
Barbastro **116 D3** 42.02N 0.07E
Barberton **126 D2** 25.48S 31.03E
Barbuda, island **137 F2**
Barcaldine **42 E4** 23.33S 145.17E
Barcellona **117 G2** 38.10N 15.15E
Barcelona, Spain **116 D3** 41.25N 2.10E
Barcelona, Venezuela **142 D5** 10.08N 64.43W
Barcelos **142 D3** 0.59S 62.58W
Barcoo River **40 E3**
Bardera **125 H2** 2.18N 42.18E
Bareilly **106 C3** 28.20N 79.24E
Barents Sea **98 C15**
Bareo **100 C3** 3.35N 115.22E
Barham **48 C2** 35.38S 144.07E
Bari **117 G3** 41.07N 16.52E
Bariloche **144 B2** 41.09S 71.18W
Barinas **142 C4** 8.36N 70.15W
Baripada **106 D3** 21.52N 86.48E
Barisan Mountains **100 A3**
Barka **124 F5** 32.49N 21.45E
Barkley, Lake **135 F2**
Barkly Tableland **80 E6**
Bar-le-Duc **116 E4** 48.46N 5.10E
Barlee, Lake **72 C3**
Barlee Range **72 C4**
Barmedman **48 D2** 34.09S 147.23E
Barn Bluff, mountain **60 C3** 41.44S 145.54E
Barnaul **118 I4** 53.21N 83.45E
Barnstaple **114 D3** 51.05N 4.04W
Baroda *see* Vadodara
Barquisimeto **142 D5** 10.03N 69.18W
Barra **143 F2** 11.06S 43.25W
Barraba **48 F4** 30.23S 150.36E
Barrancabermeja **142 C4** 7.06N 73.54W
Barranquilla **142 C5** 11.00N 74.50W
Barras **143 F3** 4.16S 42.21W
Barreiras **143 F2** 12.09S 44.58W
Barreiro **116 B2** 38.40N 9.05W
Barren Grounds, region **132 H4**
Barretos **143 F1** 20.37S 48.38W
Barrier Range **48 A4**
Barrington Tops, mountain **48 F3** 32.03S 151.24E
Barrow **132 C5** 71.16N 156.50W
Barrow Island **72 B4**
Barrow, Mount **60 E3** 41.23S 147.25E
Barrow, Point **132 C5** 71.22N 156.30W
Barrow Creek **80 D4** 21.32S 133.53E
Barschtsch, Cape **89 E3** 5.04S 145.48E
Barstow **134 C2** 34.53N 117.03W
Bartle Frere, mountain **42 E7** 17.24S 145.49E
Baru, mountain **136 D1** 8.49N 82.38W
Barunga **80 D7** 14.31S 132.52E
Barus **100 A3** 2.02N 98.20E
Baruun-Urt **102 F4** 46.42N 113.20E
Barwon River **48 E5**
Basankusu **126 B5** 1.12N 19.50E
Basel **116 E4** 47.33N 7.36E
Bashakerd Mountains **108 D3**
Bashi Channel **100 D5**
Basilan **100 D3** 6.40N 121.59E
Basilan, island **100 D3**
Basildon **114 E3** 51.34N 0.25E
Baskerville, Cape **72 D5** 17.08S 122.15E
Basra **108 C4** 30.30N 47.50E
Bass Strait **60 C5**
Bassein **107 E2** 16.46N 94.45E
Basse-Terre **137 F2** 16.00N 61.43W
Basseterre **137 F2** 17.17N 62.43W
Bastia **116 E3** 42.41N 9.26E
Bastogne **114 F3** 50.00N 5.43E
Bastrop **134 E2** 32.49N 91.54W
Bata **124 D2** 1.51N 9.49E
Batac **100 D4** 18.03N 120.36E
Batakan **100 C2** 4.03S 114.39E
Batan Islands **100 D5**
Batangas **100 D4** 13.46N 121.03E

*Entry / **Page** / **Ref.** / Lat. / Long.*

Batavia-Bivak **88 C3** 2.45S 138.24E
Batchelor **80 C8** 13.04S 131.02E
Batemans Bay **48 F2**
Batemans Bay **48 F2** 35.43S 150.10E
Bath **114 D3** 51.23N 2.22W
Bathurst, Australia **48 E3** 33.27S 149.35E
Bathurst, Canada **133 L2** 47.37N 65.40W
Bathurst Harbour **60 D1**
Bathurst Island, Australia **80 C9**
Bathurst Island, Canada **132 H5**
Batkanu **124 B2** 9.07N 12.17W
Batlow **48 E2** 35.32S 148.10E
Batna **124 D5** 35.34N 6.10E
Baton Rouge **134 E2** 30.30N 91.10W
Batouri **124 E2** 4.26N 14.27E
Battambang **100 B4** 13.06N 103.13E
Batu Islands **100 A2**
Batu, mountain **125 G2** 6.55N 39.49E
Batubetumbang **100 B2** 2.48S 106.10E
Batui **100 D2** 1.15S 122.34E
Batukelau **100 C3** 0.50N 115.00E
Batumi **108 C5** 41.37N 41.36E
Batuputih **100 C3** 1.23N 118.20E
Baubau **100 D2** 5.30S 122.37E
Bauchi **124 E3** 10.16N 9.50E
Bauer, Cape **66 E4** 32.44S 134.04E
Bauld, Cape **133 M3** 51.40N 55.25W
Bauru **143 F1** 22.19S 49.07W
Bauska **115 I4** 56.22N 24.12E
Bautzen **114 G3** 51.11N 14.29E
Baw Baw, Mount **54 D2** 37.50S 146.17E
Bawe **88 B3** 2.59S 134.39E
Bawean, island **100 C2**
Bayamo **137 E3** 20.25N 76.35W
Bayan **100 C2** 8.16S 116.28E
Bayan Har Mountains **102 D3**
Bayankhongor **102 E4** 46.07N 100.44E
Bay City **134 E1** 28.59N 96.00W
Baydar Bay **118 G6**
Bayir **108 B4** 30.46N 36.40E
Baykal, Lake **119 K4**
Bayombong **100 D4** 16.27N 121.10E
Bayonne **116 C3** 43.30N 1.28W
Bayunglincir **100 B2** 2.00S 103.39E
Baza **116 C2** 37.30N 2.45W
Beachport **66 H2** 37.29S 140.01E
Beaconsfield **60 D3** 41.12S 146.49E
Beagle Bay **72 D5**
Bear Lake **134 C3**
Beatrice **134 E3** 40.17N 96.45W
Beaudesert **42 I2** 27.59S 153.00E
Beaufort, Australia **54 B2** 37.26S 143.23E
Beaufort, U.S.A. **135 F2** 32.26N 80.40W
Beaufort Sea **132 D5**
Beaufort West **126 C1** 32.21S 22.35E
Beaumont **136 C4** 30.04N 94.06W
Beauty Point **60 D3** 41.10S 146.49E
Beauvais **114 E2** 49.26N 2.05E
Beawar **106 C3** 26.02N 74.20E
Becej **117 H4** 45.36N 20.04E
Bechar **124 C5** 31.35N 2.17W
Bedourie **42 B3** 24.22S 139.28E
Beechworth **54 D2** 36.22S 146.42E
Beecroft Peninsula **48 F2**
Beenleigh **42 I2** 27.43S 153.12E
Beersheba **108 B4** 31.15N 34.47E
Bega **48 E1** 36.41S 149.51E
Begamganj **107 E3** 22.59N 91.04E
Bei'an **103 G4** 48.16N 126.36E
Beiarn **118 B5** 67.02N 14.40E
Beihai **102 E2** 21.29N 109.10E
Beijing **103 F4** 39.55N 116.26E
Beira **126 D3** 19.49S 34.52E
Beirut **108 B4** 33.52N 35.30E
Beius **117 H4** 46.40N 22.21E
Beja **116 B2** 38.01N 7.52W
Bejaia **116 E2** 36.49N 5.03E
Bejar **116 B3** 40.24N 5.45W
Bekily **126 E2** 24.12S 45.20E
Belaga **100 C3** 2.45N 113.46E
Belarus, republic **115 J3**
Belaya River **118 F4**
Belaya Tserkov **115 K2** 49.49N 30.10E
Belcher Islands **133 K3**
Belem **143 F3** 1.27S 48.29W

Entry/***Page***/***Ref.***/*Lat.*/*Long.*

Belen, Argentina **144 C4** 27.36S 67.00W
Belen, U.S.A. **134 D2** 34.39N 106.48W
Belet Uen **125 H2** 4.42N 45.12E
Belfast **114 C3** 54.35N 5.55W
Belgaum **106 C2** 15.54N 74.36E
Belgium, kingdom **114 E3**
Belgorod **115 L3** 50.38N 36.36E
Belgorod Dnestrovskiy **115 K2** 46.10N 30.19E
Belgrade **117 H3** 44.50N 20.30E
Belgrano II, antarctic base **92 Q3** 77.52S 34.34W
Belgrano III, antarctic base **92 P3** 77.54S 45.49W
Belimbing **100 B2** 5.56S 104.37E
Belitung, island **100 B2**
Belize **136 D2** 17.29N 88.13W
Belize, independent state **136 D2**
Belkina, Cape **103 H4** 45.47N 137.43E
Bell **42 H2** 26.56S 151.27E
Bellary **106 C2** 15.11N 76.54E
Bell Bay **60 D3** 41.07S 146.52E
Belle Glade **135 F1** 26.41N 80.41W
Belle Isle, Strait of **133 M3**
Bellingen **48 G4** 30.27S 152.54E
Bellingham **134 B3** 48.45N 122.29W
Bellingshausen, antarctic base **92 P3** 62.12S 58.58W
Bellingshausen Sea **92 N3**
Bello **142 C4** 6.20N 75.41W
Belluno **116 F4** 46.08N 12.13E
Belmopan **136 D2** 17.13N 88.48W
Belogorsk **119 M4** 50.55N 128.26E
Belo Horizonte **143 F2** 19.54S 43.54W
Beloit **135 F3** 42.31N 89.04W
Beloretsk **118 F4** 53.59N 58.20E
Beltana **66 G5** 30.50S 138.20E
Beltsy **115 J2** 47.45N 27.59E
Belukha, Mount **118 I3** 49.50N 86.44E
Belyando River **40 E3**
Belyuen **80 C8** 12.33S 130.42E
Benalla **54 C2** 36.33S 145.59E
Benavente **116 B3** 42.00N 5.40W
Bend **134 B3** 44.04N 121.17W
Bende **124 D2** 5.32N 7.36E
Bender Beila **125 I2** 9.30N 50.50E
Bendigo **54 C2** 36.46S 144.17E
Benevento **117 F3** 41.08N 14.46E
Bengal, Bay of **107 E2**
Bengbu **103 F3** 32.56N 117.27E
Benghazi **124 F5** 32.07N 20.05E
Bengkulu **100 B2** 3.46S 102.16E
Bengkulu, province **100 B2**
Bengo, Bay of **126 B4**
Bengoi **88 B3** 3.01S 130.10E
Benguela **126 B3** 12.34S 13.24E
Beni River **142 D2**
Beni-Abbes **124 C4** 30.11N 2.14W
Beni Mazar **125 G4** 28.29N 30.48E
Beni Mellal **124 C5** 32.22N 6.29W
Benin, Bight of **124 D2**
Benin, people's republic **124 D3**
Benin City **124 D2** 6.19N 5.41E
Beni Suef **125 G4** 29.05N 31.05E
Benjamin Constant **142 D3** 4.23S 69.59W
Ben Nevis, mountain **114 D4** 56.48N 5.00W
Bent **108 D3** 26.16N 59.31E
Bentinck Island **42 B7**
Bentley, Mount **94 N2** 78.00S 86.00W
Bentonville **134 E2** 36.23N 94.13W
Benue River **124 D2**
Benxi **103 G4** 41.21N 123.45E
Beppu **103 H3** 33.18N 131.30E
Berat **117 G3** 40.43N 19.46E
Berau Bay **88 B3**
Berber **125 G3** 18.01N 34.00E
Berbera **125 H3** 10.28N 45.02E
Berberati **124 E2** 4.19N 15.51E
Berdichev **115 J2** 49.54N 28.39E
Berdyansk **115 L2** 46.45N 36.47E
Bereina **89 E2** 8.40S 146.30E
Berens River **132 I3** 52.22N 97.00W
Berettyoujfalu **117 H4** 47.13N 21.31E
Berezniki **118 F4** 59.26N 56.49E
Berezovo **118 G5** 63.58N 65.00E
Berga **116 D3** 42.06N 1.50E
Bergama **108 A4** 39.08N 27.10E
Bergamo **116 E4** 45.42N 9.40E
Bergen **114 F5** 60.23N 5.20E
Bergerac **116 D3** 44.50N 0.29E
Berghan Point **90 E7** 34.55S 173.33E
Berguent **124 C5** 34.03N 2.02W
Berhampore **106 D3** 24.06N 88.18E
Berhampur **106 D2** 19.21N 84.51E
Bering Sea **84 N11**
Bering Strait **132 B4**
Berkane **116 C1** 34.59N 2.20W
Berkner Island **92 P3**
Berlin, Germany **114 G3** 52.32N 13.25E
Berlin, U.S.A. **135 G3** 44.27N 71.13W
Bermejo River **144 C4**
Bermuda, island **130 M4**
Bern **116 E4** 46.57N 7.26E
Bernier Island **72 B4**
Berri **66 H3** 34.17S 140.36E
Berridale **48 E1** 36.22S 148.50E
Berrigan **48 C2** 35.40S 145.48E
Berrouaghia **116 D2** 36.07N 2.52E
Bertoua **124 E2** 4.34N 13.42E
Berwick **54 C2** 38.02S 145.21E
Berwick-upon-Tweed **114 D4** 55.46N 2 00W
Besancon **116 E4** 47.14N 6.02E
Besikama **100 D2** 9.38S 124.58E
Beswick **80 D7** 14.34S 133.07E
Betamba **126 C4** 2.17S 21.24E
Betano **100 D2** 9.11S 125.42E
Betanzos **116 B3** 43.17N 8.13W
Bethel **132 B4** 60.49N 161.49W
Bettiah **106 D3** 26.49N 84.30E
Beulah **54 B3** 35.57S 142.25E
Beykoz **117 I3** 41.09N 29.06E
Beziers **116 D3** 43.21N 3.13E
Bhadrakh **106 D3** 21.05N 86.36E
Bhavnagar **106 C3** 21.46N 72.14E
Bhilai **102 C2** 21.12N 81.26E
Bhilwara **106 C3** 25.23N 74.40E
Bhopal **106 C3** 23.17N 77.28E
Bhubaneswar **106 D3** 20.13N 85.50E
Bhuj **106 B3** 23.12N 69.54E
Bhutan, kingdom **107 E3**
Biafra, Bight of **124 D2**
Biak, Central Sulawesi **100 D2** 0.54S 122.54E
Biak, Irian Jaya **88 C3** 1.10S 136.05E
Biak, island **88 C3**
Biala Podlaska **115 I3** 52.03N 23.05E
Bialla **89 F2** 5.20S 151.01E
Bialystok **115 I3** 53.09N 23.10E
Bian River **88 D2**
Bicheno **60 F3** 41.53S 148.18E
Bicoli **101 D3** 0.35N 128.30E
Bida **124 D2** 9.06N 5.59E
Bidura, homestead **48 B2** 34.10S 143.21E
Bie Plateau **126 B3**
Biel **116 E4** 47.09N 7.16E
Bielefeld **116 E5** 52.02N 8.32E
Biella **116 E4** 45.34N 8.04E
Bielsko-Biala **117 G4** 49.50N 19.00E
Bielsk Podlaski **115 I3** 52.47N 23.11E
Bien Hoa **100 B4** 10.58N 106.50E
Bienville, Lake **133 K3**
Biga **117 I3** 40.13N 27.14E
Bigge Island **72 D6**
Biggenden **42 I3** 25.31S 152.03E
Bighorn River **134 D3**
Bihac **117 G3** 44.49N 15.53E
Bihar **106 D3** 25.13N 85.31E
Bijapur **106 C2** 16.47N 75.48E
Bikaner **106 C3** 28.01N 73.22E
Bikin River **104 G10**
Bikini, island **84 L7**
Bilaspur **106 D3** 22.03N 82.12E
Bilbao **116 C3** 43.15N 2.56W
Bilecik **117 J3** 40.10N 29.59E
Billings **134 D3** 45.47N 108.30W
Bilma **124 E3** 18.46N 12.50E
Biloela **42 H3** 24.24S 150.31E
Biloxi **135 F2** 30.24N 88.55W
Bimberi Peak **48 E2** 35.40S 148.48E
Binaiya, mountain **88 A3** 3.07S 129.25E
Binalud Mountains **108 D4**
Bindura **126 D3** 17.20S 31.21E
Bingara **48 F5** 29.52S 150.34E
Binghamton **135 G3** 42.06N 75.55W
Bingkor **100 C3** 5.26N 116.15E
Bingol **108 C4** 38.54N 40.29E
Binnaway **48 E4** 31.33S 149.25E
Bintan, island **100 B3**
Bintulu **100 C3** 3.12N 113.01E
Bintuni **88 B3** 2.07S 133.37E
Bintuni Bay **88 B3**
Bioko, island **124 D2**
Birao **124 F3** 10.11N 22.49E
Biratnagar **106 D3** 26.27N 87.17E
Birchip **54 B3** 35.59S 142.59E
Birdsville **42 B3** 25.54S 139.21E
Birhan, mountain **125 G3** 10.45N 37.55E
Birjand **108 D4** 32.55N 59.10E
Birlad **115 J2** 46.15N 27.40E
Birmingham, U.K. **114 D3** 52.30N 1.50W
Birmingham, U.S.A. **135 F2** 33.30N 86.55W
Bir-Mogrein **124 B4** 25.10N 11.35W
Birnin Kebbi **124 D3** 12.30N 4.11E
Birobidzhan **119 N3** 48.49N 132.54E
Birsk **118 F4** 55.28N 55.31E
Biscay, Bay of **116 C4**
Bishkek **118 H3** 42.53N 74.46E
Bisk **102 C5** 52.35N 85.16E
Biskra **124 D5** 34.50N 5.41E
Bislig **100 D3** 8.10N 126.18E
Bismarck **134 D3** 46.50N 100.48W
Bismarck Archipelago **89 E3**
Bismarck Sea **89 E3**
Bissagos Islands **124 B3**
Bissau **124 B3** 11.52N 15.39W
Bitlis **108 C4** 38.23N 42.04E
Bitola **117 H3** 41.01N 21.21E
Biwa, Lake **104 H4**
Bizerte **124 D5** 37.18N 9.52E
Bjelovar **117 G4** 45.54N 16.51E
Black Bluff, mountain **60 C3** 41.29S 145.57E
Black Hills **134 D3**
Black River **102 E2**
Black Sea **112 H2**
Blackall **42 E3** 24.26S 145.28E
Blackburn **114 D3** 53.45N 2.29W
Blackfoot **134 C3** 43.11N 112.23W
Blackpool **114 D3** 53.50N 3.03W
Blackstone Range **72 E3**
Black Sugarloaf, mountain **48 F4** 31.23S 151.32E
Black Volta, river **124 C3**
Blackwater **42 G4** 23.35S 148.53E
Blagoevgrad **117 H3** 42.01N 23.05E
Blagoveshchensk **103 G5** 50.19N 127.30E
Blair Athol **42 F4** 22.42S 147.33E
Blanc, Cape, Mauritania **124 B4** 20.44N 17.05W
Blanc, Cape, Tunisia **124 D5** 37.21N 9.52E
Blanc, Mount **116 E4** 45.50N 6.52E
Blanca Peak **134 D2** 37.34N 105.29W
Blanche, Lake **66 G6**
Blanco, Cape **134 B3** 42.50N 124.29W
Blanquilla, island **137 F2**
Blantyre **126 D3** 15.46S 35.00E
Blaze, Point **80 C8** 12.56S 130.08E
Blednaya, Mount **118 G6** 76.23N 65.08E
Blenheim **90 E4** 41.31S 173.57E
Blida **124 D5** 36.30N 2.50E
Blinman **66 G5** 31.06S 138.40E
Blitar **100 C2** 8.06S 112.12E
Bloemfontein **126 C2** 29.07S 26.14E
Blois **116 D4** 47.36N 1.20E
Bloomington, Illinois **135 F3** 40.29N 89.00W
Bloomington, Indiana **135 F2** 39.10N 86.31W
Bluefield **135 F2** 37.14N 81.17W
Bluefields **136 D2** 12.00N 83.49W
Blue Nile, river **125 G3**
Bluff **90 C1** 46.36S 168.21E
Bluff Knoll, mountain **72 C2** 34.22S 118.15E
Blumenau **144 E4** 26.55S 49.07W
Bo **124 B2** 7.58N 11.45W
Boac **100 D4** 13.26N 121.52E
Boa Vista **142 D4** 2.51N 60.43W
Bobo-Dioulasso **124 C3** 11.11N 4.18W
Bobonong **126 C2** 21.58S 28.26E
Bobruisk **115 J3** 53.08N 29.10E
Boca do Acre **142 D3** 8.45S 67.23W
Boddington **72 C2** 32.48S 116.28E
Bodele Depression **124 E3**
Boden **110 G4** 65.50N 21.44E
Bodmin **114 D3** 50.29N 4.43W
Bodo **118 B5** 67.18N 14.26E
Bodrum **117 I2** 37.03N 27.28E
Boende **126 C4** 0.15S 20.51E
Boffa **124 B3** 10.12N 14.02W
Boga, Lake **54 B3**
Bogalusa **135 F2** 30.43N 89.53W
Bogan River **48 D4**
Boggabilla **48 F5** 28.37S 150.22E
Boggabri **48 E4** 30.42S 150.03E
Bogia **89 E3** 4.16S 144.56E
Bogong, Mount **54 D2** 36.44S 147.18E
Bogor **100 B2** 6.34S 106.45E
Bogota **142 C4** 4.38N 74.05W
Bohol, island **100 D3**
Boigu Island **88 D2**
Boise **134 C3** 43.38N 116.12W
Bojador, Cape **124 B4** 26.09N 14.46W
Bojnurd **108 D4** 37.28N 57.20E
Bok **89 E2** 5.25S 147.13E
Boke **124 B3** 10.56N 14.18W
Bokpyin **107 E2** 11.14N 98.45E
Boku **89 G2** 6.35S 155.20E
Bolac, Lake **54 B2**
Bolangir **106 D3** 20.41N 83.30E
Bole **102 C4** 44.52N 82.06E
Bolgatanga **124 C3** 10.44N 0.53W
Bolgrad **115 J2** 45.42N 28.35E
Bolivar **144 C3** 36.15S 61.07W
Bolivar Peak **142 C4** 8.33N 71.03W
Bolivia, republic **142 D2**
Bollnas **114 H5** 61.20N 16.25E
Bolobo **126 B4** 2.10S 16.17E
Bologna **116 F3** 44.30N 11.20E
Bolshevik Island **118 K6**
Bolvadin **110 H1** 38.43N 31.02E
Bolzano **116 F4** 46.30N 11.22E
Boma **126 B4** 5.50S 13.03E
Bomaderry **48 F2** 34.51S 150.37E
Bombala **48 E1** 36.55S 149.15E
Bomberai River **88 B3**
Bomu River **124 F2**
Bonaire, island **137 F2**
Bonaparte Archipelago **72 D5**
Bon Bon, homestead **66 E5** 30.26S 135.29E
Bondo **126 C5** 3.47N 23.50E
Bongabong **100 D4** 12.44N 121.30E
Bongor **124 E3** 10.18N 15.20E
Bonin Islands **103 I2**
Bonn **114 F3** 50.44N 7.06E
Bonnie Rock **72 C2** 30.33S 118.22E
Bonobono **100 C3** 8.40N 117.32E
Bonshaw **48 F5** 29.03S 151.16E
Bontang **100 C3** 0.05N 117.31E
Bonthe **124 B2** 7.32N 12.30W
Booligal **48 C3** 33.52S 144.53E
Boonah **42 I2** 28.00S 152.41E
Boorowa **48 E2** 34.27S 148.44E
Boort **54 B2** 36.07S 143.43E
Boothby, Cape **92 C3** 67.50S 57.35E
Boothia, Gulf of **132 I5**
Boothia Peninsula **132 I5**
Bor **125 G2** 6.18N 31.34E
Borang, Cape **88 B2** 5.17S 133.05E
Boras **114 G4** 57.44N 12.55E
Borba **142 E3** 4.39S 59.35W
Borda, Cape **66 F3** 35.45S 136.35E
Bordeaux **116 C3** 44.50N 0.34W
Bordertown **66 H2** 36.18S 140.49E
Bordj Bou Arreridj **116 D2** 36.02N 4.49E
Bordj Omar Driss **124 D4** 28.04N 6.39E
Boree Plains, homestead **48 B3** 33.43S 143.22E
Borger **134 D2** 35.39N 101.24W
Borgholm **114 H4** 56.51N 16.40E

*Entry/**Page**/**Ref.**/Lat./Long.*

Borisoglebsk **118 E4** 51.23N 42.02E
Borisov **115 J3** 54.09N 28.30E
Borlange **114 H5** 60.29N 15.25E
Borneo, island **100 C2**
Bornholm, island **110 F3**
Bornu, Plain of **124 E3**
Borroloola **80 F6** 16.04S 136.18E
Borsa **117 H4** 47.39N 24.40E
Borujerd **108 C4** 33.55N 48.48E
Borzya **119 L4** 50.24N 116.35E
Bosaso **125 H3** 11.13N 49.08E
Bose **102 E2** 23.55N 106.32E
Boshan **103 F3** 36.32N 117.47E
Bosnia and Herzegovina, republic **117 G3**
Bosnia Dubica **117 G4** 45.13N 16.52E
Bosnik **88 C3** 1.09S 136.14E
Bosobolo **126 B5** 4.11N 19.55E
Bossangoa **124 E2** 6.27N 17.21E
Bossut, Cape **72 D5** 18.42S 121.37E
Bosten Lake **102 C4**
Boston **135 G3** 42.20N 71.05W
Botany Bay **48 F2**
Bothnia, Gulf of **114 I5**
Bothwell **60 E2** 42.23S 147.00E
Botosani **117 I4** 47.44N 26.41E
Botswana, republic **126 C2**
Bouafle **124 C2** 7.01N 5.47W
Bouake **124 C2** 7.42N 5.00W
Bouar **124 E2** 5.58N 15.35E
Bougainville, Cape **72 E6** 13.57S 126.05E
Bougainville, island **89 F2**
Bougouni **124 C3** 11.25N 7.28W
Boulder, Australia **72 D2** 30.47S 121.29E
Boulder, U.S.A. **134 D2** 40.02N 105.16W
Boulder City **134 C2** 35.59N 114.51W
Boulia **42 B4** 22.55S 139.54E
Boulogne **116 D5** 50.43N 1.37E
Bountiful **134 C3** 40.54N 111.54W
Bounty Islands **84 M1**
Bourg-en-Bresse **116 E4** 46.13N 5.12E
Bourges **116 D4** 47.05N 2.23E
Bourke **48 C4** 30.05S 145.56E
Bournemouth **114 D3** 50.43N 1.54W
Bou Saada **124 D5** 35.10N 4.09E
Bouvet Island **92 A2**
Bowen **42 G5** 20.01S 148.15E
Bowling Green **135 F2** 37.00N 86.29W
Boyoma Falls **126 C5** 0.18N 25.30E
Boyer **60 E2** 42.45S 147.08E
Boynton Beach **135 F1** 26.32N 80.04W
Bozeman **134 C3** 45.41N 111.00W
Bozoum **124 E2** 6.16N 16.22E
Bracke **118 B5** 62.44N 15.30E
Brad **117 H4** 46.07N 22.47E
Bradenton **135 F1** 27.29N 82.33W
Braga **116 B3** 41.32N 8.26W
Bragado **144 C3** 35.10S 60.20W
Braganca, Brazil **143 F3** 1.02S 46.46W
Braganca, Portugal **116 B3** 41.47N 6.46W
Brahmanbaria **107 E3** 23.58N 91.04E
Brahmaputra River **107 E3**
Braidwood **48 E2** 35.27S 149.48E
Braila **117 I4** 45.17N 27.58E
Brak **124 E4** 27.33N 14.17E
Branco River **142 D3**
Brandon **132 I2** 49.42N 99.53W
Brasilia **143 F2** 15.45S 47.57W
Brasov **117 I4** 45.39N 25.35E
Brassey, Mount **80 E3** 23.04S 134.55E
Bratislava **114 H2** 48.10N 17.10E
Bratsk **118 K4** 56.20N 101.50E
Bratsk Reservoir **118 K4**
Braunau **114 G2** 48.16N 13.03E
Brava **125 H2** 1.02N 44.02E
Bray **114 C3** 53.12N 6.06W
Brazil, federal republic **144 D4**
Brazilian Highlands **143 F2**
Brazos River **134 D2**
Brazzaville **126 B4** 4.14S 15.14E
Brcko **117 G3** 44.52N 18.49E
Bream Bay **90 F7**
Breda **114 E3** 51.35N 4.46E
Bremen **114 F3** 53.05N 8.48E
Bremer Bay **72 C2**
Bremerhaven **114 F3** 53.33N 8.35E

*Entry/**Page**/**Ref.**/Lat./Long.*

Bremerton **134 B3** 47.34N 122.34W
Brescia **116 F4** 45.33N 10.13E
Brest, Belarus **115 I3** 52.08N 23.40E
Brest, France **116 C4** 48.23N 4.30W
Brett, Cape **90 F7** 35.10S 174.20E
Brewarrina **48 D5** 29.57S 146.52E
Brewton **135 F2** 31.07N 87.04W
Brezina **124 D5** 33.03N 1.18E
Bria **124 F2** 6.32N 22.00E
Bridgeport **135 G3** 41.12N 73.12W
Bridgetown, Australia **72 C2** 33.58S 116.08E
Bridgetown, Barbados **137 G2** 13.06N 59.37W
Bridgewater, Australia **60 E2** 42.44S 147.14E
Bridgewater, Canada **133 L2** 44.23N 64.32W
Bridlington **114 D3** 54.05N 0.12W
Bridport **60 E3** 41.01S 147.24E
Brigham City **134 C3** 41.31N 112.01W
Bright **54 D2** 36.44S 146.58E
Brighton **114 D3** 50.50N 0.10W
Brindisi **117 G3** 40.37N 17.57E
Brisbane **42 I2** 27.28S 153.01E
Brisbane River **42 I2**
Bristol **114 D3** 51.27N 2.35W
Bristol Bay **132 B3**
British Columbia, province **132 F3**
Brive-la-Gaillarde **116 D4** 45.09N 1.32E
Brno **114 H2** 49.12N 16.40E
Broach **106 C3** 21.42N 72.58E
Broad Sound **42 G4**
Broadford **54 C2** 37.13S 145.03E
Brockville **133 K2** 44.35N 75.44W
Brodeur Peninsula **132 I5**
Broken Bay **48 F3**
Broken River **54 C2**
Broken Hill **48 A4** 31.58S 141.27E
Bronte Park **60 D2** 42.08S 146.29E
Brookhaven **135 E2** 31.36N 90.28W
Brookings **134 E3** 44.19N 96.47W
Brooks Range **132 C4**
Brookton **72 C2** 32.22S 117.00E
Broome **72 D5** 17.58S 122.14E
Brovary **115 K3** 50.30N 30.45E
Brown, antarctic base **92 O3** 64.53S 62.53W
Brown, Mount **66 G4** 32.30S 138.00E
Brown, Point **66 D4** 32.33S 133.51E
Brownfield **134 D2** 33.11N 102.16W
Brownsville, Tennessee **135 E2** 35.35N 89.15W
Brownsville, Texas **134 E1** 25.54N 97.30W
Brownwood **134 E2** 31.42N 98.59W
Bruce, Mount **72 C4** 22.36S 118.08E
Brugge **114 E3** 51.13N 3.14E
Brunei, sultanate **100 C3**
Brunette Downs, homestead **80 E5** 18.39S 135.57E
Brunner, Lake **90 D3**
Brunsbuttelkoog **114 F3** 53.54N 9.08E
Brunswick **135 F2** 31.09N 81.30W
Brunswick Bay **72 D6**
Brussels **114 E3** 50.50N 4.21E
Bruthen **55 D2** 37.43S 147.50E
Bryan, Ohio **135 F3** 41.30N 84.34W
Bryan, Texas **134 E2** 30.41N 96.24W
Bryan, Mount **66 G4** 33.26S 138.59E
Bryansk **118 D4** 53.15N 34.09E
Buala **89 G2** 8.11S 159.37E
Bucaramanga **142 C4** 7.08N 73.10W
Buccaneer Archipelago **72 D5**
Buchan **55 E2** 37.30S 148.11E
Buchanan **124 B2** 5.57N 10.02W
Bucharest **117 I3** 44.25N 26.07E
Buckalow, homestead **48 A3** 32.40S 141.14E
Buckingham Bay **80 E8**
Buckleboo **66 F4** 32.55S 136.13E
Budapest **115 H2** 47.30N 19.03E
Buenaventura **142 C4** 3.54N 77.02W
Buenos Aires **144 D3** 34.40S 58.30W
Buenos Aires, Lake **144 B2**
Bufareh **88 C3** 2.15S 138.50E
Buffalo **135 G3** 42.52N 78.55W

*Entry/**Page**/**Ref.**/Lat./Long.*

Buffalo, Mount **54 D2** 36.47S 146.45E
Buga **142 C4** 3.53N 76.17W
Buguruslan **118 F4** 53.36N 52.30E
Buin **89 G2** 6.45S 155.41E
Bujumbura **126 C4** 3.22S 29.19E
Buk **88 D2** 8.30S 142.28E
Buka **89 F2** 5.25S 154.41E
Buka Island **89 F2**
Bukama **126 C4** 9.13S 25.52E
Bukavu **126 C4** 2.30S 28.49E
Bukhara **118 G2** 39.47N 64.26E
Bukittinggi **100 B2** 0.18S 100.20E
Bukoba **126 D4** 1.20S 31.49E
Bul, Mount **108 D4** 30.48N 52.45E
Bula **88 B3** 3.07S 130.27E
Bulagan **102 E4** 48.50N 103.29E
Bulahdelah **48 G3** 32.25S 152.12E
Bulawayo **126 C2** 20.10S 28.43E
Buldan **117 I2** 38.03N 28.50E
Bulgaria, people's republic **117 H3**
Bulla **88 D2** 9.08S 141.20E
Buller, Mount **54 D2** 37.09S 146.26E
Buller River **90 D4**
Bullfinch **72 C2** 30.59S 119.06E
Bullidobu **89 E2** 9.40S 147.45E
Bulloo River **42 E2**
Bulloo Downs, homestead, Queensland **42 D1** 28.32S 142.57E
Bulloo Downs, homestead, Western Australia **72 C4** 24.00S 119.34E
Bull Shoals Lake **134 E2**
Bulman **80 E8** 13.40S 134.19E
Bulolo **89 E2** 7.12S 146.39E
Bumba **126 C5** 2.10N 22.30E
Bunbury **72 C2** 33.20S 115.38E
Bundaberg **42 I3** 24.52S 152.21E
Bundey, Mount **80 C8** 12.53S 131.36E
Bungku **100 D2** 2.33S 121.59E
Bungle Bungles, The **80 B6** 17.25S 128.23E
Bungo Channel **104 F3**
Bunia **126 D5** 1.35N 30.13E
Bura **126 D4** 1.06S 39.58E
Buraida **108 C3** 26.20N 43.59E
Burcher **48 D3** 33.31S 147.15E
Burdekin River **42 F6**
Burdwan **106 D3** 23.15N 87.52E
Bureiskiy Range **119 N4**
Bureya **119 M3** 49.49N 129.50E
Burgas **117 I3** 42.30N 27.29E
Burgersdorp **126 C1** 31.00S 26.20E
Burgos **116 C3** 42.21N 3.41W
Burgsvik **118 B4** 57.03N 18.19E
Burhaniye **117 I2** 39.29N 26.59E
Burhanpur **106 C3** 21.18N 76.08E
Burica, Point **136 D1** 8.02N 82.52W
Buriram **100 B4** 15.01N 103.06E
Burketown **42 B7** 17.45S 139.33E
Burkina Faso, republic **124 C3**
Burley **134 C3** 42.31N 113.48W
Burlington, Iowa **134 E3** 40.50N 91.07W
Burlington, North Carolina **135 G2** 36.05N 79.27W
Burlington, Vermont **135 G3** 44.28N 73.14W
Burnett River **42 H3**
Burnie **60 C3** 41.03S 145.55E
Burns **134 C3** 43.36N 119.03W
Burr, Mount **66 H2** 37.36S 140.29E
Burra **66 G4** 33.41S 138.56E
Burragorang, Lake **48 F3**
Burren Junction **48 E4** 30.06S 148.59E
Bursa **108 A5** 40.12N 29.04E
Buru, island **88 A3**
Burundi, republic **126 C4**
Burung **100 B3** 0.25N 103.32E
Bushehr **108 D3** 28.59N 50.50E
Businga **126 C5** 3.16N 20.59E
Busira River **124 E2**
Busselton **72 C2** 33.39S 115.20E
Buta **126 C5** 2.49N 24.50E
Butare **126 C4** 2.34S 29.43E
Butaritari, island **84 M6**
Butte **134 C3** 46.00N 112.31W
Butterworth **100 B3** 5.24N 100.22E
Butuan **100 D3** 8.56N 125.31E

*Entry/**Page**/**Ref.**/Lat./Long.*

Butung, island **100 D2**
Buyake Muli, passage **88 C2**
Buzau **117 I4** 45.09N 26.49E
Buzi River **126 D2**
Buzuluk **118 F4** 52.49N 52.19E
Bwagaoia **89 F1** 10.41S 152.51E
Bydgoszcz **114 H3** 53.16N 18.00E
Bylot Island **133 K5**
Byrock **48 D4** 30.40S 146.25E
Byron, Cape **48 G5** 28.38S 153.38E
Byron Bay **48 G5** 28.39S 153.37E
Byrranga Mountains **118 J6**

C

Caazapa **144 D4** 26.09S 56.25W
Cabanatuan **100 D4** 15.30N 120.58E
Cabimas **142 C5** 10.26N 71.27W
Cabinda **126 B4** 5.34S 12.12E
Caboolture **42 I2** 27.05S 152.57E
Caborca **134 C2** 30.42N 112.10W
Cabot Strait **133 L2**
Cacak **117 H3** 43.54N 20.22E
Caceres, Brazil **142 E2** 16.05S 57.40W
Caceres, Spain **116 B2** 39.29N 6.23W
Cachoeira do Sul **144 D3** 30.03S 52.52W
Cachoeiro de Itapemirim **143 F1** 20.51S 41.07W
Caconda **126 B3** 13.43S 15.03E
Cadajas **142 D3** 3.55S 62.00W
Cadibarrawirracanna, Lake **66 E6**
Cadillac **135 F3** 44.15N 85.23W
Cadiz, Philippines **100 D4** 10.57N 123.18E
Cadiz, Spain **116 B2** 36.32N 6.18W
Caen **116 C4** 49.11N 0.22W
Cagayan Islands **100 C3**
Cagayan de Oro **100 D3** 8.29N 124.40E
Cagliari **116 E2** 39.13N 9.08E
Cahors **116 D3** 44.28N 1.26E
Caicara **142 D4** 7.38N 66.10W
Caico **143 G3** 6.25S 37.04W
Caicos Islands **137 E3**
Cairn Curran Reservoir **54 B2**
Cairns **42 E7** 16.55S 145.46E
Cairo **125 G5** 30.03N 31.15E
Cajamarca **142 C3** 7.09S 78.32W
Calabar **124 D2** 4.56N 8.22E
Calabozo **142 D4** 8.58N 67.28W
Calafat **117 H3** 43.59N 22.57E
Calahorra **116 C3** 42.19N 1.58W
Calais **116 D5** 50.57N 1.52E
Calamian Group **100 C4**
Calanscio Sand Sea, desert **124 F4**
Calatayud **116 C3** 41.21N 1.39W
Calbayog **100 D4** 12.04N 124.38E
Calcutta *see* Kolkata
Caldas da Rainha **116 B2** 39.24N 9.08W
Caldera **144 B4** 27.05S 70.48W
Caldwell **134 C3** 43.39N 116.40W
Calgary **132 G3** 51.05N 114.05W
Cali **142 C4** 3.24N 76.30W
Calicut *see* Kozhikode
California, Gulf of **134 C1**
California, state **134 B2**
Callabonna, Lake **66 H6**
Callao **142 C2** 12.05S 77.08W
Caloundra **42 I2** 26.48S 153.08E
Caltanissetta **117 F2** 37.29N 14.04E
Calvados Chain, The, islands **89 F1**
Calvert Hills, homestead **80 F6** 17.14S 137.20E
Calvinia **126 B1** 31.25S 19.47E
Camaguey **136 E3** 21.25N 77.55W
Cambay, Gulf of **106 C3**
Cambodia, republic **100 B4**
Cambrai **116 D5** 50.10N 3.14E
Cambridge **60 E2** 42.50S 147.26E
Cambridge Bay **132 H4** 69.09N 105.00W
Camden **48 F2** 34.03S 150.42E
Camels Hump **80 C3** 23.46S 131.38E
Cameron Downs, homestead **42 E5** 21.23S 144.17E
Cameroon, federal republic **124 D2**
Cameta **143 F3** 2.14S 49.30W
Camfield, homestead **80 C6** 17.02S 131.17E

*Entry/**Page**/**Ref.**/Lat./Long.*

Camiranga **143 F3** 1.47S 46.15W
Camiri **142 D1** 20.03S 63.31W
Camocim **143 F3** 2.54S 40.50W
Camooweal **42 B6** 19.55S 138.07E
Campbell, Cape **90 F4** 41.44S 174.16E
Campbell Island **92 I2**
Campbellton **133 L2** 48.00N 66.41W
Campbell Town **60 E3** 41.55S 147.30E
Campeche **136 C2** 19.50N 90.30W
Campeche, Bay of **136 C3**
Camperdown **54 B1** 38.14S 143.09E
Campina Grande **143 G3** 7.15S 35.53W
Campinas **144 E4** 22.54S 47.06W
Campobasso **117 F3** 41.33N 14.39E
Campo Grande **143 E1** 20.24S 54.35W
Campo Maior **143 F3** 4.50S 42.12W
Campos **143 F1** 21.46S 41.21W
Cam Ranh **100 B4** 11.54N 109.14E
Canada, dominion **132 H3**
Canadian River **134 D2**
Canakkale **117 I3** 40.09N 26.25E
Cananea **136 A4** 31.00N 110.20W
Canary Islands **124 B4**
Canaveral, Cape **135 F1** 28.28N 80.31W
Canberra **48 E2** 35.17S 149.08E
Candala **125 H3** 11.23N 49.53E
Canea **117 H2** 35.31N 24.01E
Cangzhou **103 F3** 38.19N 116.54E
Canicatti **117 F2** 37.22N 13.51E
Cann River **55 E2**
Cannanore **106 C2** 11.53N 75.23E
Cannes **116 E3** 43.33N 7.00E
Canning Basin **72 D5**
Cann River **55 E2** 37.34S 149.09E
Canobolas, Mount **48 E3** 33.21S 148.59E
Canon City **134 D2** 38.27N 105.16W
Canopus, homestead **66 H4** 33.30S 140.42E
Canowindra **48 E3** 33.34S 148.40E
Cantabrian Mountains **112 D2**
Canterbury **114 E3** 51.17N 1.05E
Canterbury Bight **90 E2**
Canterbury Plains **90 D3**
Can Tho **100 B4** 10.03N 105.46E
Canton **135 F3** 40.48N 81.23W
Canutama **142 D3** 6.34S 64.21W
Canyon **134 D2** 34.58N 101.56W
Canyon City **134 C3** 44.24N 118.58W
Cape Barren Island **60 F4**
Cape Breton Island **133 L2**
Cape Coast **124 C2** 5.10N 1.13W
Cape Girardeau **135 F2** 37.19N 89.31W
Capella **88 D3** 4.58S 141.06E
Cape Town **126 B1** 33.56S 18.28E
Cape Verde Islands, independent state **146 B3**
Cape York Peninsula **42 D9**
Cap Haitien **137 E2** 19.47N 72.17W
Capitan Arturo Prat, antarctic base **92 O3** 62.30S 59.41W
Capoeiras Falls **142 E3** 6.34S 58.25W
Capricorn Channel **42 H5**
Capricorn Group **42 I4**
Caqueta River **142 C3**
Caracal **117 H3** 44.07N 24.18E
Caracas **142 D5** 10.35N 66.56W
Caraga **100 D3** 7.19N 126.32E
Caransebes **117 H4** 45.23N 22.13E
Caratasca Lagoon **136 D2**
Caratinga **143 F2** 19.50S 42.06W
Caravaca **116 C2** 38.06N 1.51W
Carbonia **116 E2** 39.11N 8.31E
Carcar **100 D3** 10.03N 123.36E
Carcassonne **116 D3** 43.13N 2.21E
Cardamom Hills **106 C1**
Cardenas **136 D3** 23.02N 81.12W
Cardiel, Lake **144 B2**
Cardiff **114 D3** 51.30N 3.13W
Cardwell **42 F6** 18.16S 146.01E
Carei **117 H4** 47.40N 22.28E
Carey, Lake **72 D3**
Caribbean Sea **137 E2**
Caribou **135 H3** 46.52N 68.01W
Carinda **48 D4** 30.28S 147.42E
Carlisle **114 D3** 54.54N 2.55W
Carlow **114 C3** 52.50N 6.55W
Carlsbad, California **134 C2** 33.09N 117.20W
Carlsbad, New Mexico **134 D2** 32.25N 104.14W
Carmarthen **114 D3** 51.52N 4.19W
Carmen Island **134 C1**
Carnarvon, Australia **72 B3** 24.53S 113.40E
Carnarvon, South Africa **126 C1** 30.59S 22.08E
Carnarvon Range **72 D4**
Carnegie, Lake **72 D3**
Carney Island **92 M3**
Carolina **143 F3** 7.20S 47.25W
Caroline Islands **84 J6**
Caroline Range **72 E5**
Caroni River **137 F1**
Carpathian Mountains **115 I2**
Carpentaria, Gulf of **40 D4**
Carrieton **66 G4** 32.26S 138.32E
Carson City **132 G1** 39.10N 119.46W
Cartagena, Colombia **142 C5** 10.24N 75.33W
Cartagena, Spain **116 C2** 37.36N 0.59W
Cartago **136 D1** 9.50N 83.52W
Cartwright **133 M3** 53.40N 57.00W
Caruaru **143 G3** 8.15S 35.55W
Carupano **137 F2** 10.39N 63.14W
Carwarp **54 B3** 34.28S 142.12E
Casablanca **124 C5** 33.39N 7.35W
Casa Grande **134 C2** 32.50N 111.45W
Cascade Range **134 B3**
Casey, antarctic base **92 F3** 66.17S 110.32E
Casino **48 G5** 28.50S 153.02E
Casper **134 D3** 42.50N 106.20W
Caspian Depression **118 E3**
Caspian Sea **118 E3**
Castellon de la Plana **116 D2** 39.59N 0.03W
Castelo Branco **116 B2** 39.50N 7.30W
Castelvetrano **116 F2** 37.41N 12.47E
Casterton **54 A2** 37.36S 141.24E
Castle Hill **42 E6** 19.41S 144.14E
Castlemaine **54 C2** 37.04S 144.13E
Castlereagh River **48 E4**
Castres **116 D3** 43.36N 2.14E
Castries **137 F2** 14.01N 60.59W
Castro, Brazil **144 D4** 24.46S 50.03W
Castro, Chile **144 B2** 42.30S 73.46W
Castrovillari **117 G2** 39.48N 16.12E
Caswell Sound **90 B2**
Cat Island **137 E3**
Catalao **143 F2** 18.12S 47.54W
Catamarca **144 C4** 28.28S 65.46W
Catanduva **144 E4** 21.08S 48.58W
Catania **117 G2** 37.31N 15.06E
Catanzaro **117 G2** 38.54N 16.36E
Catbalogan **100 D4** 11.46N 124.55E
Catoche, Cape **136 D3** 21.38N 87.08W
Cattle Creek, homestead **80 C6** 17.35S 131.33E
Cauayan **100 D3** 9.58N 122.37E
Cauca River **142 C4**
Caucasus Mountains **118 E3**
Cauquenes **144 B3** 35.58S 72.22W
Caura River **137 F1**
Cavite **100 D4** 14.30N 120.54E
Cawndilla Lake **48 B3**
Caxias **143 F3** 4.53S 43.20W
Caxias do Sul **144 D4** 29.14S 51.10W
Cayenne **143 E4** 4.55N 52.18W
Cayman Islands **136 D3**
Cayman Trench **136 D2**
Cayman Brac, island **136 D2**
Cazombo **126 C3** 11.54S 22.56E
Cebu **100 D4** 10.17N 123.56E
Cebu, island **100 D4**
Cecina **116 F3** 43.18N 10.31E
Cedar City **134 C2** 37.40N 113.04W
Cedar Falls **134 E3** 42.34N 92.26W
Cedar Rapids **134 E3** 41.59N 91.39W
Cedros Island **134 C1**
Ceduna **66 D4** 32.08S 133.41E
Cefalu **117 F2** 38.03N 14.03E
Cegled **115 H2** 47.10N 19.47E
Cekhira **124 E5** 34.16N 10.03E
Celje **117 G4** 46.15N 15.16E
Celle **114 G3** 52.37N 10.05E
Cenderawasih, Gulf of **88 C3**
Central Range **88 D3**
Central African Republic **124 E2**
Central Java, province **100 B2**
Central Kalimatan, province **100 C2**
Centralia **132 F2** 46.43N 122.58W
Central Makran Range **106 B3**
Central Mount Wedge **80 C3** 22.52S 131.50E
Central Russian Uplands **118 D4**
Central Sulawesi, province **100 D2**
Central Siberian Plateau **118 J5**
Cerro de Pasco **142 C2** 10.43S 76.15W
Cesis **115 J4** 57.18N 25.18E
Ceske Budejovice **114 G2** 49.00N 14.30E
Cesky Krumlov **114 G2** 48.49N 14.20E
Cesme **117 I2** 38.19N 26.20E
Cessnock **48 F3** 32.51S 151.21E
Cetinje **117 G3** 42.23N 18.55E
Ceuta **116 B2** 35.53N 5.19W
Chabjuwardoo Bay **72 B4**
Chacorao Falls **142 E3** 6.20S 58.16W
Chad, Lake **124 E3**
Chad, republic **124 E3**
Chaga **106 B3** 29.20N 64.44E
Chagda **119 N4** 58.44N 130.38E
Chagos Archipelago **146 E2**
Chah Bahar **108 E3** 25.16N 60.41E
Chaiyaphum **100 B4** 15.46N 101.55E
Chake Chake **126 D4** 5.12S 39.44E
Chalky Inlet **90 B1**
Challenger Deep **84 J7**
Cham **114 G2** 49.13N 12.41E
Chambers Bay **80 C8**
Chambers Pillar, mountain **80 D2** 24.53S 133.49E
Chambery **116 E4** 45.34N 5.55E
Chamo, Lake **125 G2**
Champlain, Lake **135 G3**
Champoton **136 C2** 19.20N 90.43W
Chanaral **144 B4** 26.23S 70.40W
Chanaran **108 D4** 36.36N 59.05E
Chandigarh **106 C4** 30.43N 76.47E
Chandler **134 C2** 33.19N 111.49W
Chandrapur **106 C2** 19.58N 79.21E
Changbai Mountains **104 D8**
Changchun **103 G4** 43.50N 125.20E
Changde **102 F2** 29.03N 111.35E
Changhua **103 G2** 24.06N 120.31E
Changji **102 C4** 44.00N 87.20E
Changjin **104 C7** 40.21N 127.20E
Changsha **102 F2** 28.10N 113.00E
Changyon **104 B6** 38.16N 125.04E
Changzhou **103 F3** 31.45N 119.58E
Channel Islands **116 C4**
Chanthaburi **100 B4** 12.35N 102.08E
Chany, Lake **118 H4**
Chaozhou **103 F2** 23.43N 116.35E
Chapaevo **118 F4** 50.12N 51.09E
Chapala, Lake **136 B3**
Chappell Islands **60 E4**
Chapra **106 D3** 25.46N 84.44E
Charcot Island **92 O3**
Chardzhou **108 E4** 39.09N 63.34E
Chari River **124 E3**
Charikar **106 B4** 35.02N 69.13E
Charleroi **114 E3** 50.25N 4.27E
Charles, Cape **130 L4** 37.08N 75.59W
Charleston, South Carolina **135 G2** 32.48N 79.58W
Charleston, West Virginia **135 F2** 38.23N 81.40W
Charleville **42 F2** 26.24S 146.15E
Charlotte **135 F2** 35.12N 80.50W
Charlotte Amalie **137 F2** 18.22N 64.56W
Charlottesville **135 G2** 38.02N 78.29W
Charlottetown **133 L2** 46.14N 63.09W
Charlton **54 B2** 36.16S 143.21E
Charters Towers **42 F5** 20.05S 146.16E
Chartres **116 D4** 48.27N 1.30E
Chascomus **144 D3** 35.34S 58.00W
Chateaubriant **116 C4** 47.43N 1.22W
Chateaudun **116 D4** 48.04N 1.20E
Chateauroux **116 D4** 46.49N 1.41E
Chatellerault **116 D4** 46.49N 0.33E
Chatham **133 L2** 47.02N 65.30W
Chatham Islands **84 N1**
Chato, Mount **144 B2** 42.35S 72.04W
Chattahoochee River **135 F2**
Chattanooga **135 F2** 35.02N 85.18W
Chaumont **116 E4** 48.07N 5.08E
Chaves **116 B3** 41.44N 7.28W
Cheb **114 G3** 50.08N 12.20E
Cheboksary **118 E4** 56.08N 47.12E
Chech Desert **124 C4**
Chechaouen **116 B2** 35.10N 5.16W
Chechon **104 D5** 37.05N 128.09E
Cheju **104 C3** 33.31N 126.29E
Cheju Island **104 C3**
Cheju Strait **104 C3**
Chekhov **115 L4** 55.07N 37.30E
Chelkar **118 F3** 47.48N 59.39E
Chelyabinsk **118 G4** 55.12N 61.25E
Chelyuskin, Cape **118 K6** 77.44N 103.55E
Chenab River **106 C4**
Chennai **106 D2** 13.05N 80.18E
Chengdu **102 E3** 30.37N 104.06E
Cherbourg **116 C4** 49.38N 1.37W
Cheremkhovo **118 K4** 53.08N 103.01E
Cherepovets **118 D4** 59.09N 37.50E
Cherkassy **115 K2** 49.27N 32.04E
Cherkessk **118 E3** 44.14N 42.05E
Chernigov **115 K3** 51.30N 31.18E
Chernovtsy **115 J2** 48.19N 25.52E
Cherokee **134 E3** 42.45N 95.32W
Cherski Range **119 N5**
Cherskiy **119 Q5** 68.45N 161.15E
Chervonograd **115 I3** 50.25N 24.10E
Chesapeake Bay **135 G2**
Chesha Bay **118 E5**
Chester **114 D3** 53.12N 2.54W
Chetumal **136 D2** 18.30N 88.17W
Cheyenne **134 D3** 41.08N 104.50W
Cheyenne River **134 D3**
Chhatarpur **106 C3** 24.52N 79.35E
Chiai **103 G2** 23.38N 120.27E
Chiang Mai **100 A4** 18.48N 98.59E
Chiang Rai **100 A4** 19.56N 99.51E
Chiba **104 J4** 35.38N 140.07E
Chibougamau **133 K2** 49.56N 74.24W
Chicago **135 F3** 41.50N 87.45W
Chiclayo **142 C3** 6.47S 79.47W
Chico **134 B2** 39.43N 121.51W
Chico River **144 B2**
Chicoutimi **133 K2** 48.26N 71.06W
Chicualacuala **126 D2** 22.06S 31.42E
Chidley, Cape **133 L4** 60.30N 64.30W
Chifeng **103 F4** 42.17N 118.56E
Chigubo **126 D2** 22.50S 33.34E
Chihli, Gulf of **103 F3**
Chihuahua **136 B3** 28.40N 106.06W
Childers **42 I3** 25.14S 152.17E
Chile, republic **144 B3**
Chillan **144 B3** 36.37S 72.10W
Chiloe Island **144 B2**
Chilpancingo **136 C2** 17.33N 99.30W
Chilung **103 G2** 25.10N 121.43E
Chilwa, Lake **126 D3**
Chimborazo, mountain **142 C3** 1.29S 78.52W
Chimbote **142 C3** 9.04S 78.34W
Chimkent **118 G3** 42.16N 69.05E
Chimoio **126 D3** 19.08S 33.29E
Chin Island **104 C4**
China, people's republic **102 E3**
Chinandega **136 D2** 12.35N 87.10W
Chincha Alta **142 C2** 13.25S 76.07W
Chinchilla **42 H2** 26.44S 150.38E
Chinde **126 D3** 18.37S 36.24E
Chindwin River **107 E3**
Chingola **126 C3** 12.31S 27.53E
Chinhae **104 D4** 35.05N 128.47E
Chinhoyi **126 D3** 17.21S 30.13E
Chinju **104 D4** 35.10N 128.06E
Chioggia **116 F4** 45.13N 12.17E
Chios **117 I2** 38.23N 26.07E
Chipata **126 D3** 13.40S 32.42E
Chiquinquira **142 C4** 5.37N 73.51W
Chirchik **102 A4** 41.28N 69.31E
Chirfa **124 E4** 20.55N 12.22E
Chiriqui, Gulf of **136 D1**

*Entry / **Page** / **Ref.** / Lat. / Long.*

Chirripo Grande, mountain **136 D1** 9.31N 83.30W
Chita **119 L4** 52.03N 113.35E
Chitato **126 C4** 7.22S 20.47E
Chitipa **126 D4** 9.41S 33.19E
Chitre **136 D1** 7.59N 80.25W
Chittagong **107 E3** 22.20N 91.48E
Chitungwiza **126 D3** 18.00S 31.06E
Chivhu **126 D3** 19.01S 30.53E
Chizha **118 E5** 67.04N 44.20E
Choiseul, island **89 G2**
Chojnice **114 H3** 53.42N 17.32E
Chokurdakh **119 O6** 70.39N 147.50E
Cholet **116 C4** 47.04N 0.53W
Choluteca **136 D2** 13.15N 87.10W
Choma **126 C3** 16.51S 27.04E
Chonan **104 C5** 36.50N 127.09E
Chon Buri **100 B4** 13.24N 100.59E
Chongjin **104 D7** 41.50N 129.55E
Chongju, North Korea **104 B6** 39.39N 125.17E
Chongju, South Korea **104 C5** 36.39N 127.27E
Chongqing **102 E2** 29.30N 106.35E
Chongup **104 C4** 35.35N 126.50E
Chonju **104 C4** 35.50N 127.05E
Chonos Archipelago **144 B2**
Chopan **106 D3** 24.32N 83.02E
Choshi **104 J4** 35.43N 140.51E
Choybalsan **102 F4** 48.02N 114.32E
Christchurch **90 D3** 43.32S 172.37E
Christmas Island, Australia **100 B1**
Christmas Island, Kiribati **84 O6**
Christmas Creek, homestead **72 E5** 18.53S 125.55E
Chubut River **144 B2**
Chugoku Mountains **104 F4**
Chukchi Sea **119 S5**
Chukot Peninsula **119 S5**
Chukot Range **119 R5**
Chula Vista **134 C2** 32.38N 117.05W
Chulucanas **142 B3** 5.08S 80.10W
Chumikan **119 N4** 54.40N 135.15E
Chunchon **104 C5** 37.53N 127.45E
Chungju **104 C5** 36.59N 127.53E
Chuquicamata **142 D1** 22.20S 68.56W
Chur **116 E4** 46.52N 9.32E
Churchill **132 I3** 58.45N 94.00W
Churchill Lake **132 H3**
Churchill Peak **132 F3** 58.19N 125.04W
Churchill River, Manitoba **132 H3**
Churchill River, Newfoundland **133 L3**
Chuxiong **102 E2** 25.03N 101.33E
Cienaga **137 E2** 11.01N 74.15W
Cienfuegos **136 D3** 22.10N 80.27W
Cieza **116 C2** 38.14N 1.25W
Cimarron River **134 D2**
Cimpina **117 I4** 45.08N 25.44E
Cimpulung **117 I4** 45.16N 25.03E
Cincinnati **135 F2** 39.10N 84.30W
Cine **117 I2** 37.37N 28.03E
Cipatuja **100 B2** 7.45S 108.02E
Cirebon **100 B2** 6.46S 108.33E
Citlaltepetl, mountain **136 C2** 19.00N 97.18W
Cittanova **117 G2** 38.21N 16.05E
Ciudad Bolivar **142 D4** 8.06N 63.36W
Ciudad Camargo **136 B3** 27.41N 105.10W
Ciudad Guzman **136 B2** 19.40N 103.30W
Ciudad Juarez **136 B4** 31.42N 106.29W
Ciudad Obregon **136 B3** 27.28N 109.59W
Ciudad Rodrigo **116 B3** 40.36N 6.33W
Ciudad Victoria **136 C3** 23.43N 99.10W
Civitavecchia **116 F3** 42.05N 11.47E
Clare **66 G4** 33.50S 138.37E
Clarence River **48 G5**
Clarence Strait **80 C8**
Clarence Town **137 E3** 23.06N 74.58W
Clarion, island **85 T7**
Clarke Island **60 F4**
Clarksdale **135 E2** 34.12N 90.33W
Clarks Hill Lake **135 F2**
Clarksville **135 F2** 36.31N 87.21W
Clear, Cape **114 C3** 51.26N 9.31W
Clearwater Lake **133 K3**
Cleburne **136 C4** 32.21N 97.24W

*Entry / **Page** / **Ref.** / Lat. / Long.*

Clermont **42 F4** 22.50S 147.38E
Clermont-Ferrand **116 D4** 45.47N 3.05E
Cleve **66 F4** 33.42S 136.30E
Cleveland, Mississippi **135 E2** 33.43N 90.46W
Cleveland, Ohio **135 F3** 41.30N 81.41W
Cleveland, Tennessee **135 F2** 35.10N 84.51W
Cleveland, Mount **134 C3** 48.55N 113.50W
Clifton Hills, homestead **66 G7** 27.01S 138.54E
Clipperton, island **85 U7**
Cloncurry **42 C5** 20.42S 140.30E
Cloquet **134 E3** 46.43N 92.28W
Cloud Peak **134 D3** 44.23N 107.11W
Clovis **134 D2** 34.24N 103.11W
Cluj **117 H4** 46.47N 23.37E
Clunes **54 B2** 37.18S 143.47E
Clutha River **90 C1**
Coaldale **134 C2** 38.02N 117.52W
Coast Mountains **132 E3**
Coats Island **133 J4**
Coats Land **94 R2**
Coatzacoalcos **136 C2** 18.10N 94.25W
Coban **136 C2** 15.28N 90.20W
Cobar **48 C4** 31.32S 145.51E
Cobberas, Mount **55 E2** 36.50S 148.09E
Cobden **54 B1** 38.20S 143.04E
Cobija **142 D2** 11.02S 68.45W
Cobourg Peninsula **80 C9**
Cobram **54 C3** 35.55S 145.39E
Coburg **114 G3** 50.15N 10.58E
Cochabamba **142 D2** 17.26S 66.10W
Cochin **106 C1** 9.56N 76.15E
Cochrane **133 J2** 49.04N 81.02W
Cockburn **66 H4** 32.05S 141.00E
Coco River **136 D2**
Cocos Islands **84 E4**
Cocos Island **142 B4**
Cocuy, mountain **142 C4** 6.30N 72.21W
Cod, Cape **135 H3** 41.48N 69.55W
Codfish Island **90 B1**
Codo **143 F3** 4.28S 43.51W
Cody **132 H2** 44.31N 109.04W
Coen **42 D9** 13.57S 143.12E
Coeur d'Alene **134 C3** 47.40N 116.46W
Coffin Bay **66 E3**
Coffs Harbour **48 G4** 30.18S 153.08E
Cognac **116 C4** 45.42N 0.19W
Cohuna **54 C3** 35.49S 144.13E
Coiba Island **136 D1**
Coig River **144 B1**
Coimbatore **106 C2** 11.00N 76.57E
Coimbra **116 B3** 40.12N 8.25W
Coin **116 C2** 36.40N 4.45W
Colac **54 B1** 38.21S 143.35E
Colatina **143 F2** 19.35S 40.37W
Colbeck, Cape **92 K3** 77.00S 158.00W
Colbinabbin **54 C2** 36.35S 144.48E
Colemerik **108 C4** 37.36N 43.45E
Coleraine, Australia **54 A2** 37.36S 141.42E
Coleraine, U.K. **114 C4** 55.08N 6.40W
Coleridge, Lake **90 D3**
Coles Bay **60 F2** 42.08S 148.18E
Colesberg **126 C1** 30.44S 25.05E
Colhue Huapi, Lake **144 C2**
Colima **136 B2** 19.14N 103.41W
Colinas **143 F3** 6.02S 44.15W
Collarenebri **48 E5** 29.33S 148.35E
Collie **72 C2** 33.22S 116.09E
Collier Bay **72 D5**
Collinsville **42 F5** 20.34S 147.51E
Colmar **116 E4** 48.05N 7.21E
Colmenar Viejo **116 C3** 40.39N 3.46W
Cologne **114 F3** 50.56N 6.57E
Colombia, republic **142 C4**
Colombo **106 C1** 6.55N 79.52E
Colon, Cuba **136 D3** 22.42N 80.54W
Colon, Panama **136 E1** 9.22N 79.54W
Colorado Plateau **134 C2**
Colorado River, Argentina **144 C3**
Colorado River, Arizona **134 C2**
Colorado River, Texas **134 D2**
Colorado, state **134 D2**
Colorado Springs **134 D2** 38.50N 104.50W

*Entry / **Page** / **Ref.** / Lat. / Long.*

Columbia, Missouri **134 E2** 38.58N 92.20W
Columbia, South Carolina **135 F2** 34.00N 81.00W
Columbia, Tennessee **135 F2** 35.37N 87.02W
Columbia, Mount **132 G3** 52.09N 117.25W
Columbia Mountains **132 F3**
Columbia River **134 B3**
Columbine, Cape **126 B1** 32.50S 17.51E
Columbus, Georgia **135 F2** 32.28N 84.59W
Columbus, Ohio **135 F2** 39.59N 83.03W
Colville River **132 C4**
Colvocoresses Bay **94 F2**
Comayagua **136 D2** 14.30N 87.39W
Comet River **42 G4**
Comilla **107 E3** 23.28N 91.10E
Commander Islands **119 Q4**
Commonwealth Hill, homestead **66 E6** 29.57S 134.10E
Communism Peak **118 H2** 38.59N 72.01E
Comodoro Rivadavia **144 C2** 45.50S 67.30W
Comorin, Cape **106 C1** 8.04N 77.35E
Comoros, independent state **126 E3**
Compiegne **116 D4** 49.25N 2.50E
Conakry **124 B2** 9.30N 13.43W
Conara Junction **60 E3** 41.50S 147.26E
Concepcion, Chile **144 B3** 36.50S 73.03W
Concepcion, Paraguay **144 D4** 23.25S 57.17W
Concepcion del Uruguay **144 D3** 32.30S 58.15W
Conception, Point **134 B2** 34.27N 120.26W
Concord **135 G3** 43.13N 71.34W
Concordia, Argentina **144 D3** 31.25S 58.02W
Concordia, U.S.A. **134 E2** 39.35N 97.39W
Condamine River **42 H2**
Condobolin **48 D3** 33.05S 147.09E
Congo, Democratic Republic of **126 C4**
Congo, people's republic **126 B4**
Congo River **126 B5**
Connecticut, state **135 H2**
Connors Range **42 G5**
Conroe **136 C4** 30.19N 95.28W
Consolacion del Sur **136 D3** 22.32N 83.32W
Constanta **117 I3** 44.12N 28.40E
Constantine **124 D5** 36.22N 6.40E
Constitucion **144 B3** 35.20S 72.28W
Consuelo Peak **42 G3** 24.57S 148.10E
Conway, Mount **80 D3** 23.45S 133.25E
Coober Pedy **66 E6** 29.01S 134.45E
Cook **66 C5** 30.37S 130.25E
Cook, Mount **90 C3** 43.37S 170.08E
Cook Strait **90 F4**
Cook Islands, independent state **84 O4**
Cooktown **42 E8** 15.29S 145.15E
Coolah **48 E4** 31.50S 149.43E
Coolamon **48 D2** 34.49S 147.12E
Coolgardie **72 D2** 30.57S 121.09E
Coolibah, homestead **80 C7** 15.33S 130.56E
Cooma **48 E1** 36.15S 149.07E
Coonabarabran **48 E4** 31.16S 149.18E
Coonalpyn **66 G3** 35.42S 139.51E
Coonamble **48 E4** 30.57S 148.24E
Coondambo **66 E5** 31.03S 135.49E
Cooper Creek **66 G6**
Coorabie **38 D2** 31.54S 132.18E
Coordewandy, homestead **72 C3** 25.36S 115.58E
Coorong, The, lake **66 G3**
Cooroy **42 I2** 26.25S 152.55E
Coos Bay **134 B3** 43.21N 124.12W
Cootamundra **48 D2** 34.38S 148.02E
Copenhagen **114 G4** 55.43N 12.34E
Copeville **66 G3** 34.48S 139.50E
Copiapo **144 B4** 27.20S 70.23W
Coppermine River **132 G4**
Coquimbo **144 B4** 29.57S 71.25W
Corabia **117 H3** 43.46N 24.31E
Coral Sea **84 K4**
Corangamite, Lake **54 B1**
Corbeta Uruguay, antarctic base **92 Q2** 59.27S 27.15W

*Entry / **Page** / **Ref.** / Lat. / Long.*

Corbin **135 F2** 36.58N 84.06W
Corcovado, Gulf of **144 B2**
Cordillo Downs, homestead **66 H7** 26.42S 140.37E
Cordoba, Argentina **144 C3** 31.25S 64.11W
Cordoba, Spain **116 C2** 37.53N 4.46W
Cordova **132 D4** 60.30N 145.44W
Corfu **117 G2** 39.38N 19.55E
Corfu, island **117 G2**
Coricudgy, Mount **48 F3** 32.50S 150.21E
Corinth **117 H2** 37.56N 22.55E
Cork **114 C3** 51.54N 8.28W
Corner Inlet **54 D1**
Corner Brook **133 M2** 48.58N 57.58W
Cornwallis Island **132 I5**
Corny Point **66 F3** 34.54S 137.00E
Coro **142 D5** 11.27N 69.41W
Corocoro **142 D2** 17.10S 68.28W
Coromandel Peninsula **90 F6**
Corona, homestead **48 A4** 31.14S 141.25E
Coronel Oviedo **144 D4** 25.24S 56.30W
Coropuna, mountain **142 C2** 15.31S 72.45W
Corowa **48 D2** 36.00S 146.23E
Corpus Christi **134 E1** 27.47N 97.26W
Corral **144 B3** 39.55S 73.30W
Corrientes **144 D4** 27.30S 58.48W
Corrientes, Cape, Argentina **144 D3** 38.06S 57.33W
Corrientes, Cape, Colombia **142 C4** 5.29N 77.36W
Corrientes, Cape, Mexico **136 B3** 20.25N 105.42W
Corryong **55 D2** 36.12S 147.54E
Corsica, island **116 E3**
Cortez **134 D2** 37.22N 108.36W
Corum **108 B5** 40.31N 34.57E
Corumba **142 E2** 19.00S 57.35W
Corunna **116 B3** 43.22N 8.24W
Coruripe **143 G2** 10.09S 36.09W
Corvallis **134 B3** 44.34N 123.17W
Cosenza **117 G2** 39.17N 16.16E
Costa Rica, republic **136 D1**
Cotopaxi, mountain **142 C3** 0.40S 78.28W
Cottage Grove **134 B3** 43.48N 123.05W
Cottbus **114 G3** 51.43N 14.21E
Couedic, Cape du **66 F2** 36.04S 136.42E
Coulman Island **92 I3**
Council Bluffs **134 E3** 41.14N 95.54W
Courtenay **132 F2** 49.40N 124.58W
Covilha **116 B3** 40.17N 7.30W
Cowan, Lake **72 D2**
Cowell **66 F4** 33.41S 136.55E
Cowes **54 C1** 38.27S 145.15E
Cowra **48 E3** 33.50S 148.41E
Cox Bight **60 D1**
Coxim **143 E2** 18.28S 54.45W
Cozumel Island **136 D3**
Cradle Mountain **60 C3** 41.41S 145.57E
Cradock **126 C1** 32.10S 25.37E
Craig **134 D3** 40.32N 107.31W
Craiova **117 H3** 44.18N 23.47E
Cranbourne **54 C1** 38.06S 145.17E
Cranbrook **132 G2** 49.29N 115.48W
Crato **143 G3** 7.10S 39.25W
Crescent, Lake **60 E2**
Crescent City **134 B3** 41.46N 124.13W
Cressy, Tasmania **60 E3** 41.41S 147.04E
Cressy, Victoria **54 B1** 38.02S 143.38E
Creston **134 E3** 41.04N 94.20W
Creswick **54 B2** 37.26S 143.54E
Crete, island **117 I1**
Crete, Sea of **125 F5**
Cretin, Cape **89 E2** 6.40S 147.53E
Crib Point **54 C1** 38.22S 145.12E
Crimea, peninsula **115 K2**
Cristobal Colon, mountain **137 E2** 10.53N 73.48W
Croatia, republic **117 G4**
Crockett **136 C4** 31.21N 95.30W
Croker, Cape **88 B1** 10.59S 132.36E
Croker Island **80 D9**
Cromwell **90 C2** 45.03S 169.14E
Crooked Island **137 E3**
Crookston **134 E3** 47.47N 96.36W

Entry / ***Page*** / ***Ref.*** / *Lat.* / *Long.*

Crookwell **48 E2** 34.28S 149.29E
Cross, Cape **126 B2** 21.49S 13.57E
Crotone **117 G2** 39.05N 17.08E
Crown Island **89 E2**
Croydon **42 D6** 18.12S 142.15E
Crozet Islands **92 C2**
Cruces **136 D3** 22.20N 80.17W
Cruz, Cape **136 E2** 19.52N 77.44W
Cruzeiro do Sul **142 C3** 7.40S 72.39W
Crystal Brook **66 G4** 33.21S 138.12E
Cuango **126 B4** 9.10S 17.59E
Cuanza River **122 E4**
Cuba, republic **136 E3**
Cucuta **142 C4** 7.55N 72.31W
Cuddalore **106 C2** 11.43N 79.46E
Cuddapah **106 C2** 14.30N 78.50E
Cue **72 C3** 27.26S 117.54E
Cuenca, Ecuador **142 C3** 2.54S 79.00W
Cuenca, Spain **116 C3** 40.04N 2.07W
Cuernavaca **136 C2** 18.57N 99.15W
Cuero **134 E1** 29.06N 97.19W
Cuiaba **143 E2** 15.35S 56.05W
Cuima **126 B3** 13.16S 15.39E
Cuito River **126 B3**
Culcairn **48 D2** 35.40S 147.02E
Culiacan **136 B3** 24.50N 107.23W
Cullera **116 C2** 39.10N 0.15W
Culver, Point **72 D2** 32.54S 124.45E
Cumana **142 D5** 10.29N 64.12W
Cumberland Sound **133 L4**
Cummins **66 E3** 34.16S 135.44E
Cunani **143 E4** 2.55N 51.07W
Cunene River **122 E3**
Cuneo **116 E3** 44.24N 7.33E
Cunnamulla **42 E1** 28.04S 145.40E
Curacao, island **142 D5**
Curanilahue **144 B3** 37.29S 73.28W
Curico **144 B3** 35.00S 71.15W
Curitiba **144 E4** 25.24S 49.16W
Curnamona, homestead **66 G5** 31.39S 139.33E
Curoca River **126 B3**
Curraweena, homestead **48 C4** 30.43S 145.52E
Currie **60 A5** 39.56S 143.52E
Curtin Springs, homestead **80 C2** 25.19S 131.45E
Curtis Group **60 D5**
Curtis Island, Australia **42 H4**
Curtis Island, New Zealand **84 N2**
Curtis, Port **42 H4**
Curuzu Cuatia **144 D4** 29.50S 58.05W
Cuttack **106 D3** 20.26N 85.56E
Cuvelai **126 B3** 15.44S 15.50E
Cuzco **142 C2** 13.32S 71.57W
Cygnet **60 E1** 43.10S 147.05E
Cyprus, republic **110 H1**
Czech, republic **114 H2**
Czestochowa **117 G5** 50.49N 19.07E

D

Daba Mountains **102 E3**
Dabola **124 B3** 10.48N 11.02W
Dacca **107 E3** 23.42N 90.22E
Dahlak Archipelago **125 H3**
Dai Island **89 H2**
Daimiel **116 C2** 39.04N 3.37W
Dairut **125 G4** 27.34N 30.48E
Dajarra **42 B5** 21.42S 139.31E
Dakar **124 B3** 14.38N 17.27W
Dakhla **124 B4** 23.43N 15.57W
Dakshin Gangotri, antarctic base **92 A3** 70.05S 12.00E
Dalaba **124 B3** 10.47N 12.12W
Dalandzadagad **102 E4** 43.34N 104.20E
Da Lat **100 B4** 11.56N 108.25E
Dalbandin **106 B3** 28.56N 64.30E
Dalby **42 H2** 27.11S 151.15E
Dalgety Downs, homestead **72 C3** 25.17S 116.12E
Dalhart **134 D2** 36.05N 102.31W
Dalhousie **133 L2** 48.03N 66.22W
Dali **102 D2** 25.45N 100.06E
Dallas **134 E2** 32.47N 96.48W
Dalma Island **108 D3**
Daloa **124 C2** 6.56N 6.28W
Dalrymple, Lake **42 F5**
Daltonganj **106 D3** 24.02N 84.07E
Dalwallinu **72 C2** 30.17S 116.40E
Daly River **80 C8**
Daly River **38 D4** 13.45S 130.41E
Daly Waters **80 D6** 16.16S 133.22E
Daman **106 C3** 20.25N 72.58E
Damanhur **108 B4** 31.03N 30.28E
Damar, island **88 A2**
Damascus **108 B4** 33.30N 36.19E
Damavand Peak **108 D4** 35.56N 52.08E
Damba **126 B4** 6.44S 15.29E
Damghan **108 D4** 36.09N 54.22E
Dammam **108 D3** 26.26N 50.07E
Dampier **72 C4** 20.39S 116.43E
Dampier Land **72 D5**
Dampier Strait **88 B3**
Danakil Depression **125 G3**
Da Nang **100 B4** 16.04N 108.14E
Dandong **103 G3** 40.08N 124.24E
Danger Point **126 B1** 34.37S 19.17E
Dangila **125 G3** 11.18N 36.56E
Dannevirke **90 G4** 40.12S 176.06E
Danube River **116 F4**
Danville **135 G2** 36.34N 79.25W
Daqing **103 G4** 46.37N 124.59E
Darbhanga **106 D3** 26.10N 85.54E
Dar-es-Salaam **126 D4** 6.51S 39.18E
Dareton **48 B2** 34.05S 142.04E
Darfield **90 E3** 43.29S 172.07E
Dargaville **90 E7** 35.56S 173.52E
Darhan **102 E4** 49.30N 105.58E
Darien, Gulf of **137 E1**
Darjeeling **102 C2** 27.02N 88.20E
Darling River **48 B4**
Darlington Point **48 C2** 34.34S 146.00E
Darmstadt **114 F2** 49.52N 8.39E
Darnley, Cape **92 D3** 67.16S 69.53E
Dartmoor **54 A2** 37.55S 141.16E
Dartmouth **133 L2** 44.40N 63.35W
Dartmouth, Lake **55 D2**
Daru **88 D2** 9.05S 143.10E
Darvel Bay **100 C3**
Darwin **80 C8** 12.27S 130.50E
Darwin, Mount **144 C1** 54.45S 69.20W
Darwin, Port **80 C8**
Date **104 J8** 42.20N 140.53E
Datong **102 F4** 40.12N 113.12E
Datu, Cape **100 B3** 2.06N 109.37E
Datumakuta **100 C3** 2.30N 117.55E
Daugavpils **115 J4** 55.52N 26.31E
Daulat Yar **106 B4** 34.33N 65.46E
Davangere **106 C2** 14.30N 75.52E
Davao **100 D3** 7.05N 125.38E
Davenport Range **40 D3**
Davey, Port **60 C1**
David **136 D1** 8.26N 82.26W
Davis, antarctic base **92 D3** 68.35S 77.58E
Davis Sea **92 E2**
Davis Strait **133 M4**
Dawa River **125 G2**
Dawson **132 E4** 64.04N 139.24W
Dawson River **40 F3**
Dawson Creek **132 F3** 55.46N 120.14W
Dax **116 C3** 43.43N 1.03W
Daxue Mountains **102 E3**
Daylesford **54 C2** 37.21S 144.09E
Daytona Beach **135 F1** 29.11N 81.01W
Dayu **100 C2** 2.00S 115.00E
De Aar **126 C1** 30.40S 24.01E
Dead Sea **108 B4**
Deakin **72 E2** 30.46S 128.58E
Deal Island **60 E5**
Dean Funes **144 C3** 30.25S 64.22W
Death Valley **134 C2**
Deblin **115 I3** 51.34N 21.50E
Debra Tabor **125 G3** 11.50N 38.26E
Debrecen **115 I2** 47.30N 21.37E
Debre Markos **125 G3** 10.19N 37.41E
Decamere **125 G3** 15.06N 39.00E
Decatur **135 F2** 39.51N 88.57W
Deccan Plateau **106 C3**
Dee River **60 D2**
Deering **132 B4** 66.04N 162.50W
Deering, Mount **72 E3** 25.06S 128.55E
De Grey River **72 C4**
Dehra Dun **106 C4** 30.19N 78.03E
Deir-ez-Zor **108 C4** 35.20N 40.09E
Dej **117 H4** 47.08N 23.55E
Delagoa Bay **126 D2**
De Land **136 D3** 29.02N 81.18W
Delano **134 C2** 35 45N 119.16W
Delano Peak **134 C2** 38.23N 112.22W
Delaware Bay **135 G2**
Delaware, state **135 H2**
Delegate **48 E1** 37.03S 148.57E
Delhi **135 G3** 42.17N 74.57W
Delicias **136 B3** 28.10N 105.30W
Deloraine **60 D3** 41.32S 146.39E
Del Rio **134 D1** 29.23N 100.56W
Dembidolo **125 G2** 8.34N 34.50E
Deming **134 D2** 32.17N 107.46W
Demopolis **136 D4** 32.31N 87.50W
Demta **88 D3** 2.22S 140.06E
Denau **118 G2** 38.20N 67.54E
Dendi, mountain **125 G2** 8.50N 37.43E
Denham **72 B3** 25 56S 113.32E
Den Helder **114 E3** 52.58N 4.46E
Denial Bay **66 D4**
Deniliquin **48 C2** 35.32S 144.57E
Denizli **108 A4** 37.46N 29.05E
Denman **48 F3** 32.23S 150.41E
Denmark **72 C2** 34.58S 117.21E
Denmark, kingdom **114 F4**
Denmark Strait **133 O4**
Denpasar **100 C2** 8.40S 115.14E
Denton **134 E2** 33.14N 97.18W
D'Entrecasteaux Channel **60 D1**
D'Entrecasteaux Islands **89 F2**
D'Entrecasteaux, Point **72 C2** 34.50S 116.00E
D'Entrecasteaux Reefs **84 K4**
Denver **134 D2** 39.45N 105.00W
Derbent **108 C5** 42.03N 48.18E
Derby, Tasmania **60 E3** 41.09S 147.47E
Derby, U.K. **114 D3** 52.55N 1.30W
Derby, Western Australia **72 D5** 17.19S 123.38E
Derj **124 E5** 30.10N 10.29E
Derna **124 F5** 32.46N 22.39E
Dernberg, Cape **126 B2** 27.45S 15.33E
Derrinallum **54 B2** 37.57S 143.13E
Derwent, River **60 D2**
Derwent Bridge **60 D2** 42.08S 146.14E
Dese **125 G3** 11.05N 39.40E
Deseado River **144 C2**
Des Moines **134 E3** 41.35N 93.35W
Dessau **114 G3** 51.51N 12.15E
Dete **126 C3** 18.38S 26.50E
Detroit **135 F3** 42.23N 83.05W
Deva **117 H4** 45.53N 22.55E
Devil's Island **143 E4**
Devils Lake **134 E3** 48.08N 98.50W
Devils Marbles, mountain **80 E4** 20.34S 134.16E
Devon Island **132 I5**
Devonport **60 D3** 41.09S 146.16E
Dey Dey, Lake **66 C6**
Deyong, Cape **88 C2** 6.56S 138.33E
Dezful **108 C4** 32.23N 48.28E
Dhaba **108 B3** 27.19N 35.46E
Dhahran **108 D3** 26.18N 50.05E
Dhamtari **106 D3** 20.43N 81.36E
Dhanbad **106 D3** 23.47N 86.32E
Dhenkanal **106 D3** 20.40N 85.39E
Dhoraji **106 C3** 21.42N 70.32E
Dhulia **106 C3** 20.54N 74.47E
Diamantina **143 F2** 18.17S 43.37W
Diamantina River **42 C3**
Diamond Head **48 G4** 31.44S 152.48E
Dibrugarh **107 E3** 27.29N 94.56E
Dickinson **134 D3** 46.52N 102.49W
Dieppe **116 D4** 49.55N 1.05E
Diffa **124 E3** 13.28N 12.35E
Digne **116 E3** 44.05N 6.14E
Digoin **116 D4** 46.29N 3.54E
Digul River **88 D2**
Dijon **116 D4** 47.20N 5.02E
Dikson **118 I6** 73.32N 80.39E
Dikwa **124 E3** 12.01N 13.55E
Dilam **108 C3** 23.59N 47.06E
Dili **100 D2** 8.35S 125.35E
Dillingham **132 C3** 59.03N 158.30W
Dimboola **54 B2** 36.27S 142.02E
Dinagat, island **100 D4**
Dinan **116 C4** 48.27N 2.02W
Dinaric Alps **117 G3**
Dindigul **106 C2** 10.23N 78.00E
Dipolog **100 D3** 8.34N 123.28E
Dire Dawa **125 H2** 9.35N 41.50E
Dirk Hartog Island **72 B3**
Dirranbandi **42 G1** 28.35S 148.14E
Disappointment Islands **85 Q4**
Disappointment, Lake **72 D4**
Disaster Bay **48 F1**
Discovery Bay **54 A1**
Disko Bay **133 M4**
Disko Island **133 M5**
Dispur **107 E3** 26.03N 91.52E
Diu **106 C3** 20.41N 71.03E
Diwaniyah **108 C4** 32.00N 44.57E
Diyarbakir **108 C4** 37.55N 40.14E
Dja River **124 E2**
Djambala **126 B4** 2.32S 14.43E
Djanet **124 D4** 24.34N 9.30E
Djelfa **124 D5** 34.43N 3.14E
Djerid, Lake **124 D5**
Djibouti **125 H3** 11.36N 43.09E
Djibouti, republic **125 H3**
Djougou **124 D2** 9.40N 1.47E
Dmitri Laptev Strait **119 N6**
Dmitrov **115 L4** 56.21N 37.34E
Dneprodzerzhinsk **115 K2** 48.30N 34.37E
Dnepropetrovsk **115 L2** 48.29N 35.00E
Dnieper River **115 K3**
Dniester River **115 J2**
Doberai Peninsula **88 B3**
Dobo **88 B2** 5.48S 134.13E
Dodge City **134 D2** 37.45N 100.02W
Dodoma **126 D4** 6.10S 35.40E
Dogo, island **104 F5**
Doha **108 D3** 25.15N 51.36E
Dolak, island **88 C2**
Dole **116 E4** 47.05N 5.30E
Dolo **125 H2** 4.11N 42.03E
Dolores **144 D3** 36.23S 57.44W
Dolphin Head **126 B2** 25.43S 14.50E
Dom, mountain **88 C3** 2.40S 136.53E
Domazlice **114 G2** 49.26N 12.57E
Dombey, Cape **80 B8** 13.53S 129.43E
Dominica, republic **137 F2**
Dominican Republic **137 E3**
Don River **118 E3**
Donald **54 B2** 36.22S 142.59E
Dondra Head **106 D1** 5.55N 80.35E
Donets River **115 L2**
Donetsk **115 L2** 48.00N 37.50E
Dongara **72 C3** 29.15S 114.56E
Dongchuan **102 E2** 26.10N 103.02E
Donggala **100 C2** 0.38S 119.45E
Dongguan **102 F2** 23.03N 113.43E
Dong Hoi **100 B4** 17.32N 106.35E
Dongola **125 G3** 19.10N 30.27E
Dongsheng **102 F3** 39.51N 110.00E
Dongting Lake **102 F2**
Donnybrook **72 C2** 33.35S 115.49E
Dora, Lake **72 D4**
Dordrecht **116 D5** 51.48N 4.40E
Dorgon Lake **102 D4**
Dori **124 C3** 14.03N 0.02W
Dorohoi **117 I4** 47.57N 26.31E
Dorre Island **72 B3**
Dorrigo **48 G4** 30.20S 152.43E
Dortmund **114 F3** 51.32N 7.27E
Dothan **135 F2** 31.12N 85.25W
Douala **124 D2** 4.04N 9.43E
Douarnenez **116 C4** 48.05N 4.20W
Double Island Point **42 I3** 25.56S 153.11E
Doubtful Sound **90 B2**
Doubtless Bay **90 E7**
Douglas, U.K. **114 D3** 54.09N 4.29W
Douglas, U.S.A. **136 D4** 31.30N 82.54W

*Entry / **Page** / **Ref.** / Lat. / Long.*

Douro River **116 B3**
Dover, Australia **60 D1** 43.19S 147.01E
Dover, U.K. **114 E3** 51.08N 1.19E
Dozen, island **104 F5**
Drake Passage **92 P2**
Drakensberg Mountains **122 F1**
Drama **117 H3** 41.10N 24.11E
Drammen **114 G4** 59.45N 10.15E
Dresden **114 G3** 51.03N 13.45E
Drogheda **114 C3** 53.43N 6.21W
Drogobych **115 I2** 49.20N 23.30E
Dromedary, Cape **48 F1** 36.18S 150.08E
Drouin **54 C1** 38.08S 145.51E
Drumheller **132 G3** 51.30N 112.46W
Drummondville **135 G3** 45.52N 72.30W
Druzhina **119 O5** 68.11N 145.19E
Dryden **132 I2** 49.48N 92.48W
Drysdale River **72 E5**
Duarte Peak **137 E2** 19.02N 70.59W
Dubai **108 D3** 25.14N 55.17E
Dubawnt Lake **132 H4**
Dubawnt River **132 H4**
Dubbo **48 E3** 32.15S 148.37E
Dublin, Ireland **114 C3** 53.20N 6.15W
Dublin, U.S.A. **135 F2** 32.31N 82.54W
Dubna **115 L3** 54.09N 36.59E
Dubno **115 J3** 50.28N 25.40E
Dubrovnik **117 G3** 42.40N 18.07E
Dubuque **134 E3** 42.31N 90.41W
Duchesne **134 C3** 40.10N 110.24W
Duchess **42 B5** 21.21S 139.52E
Ducie Island **85 S3**
Dudinka **118 I5** 69.27N 86.13E
Dudley **114 D3** 52.30N 2.05W
Duisburg **114 F3** 51.26N 6.45E
Duke of Gloucester Islands **85 P3**
Dukhan **108 D3** 25.25N 50.48E
Dukou **102 E2** 26.30N 101.45E
Duluth **134 E3** 46.45N 92.10W
Dumaguete **100 D3** 9.20N 123.18E
Dumaresq River **48 F5**
Dumas **134 D2** 35.51N 101.57W
Dumfries **114 D4** 55.04N 3.37W
Dumont d'Urville, antarctic base **92 H3** 68.40S 140.01E
Dunafoldvar **117 G4** 46.48N 18.55E
Duncan **134 E2** 34.30N 97.57W
Dundalk **114 C3** 54.01N 6.25W
Dundas **133 L5** 76.35N 68.30W
Dundas, Lake **72 D2**
Dundas Strait **80 C9**
Dundee, South Africa **126 D2** 28.10S 30.15E
Dundee, U.K. **114 D4** 56.28N 3.00W
Dunedin **90 D2** 45.53S 170.31E
Dungog **48 F3** 32.24S 151.45E
Dungun **100 B3** 4.45N 103.24E
Dunk Island **42 F7**
Dunkeld **54 B2** 37.39S 142.21E
Dunkirk **116 D5** 51.02N 2.23E
Dun Laoghaire **114 C3** 53.17N 6.08W
Dunmarra, homestead **80 D6** 16.42S 133.25E
Dunolly **54 B2** 36.52S 143.44E
Durack River **72 E5**
Durango, Mexico **136 B3** 24.01N 104.40W
Durango, U.S.A. **134 D2** 37.15N 107.55W
Durban **126 D2** 29.53S 31.00E
Durham, U.K. **114 D3** 54.47N 1.34W
Durham, U.S.A. **135 G2** 36.00N 78.54W
Durres **117 G3** 41.18N 19.28E
Dursunbey **117 I2** 39.35N 28.37E
D'Urville Island **90 E4**
Dushanbe **118 G2** 38.38N 68.51E
Dusky Sound **90 B2**
Dusseldorf **114 F3** 51.13N 6.47E
Duyfken Point **42 C9** 12.34S 141.35E
Duyun **102 E2** 26.16N 107.29E
Dyke Ackland Bay **89 E2**
Dysart **42 G4** 22.37S 148.21E
Dzerzhinsk **115 M4** 56.15N 43.30E
Dzhambul **118 H3** 42.50N 71.25E
Dzhankoi **115 K2** 45.40N 34.20E
Dzhizak **118 G3** 40.06N 67.45E
Dzugdzhur Range **119 N4**

E

Eagle Point **89 E1** 10.33S 149.51E
Eaglehawk Neck **60 E1** 43.01S 147.55E
Eagle Pass **134 D1** 28.44N 100.31W
East Cape, New Zealand **90 H6** 37.41S 178.33E
East Cape, P.N.G. **89 F3** 4.15S 153.00E
East China Sea **103 G2**
Easter Island **85 T3**
Eastern Ghats, mountains **106 D2**
Eastern Sierra Madre, mountains **136 B3**
East Falkland, island **144 D1**
East Hampton **133 K2** 40.58N 72.11W
East Java, province **100 C2**
East Kalimantan, province **100 C3**
East Korea Bay **104 C6**
East London **126 C1** 33.00S 27.54E
Eastmain **133 K3** 52.10N 78.30W
Eastmain River **133 K3**
East Nusa Tenggara, province **100 D2**
East Timor, republic **100 D2**
East Siberian Sea **119 P6**
East Sister Island **60 E5**
Eau Claire **134 E3** 44.50N 91.30W
Eauripik, island **101 F3**
Eberswalde **114 G3** 52.50N 13.50E
Ebolowa **124 E2** 2.56N 11.11E
Ebon, island **84 L6**
Ebro River **116 C3**
Ech Cheliff **124 D5** 36.11N 1.21E
Echo, Lake **60 D2**
Echuca **54 C2** 36.08S 144.45E
Ecuador, republic **142 B3**
Edd **125 H3** 13.57N 41.38E
Ed Damer **125 G3** 17.37N 33.59E
Ed Dueim **125 G3** 14.00N 32.20E
Eddystone Point **60 F4** 41.00S 148.21E
Edea **124 E2** 3.47N 10.13E
Eden **48 F1** 37.04S 149.55E
Edenhope **66 H2** 37.02S 141.17E
Edessa **117 H3** 40.48N 22.03E
Edgeworth David **92 F3** 65.50S 102.00E
Edinburgh **114 D4** 55.57N 3.13W
Edirne **108 A5** 41.40N 26.34E
Edithburgh **66 F3** 35.05S 137.45E
Edmond **134 E2** 35.40N 97.30W
Edmonton **132 G3** 53.34N 113.25W
Edmundston **135 H3** 47.22N 68.20W
Edremit **117 I2** 39.34N 27.01E
Edson **132 G3** 53.36N 116.28W
Edward, Lake **126 C4**
Edward River, homestead **42 C8** 14.54S 141.36E
Edward VII Land **94 K2**
Efate, island **84 L4**
Efogi **89 E2** 9.10S 147.40E
Egea de Los Caballeros **116 C3** 42.07N 1.05W
Eger **115 I2** 47.53N 20.28E
Egersund **114 F4** 58.27N 6.01E
Egerton, Mount **72 C4** 24.46S 117.42E
Egmont, Cape **90 E5** 39.17S 173.45E
Egmont, Mount **90 E5** 39.18S 174.04E
Egypt, Arab republic **125 F4**
Eiao, island **85 Q5**
Eidsvoll **114 G5** 60.19N 11.14E
Eighty Mile Beach **72 D5**
Eil **125 H2** 8.00N 49.49E
Eildon, Lake **54 D2**
Eisenach **114 G3** 50.59N 10.19E
Ekonda **118 K5** 65.59N 103.55E
El Aaiun **124 B4** 27.10N 13.11W
El Alamein **125 F5** 30.48N 28.58E
Elath **108 B3** 29.33N 34.57E
Elato, island **101 F3**
El Atrun Oasis **125 F3** 18.12N 26.40E
Elazig **108 B4** 38.41N 39.14E
Elba, island **116 F3**
El Banco **142 C4** 9.04N 73.59W
Elbasan **117 H3** 41.07N 20.05E
El Bayadh **116 D1** 33.40N 1.00E
Elbe River **114 F3**
Elbert, Mount **134 D2** 39.05N 106.27W
Elblag **114 H3** 54.10N 19.25E
Elbrus, Mount **112 I2** 43.21N 42.29E
Elburz Mountains **108 D4**
El Cajon **134 C2** 32.48N 116.58W
El Callao **142 D4** 7.18N 61.50W
Elche **116 C2** 38.16N 0.41W
El Djem **124 E5** 35.15N 10.15E
Eldon Peak **60 C3** 41.58S 145.44E
El Dorado **134 E2** 37.51N 96.52W
Elektrostal **115 L4** 55.46N 38.30E
El Encanto **142 C3** 1.37S 73.12W
Eleuthera, island **137 E3**
El Ferrol **116 B3** 43.29N 8.14W
Elgin **114 D4** 57.39N 3.20W
Elgin Downs, homestead **42 F4** 22.00S 146.55E
El Gir **125 F3** 19.50N 28.18E
El Golea **124 D5** 30.35N 2.51E
Elgon, Mount **126 D5** 1.07N 34.35E
El Haseke **108 C4** 36.29N 40.45E
Eliase **88 B2** 8.21S 130.48E
Elista **118 E3** 46.18N 44.14E
Elizabeth City **135 G2** 36.18N 76.16W
El Jadida **116 B1** 33.19N 8.35W
El Jauf **124 F4** 24.09N 23.19E
Elk **115 I3** 53.51N 22.20E
Elkedra, homestead **80 E4** 21.10S 135.27E
El Kharga **125 G4** 25.27N 30.32E
Elko **134 C3** 40.50N 115.46W
Ellery, Mount **55 E2** 37.24S 148.47E
Ellesmere Island **132 J5**
Ellesmere, Lake **90 E3**
Elliot Lake **135 F3** 46.24N 82.41W
Elliott **80 D6** 17.33S 133.32E
Elliott Bay **60 C1**
Elliston **66 E4** 33.39S 134.53E
Ellsworth Land **94 N2**
El Milia **116 E2** 36.48N 6.14E
Elmore **54 C2** 36.30S 144.40E
El Nido **100 C4** 11.14N 119.25E
El Oued **124 D5** 33.30N 6.45E
El Paso **134 D2** 31.45N 106.30W
El Salvador, republic **136 C2**
Elsey, homestead **80 D7** 14.58S 133.20E
El Shab **125 F4** 22.34N 29.52E
Eltham **90 F5** 39.26S 174.18E
El Tigre **142 D4** 8.44N 64.18W
El Tur **108 B3** 28.14N 33.36E
Eluru **106 D2** 16.45N 81.10E
Elverum **114 G5** 60.54N 11.33E
El Wak **126 E5** 2.48N 40.55E
Ely **134 C2** 39.15N 114.53W
Emden **114 F3** 53.23N 7.13E
Emerald, Queensland **42 G4** 23.30S 148.10E
Emerald, Victoria **54 C2** 37.56S 145.26E
Emeti **88 D2** 7.52S 143.15E
Emi Koussi, mountain **124 E3** 19.52N 18.31E
Emirau Island **89 E3**
Emita **60 E5** 40.00S 147.55E
Emmaville **48 F5** 29.27S 151.36E
Emmen **114 F3** 52.47N 6.55E
Empedrado **144 D4** 27.50S 58.46W
Emperor Seamounts **84 L9**
Emporia **134 E2** 38.24N 96.10W
Empress Augusta Bay **89 F2**
Emu Park **42 H4** 23.16S 150.50E
Encarnacion **144 D4** 27.20S 55.50W
Encounter Bay **66 G3**
Ende **100 D2** 8.51S 121.40E
Endeavour Strait **42 C10**
Enderbury, island **84 N5**
Enderby Land **94 C2**
Engano, Cape **100 D4** 18.35N 122.13E
Englewood **134 D2** 39.39N 105.00W
English Channel **114 E3**
English Company's Islands, The **80 F9**
Enid **134 E2** 36.24N 97.54W
Eniwetok, island **84 L7**
En Nahud **125 F3** 12.41N 28.28E
Ennedi Plateau **124 F3**
Ennis **114 C3** 52.50N 8.59W
Enniskillen **114 C3** 54.21N 7.38W
Ennisvale, homestead **48 A3** 33.13S 141.10E
Enschede **114 F3** 52.13N 6.55E
Ensenada **134 C2** 31.53N 116.35W
Entebbe **126 D5** 0.05N 32.29E
Enugu **124 D2** 6.20N 7.29E
Epernay **116 D4** 49.02N 3.58E
Epinal **116 E4** 48.10N 6.28E
Eppalock, Lake **54 C2**
Equatorial Guinea, republic **124 D2**
Eran **100 C3** 9.04N 117.44E
Erave **89 D2** 6.37S 143.54E
Erbil **108 C4** 36.11N 44.01E
Erciyas, Mount **108 B4** 38.32N 35.27E
Erebus, Mount **94 I2** 77.40S 167.20E
Erenhot **102 F4** 43.50N 112.00E
Erfurt **114 G3** 50.58N 11.02E
Erie **135 F3** 42.07N 80.05W
Erie, Lake **135 F3**
Erigavo **125 H3** 10.40N 47.20E
Erimo, Cape **104 K7** 41.55N 143.13E
Eritrea, republic **125 G3**
Erlangen **114 G2** 49.36N 11.02E
Erldunda, homestead **80 D2** 25.13S 133.11E
Ernabella **66 D7** 26.17S 132.08E
Ernest Giles Range **72 D3**
Erode **106 C2** 11.21N 77.43E
Eromanga **42 D2** 26.40S 143.16E
Errol **133 K2** 44.47N 71.10W
Erromango, island **84 M4**
Er Rosieres **125 G3** 11.52N 34.23E
Erzincan **108 B4** 39.44N 39.30E
Erzurum **108 C4** 39.57N 41.17E
Esa'ala **89 F2** 9.45S 150.50E
Esan, Cape **104 J7** 41.49N 141.12E
Esashi **104 K9** 44.58N 142.38E
Esbjerg **114 F4** 55.28N 8.28E
Escanaba **135 F3** 45.47N 87.04W
Escondido **134 C2** 33.07N 117.07W
Escuinapa **136 B3** 22.49N 105.44W
Escuintla **136 C2** 14.18N 90.47W
Esfahan **108 D4** 32.41N 51.41E
Esk **42 I2** 27.14S 152.25E
Eskisehir **108 B4** 39.46N 30.30E
Esmeraldas **142 C4** 0.56N 79.40W
Espanola **134 D2** 35.59N 106.05W
Esperance **72 D2** 33.52S 121.54E
Esperance Bay **72 D2**
Esperanza, antarctic base **92 P3** 63.24S 56.59W
Espiritu Santo, island **84 L4**
Espoo **115 I5** 60.10N 24.42E
Essaouira **124 C5** 31.30N 9.48W
Essen **114 F3** 51.27N 6.57E
Essequibo River **142 E4**
Es Sider **124 E5** 30.38N 18.20E
Es Suweida **108 B4** 32.43N 36.33E
Estella **116 C3** 42.41N 2.02W
Estepona **116 B2** 36.26N 5.09W
Estevan **132 H2** 49.11N 102.57W
Estherville **134 E3** 43.28N 94.49W
Estonia, republic **115 I4**
Ethiopia, independent state **125 G2**
Ethiopian Highlands **125 H2**
Etna, Mount **117 F2** 37.45N 15.00E
Etosha Pan, lake **126 B3**
Eucla **72 E2** 31.41S 128.52E
Eucumbene, Lake **48 E2**
Eudunda **66 G3** 34.11S 139.05E
Eufaula **136 D4** 31.55N 85.10W
Eugene **134 B3** 44.03N 123.04W
Eugenia, Point **134 C1** 27.50N 115.05W
Eugowra **48 E3** 33.26S 148.23E
Euphrates River **108 C4**
Eureka, California **134 B3** 40.49N 124.10W
Eureka, Utah **134 C2** 39.57N 112.09W
Euroa **54 C2** 36.45S 145.34E
Europa, island **122 H2**
Eustis **135 F1** 28.53N 81.42W
Eva Downs, homestead **80 E5** 18.00S 134.52E
Evandale **60 E3** 41.34S 147.15E
Evans, Lake **133 K3**
Evans Head **48 G5** 29.07S 153.26E
Evelyn, Mount **80 D8** 13.36S 132.54E
Evensk **119 P5** 61.56N 159.30E

Entry/***Page***/***Ref.***/*Lat.*/*Long.*

Everard, Lake **66 E5**
Everard, Mount **66 D7** 26.15S 132.04E
Everest, Mount **106 D3** 27.59N 86.56E
Everett **134 B3** 47.59N 122.11W
Evesham **116 C5** 52.06N 1.56W
Evora **116 B2** 38.34N 7.54W
Evreux **116 D4** 49.03N 1.11E
Exeter **114 D3** 50.43N 3.31W
Exmouth **72 B4** 21.55S 114.08E
Exmouth Gulf **72 B4**
Expedition Range **42 G3**
Eyasi, Lake **126 D4**
Eyre Peninsula **66 E4**
Eyre, Lake, (North) **66 F6**
Eyre, Lake, (South) **66 F6**
Ezine **108 A4** 39.46N 26.20E

F

Fada **124 F3** 17.14N 21.32E
Fagaras **117 H4** 45.50N 24.59E
Fairbanks **132 D4** 64.50N 147.50W
Fairmont **135 F2** 39.28N 80.08W
Fais, island **101 F3**
Faisalabad **106 C4** 31.25N 73.09E
Faith **134 D3** 45.02N 102.03W
Faizabad **102 C2** 26.46N 82.08E
Fakfak **88 B3** 2.55S 132.17E
Falam **107 E3** 22.58N 93.45E
Falcon Reservoir **136 C3**
Falkenberg **114 G4** 56.55N 12.30E
Falkland Islands **144 D1**
Falkland Sound **144 C1**
Fallon **134 C2** 39.29N 118.47W
False Cape **88 C2** 8.26S 137.25E
Falun **114 H5** 60.37N 15.40E
Fanning Island **84 O6**
Fan si pan, mountain **102 E2** 22.19N 103.46E
Faraday, antarctic base **92 O3** 65.15S 64.16W
Farafra Oasis **125 F4** 27.03N 28.17E
Farah **106 B4** 32.23N 62.08E
Farallon de Medinilla, island **101 F4**
Farallon de Pajaros, island **101 F5**
Farasan Islands **108 C2**
Faraulep, island **101 F3**
Faraway Hill, homestead **66 G4** 33.08S 139.46E
Farewell, Cape, Greenland **133 N3** 59.45N 43.30W
Farewell, Cape, New Zealand **90 E4** 40.30S 172.43E
Farewell Spit **90 E4**
Fargo **134 E3** 46.52N 96.49W
Faribault **132 I2** 44.19N 93.15W
Farina **66 G5** 30.04S 138.17E
Farmington **134 D2** 36.40N 108.10W
Faro, Brazil **142 E3** 2.10S 56.39W
Faro, Portugal **116 B2** 37.01N 7.56W
Faroe Islands **114 C5**
Farquhar, Cape **72 B4** 23.38S 113.37E
Farrukhabad **106 C3** 27.23N 79.35E
Farwell Island **92 N3**
Fasano **117 G3** 40.50N 17.21E
Fastov **115 J3** 50.08N 29.59E
Fatehpur **106 D3** 25.56N 80.55E
Fatigue, Mount **54 D1** 38.34S 146.19E
Fatuhiva, island **85 R4**
Faya-Largeau **124 E3** 17.58N 19.06E
Fayetteville, Arkansas **134 E2** 36.03N 94.10W
Fayetteville, North Carolina **135 G2** 35.03N 78.53W
Fderick **124 B4** 22.40N 12.41W
Fear, Cape **135 G2** 33.51N 77.59W
Feathertop, Mount **54 D2** 36.54S 147.09E
Feilding **90 F4** 40.12S 175.35E
Feira de Santana **143 G2** 12.17S 38.53W
Felanitx **116 D2** 39.28N 3.08E
Fen River **102 F3**
Fengcheng **103 G4** 40.27N 124.02E
Feni Islands **89 F3**
Feodosiya **115 L2** 45.03N 35.23E
Ferdows **108 D4** 34.00N 58.09E
Fergus Falls **134 E3** 46.18N 96.07W

Entry/***Page***/***Ref.***/*Lat.*/*Long.*

Fergusson Island **89 F2**
Ferkessedougou **124 C2** 9.30N 5.10W
Fermo **116 F3** 43.09N 13.44E
Fernie **134 C3** 49.30N 115.00W
Ferrara **116 F3** 44.50N 11.38E
Ferrat, Cape **112 D1** 35.54N 0.25W
Fetesti **117 I3** 44.22N 27.51E
Fethiye **108 A4** 36.37N 29.07E
Fez **124 C5** 34.05N 5.00W
Field West, Mount **60 D2** 42.39S 146.31E
Figueira da Foz **116 B3** 40.09N 8.51W
Figueras **116 D3** 42.16N 2.57E
Fiji, dominion **84 M4**
Fingal **60 E3** 41.39S 147.59E
Finisterre, Cape **116 B3** 42.52N 9.16W
Finisterre Range **89 E2**
Finke **80 E2** 25.35S 134.34E
Finke River **80 D2**
Finland, Gulf of **115 J4**
Finland, republic **115 I5**
Finley **48 C2** 35.39S 145.34E
Finnis, Cape **66 E4** 33.37S 134.50E
Finschhafen **89 E2** 6.35S 147.51E
Fiordland, region **90 B2**
Firminy **116 D4** 45.23N 4.18E
Fish River **126 B2**
Fitzgerald **135 F2** 31.43N 83.16W
Fitzroy, Mount **144 B2** 49.16S 73.03W
Fitzroy River **72 D5**
Fitzroy Crossing **72 E5** 18.11S 125.36E
Flagstaff **134 C2** 35.15N 111.40W
Flamingo Bay **88 C2**
Flathead Lake **134 C3**
Flattery, Cape, Australia **42 E8** 14.58S 145.21E
Flattery, Cape, U.S.A. **134 B3** 48.24N 124.43W
Flensburg **114 F3** 54.47N 9.27E
Flinders Bay **72 B2**
Flinders Island **60 F5**
Flinders Ranges **66 G4**
Flinders Reefs **42 G7**
Flinders River **42 C6**
Flin Flon **132 H3** 54.47N 101.51W
Flint **135 F3** 43.03N 83.40W
Flint, island **85 P4**
Florence, Alabama **135 F2** 34.48N 87.40W
Florence, Arizona **134 C2** 33.02N 111.22W
Florence, Italy **116 F3** 43.47N 11.15E
Florence, South Carolina **135 G2** 34.12N 79.44W
Florencia **142 C4** 1.37N 75.37W
Flores **136 D2** 16.58N 89.50W
Flores, island **100 D2**
Flores Sea **100 C2**
Florianopolis **144 E4** 27.35S 48.31W
Florida **144 D3** 34.04S 56.14W
Florida Islands **89 H2**
Florida Keys, islands **135 F1**
Florida, state **135 F1**
Florida Strait **136 D3**
Fluk **88 A3** 1.39S 127.44E
Flushing **114 E3** 51.27N 3.35E
Fly River **88 D2**
Flying Fish, Cape **92 M3** 71.53S 100.50W
Foca **117 G3** 43.30N 18.48E
Focsani **117 I4** 45.41N 27.12E
Foggia **117 G3** 41.28N 15.33E
Fohnsdorf **114 G2** 47.13N 14.40E
Foligno **116 F3** 42.57N 12.43E
Fonseca, Gulf of **136 D2**
Fonte Boa **142 D3** 2.33S 65.59W
Forbes **48 D3** 33.23S 148.01E
Ford, Cape **40 C4** 13.26S 129.55E
Forestier, Cape **60 F2** 42.11S 148.21E
Forestier Peninsula **60 F2**
Forfar **114 D4** 56.38N 2.54W
Forli **116 F3** 44.13N 12.02E
Formosa **144 D4** 26.07S 58.14W
Formosa Bay **126 E4**
Forrest **54 B1** 38.31S 143.43E
Forrest City **135 E2** 35.01N 90.48W
Forsayth **42 D6** 18.35S 143.36E
Forsnas **118 B5** 66.16N 18.35E
Forssa **115 I5** 60.49N 23.40E
Forst **114 G3** 51.46N 14.39E

Entry/***Page***/***Ref.***/*Lat.*/*Long.*

Forster **48 G3** 32.11S 152.31E
Fort Albany **133 J3** 52.15N 81.35W
Fortaleza **143 G3** 3.45S 38.35W
Fort Bragg **134 B2** 39.29N 123.46W
Fort Collins **134 D3** 40.36N 105.04W
Fort-de-France **137 F2** 14.36N 61.05W
Fortescue River **72 C4**
Fort Frances **132 I2** 48.37N 93.23W
Fort George **133 K3** 53.50N 79.01W
Forth, River **60 D3**
Fort Lauderdale **135 F1** 26.08N 80.08W
Fort McMurray **132 G3** 56.40N 111.07W
Fort Nelson **132 F3** 58.48N 122.44W
Fort Olimpo **142 E1** 21.02S 57.51W
Fort Peck Reservoir **134 D3**
Fort Pierce **135 F1** 27.28N 80.20W
Fort St John **132 F3** 56.14N 120.55W
Fort Severn **132 J3** 56.00N 87.40W
Fort Shevchenko **118 F3** 44.31N 50.15E
Fort Simpson **132 F4** 61.52N 121.15W
Fort Smith, Canada **132 G4** 60.01N 111.55W
Fort Smith, U.S.A. **134 E2** 35.22N 94.27W
Fort Stockton **134 D2** 30.54N 102.54W
Fortuna **134 B3** 40.36N 124.10W
Fort Walton Beach **135 F2** 30.25N 86.38W
Fort Wayne **135 F3** 41.05N 85.08W
Fort William **114 C4** 56.49N 5.07W
Fort Worth **134 E2** 32.45N 97.20W
Fort Yukon **132 D4** 66.35N 145.20W
Foster **54 D1** 38.40S 146.13E
Fougeres **116 C4** 48.21N 1.12W
Foul Bay **125 G4**
Foulwind, Cape **90 D4** 41.45S 171.28E
Foumban **126 B5** 5.43N 10.50E
Four Archers, The, mountain **80 E7** 15.28S 135.25E
Fourcroy, Cape **80 C9** 11.48S 131.01E
Fouta Djallon, mountains **124 B3**
Foveaux Strait **90 B1**
Fowlers Bay **66 D4**
Fowlers Bay **66 D5** 31.59S 132.26E
Fox Glacier **90 C3**
Foxe Basin, bay **133 K4**
Foxton **90 F4** 40.23S 175.17E
Foz do Cunene **126 B3** 17.15S 11.48E
Franca **143 F1** 20.33S 47.27W
France, republic **116 D4**
Franceville **126 B4** 1.40S 13.31E
Francis Case, Lake **134 D3**
Francistown **126 C2** 21.11S 27.32E
Frankfort **135 F2** 38.11N 84.53W
Frankfurt **114 F3** 50.06N 8.41E
Frankland River **60 B3**
Franklin **60 D1** 43.06S 147.00E
Franklin, Mount **48 E2** 35.29S 148.47E
Franklin River **60 C2**
Franklin D. Roosevelt Lake **134 C3**
Franstontein **126 B2** 20.12S 15.01E
Franz Josef Glacier **90 C3**
Franz Josef Land, islands **118 E6**
Fraser Island **42 I3**
Fraser, Mount **72 C3** 25.38S 118.23E
Fraser River **132 F3**
Fraserburgh **114 D4** 57.42N 2.00W
Fredericia **114 F4** 55.34N 9.47E
Fredericton **133 L2** 45.57N 66.40W
Frederikshab **133 N4** 62.00N 49.47W
Frederikshavn **114 G4** 57.28N 10.33E
Freeling, Mount **80 D3** 22.35S 133.06E
Freetown **124 B2** 8.30N 13.17W
Freiburg **114 F2** 48.00N 7.52E
Fremantle **72 C2** 32.03S 115.44E
French Guiana, dependency **143 E4**
Frenchmans Cap, mountain **60 C2** 42.16S 145.50E
French Polynesia, dependency **85 Q4**
Frenda **116 D2** 35.04N 1.03E
Fresno **134 C2** 36.41N 119.47W
Freycinet Peninsula **60 F2**
Fria, Cape **126 B3** 18.30S 12.01E
Friedrichshafen **114 F2** 47.39N 9.30E
Frio, Cape **140 G4** 22.59S 42.00W
Frome, Lake **66 H5**
Frome Downs, homestead **66 G5** 31.13S 139.46E

Entry/***Page***/***Ref.***/*Lat.*/*Long.*

Fuengirola **116 C2** 36.32N 4.38W
Fuerteventura, island **124 B4**
Fujairah **108 D3** 25.00N 56.00E
Fuji, Mount **104 I4** 35.23N 138.42E
Fukuchiyama **104 G4** 35.19N 135.08E
Fukue Island **104 D3**
Fukui **104 H5** 36.04N 136.12E
Fukuoka **104 E3** 33.39N 130.21E
Fukushima **104 J5** 37.44N 140.28E
Fulda **116 E5** 50.33N 9.41E
Funafuti **84 N5** 8.34S 178.59E
Funafuti, island **84 M5**
Funchal **124 B5** 32.40N 16.55W
Fundy, Bay of **133 L2**
Funtua **124 D3** 11.34N 7.17E
Furg **108 D3** 28.19N 55.10E
Furneaux Group **60 E5**
Furukawa **104 J6** 38.34N 140.56E
Fushun **103 G4** 41.51N 123.53E
Futuna Islands **84 M4**
Fuxin **103 G4** 42.04N 121.39E
Fuzhou, Fujian **103 F2** 26.09N 119.17E
Fuzhou, Jiangxi **103 F2** 28.03N 116.15E
F.Y.R.O.M. *see* Macedonia

G

Gabes **124 E5** 33.52N 10.06E
Gabes, Gulf of **124 E5**
Gabo Island **55 E2**
Gabon, republic **126 B4**
Gaborone **126 C2** 24.45S 25.55E
Gadag **106 C2** 15.26N 75.42E
Gadsden **135 F2** 34.00N 86.00W
Gaferut, island **101 F3**
Gafsa **124 D5** 34.28N 8.43E
Gagnoa **124 C2** 6.04N 5.55W
Gainesville **135 F1** 29.37N 82.21W
Gairdner, Lake **66 E5**
Galadi **125 H2** 6.56N 46.21E
Galana River **125 G1**
Galapagos Islands **142 A4**
Galati **117 I4** 45.27N 28.02E
Galera Point **137 F2** 10.49N 60.54W
Galilee, Lake **42 E4**
Galiwinku **80 E8** 12.02S 135.34E
Galkayu **125 H2** 6.47N 47.21E
Galle **106 D1** 6.01N 80.13E
Gallegos **144 C1** 51.35S 69.15W
Gallinas, Point **142 C5** 12.27N 71.44W
Gallipoli **108 A5** 40.25N 26.41E
Gallivare **118 C5** 67.10N 20.40E
Gallup **134 D2** 35.30N 108.45W
Galveston **134 E1** 29.17N 94.48W
Galway **114 C3** 53.16N 9.03W
Gambela **125 G2** 8.15N 34.35E
Gambia, republic **124 B3**
Gambia River **124 B3**
Gambier Islands **66 F3**
Gamboma **126 B4** 1.50S 15.58E
Gan River **103 F2**
Gander **133 M2** 48.58N 54.34W
Gandia **116 C2** 38.59N 0.11W
Ganganagar **106 C3** 29.54N 73.56E
Ganges, Mouths of the **107 D3**
Ganges River **106 D3**
Gangtok **106 D3** 27.20N 88.39E
Gannett Peak **134 D3** 43.10N 109.38W
Ganta **124 C2** 7.15N 8.59W
Gantheaume Bay **72 B3**
Gantheaume Point **72 D5** 17.59S 122.11E
Ganzhou **103 F2** 25.52N 114.51E
Gao **124 C3** 16.19N 0.09W
Gaoyou Lake **103 F3**
Gap **116 E3** 44.33N 6.05E
Gapuwiyak **80 E8** 12.25S 135.43E
Garanhuns **143 G3** 8.53S 36.28W
Garantah **100 C2** 9.02S 117.11E
Gardiner, Mount **80 D3** 22.10S 132.38E
Gardner Pinnacles, islands **84 O8**
Gardo **125 H2** 9.30N 49.06E
Garissa **126 D4** 0.27S 39.39E
Garonne River **116 D3**
Garoua **124 E2** 9.17N 13.22E
Garwa **106 D3** 24.14N 83.51E
Gary **135 F3** 41.34N 87.20W

*Entry/**Page**/**Ref.**/Lat./Long.*

Gasan-Kuli **108 D4** 37.31N 53.59E
Gascoyne, Mount **72 C4** 24.57S 116.38E
Gascoyne River **72 B4**
Gascoyne Junction **72 C3** 25.03S 115.12E
Gaspe **133 L2** 48.50N 64.30W
Gaspe, Cape **133 L2** 48.46N 64.11W
Gatchina **115 K4** 59.32N 30.09E
Gatton **42 I2** 27.34S 152.17E
Gauhati **107 E3** 26.10N 91.45E
Gavater **108 E3** 25.10N 61.31E
Gavle **114 H5** 60.41N 17.10E
Gawler **66 G3** 34.36S 138.45E
Gaya **106 D3** 24.48N 85.00E
Gayndah **42 H3** 25.37S 151.37E
Gaza **108 B4** 31.30N 34.28E
Gazelle Peninsula **89 F3**
Gaziantep **108 B4** 37.04N 37.21E
Gboko **124 D2** 7.15N 9.09E
Gdansk **114 H3** 54.22N 18.41E
Gdynia **114 H3** 54.31N 18.30E
Gebe **88 A3**
Gebeit **125 G3** 18.58N 36.48E
Gedaref **125 G3** 14.01N 35.24E
Gediz **117 I2** 39.04N 29.25E
Geelong **54 C1** 38.10S 144.21E
Geelvink Channel **72 B3**
Geeveston **60 D1** 43.10S 146.55E
Geidam **124 E3** 12.55N 11.55E
Gejiu **102 E2** 23.25N 103.05E
Gela **117 F2** 37.04N 14.15E
Gemena **126 B5** 3.13N 19.48E
Gemlik **117 I3** 40.26N 29.10E
Geneina **124 F3** 13.27N 22.30E
General Alvarado **144 D3** 38.15S 57.50W
General Bernardo O'Higgins, antarctic base **92 P3** 63.19S 57.54W
General Carrera, Lake **144 B2**
General Pico **144 C3** 35.43S 63.45W
General Roca **144 C3** 39.00S 67.35W
General Santos **100 D3** 6.05N 125.15E
Geneva **116 E4** 46.13N 6.09E
Genoa, Australia **55 E2** 37.29S 149.35E
Genoa, Italy **116 E3** 44.24N 8.56E
Geographe Bay **72 B2**
Geographe Channel **72 B4**
George, Lake, Australia **48 E2**
George, Lake, U.S.A. **135 F1**
George River **133 L3**
George Town, Australia **60 D3** 41.06S 146.49E
George Town, Bahamas **137 E3** 23.33N 75.47W
George Town, Malaysia **100 B3** 5.26N 100.16E
Georgetown, Australia **42 D6** 18.17S 143.33E
Georgetown, Gambia **124 B3** 13.31N 14.50W
Georgetown, Guyana **142 E4** 6.46N 58.10W
Georgetown, U.S.A. **135 G2** 33.23N 79.18W
George V Land **94 H2**
Georg Forster, antarctic base **92 A3** 70.46S 11.51E
Georgia, republic **108 C5**
Georgia, state **135 F2**
Georgian Bay **135 F3**
Georgina River **42 B5**
Georgina Downs, homestead **80 F4** 21.08S 137.40E
Georg von Neumayer, antarctic base **92 R3** 70.37S 8.22W
Gera **114 G3** 50.51N 12.11E
Geraldine **90 D2** 44.05S 171.15E
Geraldton, Australia **72 B3** 28.46S 114.37E
Geraldton, Canada **133 J2** 49.44N 86.59W
Geranium **66 H3** 35.23S 140.10E
Gerlachovka, mountain **115 I2** 49.09N 20.05E
Germany, republic **114 F3**
Gerona **116 D3** 41.59N 2.49E
Gestro River **125 H2**
Getafe **116 C3** 40.18N 3.44W
Gettysburg **135 G2** 39.50N 77.16W
Getz Ice Shelf **92 L3**

*Entry/**Page**/**Ref.**/Lat./Long.*

Ghadames **124 D5** 30.08N 9.30E
Ghaghe, island **89 G2**
Ghana, republic **124 C2**
Ghanzi **126 C2** 21.34S 21.42E
Ghardaia **124 D5** 32.20N 3.40E
Ghat **124 E4** 24.59N 10.11E
Ghent **114 E3** 51.02N 3.42E
Gheorgheni **117 I4** 46.43N 25.36E
Ghriss **116 D2** 35.16N 0.14E
Gibb River, homestead **72 E5** 16.26S 126.26E
Gibraltar **116 B2** 36.09N 5.21W
Gibraltar, Strait of **116 B2**
Gibson Desert **72 D4**
Gidgealpa **66 H7** 27.57S 140.03E
Gien **116 D4** 47.41N 2.37E
Giessen **116 E5** 50.35N 8.42E
Gifu **104 H4** 35.27N 136.46E
Gijon **116 B3** 43.32N 5.40W
Gila Bend **134 C2** 32.56N 112.42W
Gilbert River **42 C7**
Giles Meteorological Station **72 E3** 25.02S 128.18E
Gilgandra **48 E4** 31.43S 148.39E
Gilgit **106 C4** 35.54N 74.20E
Gillam **132 I3** 56.25N 94.45W
Gilles, Lake **66 F4**
Gillette **134 D3** 44.18N 105.30W
Gillock Island **92 D3**
Gilruth, Mount **80 D8** 13.02S 133.05E
Giluwe, Mount **101 F2** 6.03S 143.53E
Gimbala, mountain **125 F3** 12.59N 24.20E
Ginir **125 H2** 7.06N 40.40E
Girga **125 G4** 26.17N 31.58E
Girgarre **54 C2** 36.24S 144.59E
Girgir, Cape **89 D3** 3.50S 144.30E
Girishk **106 B4** 31.50N 64.35E
Gisborne **90 H5** 38.39S 178.01E
Giza **125 G5** 30.01N 31.12E
Gizhiga **119 Q5** 62.00N 160.34E
Gizo **89 G2** 8.04S 156.45E
Gizycko **115 I3** 54.03N 21.48E
Gjirokaster **117 H3** 40.05N 20.10E
Gjovik **114 G5** 60.47N 10.41E
Glace Bay **133 M2** 46.11N 59.58W
Gladstone, Queensland **42 H4** 23.51S 151.16E
Gladstone, South Australia **66 G4** 33.17S 138.22E
Gladstone, Tasmania **60 F4** 40.58S 148.01E
Glasgow, U.K. **114 D4** 55.53N 4.15W
Glasgow, U.S.A. **134 D3** 48.12N 106.37W
Glazov **118 F4** 58.09N 52.42E
Glendale **134 C2** 33.30N 112.15W
Glenelg River **54 A1**
Glengarry Range **72 C3**
Glen Innes **48 F5** 29.44S 151.44E
Glenmaggie, Lake **54 D2**
Glenmorgan **42 G2** 27.15S 149.41E
Glennie Group **54 D1**
Glenorchy, homestead **66 G5** 31.55S 139.47E
Glenroy, homestead **72 E5** 17.22S 126.06E
Glittertind, mountain **114 F5** 61.40N 8.32E
Gliwice **114 H3** 50.20N 18.40E
Glogow **114 H3** 51.40N 16.06E
Gloucester, Australia **48 G3** 32.01S 151.58E
Gloucester, U.K. **114 D3** 51.53N 2.14W
Gloucester, Cape **89 E2** 5.27S 148.25E
Gluepot, homestead **66 H4** 33.46S 140.08E
Gnowangerup **72 C2** 33.56S 118.00E
Goa **106 C2** 15.31N 73.56E
Goba **125 G2** 7.01N 39.59E
Gobabis **126 B2** 22.30S 18.58E
Gobi Desert **102 E4**
Godavari River **106 C2**
Godda **106 D3** 24.50N 87.13E
Godhavn **133 M4** 69.18N 53.40W
God's Lake **132 I3**
Godthab **133 M4** 64.15N 51.35W
Godwin Austen, K2, mountain **106 C4** 35.53N 76.30E
Goiania **143 F2** 16.43S 49.18W

*Entry/**Page**/**Ref.**/Lat./Long.*

Goldap **115 I3** 54.19N 22.19E
Gold Coast **42 I1** 28.05S 153.26E
Golden Bay **90 E4**
Goldfield **134 C2** 37.42N 117.15W
Golmud **102 D3** 36.22N 94.55E
Gombi **124 E3** 10.10N 12.45E
Gomel **115 K3** 52.25N 31.00E
Gomogomo **88 B2** 6.39S 134.43E
Gonaives **137 E2** 19.29N 72.42W
Gonave Island **137 E2**
Gonda **106 D3** 27.08N 81.58E
Gonder **125 G3** 12.39N 37.29E
Gondia **106 D3** 21.23N 80.14E
Gonen **117 I3** 40.06N 27.39E
Gongga Mountain **102 E2** 29.34N 101.53E
Gongzhuling **103 G4** 43.30N 124.48E
Goodenough, Cape **92 G3** 66.15S 126.35E
Goodenough Island **89 F2**
Good Hope, Cape of **126 B1** 34.20S 18.25E
Goodland **134 D2** 39.20N 101.43W
Goodooga **48 D5** 29.07S 147.27E
Goolwa **66 G3** 35.31S 138.45E
Goondiwindi **42 H1** 28.30S 150.17E
Goonyella **42 F5** 21.48S 147.58E
Goose Lake **134 B3**
Goose Bay-Happy Valley **133 L3** 53.15N 60.20W
Gorakhpur **106 D3** 26.45N 83.23E
Gorda, Point **138 C8** 14.23N 83.10W
Gordon **132 D4** 69.39N 141.18W
Gordon, Lake **60 D2**
Gordon River **60 C2**
Gordonvale **42 E7** 17.06S 145.47E
Gore, Ethiopia **125 G2** 8.10N 35.29E
Gore, New Zealand **90 C1** 46.06S 168.56E
Gorgan **108 D4** 36.50N 54.29E
Gorki **115 K3** 54.17N 30.59E
Gormanston **60 C2** 42.08S 145.36E
Goroka **89 E2** 6.05S 145.23E
Goroke **54 A2** 36.43S 141.30E
Gorong, island **88 B3**
Gorontalo **100 D3** 0.33N 123.05E
Gorzow Wielkopolski **114 H3** 52.42N 15.12E
Goschen Strait **89 F1**
Gosford **48 F3** 33.26S 151.20E
Gostyn **118 B4** 51.52N 17.00E
Goteborg **114 G4** 57.45N 12.00E
Gotha **114 G3** 50.57N 10.43E
Gotland, island **114 H4**
Goto Archipelago **104 D3**
Gottingen **114 F3** 51.32N 9.57E
Gough Island **146 B1**
Goulburn **48 E2** 34.45S 149.43E
Goulburn Islands **80 D9**
Goulburn River, New South Wales **48 E3**
Goulburn River, Victoria **54 C2**
Gould, Mount **72 C3** 25.48S 117.20E
Goundam **124 C3** 16.27N 3.39W
Goure **124 E3** 13.59N 10.15E
Gove Peninsula **80 F8**
Governador Valadares **143 F2** 18.51S 41.57W
Governor's Harbour **137 E3** 25.13N 76.15W
Goya **144 D4** 29.10S 59.15W
Goyder River **80 E8**
Graaff Reinet **126 C1** 32.15S 24.32E
Gracias a Dios, Cape **130 K2** 15.00N 83.10W
Gradaus **143 E3** 7.43S 51.10W
Grafton **48 G5** 29.40S 152.56E
Grafton, Cape **42 E7** 16.52S 145.55E
Graham Land **94 O2**
Grahamstown **126 C1** 33.19S 26.32E
Grampians, The, mountains **54 B2**
Granada, Nicaragua **136 D2** 11.58N 85.59W
Granada, Spain **116 C2** 37.10N 3.35W
Gran Chaco, region **144 C4**
Grand Canyon **134 C2**
Grand Bahama, island **136 E3**
Grand Bassam **124 C2** 5.14N 3.45W
Grand Canary, island **124 B4**
Grand Cayman, island **136 D2**

*Entry/**Page**/**Ref.**/Lat./Long.*

Grande River **143 E2**
Grande Prairie **132 G3** 55.10N 118.52W
Grand Forks **134 E3** 47.57N 97.05W
Grandioznyy, mountain **118 J4** 53.54N 96.09E
Grand Junction **134 D2** 39.04N 108.33W
Grand Rapids, Michigan **135 F3** 42.57N 85.40W
Grand Rapids, Minnesota **134 E3** 47.13N 93.31W
Grand Teton, mountain **134 C3** 43.45N 110.50W
Granite Peak **134 D3** 45.10N 109.50W
Granites, The, mountain **80 C4** 20.34S 130.21E
Grants Pass **134 B3** 42.26N 123.20W
Granville **116 C4** 48.50N 1.35W
Grassy **60 B4** 40.04S 144.04E
Graz **114 H2** 47.05N 15.22E
Great Lake **60 D3**
Great Sand Sea, desert **125 F4**
Great Abaco, island **137 E3**
Great Antrim Plateau **72 E5**
Great Australian Bight **40 C2**
Great Barrier Island **90 F6**
Great Barrier Reef **42 F7**
Great Bear Lake **132 F4**
Great Britain, island **112 D3**
Great Dividing Range **40 E2**
Great Eastern Desert **124 D5**
Greater Antarctica, region **94 G2**
Greater Antilles, islands **130 K2**
Greater Hinggan Mountains **103 G5**
Great Exuma, island **137 E3**
Great Falls **134 C3** 47.30N 111.16W
Great Inagua, island **137 E3**
Great Kai, island **88 B2**
Great Karoo, region **126 B1**
Great Oyster Bay **60 F2**
Great Rift Valley **122 G5**
Great Salt Lake **134 C3**
Great Sandy Desert **72 D4**
Great Slave Lake **132 G4**
Great Victoria Desert **72 E3**
Great Western Desert **124 D5**
Greece, republic **117 H2**
Greeley **134 D3** 40.26N 104.43W
Green River **134 C2**
Green Bay **135 F3** 44.32N 88.00W
Greenland, dependency **133 N5**
Greenock **114 D4** 55.57N 4.45W
Greenough River **72 B3**
Green River, P.N.G. **88 D3** 3.54S 141.08E
Green River, Utah **134 C2** 39.00N 110.10W
Green River, Wyoming **134 D3** 41.32N 109.26W
Greensboro **135 G2** 36.03N 79.50W
Greenville, Alabama **136 D4** 31.50N 86.40W
Greenville, Liberia **124 C2** 5.01N 9.03W
Greenville, Mississippi **134 E2** 33.23N 91.03W
Greenville, South Carolina **135 F2** 34.52N 82.25W
Greenwood, Mount **80 C8** 13.45S 130.05E
Gregory, Lake **66 G6**
Gregory Lake **72 E5**
Gregory Range **42 D6**
Greifswald **114 G3** 54.06N 13.24E
Grena **114 G4** 56.25N 10.53E
Grenada, independent state **137 F2**
Grenadines, islands **137 F2**
Grenfell **48 E3** 33.54S 148.10E
Grenoble **116 E4** 45.11N 5.43E
Grenville, Cape **42 D10** 12.00S 143.13E
Grey Range **42 C1**
Greymouth **90 D3** 42.27S 171.12E
Griffin **135 F2** 33.15N 84.17W
Griffith **48 C2** 34.17S 146.02E
Grim, Cape **60 B4** 40.41S 144.41E
Grimstad **114 F4** 58.20N 8.35E
Griquatown **126 C2** 28.51S 23.15E
Grodno **115 I3** 53.40N 23.50E
Groningen **114 F3** 53.13N 6.35E
Groote Eylandt, island **80 F7**
Grootfontein **126 B3** 19.32S 18.05E

*Entry/**Page**/**Ref.**/Lat./Long.*

Gros Morne, mountain **133 M2** 49.36N 57.47W
Grosseto **116 F3** 42.46N 11.07E
Grossglockner, mountain **114 G2** 47.05N 12.44E
Grozny **118 E3** 43.21N 45.42E
Grudziadz **114 H3** 53.29N 18.45E
Grytviken, antarctic base **92 Q2** 54.00S 37.16W
Guadalajara **136 B3** 20.30N 103.20W
Guadalcanal, island **89 G2**
Guadalquivir River **116 B2**
Guadalupe, island **130 H3**
Guadeloupe, island **137 F2**
Guadix **116 C2** 37.19N 3.08W
Guajira Peninsula **142 C5**
Guam, island **101 F4**
Guanambi **143 F2** 14.14S 42.44W
Guanare **142 D4** 9.04N 69.45W
Guanghua **102 F3** 32.24N 111.40E
Guangzhou **102 F2** 23.08N 113.20E
Guantanamo **137 E3** 20.09N 75.14W
Guapore River **142 D2**
Guarapuava **144 D4** 25.22S 51.28W
Guarda **116 B3** 40.32N 7.17W
Guasipati **142 D4** 7.28N 61.58W
Guasopa **89 F2** 9.15S 152.55E
Guatemala, republic **136 C2**
Guatemala City **136 C2** 14.38N 90.22W
Guaviare **142 C4** 2.35N 72.38W
Guaviare River **142 C4**
Guayaquil **142 C3** 2.13S 79.54W
Guayaquil, Gulf of **142 B3**
Guaymas **134 C1** 27.59N 110.54W
Gubkin **115 L3** 51.17N 37.32E
Guelma **116 E2** 36.29N 7.25E
Guelph **135 F3** 43.34N 80.16W
Gueret **116 D4** 46.10N 1.52E
Guernica **116 C3** 43.19N 2.40W
Guerrero **136 B3** 28.33N 107.28W
Gughe, mountain **125 G2** 6.16N 37.28E
Guguan, island **101 F4**
Guiana Highlands **142 D4**
Guilin **102 F2** 25.21N 110.11E
Guinea, Gulf of **124 D2**
Guinea Highlands **124 C2**
Guinea, republic **124 B2**
Guinea Bissau, independent state **124 B3**
Guines **136 D3** 22.50N 82.20W
Guiria **137 F2** 10.37N 62.21W
Guiyang **102 E2** 26.35N 106.40E
Gujranwala **106 C4** 32.06N 74.11E
Gujrat **106 C4** 32.35N 74.06E
Gulbarga **106 C2** 17.22N 76.47E
Gulbene **115 J4** 57.15N 26.42E
Gulfport **135 F2** 30.21N 89.08W
Gulgong **48 E3** 32.22S 149.32E
Gulu **126 D5** 2.46N 32.21E
Gums, The, homestead **66 G4** 33.51S 139.21E
Guna, mountain **125 G3** 11.40N 38.20E
Gunbower **54 C3** 35.58S 144.22E
Gundagai **48 E2** 35.04S 148.07E
Gunderbooka, Mount **48 C4** 30.36S 145.41E
Gungu **126 B4** 5.43S 19.20E
Gunnedah **48 F4** 30.59S 150.15E
Gunson, Mount **66 F5** 31.26S 137.06E
Guntur **106 D2** 16.20N 80.27E
Gunungsugih **100 B2** 5.00S 105.16E
Gurue **126 D3** 15.30S 36.58E
Guryev **118 F3** 47.08N 51.59E
Gusau **124 D3** 12.12N 6.40E
Gusev **115 I3** 54.32N 22.12E
Guyana, republic **142 E4**
Guymon **134 D2** 36.42N 101.30W
Guyra **48 F4** 30.14S 151.40E
Gwabegar **48 E4** 30.37S 148.58E
Gwadar **106 B3** 25.09N 62.21E
Gwalia **72 D3** 28.55S 121.20E
Gwalior **106 C3** 26.12N 78.09E
Gwanda **126 C2** 21.00S 29.02E
Gwaram **124 D3** 11.15N 9.52E
Gweru **120 G3** 19.25S 29.50E
Gwydir River **48 E5**

*Entry/**Page**/**Ref.**/Lat./Long.*

Gyanja **108 C5** 40.39N 46.20E
Gympie **42 I2** 26.11S 152.40E
Gyor **114 H2** 47.41N 17.40E
Gypsumville **132 I3** 51.47N 98.38W

H

Haakon VII Sea **92 B3**
Haast Bluff **80 C3** 23.27S 131.53E
Haast Pass **90 C2** 44.07S 169.22E
Haboro **104 J9** 44.23N 141.43E
Hachijo Island **104 I3**
Hachinohe **104 J7** 40.30N 141.30E
Hack, Mount **66 G5** 30.46S 138.48E
Hadd, Cape **106 A3** 22.32N 59.48E
Hadejia **124 E3** 12.30N 10.03E
Haderslev **114 F4** 55.15N 9.30E
Hadjout **116 D2** 36.33N 2.20E
Haeju **104 B6** 38.04N 125.40E
Hafun **125 I3** 10.27N 51.15E
Hafun, Cape **125 I3** 10.29N 51.20E
Hagerstown **135 G2** 39.39N 77.44W
Haifa **108 B4** 32.49N 34.59E
Haikou **102 F2** 20.05N 110.25E
Hail **108 C3** 27.31N 41.45E
Hailar **103 F4** 49.15N 119.41E
Hainan, island **102 E1**
Haiphong **100 B5** 20.50N 106.41E
Haiti, republic **137 E2**
Hajduboszormeny **117 H4** 47.40N 21.29E
Hakodate **104 J7** 41.46N 140.44E
Halaib **125 G4** 22.12N 36.35E
Halberstadt **114 G3** 51.54N 11.04E
Hale, Mount **72 C3** 26.03S 117.15E
Hale River **80 E2**
Halfmoon Bay **90 C1** 46.54S 168.08E
Halifax **133 L2** 44.38N 63.35W
Halifax Bay **42 F6**
Hall Islands **101 G3**
Hall Point **72 D5** 15.40S 124.22E
Halle **114 G3** 51.28N 11.58E
Halley, antarctic base **92 Q3** 75.35S 26.40W
Halls Creek **72 E5** 18.14S 127.40E
Halls Gap **54 B2** 37.08S 142.31E
Halmahera, island **100 D3**
Halmstad **114 G4** 56.41N 12.55E
Halsingborg **114 G4** 56.05N 12.45E
Haltiatunturi, mountain **112 G4** 69.17N 21.21E
Hama **108 B4** 35.09N 36.44E
Hamada **104 F4** 34.56N 132.04E
Hamadan **108 C4** 34.46N 48.35E
Hamamatsu **104 H4** 34.42N 137.42E
Hamar **114 G5** 60.57N 10.55E
Hambantota **106 D1** 6.07N 81.07E
Hamburg **114 G3** 53.33N 10.00E
Hameenlinna **115 I5** 61.00N 24.25E
Hamersley Range **72 C4**
Hamhung **104 C6** 39.54N 127.35E
Hami **102 D4** 42.37N 93.32E
Hamilton, Canada **133 J2** 43.15N 79.50W
Hamilton, New Zealand **90 F6** 37.47S 175.17E
Hamilton, Tasmania **60 D2** 42.33S 146.50E
Hamilton, U.S.A. **135 F2** 39.23N 84.33W
Hamilton, Victoria **54 B2** 37.45S 142.02E
Hamilton Inlet **133 M3**
Hamilton Downs, homestead **80 D3** 23.30S 133.16E
Hammamet, Gulf of **124 E5**
Hammerfest **110 G5** 70.40N 23.44E
Hampton **135 G2** 37.02N 76.23W
Hampton Tableland **72 E2**
Hamuku **88 C3** 3.23S 135.09E
Han River, China **102 E3**
Han River, South Korea **104 C5**
Hanamaki **104 J6** 39.25N 141.04E
Handan **102 F3** 36.37N 114.25E
Hanford **134 C2** 36.20N 119.38W
Hango **115 I4** 59.50N 23.00E
Hangzhou **103 G3** 30.18N 120.07E
Hann, Mount **72 E5** 15.50S 125.49E
Hannover **114 F3** 52.23N 9.44E
Hanoi **100 B5** 21.01N 105.52E
Hanover **133 K2** 43.42N 72.17W

*Entry/**Page**/**Ref.**/Lat./Long.*

Hanpan, Cape **89 F3** 5.00S 154.37E
Hao, island **85 Q4**
Har **88 B2** 5.21S 133.10E
Harare **126 D3** 17.43S 31.05E
Harbin **103 G4** 45.45N 126.41E
Hardwar **106 C3** 29.58N 78.09E
Hardy, Mount **80 C3** 22.08S 131.31E
Harer **125 H2** 9.20N 42.10E
Hargeysa **125 H2** 9.31N 44.02E
Hari River **108 E4**
Harkiko **108 B2** 15.32N 39.28E
Harlingen, Netherlands **114 F3** 53.10N 5.25E
Harlingen, U.S.A. **134 E1** 26.12N 97.43W
Harnosand **114 H5** 62.37N 17.55E
Harper **124 C2** 4.25N 7.43W
Harricanaw River **133 K3**
Harrietville **54 D2** 36.54S 147.04E
Harrisburg, Illinois **135 F2** 37.40N 88.10W
Harrisburg, Pennsylvania **135 G3** 40.17N 76.54W
Harrismith **126 C2** 28.16S 29.08E
Harrisonburg **135 G2** 38.27N 78.54W
Harstad **118 B5** 68.48N 16.30E
Hart, Mount **72 E5** 16.55S 125.04E
Hartford **135 G3** 41.45N 72.42W
Harts Range **80 E3** 22.59S 134.55E
Harvey **72 C2** 33.05S 115.54E
Haslam **66 E4** 32.30S 134.13E
Hassleholm **114 G4** 56.09N 13.45E
Hastings, New Zealand **90 G5** 39.38S 176.50E
Hastings, Tasmania **60 D1** 43.25S 146.55E
Hastings, U.S.A. **134 E3** 40.37N 98.22W
Hastings, Victoria **54 C1** 38.18S 145.11E
Hatches Creek, homestead **80 E4** 20.57S 135.13E
Hateg **117 H4** 45.35N 22.57E
Hattah **54 B3** 34.46S 142.16E
Hattah Lakes **54 E3**
Hatteras, Cape **135 G2** 35.14N 75.31W
Hattiesburg **135 F2** 31.20N 89.19W
Hat Yai **100 B3** 7.00N 100.28E
Haugesund **114 F4** 59.25N 5.16E
Hauraki Gulf **90 F6**
Hauroko, Lake **90 B2**
Havana **136 D3** 23.07N 82.25W
Havre **134 D3** 48.34N 109.40W
Hawaii, island **84 P7**
Hawaiian Islands **84 O8**
Hawea, Lake **90 C2**
Hawera **90 F5** 39.35S 174.17E
Hawick **114 D4** 55.25N 2.47W
Hawke Bay **90 G5**
Hawke, Cape **48 G3** 32.12S 152.33E
Hawker **66 G5** 31.53S 138.25E
Hawkesdale **54 B1** 38.07S 142.20E
Hawthorne **134 C2** 38.31N 118.37W
Hay **48 C2** 34.30S 144.51E
Hay Point **42 G5** 21.16S 149.13E
Hay River **80 F3**
Hay River **132 G4** 60.51N 115.42W
Hazaran, Mount **108 D3** 29.32N 57.22E
Hazaribagh **102 C2** 24.00N 85.23E
Head of Bight, bay **66 C5**
Healesville **54 C2** 37.39S 145.32E
Heard Island **146 E1**
Hearst **133 J2** 49.42N 83.40W
Heathcote **54 C2** 36.55S 144.42E
Hede **118 B5** 62.27N 13.30E
Hedemora **114 H5** 60.13N 16.00E
Hefei **103 F3** 31.55N 117.18E
Hegang **103 H4** 47.36N 130.30E
Heidelberg **114 F2** 49.25N 8.42E
Heihe **103 G5** 50.16N 127.25E
Heilbron **126 C2** 27.17S 27.58E
Heilbronn **114 F2** 49.09N 9.16E
Heinola **115 J5** 61.13N 26.05E
Hekla, mountain **133 Q4** 64.00N 19.41W
Helen, island **101 E3**
Helena **134 C3** 46.35N 112.00W
Helensburgh **48 F2** 34.11S 150.59E
Helensville **90 F6** 36.40S 174.27E
Hellin **116 C2** 38.31N 1.43W
Helmeringhausen **126 B2** 25.51S 16.59E

*Entry/**Page**/**Ref.**/Lat./Long.*

Helsingor **114 G4** 56.03N 12.38E
Helsinki **115 J5** 60.08N 25.00E
Henbury, homestead **80 D2** 24.33S 133.14E
Henderson Island **85 S3**
Hendorabi, island **125 I4**
Hengduan Mountains **107 E3**
Hengyang **102 F2** 26.59N 112.22E
Henzada **107 E2** 17.36N 95.26E
Herat **106 B4** 34.20N 62.10E
Herberton **42 E7** 17.23S 145.23E
Hercules Bay **89 E2**
Hereford **134 D2** 34.49N 102.25W
Hermannsburg **80 D3** 23.57S 132.46E
Hermiston **134 C3** 45.50N 119.18W
Hermit Islands **89 D3**
Hermosillo **134 C1** 29.15N 110.59W
Herning **114 F4** 56.08N 8.59E
Heron Island **42 H4**
Heron Bay **133 J2** 48.41N 86.28W
Herrick **60 E3** 41.06S 147.52E
Hervey Bay **42 I3**
Hesso **66 F4** 32.08S 137.27E
Heyfield **54 D2** 37.59S 146.47E
Heywood **54 A1** 38.08S 141.38E
Hialeah **135 F1** 25.49N 80.17W
Hibbing **134 E3** 47.25N 92.55W
Hibbs, Point **60 C2** 42.37S 145.15E
Hicks, Point **55 E2** 37.48S 149.16E
Hidaka Range **104 K8**
High Atlas, mountains **124 C5**
Highfield Point **60 C4** 40.44S 145.17E
High Point **135 G2** 36.03N 79.50W
High River **132 G3** 50.40N 113.47W
Hiiumaa, island **115 I4**
Hikurangi, mountain **90 H6** 37.55S 178.04E
Hildesheim **116 F5** 52.09N 9.58E
Hilla **108 C4** 32.28N 44.29E
Hillsboro, Oregon **134 B3** 45.31N 122.59W
Hillsboro, Texas **136 C4** 32.01N 97.08W
Hillston **48 C3** 33.30S 145.33E
Himalayas, mountains **106 D3**
Himeji **104 G4** 34.50N 134.40E
Hinchinbrook Island **42 F6**
Hindiya **108 C4** 32.32N 44.14E
Hindmarsh, Lake **54 A2**
Hindu Kush, mountains **106 B4**
Hirado **104 D3** 33.22N 129.31E
Hirosaki **104 J7** 40.34N 140.28E
Hiroshima **104 F4** 34.23N 132.27E
Hirsova **117 I3** 44.41N 27.56E
Hirtshals **114 F4** 57.35N 9.58E
Hisiu **89 E2** 9.02S 146.48E
Hispaniola, island **137 E2**
Hissar **102 B2** 29.10N 75.45E
Hit **108 C4** 33.38N 42.49E
Hitachi **104 J5** 36.35N 140.40E
Hitoyoshi **104 E3** 32.12N 130.48E
Hivaoa, island **85 R5**
Hjorring **114 F4** 57.28N 9.59E
Hkakabo Razi, mountain **107 E3** 28.17N 97.46E
Hlako Kangri, mountain **106 D3** 28.48N 87.40E
Ho **124 D2** 6.38N 0.38E
Hobart **60 E2** 42.50S 147.15E
Hobbs **134 D2** 32.42N 103.08W
Ho Chi Minh City **100 B4** 10.46N 106.43E
Hodaka, mountain **104 H5** 36.21N 137.39E
Hodeida **108 C2** 14.48N 42.57E
Hodgson Downs, homestead **80 E7** 15.13S 134.05E
Hoeryong **104 D8** 42.29N 129.45E
Hof **114 G3** 50.19N 11.56E
Hofuf **108 C3** 25.20N 49.34E
Hogan Group **60 E5**
Hohhot **102 F4** 40.49N 111.37E
Hokianga Harbour **90 E7**
Hokitika **90 D3** 42.43S 170.58E
Hokkaido, island **104 K8**
Holbrook, Australia **48 D2** 35.44S 147.19E
Holbrook, U.S.A. **134 C2** 34.54N 110.11W
Holguin **137 E3** 20.54N 76.15W
Holla, Mount **125 G2** 7.00N 36.52E
Hollywood **135 F1** 26.01N 80.09W
Holmes Reefs **42 G7**

*Entry / **Page** / **Ref.** / Lat. / Long.*

Home Hill **42 F6** 19.40S 147.25E
Homer **132 C3** 59.40N 151.37W
Homestead **135 F1** 25.29N 80.29W
Homs, Libya **124 E5** 32.39N 14.15E
Homs, Syria **108 B4** 34.44N 36.43E
Hondeklipbaai **126 B1** 30.20S 17.18E
Hondo **104 E3** 32.28N 130.12E
Honduras, Gulf of **136 D2**
Honduras, republic **136 D2**
Hong Kong **103 F2** 22.16N 114.13E
Hongze Lake **103 F3**
Honiara **89 H2** 9.28S 159.52E
Honjo **104 I6** 39.22N 140.02E
Honolulu **84 P8** 21.19N 157.50W
Honshu, island **104 J5**
Hood Point **72 C2** 34.23S 119.33E
Hook Island **42 G5**
Hope **134 E2** 33.41N 93.36W
Hope, Point **132 B4** 68.20N 166.49W
Hopedale **133 L3** 55.30N 60.10W
Hopetoun, Victoria **54 B3** 35.44S 142.22E
Hopetoun, Western Australia **72 D2** 33.57S 120.08E
Hopkins River **54 B2**
Hopkinsville **135 F2** 36.50N 87.30W
Hordio **125 I3** 10.36N 51.08E
Hormuz, Strait of **108 D3**
Horn, Cape **144 C1** 56.00S 67.15W
Horn Cape **133 P4** 66.28N 22.27W
Horoshiri Peak **104 K8** 42.43N 142.40E
Horsens **114 F4** 55.53N 9.53E
Horsham **54 B2** 36.43S 142.12E
Horten **114 G4** 59.25N 10.30E
Hose Mountains **100 C3**
Hoskins **89 F2** 5.30S 150.27E
Hospet **102 B1** 15.16N 76.20E
Hosseina **125 G2** 7.30N 37.47E
Hotan **102 C3** 37.07N 79.57E
Hotham, Mount **54 D2** 36.59S 147.08E
Hot Springs **134 E2** 34.30N 93.02W
Houghton **135 F3** 47.06N 88.34W
Houlton **135 H3** 46.09N 67.50W
Houma, China **102 F3** 35.36N 111.15E
Houma, U.S.A. **135 E1** 29.35N 90.44W
Houston **134 E1** 29.45N 95.25W
Houtman Abrolhos, islands **72 B3**
Hovd **102 D4** 48.00N 91.43E
Hovsgol, Lake **102 E5**
Howe, Cape **48 F1** 37.30S 149.59E
Howe, West Cape **72 C1** 35.08S 117.36E
Howland, island **84 N6**
Howrah **106 D3** 22.35N 88.20E
Hradec Kralove **117 G5** 50.13N 15.50E
Hrubieszow **115 I3** 50.49N 23.53E
Huacho **142 C2** 11.14S 77.35W
Huaibei **103 F3** 34.00N 116.48E
Huainan **103 F3** 32.41N 117.06E
Huambo **126 B3** 12.47S 15.44E
Huancayo **142 C2** 12.05S 75.12W
Huangshi **103 F3** 30.13N 115.05E
Huanren **104 B7** 41.14N 125.22E
Huanuco **142 C3** 9.55S 76.14W
Huascaran, mountain **142 C3** 9.08S 77.36W
Huasco **144 B4** 28.30S 71.15W
Hubli **106 C2** 15.20N 75.14E
Hudson Bay **132 J3**
Hudson River **130 L5**
Hudson Strait **133 K4**
Hue **100 B4** 16.28N 107.35E
Huelva **116 B2** 37.15N 6.56W
Huesca **116 C3** 42.08N 0.25W
Hughenden **42 E5** 20.51S 144.12E
Hughes **66 B5** 30.43S 129.31E
Huiarau Range **90 G5**
Huila, mountain **142 C4** 3.00N 75.59W
Hulan River **104 C10**
Hull, Canada **133 K2** 45.26N 75.45W
Hull, U.K. **114 D3** 53.45N 0.20W
Hulun Lake **103 F4**
Humaita **142 D3** 7.33S 63.01W
Humboldt River **134 C3**
Hume, Lake **54 D2**
Humedan **108 D3** 25.27N 59.42E
Humenne **115 I2** 48.57N 21.55E
Humphreys Peak **134 C2** 35.21N 111.41W
Hun River **104 B7**
Huna Bay **133 P4**
Hungary, people's republic **115 H2**
Hungnam **104 C6** 39.49N 127.40E
Hunter Island **60 B4**
Hunter River **48 F3**
Huntington **135 F2** 38.24N 82.26W
Huntly **90 F6** 37.33S 175.10E
Huntsville, Alabama **135 F2** 34.44N 86.35W
Huntsville, Canada **133 K2** 45.20N 79.14W
Huntsville, Texas **136 C4** 30.43N 95.34W
Huon Gulf **89 E2**
Huon Peninsula **89 E2**
Huon River **60 D1**
Huonville **60 E1** 43.02S 147.04E
Huron **134 E3** 44.22N 98.12W
Huron, Lake **135 F3**
Hurstbridge **54 C2** 37.38S 145.12E
Husi **115 J2** 46.40N 28.05E
Huskvarna **114 G4** 57.47N 14.15E
Hussein-Dey **124 D5** 36.46N 2.55E
Husum **114 F3** 54.29N 9.04E
Hutchinson **134 E2** 38.03N 97.56W
Hwang River **103 F3**
Hwange **126 C3** 18.20S 26.25E
Hyden **72 C2** 32.26S 118.52E
Hyderabad, India **106 C2** 17.22N 78.26E
Hyderabad, Pakistan **106 B3** 25.23N 68.24E
Hyesan **104 D7** 41.22N 128.14E
Hyuga **104 E3** 32.25N 131.38E
Hyvinkaa **115 I5** 60.37N 24.50E

I

Ialibu **89 D2** 6.17S 144.00E
Iasi **117 I4** 47.09N 27.38E
Ibadan **124 D2** 7.23N 3.56E
Ibague **142 C4** 4.25N 75.14W
Ibarra **142 C4** 0.23N 78.05W
Iberian Peninsula **116 B2**
Ibiza **116 D2** 38.54N 1.26E
Ibiza, island **116 D2**
Ibonma **88 B3** 3.29S 133.30E
Ibra **108 D3** 22.43N 58.32E
Ibri **108 D3** 23.15N 56.16E
Ibu **100 D3** 1.30N 127.33E
Ica **142 C2** 14.05S 75.43W
Iceland, republic **110 C4**
Ichalkaranji **106 C2** 16.40N 74.33E
Ichinomiya **104 H4** 35.18N 136.48E
Ichinoseki **104 J6** 38.56N 141.08E
Idabel **134 E2** 33.54N 94.50W
Idah **124 D2** 7.05N 6.45E
Idaho, state **134 C3**
Idaho Falls **134 C3** 43.30N 112.01W
Idfu **125 G4** 24.58N 32.52E
Idlib **108 B4** 35.57N 36.38E
Igarka **118 I5** 67.31N 86.33E
Iglesias **116 E2** 39.19N 8.32E
Igli **124 C5** 30.28N 2.19W
Igom **88 B3** 2.00S 130.03E
Iguala **136 C2** 18.21N 99.31W
Igualada **116 D3** 41.35N 1.37E
Iguatu **143 G3** 6.22S 39.20W
Iguidi Desert **124 C4**
Ihosy **126 E2** 22.23S 46.09E
Ihu **89 E2** 7.55S 145.25E
Iisalmi **115 J5** 63.32N 27.15E
Ikela **126 C4** 1.06S 23.06E
Ikot Ekpene **124 D2** 5.10N 7.43E
Ilagan **100 D4** 17.07N 121.53E
Ilawa **114 H3** 53.35N 19.33E
Ilebo **126 C4** 4.20S 20.35E
Ilesha **124 D2** 7.39N 4.38E
Ilheus **143 G2** 14.50S 39.06W
Ili River **102 B4**
Iligan **100 D3** 8.12N 124.13E
Illapel **144 B3** 31.40S 71.13W
Illimani, mountain **142 D2** 16.37S 67.48W
Illinois, state **135 F2**
Illo **124 D3** 11.30N 3.40E
Ilo **142 C2** 17.38S 71.20W
Iloilo **100 D4** 10.41N 122.33E
Ilorin **124 D2** 8.32N 4.34E
Ilychovsk **115 K2** 46.19N 30.32E
Iman River **104 G9**
Imatra **115 J5** 61.14N 28.50E
Imi **125 H2** 6.28N 42.10E
Imjin River **104 C6**
Imlay, Mount **48 E1** 37.11S 149.44E
Imonda **88 D3** 3.20S 141.10E
Imperia **116 E3** 43.52N 8.01E
Impfondo **126 B5** 1.36N 18.00E
Imphal **107 E3** 24.47N 93.55E
Ina **104 H4** 35.52N 137.58E
In Amenas **124 D4** 28.05N 9.23E
Inanwatan **88 B3** 2.09S 132.14E
Inari **118 C5** 68.54N 27.05E
Inawashiro, Lake **104 J5**
Inca **116 D2** 39.43N 2.54E
Incahuasi, mountain **144 C4** 27.03S 68.20W
Inchon **104 C5** 37.30N 126.38E
India, republic **106 C3**
Indian Ocean **146 E2**
Indiana, state **135 F2**
Indianapolis **135 F2** 39.45N 86.10W
Indiga **118 E5** 67.40N 49.00E
Indigirka River **119 O5**
Indonesia, republic **100 C2**
Indore **106 C3** 22.42N 75.54E
Indus, Mouths of the **106 B3**
Indus River **106 B3**
Infiernollo, Lake **136 B2**
Ingham **42 F6** 18.39S 146.10E
Inglewood, New Zealand **90 F5** 39.09S 174.12E
Inglewood, Queensland **42 H1** 28.25S 151.02E
Inglewood, Victoria **54 B2** 36.35S 143.52E
Ingolstadt **114 G2** 48.46N 11.27E
Inhambane **126 D2** 23.51S 35.29E
Inirida **142 D4** 3.55N 67.52W
Inland Sea **104 F4**
Innamincka, homestead **66 H7** 27.44S 140.46E
Innisfail **42 F7** 17.32S 146.01E
Innsbruck **114 G2** 47.17N 11.25E
Inongo **126 B4** 1.55S 18.20E
Inscription, Cape **72 B3** 25.29S 112.58E
Inta **118 G5** 66.04N 60.01E
Inubo, Cape **104 J4** 35.41N 140.52E
Inuvik **132 E4** 68.16N 133.40W
Invercargill **90 C1** 46.25S 168.21E
Inverell **48 F5** 29.46S 151.07E
Inverness **114 D4** 57.27N 4.15W
Investigator Group **66 E4**
Investigator Strait **66 F3**
Ioannina **117 H2** 39.40N 20.51E
Ioma **89 E2** 8.22S 147.49E
Ionian Sea **117 G2**
Iowa, state **134 E3**
Iowa City **134 E3** 41.39N 91.31W
Ipiau **143 G2** 14.07S 39.43W
Ipoh **100 B3** 4.36N 101.02E
Ipsala **117 I3** 40.56N 26.23E
Ipswich, Australia **42 I2** 27.37S 152.47E
Ipswich, U.K. **114 E3** 52.04N 1.10E
Ipuh **100 B2** 2.58S 101.28E
Iqaluit **133 L4** 63.45N 68.30W
Iquique **142 C1** 20.15S 70.08W
Iquitos **142 C3** 3.51S 73.13W
Iraklion **117 I2** 35.20N 25.08E
Iran, Islamic republic **108 D4**
Iran, Plateau of **108 D4**
Irapuato **136 B3** 20.40N 101.40W
Iraq, socialist republic **108 C4**
Irbid **108 B4** 32.33N 35.51E
Ireland, Republic of **114 B3**
Iri **104 C4** 35.59N 126.55E
Iringa **126 D4** 7.49S 35.39E
Iriri River **143 E3**
Irish Sea **114 C3**
Irkutsk **118 K4** 52.18N 104.15E
Iron Baron **66 F4** 33.00S 137.10E
Iron Knob **66 F4** 32.44S 137.09E
Iron Mountain **135 F3** 45.51N 88.03W
Irrawaddy, Mouths of the **107 E2**
Irrawaddy River **107 E2**
Irtysh River **118 H4**
Isa **124 D3** 13.11N 6.24E
Isaac River **42 G4**
Isabella, island **142 A3**
Isafjordhur **133 P4** 66.05N 23.08W
Ise **104 H4** 34.29N 136.41E
Ise Bay **104 H4**
Isernia **117 F3** 41.35N 14.14E
Isherim, Mount **118 F5** 61.07N 59.08E
Ishikari Bay **104 J8**
Ishim **118 G4** 56.10N 69.30E
Ishim River **118 G4**
Ishinomaki **104 J6** 38.25N 141.18E
Ishpeming **135 F3** 46.29N 87.40W
Isiro **126 C5** 2.50N 27.40E
Iskenderun **108 B4** 36.37N 36.08E
Islamabad **106 C4** 33.42N 73.09E
Island Lagoon **66 F5**
Islands, Bay of **90 F7**
Ismail **115 J2** 45.21N 28.51E
Ismailia **108 B4** 30.36N 32.15E
Isna **125 G4** 25.16N 32.30E
Isparta **108 B4** 37.46N 30.32E
Israel, republic **108 B4**
Issyk, Lake **118 H3**
Istanbul **108 A5** 41.02N 28.57E
Itabuna **143 G2** 14.48S 39.18W
Italy, republic **117 G3**
Itanagar **107 E3** 27.02N 93.38E
Ithaca **135 G3** 42.26N 76.30W
Itigi **126 D4** 5.41S 34.30E
Iturup, island **119 O3**
Ivanhoe **48 C3** 32.54S 144.18E
Ivano-Frankovsk **115 I2** 48.53N 24.46E
Ivanovo **115 M4** 57.00N 41.00E
Ivdel **110 K4** 60.45N 60.30E
Ivory Coast, republic **124 C2**
Iwaki **104 J5** 36.58N 140.56E
Iwakuni **104 F4** 34.10N 132.09E
Iwamizawa **104 J8** 43.12N 141.47E
Iwate, mountain **104 J6** 39.53N 140.59E
Iwewi **89 D2** 7.05S 143.56E
Iwo Jima, island **84 J8**
Iwupataka **80 D3** 23.48S 133.35E
Izabal, Lake **136 D2**
Izhevsk **118 F4** 56.49N 53.11E
Izmir **108 A4** 38.25N 27.10E
Izmit **108 A5** 40.47N 29.55E
Iznik **117 I3** 40.27N 29.43E
Izumo **104 F4** 35.23N 132.43E
Izushichito Islands **104 I4**

J

Jabalpur **106 C3** 23.10N 79.59E
Jabiru **80 D8** 12.40S 132.53E
Jaca **116 C3** 42.34N 0.33W
Jackson, Mississippi **135 E2** 32.20N 90.11W
Jackson, Missouri **135 F2** 37.23N 89.40W
Jackson, Tennessee **135 F2** 35.37N 88.50W
Jackson Bay **90 C3**
Jackson Head **90 C3** 43.58S 168.37E
Jackson, Mount **72 C2** 30.15S 119.16E
Jackson, Port **48 F3**
Jacksonville, Florida **135 F2** 30.20N 81.40W
Jacksonville, North Carolina **135 G2** 34.45N 77.26W
Jacksonville Beach **135 F2** 30.18N 81.24W
Jacmel **137 E2** 18.18N 72.32W
Jacobina **143 F2** 11.13S 40.30W
Jacques Cartier Passage **133 L3**
Jacquinot Bay **89 F2**
Jaen **116 C2** 37.46N 3.48W
Jaffna **106 D1** 9.40N 80.01E
Jagdalpur **106 D2** 19.04N 82.05E
Jagersfontein **126 C2** 29.46S 25.25E
Jaguarao **144 D3** 32.30S 53.25W
Jagungal, Mount **48 E1** 36.09S 148.23E
Jaipur **106 C3** 26.53N 75.50E
Jaisalmer **106 C3** 26.52N 70.55E
Jajpur **106 D3** 20.50N 86.25E
Jakarta **100 B2** 6.08S 106.45E
Jalapa Enriquez **136 C2** 19.32N 96.56W
Jalandhar **106 C4** 31.18N 75.40E
Jalna **106 C2** 19.50N 75.58E
Jalpaiguri **107 D3** 26.30N 88.50E
Jaluit, island **84 L6**

*Entry / **Page** / **Ref.** / Lat. / Long.*

Jamaari **124 D3** 11.42N 9.54E
Jamaica, independent state **136 E2**
Jamama **125 H1** 0.01N 42.44E
Jambi, province **100 B2**
Jambuike **88 D2** 7.24S 140.06E
James Bay **133 J3**
James River **134 E3**
Jamestown, Australia **66 G4** 33.12S 138.36E
Jamestown, U.S.A. **134 E3** 46.54N 98.42W
Jamieson **54 D2** 37.18S 146.08E
Jammu **106 C4** 32.43N 74.54E
Jamnagar **106 C3** 22.28N 70.06E
Jamsa **115 J5** 61.51N 25.10E
Jamshedpur **106 D3** 22.47N 86.12E
Jandowae **42 H2** 26.47S 151.07E
Januaria **143 F2** 15.28S 44.23W
Japan, independent state **104 J5**
Japan, Sea of **104 G7**
Japan Trench **84 J8**
Japura River **142 D3**
Jarocin **114 H3** 51.59N 17.30E
Jarrahdale **72 C2** 32.20S 116.07E
Jarvis, island **84 P5**
Jask **108 D3** 25.40N 57.46E
Jativa **116 C2** 39.00N 0.32W
Jauf **108 B3** 29.47N 39.52E
Jaunpur **106 D3** 25.44N 82.41E
Java, island **100 C2**
Java Sea **100 C2**
Java Trench **84 F5**
Jawi **100 B2** 0.47S 109.15E
Jaya Peak **88 C3** 4.05S 137.09E
Jayapura **88 D3** 2.37S 140.39E
Jayawijaya Range **88 C3**
Jayrud **108 B4** 33.48N 36.44E
Jeble **108 B4** 35.22N 35.56E
Jefferson City **134 E2** 38.33N 92.10W
Jekabpils **115 J4** 56.28N 25.58E
Jelgava **115 I4** 56.39N 23.40E
Jena **114 G3** 50.56N 11.35E
Jeparit **54 A2** 36.09S 141.59E
Jequie **143 F2** 13.52S 40.06W
Jeremoabo **143 G2** 10.04S 38.23W
Jerez de la Frontera **116 B2** 36.41N 6.08W
Jericho **38 E3** 23.36S 146.07E
Jerilderie **48 C2** 35.22S 145.43E
Jerusalem **108 B4** 31.47N 35.13E
Jervis Bay **48 F2**
Jessore **107 D3** 23.10N 89.12E
Jhansi **102 B2** 25.27N 78.34E
Jhelum River **106 C4**
Jiamusi **103 H4** 46.59N 130.29E
Jianou **103 F2** 27.04N 118.20E
Jiayuguan **102 D3** 39.47N 98.14E
Jidda **108 B3** 21.30N 39.10E
Jiggalong **72 D4** 23.21S 120.47E
Jihlava **114 H2** 49.24N 15.35E
Jijiga **125 H2** 9.22N 42.50E
Jilin **103 G4** 43.53N 126.35E
Jima **125 G2** 7.39N 36.47E
Jinan **103 F3** 36.41N 117.00E
Jindabyne **48 E1** 36.25S 148.37E
Jingdezhen **103 F2** 29.17N 117.12E
Jingpo Lake **104 D8**
Jining **103 F3** 35.25N 116.40E
Jinja **126 D5** 0.27N 33.14E
Jinka **80 F3** 22.55S 135.38E
Jinxi **103 G4** 40.46N 120.47E
Jinzhou **103 G4** 41.07N 121.06E
Jiuquan **102 D3** 39.47N 98.30E
Jixi **103 H4** 45.17N 131.00E
Joao Pessoa **143 G3** 7.06S 34.53W
Jodhpur **106 C3** 26.18N 73.08E
Joensuu **115 J5** 62.36N 29.45E
Joghatay **108 D4** 36.34N 57.00E
Johannesburg **126 C2** 26.10S 28.02E
Johnston Atoll **84 O7**
Johnston, Lake **72 D2**
Johor Baharu **100 B3** 1.28N 103.46E
Joinvile **144 E4** 26.20S 48.49W
Joliet **135 F3** 41.32N 88.05W
Joliette **133 K2** 46.02N 73.27W
Jolo **100 D3** 6.03N 121.01E
Jolo, island **100 D3**
Jonkoping **114 G4** 57.45N 14.10E
Jonquiere **133 K2** 48.25N 71.16W
Joplin **134 E2** 37.04N 94.31W
Jordan, kingdom **108 B4**
Jos **124 D2** 9.54N 8.53E
Jos Plateau **124 D3**
Joseph Bonaparte Gulf **80 B8**
Juan de Fuca Strait **132 F2**
Juan Fernandez Islands **85 W2**
Juazeiro **143 F3** 9.25S 40.30W
Juba **125 G2** 4.50N 31.35E
Juba River **125 H2**
Juiz de Fora **143 F1** 21.47S 43.23W
Jujuy **144 C4** 24.10S 65.18W
Juliaca **142 C2** 15.29S 70.09W
Julia Creek **42 C5** 20.40S 141.40E
Jullundur *see* Jalandhar
Jundah **42 D3** 24.50S 143.04E
Juneau **132 E3** 58.20N 134.20W
Junee **48 D2** 34.52S 147.35E
Jurua River **142 C3**
Juruena River **142 E2**
Jutiapa **136 C2** 14.17N 89.50W
Jyvaskyla **115 J5** 62.16N 25.50E

K

Kabala **124 B2** 9.40N 11.36W
Kabalo **126 C4** 6.02S 26.55E
Kabaman **89 F3** 4.36S 152.43E
Kabarole **126 D5** 0.40N 30.17E
Kabong **100 C3** 1.50N 111.07E
Kabul **106 B4** 34.31N 69.12E
Kabuli **89 E3** 2.10S 146.40E
Kabwe **126 C3** 14.29S 28.25E
Kadavu, island **84 M4**
Kadina **66 F4** 33.58S 137.43E
Kadoma **126 C3** 18.16S 29.55E
Kaduna **124 D3** 10.28N 7.25E
Kaduna River **124 D3**
Kaedi **124 B3** 16.12N 13.32W
Kaesong **104 C5** 37.59N 126.30E
Kafan **108 C4** 39.11N 46.22E
Kafiau, island **88 A3**
Kafue River **126 C3**
Kagoshima **104 E2** 31.37N 130.32E
Kagoshima Bay **104 E2**
Kagua **89 D2** 6.25S 143.50E
Kagul **115 J2** 45.58N 28.10E
Kai Islands **88 B2**
Kaiapit **89 E2** 6.15S 146.15E
Kaiapoi **90 E3** 43.23S 172.39E
Kaifeng **107 G4** 34.47N 114.20E
Kaikohe **90 E7** 35.24S 173.48E
Kaikoura **90 E3** 42.25S 173.41E
Kaikoura Range **90 E3**
Kailahun **124 B2** 8.21N 10.21W
Kaimana **88 B3** 3.39S 133.44E
Kainji Reservoir **124 D3**
Kaipara Harbour **90 F6**
Kairouan **116 E2** 35.42N 10.01E
Kairuku **89 E2** 8.50S 146.30E
Kaitaia **90 E7** 35.07S 173.14E
Kaiyuan **102 E2** 23.42N 103.09E
Kajaani **115 J5** 64.12N 27.45E
Kajabbi **42 C5** 20.02S 140.02E
Kakali **100 C2** 1.57S 119.22E
Kakinada **106 D2** 16.59N 82.20E
Kalabo **126 C3** 15.00S 22.37E
Kalahari Desert **126 B2**
Kalamata **117 H2** 37.02N 22.07E
Kalamazoo **135 F3** 42.17N 85.36W
Kalangadoo **66 H2** 37.34S 140.42E
Kalannie **72 C2** 30.22S 117.07E
Kalat **106 B3** 29.01N 66.38E
Kalbarri **72 B3** 27.42S 114.12E
Kalemie **126 C4** 5.57S 29.10E
Kalgoorlie **72 D2** 30.45S 121.28E
Kalianda **100 B2** 5.43S 105.34E
Kalimantan, states **100 C2**
Kaliningrad **115 I3** 54.40N 20.30E
Kalispell **134 C3** 48.12N 114.19W
Kalisz **114 H3** 51.46N 18.02E
Kalkarindji **80 C6** 17.27S 130.50E
Kalmar **114 H4** 56.39N 16.20E
Kalmykovo **118 F3** 49.02N 51.55E
Kaltukatjara **80 B2** 24.50S 129.05E
Kaluga **115 L3** 54.31N 36.16E
Kalumburu **72 E6** 14.18S 126.39E
Kalush **115 I2** 49.02N 24.20E
Kalutara **106 C1** 6.35N 79.59E
Kama River **118 F4**
Kamaishi **104 J6** 39.18N 141.52E
Kambalda **72 D2** 31.11S 121.40E
Kambang **100 B2** 1.40S 100.42E
Kamchatka Peninsula **119 P4**
Kamenets Podolski **115 J2** 48.40N 26.36E
Kamensk-Uralski **118 G4** 56.29N 61.49E
Kamet, mountain **106 C4** 30.54N 79.37E
Kamina **126 C4** 8.46S 25.00E
Kamishli **108 C4** 37.03N 41.15E
Kamloops **132 F3** 50.39N 120.24W
Kampa **100 B2** 1.46S 105.26E
Kampala **126 D5** 0.19N 32.35E
Kampar **100 B3** 4.17N 101.08E
Kampong Cham **100 B4** 11.59N 105.26E
Kampong Chhnang **100 B4** 12.16N 104.39E
Kampot **100 B4** 10.37N 104.11E
Kamrau Bay **88 B3**
Kamyshin **118 E4** 50.05N 45.24E
Kananga **126 C4** 5.53S 22.26E
Kanazawa **104 H5** 36.35N 136.38E
Kanchipuram **106 C2** 12.50N 79.44E
Kandahar **106 B4** 31.36N 65.47E
Kandalaksha **118 D5** 67.09N 32.31E
Kandi **124 D3** 11.05N 2.59E
Kandos **48 E3** 32.53S 149.59E
Kandrian **89 E2** 6.14S 149.32E
Kandy **106 D1** 7.17N 80.40E
Kangaroo Island **66 F3**
Kangean Islands **100 C2**
Kanggye **104 C7** 40.55N 126.40E
Kangnung **104 D5** 37.48N 127.52E
Kanin, Cape **118 E5** 68.38N 43.20E
Kanin Peninsula **118 E5**
Kaniva **54 A2** 36.23S 141.15E
Kankaanpaa **115 I5** 61.47N 22.25E
Kankakee **135 F3** 41.08N 87.52W
Kankan **124 C3** 10.22N 9.11W
Kanker **106 D3** 20.17N 81.30E
Kannauj **102 B2** 27.02N 79.56E
Kano **124 D3** 12.00N 8.31E
Kanoya **104 E2** 31.22N 130.50E
Kanpur **106 D3** 26.27N 80.14E
Kansas River **134 E2**
Kansas, state **134 D2**
Kansas City **134 E2** 39.05N 94.37W
Kansong **104 D6** 38.20N 128.28E
Kanton, island **84 N5**
Kaohsiung **103 G2** 22.36N 120.17E
Kaolack **124 B3** 14.09N 16.08W
Kaoma **126 C3** 14.50S 24.48E
Kapingamarangi, island **84 K6**
Kapiti Island **90 F4**
Kapsan **104 D7** 40.59N 128.22E
Kapuas River **100 C3**
Kapunda **66 G3** 34.21S 138.55E
Kapuskasing **133 J2** 49.25N 82.26W
Kaputar, Mount **48 F4** 30.16S 150.10E
Kara Desert **118 F2**
Kara Sea **118 H6**
Kara Strait **118 F5**
Kara Bogaz Gol Bay **118 F3**
Karabuk **108 B5** 41.12N 32.36E
Karachi **106 B3** 24.51N 67.02E
Karaganda **118 H3** 49.53N 73.07E
Karagin Island **119 Q4**
Karaikudi **106 C2** 10.04N 78.46E
Karaj **108 D4** 35.48N 50.58E
Karakelong, island **100 D3**
Karakoram Range **102 B3**
Karakumski Canal **108 D4**
Karakuwisa **126 B3** 18.56S 19.40E
Karalundi **72 C3** 26.08S 118.41E
Karaman **108 B4** 37.11N 33.13E
Karamay **102 C4** 45.48N 84.30E
Karamea Bight **90 D4**
Karas, island **88 B3**
Karasburg **126 B2** 28.00S 18.43E
Karasjok **118 C5** 69.27N 25.30E
Karasu **117 J3** 41.07N 30.37E
Karatsu **104 D3** 33.28N 129.58E
Karaul **118 I6** 70.08N 83.13E
Karbala **108 C4** 32.37N 44.03E
Karditsa **117 H2** 39.22N 21.55E
Kargopol **118 D5** 61.32N 38.59E
Kariba, Lake **126 C3**
Karibib **126 B2** 21.59S 15.51E
Karimata, island **100 B2**
Kariz **108 E4** 34.50N 60.49E
Kar Kar Island **89 E3**
Karl Marx Stadt **114 G3** 50.50N 12.55E
Karlovac **117 G4** 45.30N 15.34E
Karlovo **117 H3** 42.38N 24.49E
Karlskrona **114 H4** 56.10N 15.35E
Karlsruhe **114 F2** 49.00N 8.24E
Karlstad **114 G4** 59.24N 13.32E
Karnal **106 C3** 29.41N 76.58E
Karnobat **117 I3** 42.39N 26.59E
Karonga **126 D4** 9.56S 33.56E
Karonie **72 D2** 30.58S 122.32E
Karpathos, island **117 I2**
Karratha **72 C4** 20.44S 116.52E
Kars **108 C5** 40.35N 43.05E
Karshi **118 G2** 38.53N 65.45E
Kartal **117 I3** 40.54N 29.12E
Karufa **88 B3** 3.50S 133.25E
Karumba **42 C7** 17.29S 140.50E
Karuni **100 C2** 9.25S 119.15E
Kasai River **126 B4**
Kasama **126 D3** 10.10S 31.11E
Kasempa **126 C3** 13.28S 25.48E
Kashan **108 D4** 33.59N 51.29E
Kashi **102 B3** 39.29N 76.02E
Kashiwazaki **104 I5** 37.22N 138.33E
Kashmar **108 D4** 35.13N 58.25E
Kasimov **115 M3** 54.55N 41.25E
Kasiui, island **88 B3**
Kasongo **126 C4** 4.32S 26.33E
Kassala **125 G3** 15.24N 36.25E
Kassel **114 F3** 51.18N 9.30E
Kastamonu **118 D3** 41.22N 33.47E
Kastoria **117 H3** 40.33N 21.15E
Kasur **106 C4** 31.07N 74.30E
Kata Tjuta **80 C2** 25.18S 130.44E
Katanning **72 C2** 33.41S 117.33E
Katerine **117 H3** 40.15N 22.30E
Katherine **80 D7** 14.28S 132.16E
Katherine River **80 D7**
Kathmandu **106 D3** 27.42N 85.19E
Katihar **102 C2** 25.33N 87.34E
Katiola **124 C2** 8.11N 5.04W
Katoomba **48 F3** 33.43S 150.19E
Katowice **114 H3** 50.15N 18.59E
Katrineholm **114 H4** 58.59N 16.15E
Katsina **124 D3** 13.00N 7.32E
Kattakurgan **118 G2** 39.54N 66.13E
Kattegat, passage **114 G4**
Kau **100 D3** 1.11N 127.54E
Kaunas **115 I3** 54.52N 23.55E
Kaura Namoda **124 D3** 12.39N 6.38E
Kautokeino **118 C5** 69.00N 23.06E
Kavaje **117 G3** 41.11N 19.33E
Kavalla **117 H3** 40.56N 24.24E
Kavieng **89 F3** 2.34S 150.48E
Kavir Desert **108 D4**
Kawhia Harbour **90 F5**
Kayeli **88 A3** 3.25S 127.07E
Kayes **124 B3** 14.26N 11.28W
Kayseri **108 B4** 38.42N 35.28E
Kazakh Uplands **118 H3**
Kazakhstan, republic **118 G3**
Kazan **118 E4** 55.45N 49.10E
Kazan River **132 H4**
Kazandzhik **118 F2** 39.15N 55.31E
Kazanluk **117 I3** 42.37N 25.23E
Kazerun **108 D3** 29.37N 51.38E
Kebnekaise, mountain **112 F4** 67.55N 18.35E
Kecskemet **115 H2** 46.56N 19.43E
Kediri **100 C2** 7.45S 112.01E
Keer-weer, Cape **42 C9** 13.55S 141.28E
Keetmanshoop **126 B2** 26.36S 18.08E
Keith **66 H2** 36.06S 140.22E
Kelang **100 B3** 3.01N 101.25E
Kellerberrin **72 C2** 31.38S 117.43E

*Entry/**Page**/**Ref.**/Lat./Long.*

Kelo **124 E2** 9.21N 15.50E
Kelowna **132 G2** 49.50N 119.29W
Kelso **134 B3** 46.09N 122.54W
Kem **118 D5** 64.58N 34.39E
Kemano **132 F3** 53.39N 127.58W
Kembe **124 F2** 4.29N 21.53E
Kemerovo **118 I4** 55.25N 86.05E
Kemp Land **94 D2**
Kempsey **48 G4** 31.05S 152.50E
Kempten **118 B3** 47.44N 10.19E
Kenai **132 C4** 60.35N 151.19W
Kendal **114 D3** 54.20N 2.45W
Kendari **100 D2** 3.57S 122.36E
Kendawangan **100 C2** 2.32S 110.13E
Kenema **124 B2** 7.57N 11.11W
Kenge **126 B4** 4.56S 17.04E
Kenitra **124 C5** 34.20N 6.34W
Kennedy Range **72 C4**
Kennett **135 E2** 36.15N 90.04W
Kenora **132 I2** 49.47N 94.26W
Kenosha **135 F3** 42.34N 87.50W
Kent Group **60 E5**
Kentucky Lake **135 F2**
Kentucky, state **135 F2**
Kenya, Mount **126 D4** 0.10S 37.19E
Kenya, republic **126 D5**
Kepi **88 C2** 6.34S 139.22E
Keppel Bay **42 H4**
Kerang **54 B3** 35.44S 143.55E
Keraudren, Cape **72 C5** 19.57S 119.45E
Keravat **89 F3** 4.20S 152.00E
Kerch **115 L2** 45.22N 36.27E
Kerema **89 E2** 7.59S 145.46E
Keren **125 G3** 15.46N 38.28E
Kerguelen Islands **146 E1**
Kericho **126 D4** 0.22S 35.19E
Kerikeri **90 E7** 35.13S 173.57E
Kerinci, mountain **100 B2** 1.43S 101.15E
Kermadec Islands **84 N3**
Kermadec Trench **84 N2**
Kerman **108 D4** 30.18N 57.05E
Kermanshah **110 I1** 34.19N 47.04E
Kermit **134 D2** 31.49N 103.07W
Kerrville **136 C4** 30.03N 99.09W
Kerulen River **102 F4**
Kerzaz **124 C4** 29.27N 1.27W
Kesennuma **104 J6** 38.54N 141.34E
Kestenga **118 D5** 65.52N 31.45E
Ketaun **100 B2** 3.19S 101.49E
Ketchikan **132 E3** 55.25N 131.40W
Ketchum **134 C3** 43.41N 114.23W
Keweenaw Bay **135 F3**
Key West **135 F1** 24.34N 81.48W
Khabarovsk **119 N3** 48.32N 135.08E
Khaibar **108 B3** 25.48N 39.12E
Khalkis **117 H2** 38.28N 23.36E
Khamis Mushait **108 C2** 18.19N 42.45E
Khanaqin **108 C4** 34.22N 45.22E
Khancoban **48 E1** 36.13S 148.08E
Khandwa **106 C3** 21.49N 76.23E
Khandyga **119 N5** 62.42N 135.35E
Khangai Mountains **102 D4**
Khanh Hung **100 B3** 9.36N 105.59E
Khanka, Lake **119 N3**
Khanpur **106 C3** 28.38N 70.40E
Khanty-Mansisk **118 G5** 61.01N 69.00E
Kharkov **115 L2** 50.00N 36.15E
Kharovsk **115 M5** 59.67N 40.07E
Khartoum **125 G3** 15.33N 32.32E
Khaskovo **117 I3** 41.57N 25.32E
Khatanga **118 K6** 71.59N 102.31E
Khatanga River **118 K6**
Khemis **116 D2** 36.15N 2.18E
Kherson **115 K2** 46.39N 32.38E
Kheta River **118 J6**
Khiva **118 G3** 41.25N 60.49E
Khmelnitski **115 J2** 49.25N 26.59E
Khon Kaen **100 B4** 16.25N 102.50E
Khorog **118 H2** 37.32N 71.32E
Khorramabad **108 C4** 33.29N 48.21E
Khorramshahr **108 C4** 30.25N 48.09E
Khouribga **124 C5** 32.54N 6.57W
Khuaf **108 E4** 34.32N 60.05E
Khujand **118 G2** 40.14N 69.40E
Khulna **107 D3** 22.49N 89.34E
Khurda **106 D3** 20.10N 85.42E
Khurma **108 C3** 21.55N 42.02E
Khust **115 I2** 48.11N 23.19E
Khvor **108 D4** 33.47N 55.03E
Khvoy **108 C4** 38.32N 45.02E
Khyber Pass **106 C4** 34.06N 71.05E
Kiama **48 F2** 34.40S 150.51E
Kibombo **126 C4** 3.54S 25.55E
Kicking Horse Pass **132 G3** 51.28N 116.23W
Kidnappers, Cape **90 G5** 39.38S 177.06E
Kiel **114 G3** 54.20N 10.08E
Kielce **115 I3** 50.51N 20.39E
Kieta **89 G2** 6.13S 155.38E
Kiev **115 K3** 50.25N 30.30E
Kiewa River **54 D2**
Kiffa **124 B3** 16.38N 11.28W
Kifri **108 C4** 34.42N 44.58E
Kigali **126 D4** 1.59S 30.05E
Kigoma **126 C4** 4.52S 29.36E
Kii Channel **104 G3**
Kikinda **117 H4** 45.50N 20.30E
Kikori **89 D2** 7.25S 144.15E
Kikwit **126 B4** 5.02S 18.51E
Kilchu **104 D7** 40.59N 129.23E
Kilimanjaro, mountain **126 D4** 3.02S 37.20E
Kiliya **117 I4** 45.30N 29.16E
Kilkis **117 H3** 40.59N 22.52E
Killarney, homestead **80 C6** 16.21S 131.52E
Killeen **134 E2** 31.08N 97.44W
Kilmarnock **114 D4** 55.37N 4.30W
Kilmore **54 C2** 37.18S 144.57E
Kilwa Kivinje **126 D4** 9.00S 39.30E
Kimba **66 F4** 33.08S 136.25E
Kimbe **89 F2** 5.33S 150.09E
Kimbe Bay **89 F2**
Kimberley, Canada **134 C3** 49.40N 115.58W
Kimberley, South Africa **126 C2** 28.45S 24.46E
Kimberley, region **72 E5**
Kimberley Downs, homestead **72 D5** 17.24S 124.22E
Kimchon **104 D5** 36.07N 128.08E
Kimmirut **133 L4** 62.50N 69.50W
Kimovsk **115 L3** 53.58N 38.35E
Kimry **115 L4** 56.51N 37.20E
Kinabalu, mountain **100 C3** 6.03N 116.32E
Kindia **124 B3** 10.03N 12.49W
Kindu **126 C4** 3.00S 25.55E
Kineshma **115 M4** 57.28N 42.08E
King Island **60 B5**
King, Lake **54 D2**
King River **60 C2**
King Sound **72 D5**
Kingaroy **42 H2** 26.32S 151.50E
King Frederick VI Coast **133 O4**
King George Islands **85 P4**
King George Sound **72 C1**
King Leopold Range **72 D5**
Kingman Reef **84 O6**
Kingoonya **66 E5** 30.55S 135.19E
Kingscote **66 F3** 35.39S 137.38E
Kingston, Australia **60 E2** 42.59S 147.20E
Kingston, Canada **133 K2** 44.14N 76.30W
Kingston, Jamaica **137 E2** 17.58N 76.48W
Kingston South East **66 G2** 36.50S 139.51E
Kingstown **137 F2** 13.12N 61.14W
Kingsville **134 E1** 27.32N 97.53W
King William Island **132 I4**
King William, Lake **60 D2**
King William's Town **126 C1** 32.53S 27.24E
Kinshasa **126 B4** 4.18S 15.18E
Kipembawe **126 D4** 7.40S 33.25E
Kipushi **126 C3** 11.46S 27.15E
Kirakira **89 H1** 10.30S 161.55E
Kirensk **119 K4** 57.45N 108.02E
Kirghiz Steppe **118 F3**
Kirgizia, republic **118 H3**
Kiribati, independent state **84 N6**
Kirikkale **108 B4** 39.51N 33.32E
Kiriwina Island **89 F2**
Kirkcaldy **114 D4** 56.07N 3.10W
Kirkland Lake **133 J2** 48.10N 80.02W
Kirklareli **117 I3** 41.45N 27.12E
Kirkpatrick, Mount **94 I3** 84.25S 165.30E
Kirksville **134 E3** 40.12N 92.35W
Kirkuk **108 C4** 35.28N 44.26E
Kirkwood **126 C1** 33.24S 25.26E
Kirov **118 E4** 58.38N 49.38E
Kirova Bay **108 C4**
Kirovakan **108 C5** 40.49N 44.30E
Kirovograd **115 K2** 48.31N 32.15E
Kirsehir **108 B4** 39.09N 34.08E
Kirthar Range **106 B3**
Kiruna **118 C5** 67.53N 20.15E
Kisangani **126 C5** 0.33N 25.14E
Kishi **124 D2** 9.06N 3.56E
Kishinev **115 J2** 47.00N 28.50E
Kiskunhalas **115 H2** 46.26N 19.29E
Kismayu **125 H1** 0.25S 42.31E
Kissidougou **124 B2** 9.15N 10.08W
Kisumu **126 D4** 0.03S 34.47E
Kisvarda **115 I2** 48.13N 22.03E
Kita **124 C3** 13.04N 9.29W
Kita Iwo, island **84 J8**
Kitakami River **104 J6**
Kitakyushu **104 E3** 33.52N 130.49E
Kitale **126 D5** 1.01N 35.01E
Kitami **104 K8** 43.51N 143.54E
Kitchener, Australia **72 D2** 31.02S 124.11E
Kitchener, Canada **133 J2** 43.27N 80.30W
Kitgum **126 D5** 3.17N 33.04E
Kitimat **132 F3** 54.05N 128.38W
Kitwe **126 C3** 12.48S 28.14E
Kiunga **88 D2** 6.07S 141.18E
Kivu, Lake **126 C4**
Kiwai Island **89 D2**
Kizil River **112 H1**
Kizyl-Arvat **118 F2** 39.00N 56.23E
Kjolen Mountains **112 F4**
Klagenfurt **114 G2** 46.38N 14.20E
Klaipeda **115 I4** 55.43N 21.07E
Klamath Falls **134 B3** 42.12N 121.47W
Klerksdorp **126 C2** 26.52S 26.39E
Klin **115 L4** 56.02N 36.45E
Klintsy **115 K3** 52.45N 32.15E
Klodzko **114 H3** 50.28N 16.40E
Kluang **100 B3** 2.01N 103.18E
Klyuchevskaya Sopka, mountain **119 Q4** 56.03N 160.38E
Knob, Cape **72 C2** 34.32S 119.15E
Knoxville **135 F2** 36.00N 83.57W
Koba **100 B2** 2.30S 106.26E
Kobe **104 G4** 34.40N 135.12E
Kobi **88 A3** 2.59S 129.54E
Koblenz **114 F3** 50.21N 7.36E
Kobrin **115 I3** 52.16N 24.22E
Kobroor, island **88 B2**
Kochi **104 F3** 33.33N 133.52E
Kodiak **132 C3** 57.49N 152.30W
Kodiak Island **132 C3**
Koforidua **124 C2** 6.01N 0.12W
Kofu **104 I4** 35.42N 138.34E
Kohat **106 C4** 33.37N 71.30E
Kohima **107 E3** 25.40N 94.08E
Kohsan **106 B4** 34.40N 61.11E
Koje Island **104 D4**
Kojonup **72 C2** 33.50S 117.09E
Kok **100 C3** 1.48N 113.41E
Kokkola **115 I5** 63.50N 23.10E
Kokoda **89 E2** 8.55S 147.44E
Kokomo **135 F3** 40.30N 86.09W
Kokopo **89 F3** 4.20S 152.18E
Kokstad **126 C1** 30.32S 29.25E
Kola, island **88 B2**
Kola Peninsula **118 D5**
Kolaka **100 D2** 4.04S 121.38E
Kolar **106 C2** 13.10N 78.10E
Kolding **114 F4** 55.29N 9.30E
Kolguyev, island **118 E5**
Kolhapur **106 C2** 16.40N 74.20E
Kolkata **106 D3** 22.30N 88.20E
Kolo **114 H3** 52.12N 18.37E
Kolobrzeg **114 H3** 54.10N 15.35E
Kolombangara, island **89 G2**
Kolomna **115 L4** 55.05N 38.45E
Kolomyia **115 I2** 48.31N 25.00E
Koluszki **117 G5** 51.44N 19.50E
Kolwezi **126 C3** 10.44S 25.28E
Kolyma Range **119 P5**
Kolyma River **119 P5**
Komarno **117 G4** 47.46N 18.05E
Komatsu **104 H5** 36.25N 136.27E
Komo **88 D2** 6.04S 142.52E
Kom Ombo **125 G4** 24.26N 32.57E
Komoran Island **88 C2**
Komotini **117 I3** 41.06N 25.25E
Kompong Som **100 B4** 11.03N 103.41E
Komsomolsk-na-Amure **119 N4** 50.32N 136.59E
Konda **88 B3** 1.34S 131.55E
Kondinin **72 C2** 32.30S 118.16E
Kongolo **126 C4** 5.20S 27.00E
Kongsberg **114 F4** 59.42N 9.39E
Konin **114 H3** 52.12N 18.12E
Konjic **117 G3** 43.40N 17.59E
Konogogo **89 F3** 3.29S 152.06E
Konos **89 F3** 3.10S 151.45E
Konotop **115 K3** 51.15N 33.14E
Konskie **117 H5** 51.12N 20.23E
Konstantinovka **115 L2** 48.33N 37.45E
Konya **108 B4** 37.51N 32.30E
Konzhakovskiy, Mount **118 F4** 59.40N 59.05E
Kooloonong **54 B3** 34.53S 143.08E
Koolpinyah, homestead **80 C8** 12.23S 131.11E
Koolyanobbing **72 C2** 30.49S 119.31E
Koombooloomba, Lake **42 E7**
Koonalda, homestead **66 B5** 31.27S 129.51E
Koonawarra, homestead **48 B4** 30.55S 142.17E
Koondrook **54 C3** 35.38S 144.08E
Koonibba **66 D5** 31.54S 133.25E
Korce **117 H3** 40.38N 20.44E
Korea Strait **104 D3**
Korhogo **124 C2** 9.22N 5.31W
Korindina, homestead **48 B5** 29.40S 143.38E
Koriyama **104 J5** 37.23N 140.22E
Korla **102 C4** 41.48N 86.10E
Korogoro Point **48 G4** 31.05S 153.02E
Koroit **54 B1** 38.18S 142.22E
Koronadal **100 D3** 6.10N 125.10E
Koror **101 E3** 7.21N 134.31E
Koror, island **101 E3**
Korosten **115 J3** 51.00N 28.30E
Korsakov **104 K10** 46.36N 142.50E
Korumburra **54 C1** 38.26S 145.49E
Koryak Range **119 Q5**
Kos, island **108 A4**
Koscian **114 H3** 52.05N 16.38E
Kosciuszko, Mount **48 E1** 36.27S 148.16E
Koshiki Archipelago **104 D2**
Kosice **115 I2** 48.44N 21.15E
Koslan **118 E5** 63.29N 48.59E
Kosong **104 D6** 38.36N 128.21E
Kosovo **117 H3**
Kosrae, island **84 L6**
Kosti **125 G3** 13.11N 32.38E
Kostroma **115 M4** 57.46N 40.59E
Koszalin **114 H3** 54.10N 16.10E
Kotabaharu **100 C2** 0.51S 111.32E
Kota Bharu **100 B3** 6.07N 102.15E
Kotabumi **100 B2** 4.52S 104.59E
Kotah **106 C3** 25.11N 75.58E
Kota Kinabalu **100 C3** 5.59N 116.04E
Kotka **115 J5** 60.26N 26.55E
Kotlas **118 E5** 61.15N 46.35E
Kotovsk **115 J2** 47.42N 29.30E
Kotto River **124 F2**
Kotuy River **118 J5**
Kotzebue **132 B4** 66.51N 162.40W
Kotzebue Sound **119 T5**
Koudougou **124 C3** 12.15N 2.23W
Koulen **100 B4** 13.49N 104.43E
Kounradski **118 H3** 46.58N 74.59E
Kouroussa **124 C3** 10.40N 9.50W
Kouvola **115 J5** 60.54N 26.45E
Kovel **115 I3** 51.12N 24.48E
Kovik **133 K4** 61.40N 77.40W
Kovrov **115 M4** 56.23N 41.21E
Kow Swamp **54 C3**
Kowanyama, homestead **42 C8** 15.25S 141.58E

Entry / ***Page*** / ***Ref.*** / *Lat.* / *Long.*

Koyp, Mount **118 F5** 62.05N 59.09E
Koyukuk River **132 C4**
Kozani **117 H3** 40.18N 21.48E
Kozhikode **106 C2** 18.56N 72.51E
Kra, Isthmus of **107 E1**
Kragero **114 F4** 58.54N 9.25E
Krakow **115 I2** 50.03N 19.55E
Kraljevo **117 H3** 43.44N 20.41E
Kramatorsk **115 L2** 48.43N 37.33E
Kramfors **114 H5** 62.55N 17.50E
Kranj **117 F4** 46.15N 14.20E
Krasnodar **118 D3** 45.02N 39.00E
Krasnograd **115 L2** 49.22N 35.28E
Krasnoturinsk **118 G4** 59.46N 60.10E
Krasnouralsk **118 G4** 58.25N 60.00E
Krasnovodsk **118 F2** 40.00N 53.00E
Krasnovodsk Gulf **108 D4**
Krasnoyarsk **118 J4** 56.05N 92.46E
Krasnystaw **115 I3** 51.00N 23.10E
Kratie **100 B4** 12.30N 106.03E
Krau **88 D3** 3.22S 140.04E
Kremenchug **115 K2** 49.04N 33.25E
Kremenets **115 J2** 50.05N 25.48E
Krems **114 H2** 48.25N 15.36E
Kribi **124 D2** 2.56N 9.56E
Krishna River **106 C2**
Kristiansand **114 F4** 58.08N 8.01E
Kristiansund **114 F5** 63.06N 7.58E
Kristinehamn **114 G4** 59.17N 14.09E
Krivoi Rog **115 K2** 47.55N 33.24E
Krolevets **115 K3** 51.34N 33.24E
Kroonstad **126 C2** 27.40S 27.15E
Krosno **115 I2** 49.40N 21.46E
Krotoszyn **114 H3** 51.41N 17.27E
Krui **100 B2** 5.13S 103.56E
Krusevac **117 H3** 43.34N 21.20E
Ksar-el-Boukhari **116 D2** 35.55N 2.47E
Ksar-el-Kebir **116 B2** 35.04N 5.56W
Kualakurun **100 C2** 1.10S 113.54E
Kuala Lumpur **100 B3** 3.08N 101.42E
Kualapembuang **100 C2** 3.28S 112.35E
Kuala Selangor **100 B3** 3.24N 101.12E
Kuala Terengganu **100 B3** 5.20N 103.07E
Kualatungkal **100 B2** 0.50S 103.28E
Kuamut **100 C3** 5.08N 117.30E
Kuantan **100 B3** 3.50N 103.19E
Kuba **108 C5** 41.23N 48.33E
Kubumesaai **100 C3** 1.26N 115.05E
Kuching **100 C3** 1.32N 110.20E
Kudat **100 C3** 6.54N 116.47E
Kudgee, homestead **48 A3** 32.42S 141.40E
Kugluktuk **132 G4** 67.49N 115.12W
Kuhmo **115 J5** 64.04N 29.30E
Kui **89 E2** 7.30S 147.16E
Kuibyshev Reservoir **112 I3**
Kuji **104 J7** 40.10N 141.47E
Kukipi **89 E2** 8.13S 146.09E
Kukudu **89 G2** 7.58S 156.56E
Kula **117 I2** 38.33N 28.38E
Kula Kangri, mountain **107 E3** 28.15N 90.34E
Kuldiga **115 I4** 56.58N 21.58E
Kulgera, homestead **80 D2** 25.50S 133.18E
Kulwin **54 B3** 35.02S 142.40E
Kulyab **106 B4** 37.55N 69.47E
Kumagaya **104 I5** 36.09N 139.22E
Kumairi **108 C5** 40.47N 43.49E
Kumamba Islands **88 C3**
Kumamoto **104 E3** 32.50N 130.42E
Kumanovo **117 H3** 42.07N 21.40E
Kumasi **124 C2** 6.45N 1.35W
Kumbo **124 E2** 6.10N 10.41E
Kumo **124 E3** 10.03N 11.13E
Kunashir, island **119 O3**
Kundiawa **89 E2** 6.01S 144.58E
Kunduz **106 B4** 36.47N 68.51E
Kunlun Mountains **102 C3**
Kunming **102 E2** 25.04N 102.41E
Kunsan **104 C4** 35.57N 126.42E
Kunua **89 F2** 5.45S 154.45E
Kununurra **72 E5** 15.47S 128.44E
Kuop, island **101 G3**
Kuopio **115 J5** 62.54N 27.40E
Kupang **100 D1** 10.13S 123.38E
Kupiano **89 E1** 10.06S 148.12E

Entry / ***Page*** / ***Ref.*** / *Lat.* / *Long.*

Kuqa **102 C4** 41.43N 82.58E
Kur, island **88 B2**
Kura River **108 C5**
Kurashiki **104 F4** 34.36N 133.43E
Kure **104 F4** 34.14N 132.32E
Kurgan **118 G4** 55.30N 65.20E
Kurgan Tyube **106 B4** 37.52N 68.47E
Kurikka **115 I5** 62.36N 22.25E
Kuril Islands **119 P3**
Kuril Trench **84 K10**
Kuring Kuru **126 B3** 17.38S 18.39E
Kurnool **106 C2** 15.51N 78.01E
Kurri Kurri **48 F3** 32.49S 151.30E
Kursk **115 L3** 51.45N 36.14E
Kuruman **126 C2** 27.28S 23.27E
Kurume **104 E3** 33.20N 130.29E
Kurunegala **106 D1** 7.28N 80.23E
Kushiro **104 L8** 42.58N 144.24E
Kushka **106 B4** 35.14N 62.15E
Kuskokwim River **132 B4**
Kustanai **118 G4** 53.15N 63.40E
Kuta **124 D2** 9.50N 6.40E
Kutahya **108 A4** 39.25N 29.56E
Kutch, Gulf of **106 B3**
Kutno **117 G5** 52.13N 19.20E
Kuujjuaq **133 L3** 58.10N 68.15W
Kuujjuarapik **133 K3** 55.20N 77.50W
Kuusamo **118 C5** 65.57N 29.15E
Kuwait **108 C3** 29.20N 48.00E
Kuwait, Arab republic **108 C3**
Kuytun **102 C4** 44.30N 85.00E
Kwajalein, island **84 L6**
Kwakoegron **143 E4** 5.14N 55.20W
Kwando River **126 C3**
Kwangju **104 C4** 35.07N 126.52E
Kwango River **126 B4**
Kwei River **102 F2**
Kwekwe **126 C3** 18.55S 29.51E
Kwenge River **126 B4**
Kwidzyn **114 H3** 53.44N 18.53E
Kwikila **89 E2** 9.51S 147.43E
Kwilu River **126 B4**
Kwinana **72 C2** 32.14S 115.46E
Kwoka, mountain **88 B3** 0.40S 132.24E
Kyabram **54 C2** 36.19S 145.03E
Kyancutta **66 E4** 33.08S 135.33E
Kyneton **54 C2** 37.15S 144.27E
Kynuna **42 C5** 21.35S 141.55E
Kyoga, Lake **126 D5**
Kyogle **48 G5** 28.38S 153.00E
Kyongju **104 D4** 35.52N 129.15E
Kyoto **104 G4** 35.02N 135.45E
Kyushu, island **104 E2**
Kyustendil **117 H3** 42.16N 22.40E
Kyzyl **118 J4** 51.45N 94.28E
Kyzyl Desert **118 G3**
Kzyl-Orda **118 G3** 44.52N 65.28E

L

Labe **124 B3** 11.17N 12.11W
Labrador, region **133 L3**
Labrador Sea **133 M4**
Labrador City **133 L3** 52.56N 66.52W
Labrea **142 D3** 7.20S 64.46W
Labuan **100 B2** 6.25S 105.49E
Labuhanbilik **100 B3** 2.30N 100.10E
Laccadive Islands **106 C2**
La Ceiba **136 D2** 15.45N 86.45W
Lacepede Bay **66 G2**
Lachlan River **48 D3**
La Chorrera **142 C4** 8.53N 79.46W
La Ciotat **116 E3** 43.10N 5.37E
Lacombe **132 G3** 52.30N 113.42W
Ladoga, Lake **115 K5**
Lady Barron **60 F4** 40.13S 148.15E
Lady Elliot Island **42 I3**
Lady Musgrave Island **42 I4**
Ladysmith **126 C2** 28.34S 29.47E
Lae **89 E2** 6.45S 147.00E
Laefu **89 F3** 3.00S 151.18E
La Estrada **116 B3** 42.40N 8.30W
Lafayette, Indiana **135 F3** 40.25N 86.54W
Lafayette, Louisiana **134 E2** 30.12N 92.18W

Entry / ***Page*** / ***Ref.*** / *Lat.* / *Long.*

Lafia **124 D2** 8.30N 8.34E
Lages **144 D4** 27.48S 50.20W
Laghouat **124 D5** 33.49N 2.55E
Lagos, Nigeria **124 D2** 6.27N 3.28E
Lagos, Portugal **116 B2** 37.05N 8.40W
La Grande **134 C3** 45.21N 118.05W
Laguna **144 E4** 28.29S 48.45W
Lahat **100 B2** 3.46S 103.32E
Lahej **108 C2** 13.02N 44.54E
Lahewa **100 A3** 1.23N 97.08E
Lahore **106 C4** 31.34N 74.22E
Lahti **115 J5** 61.00N 25.40E
Lai **124 E2** 9.22N 16.14E
Laila **108 C3** 22.16N 46.45E
Laiwui **88 A3** 1.20S 127.39E
Lajamanu, homestead **80 C5** 18.20S 130.38E
La Junta **134 D2** 37.59N 103.35W
Lake Cargelligo **48 D3** 33.18S 146.22E
Lake Charles **134 E2** 30.13N 93.13W
Lake Havasu City **134 C2** 34.28N 114.20W
Lakeland **135 F1** 28.02N 81.59W
Lake Murray **88 D2** 6.50S 141.25E
Lake Nash, homestead **80 F4** 20.58S 137.56E
Lake Providence **136 C4** 32.48N 91.10W
Lakes Entrance **55 E2** 37.53S 148.00E
Lake Stewart, homestead **48 A5** 29.16S 141.15E
Lakshadweep, islands **106 C2**
Lalago **126 D4** 3.31S 33.59E
Laleham **42 G4** 23.58S 148.48E
Lalin River **104 B9**
Lambarene **124 E1** 0.41S 10.13E
Lambert, Cape **89 F3** 4.13S 151.30E
Lambu **89 F3** 3.18S 151.43E
Lamego **116 B3** 41.05N 7.49W
Lameroo **66 H3** 35.20S 140.31E
Lamesa **134 D2** 32.43N 101.57W
Lamia **117 H2** 38.55N 22.26E
Lamitan **100 D3** 6.39N 122.09E
Lampang **100 A4** 18.16N 99.30E
Lampung, province **100 B2**
Lamu **126 E4** 2.17S 40.54E
Lanbi island **107 E2**
Lancaster, U.K. **114 D3** 54.03N 2.48W
Lancaster, U.S.A. **135 G2** 40.01N 76.19W
Lancelin **72 C2** 31.01S 115.20E
Lander **134 D3** 42.49N 108.44W
Land's End, cape **114 C2** 50.03N 5.44W
Landshut **114 G2** 48.31N 12.10E
Lang Mountains **102 E4**
Langdon, Port **80 F8**
Langgapayung **100 B3** 1.44N 99.59E
Langkawi, island **100 A3**
Langley **134 B3** 49 07N 122.39W
Langres **114 F2** 47 53N 5.20E
Lannion **116 C4** 48 44N 3.27W
Lansing **135 F3** 42.44N 84.34W
Lanuza **100 D3** 9.12N 126.04E
Lanzarote, island **124 B4**
Lanzhou **102 E3** 36.01N 103.45E
Laoag **100 D4** 18.14N 120.36E
Laon **116 D4** 49.34N 3.37E
La Oroya **142 C2** 11.36S 75.54W
Laos, people's democratic republic **100 B5**
La Palma **142 C4** 8.22N 78.08W
La Palma, island **124 B4**
La Paragua **142 D4** 6.53N 63.22W
La Paz, Argentina **144 D3** 30.45S 59.36W
La Paz, Bolivia **142 D2** 16.30S 68.10W
La Paz, Mexico **134 C1** 24.00N 110.15W
La Percuse Strait **104 J9**
Lapland, region **112 G4**
La Plata **144 D3** 34.52S 57.55W
Lappeenranta **115 J5** 61.04N 28.15E
Laptev Sea **119 M6**
Laptz, Mount **72 E5** 17.41S 126.43E
La Quiaca **142 D1** 22.06S 65.37W
Larache **116 B2** 35.12N 6.10W
Laramie **134 D3** 41.20N 105.38W
Larantuka **100 D2** 8.20S 123.00E
Larat, island **88 B2**
Laredo, Spain **116 C3** 43.24N 3.24W

Entry / ***Page*** / ***Ref.*** / *Lat.* / *Long.*

Laredo, U.S.A. **134 E1** 27.32N 99.22W
Largo **135 F1** 27.54N 82.47W
Lariang **100 C2** 1.27S 119.19E
La Rioja **144 C4** 29.26S 66.50W
Larissa **117 H2** 39.38N 22.25E
La Rochelle **116 C4** 46.10N 1.10W
La Roche-sur-Yon **116 C4** 46.40N 1.25W
La Roda **116 C2** 39.13N 2.10W
Larrimah **80 D7** 15.35S 133.13E
Larsen Ice Shelf **92 P3**
La Salle **133 K2** 45.26N 73.40W
La Sarre **133 K2** 48.49N 79.12W
Las Cruces **134 D2** 32.18N 106.47W
La Serena **144 B4** 29.54S 71.18W
Las Flores **144 D3** 36.03S 59.08W
Lashio **107 E3** 22.58N 97.48E
Las Palmas **124 B4** 28.08N 15.27W
La Spezia **116 E3** 44.07N 9.48E
Las Piedras **144 D3** 34.42S 56.14W
Las Tablas **136 D1** 7.49N 80.16W
Lastoursville **126 B4** 0.50S 12.03E
Las Vegas, Nevada **134 C2** 36.12N 115.10W
Las Vegas, New Mexico **134 D2** 35.36N 105.13W
Latakia **108 B4** 35.31N 35.47E
Latina **116 F3** 41.28N 12.53E
Latouche Treville, Cape **72 D5** 18.27S 121.50E
La Trobe, Mount **54 D1** 39.01S 146.22E
La Trobe River **54 C2**
Latrobe **60 D3** 41.14S 146.24E
La Tuque **133 K2** 47.26N 72.47W
Latur **106 C2** 18.24N 76.34E
Latvia, republic **115 I4**
Laughlen, Mount **80 E3** 23.22S 134.23E
Laun **100 A4** 10.10N 98.45E
Launceston **60 E3** 41.27S 147.10E
La Union **144 B2** 40.15S 73.02W
Laura **42 E8** 15.34S 144.27E
Laurel **135 F2** 31.41N 89.09W
Laurieton **48 G4** 31.40S 152.46E
Lausanne **116 E4** 46.32N 6.39E
Laut, island **100 C2**
Lautaro **144 B3** 38.30S 72.30W
Laval **116 C4** 48.04N 0.45W
La Vega **137 E2** 19.15N 70.33W
L'Averdy, Cape **89 G2** 5.33S 155.04E
Lavers Hill **54 B1** 38.40S 143.24E
Laverton **72 D3** 28.38S 122.24E
Lawrence **135 G3** 42.41N 71.12W
Lawton **134 E2** 34.36N 98.25W
Leaf River **133 K3**
Leake, Lake **60 E3**
Lebadea **117 H2** 38.26N 22.53E
Lebanon, republic **108 B4**
Lebu **144 B3** 37.38S 73.43W
Lecce **117 G3** 40.21N 18.11E
Le Creusot **116 D4** 46.48N 4.27E
Ledyanaya, Mount **119 R5** 61.49N 171.39E
Leech Lake **134 E3**
Leeds **114 D3** 53.50N 1.35W
Leesburg **135 F1** 28.49N 81.54W
Leeton **48 D2** 34.33S 146.24E
Leeuwin, Cape **72 B2** 34.22S 115.08E
Leeward Islands **137 G2**
Lefroy, Lake **72 D2**
Legazpi **100 D4** 13.10N 123.45E
Legges Tor, mountain **60 E3** 41.32S 147.41E
Leghorn **116 F3** 43.33N 10.18E
Legnica **114 H3** 51.12N 16.10E
Leh **106 C4** 34.09N 77.35E
Le Havre **116 D4** 49.30N 0.06E
Leichhardt Range **42 F5**
Leichhardt River **42 B6**
Leigh Creek **66 G5** 30.29S 138.25E
Leipzig **114 G3** 51.20N 12.20E
Leiria **116 B2** 39.45N 8.49W
Leksula **88 A3** 3.46S 126.35E
Lelingluan **88 B2** 7.09S 131.43E
Le Mans **116 D4** 48.00N 0.12E
Lemnos, island **117 I2**
Lemutan **100 C3** 3.02N 115.50E
Lemvig **114 F4** 56.33N 8.20E

*Entry/**Page**/**Ref.**/Lat./Long.*

Lena, Delta of the **119 M6**
Lena River **119 M5**
Lengshuijiang **102 F2** 27.40N 111.26E
Leningradskaya, antarctic base **92 I3** 69.30S 159.23E
Lenkoran **108 C4** 38.45N 48.50E
Lenmalu **88 B3** 1.45S 130.15E
Lens **116 D5** 50.26N 2.50E
Lensk **119 L5** 60.48N 114.55E
Leon, Mexico **136 B3** 21.10N 101.42W
Leon, Nicaragua **136 D2** 12.24N 86.52W
Leon, Spain **116 B3** 42.35N 5.34W
Leongatha **54 C1** 38.29S 145.57E
Leonora **72 D3** 28.53S 121.20E
Le Puy **116 D4** 45.03N 3.53E
Lerida **116 D3** 41.37N 0.38E
Lesbos, island **117 I2**
Leskovac **117 H3** 43.00N 21.57E
Lesosibirsk **118 J4** 58.13N 92.30E
Lesotho, kingdom **126 C2**
Les Sables d'Olonne **116 C4** 46.30N 1.47W
Lesser Antarctica, region **94 L2**
Lesser Kai, island **88 B2**
Lesser Slave Lake **132 G3**
Lethbridge **132 G2** 49.43N 112.48W
Leti Islands **88 A2**
Leticia **142 D3** 4.09S 69.57W
Letlhakane **126 C2** 21.20S 25.32E
Leuser, mountain **100 A3** 3.46N 97.12E
Levanger **114 G5** 63.44N 11.20E
Levelland **134 D2** 33.35N 102.23W
Leveque, Cape **72 D5** 16.24S 122.55E
Levice **117 G4** 48.14N 18.35E
Levin **90 F4** 40.37S 175.17E
Levis **135 G3** 46.47N 71.12W
Lewis Range **72 E5**
Lewiston, Idaho **134 C3** 46.24N 116.59W
Lewiston, Maine **135 G3** 44.06N 70.14W
Lewistown **134 D3** 47.04N 109.26W
Lexington **135 F2** 38.02N 84.30W
Leyte, island **100 D4**
Lhasa **107 E3** 29.41N 91.10E
Lianyungang **103 F3** 34.37N 119.10E
Liao River **103 F4**
Liaoyang **103 G4** 41.16N 123.12E
Liaoyuan **103 G4** 42.53N 125.10E
Liard River **132 E4**
Libenge **126 B5** 3.39N 18.39E
Liberal **134 D2** 37.03N 100.56W
Liberec **114 H3** 50.48N 15.05E
Liberia, republic **124 C2**
Libourne **116 C3** 44.55N 0.14W
Libreville **126 A5** 0.30N 9.25E
Libya, Arab republic **124 E4**
Libyan Desert **124 F4**
Licata **117 F2** 37.07N 13.57E
Lichinga **126 D3** 13.19S 35.13E
Lida **115 J3** 53.50N 25.19E
Liebig, Mount **80 C3** 23.18S 131.22E
Liechtenstein, principality **116 E4**
Liege **114 F3** 50.38N 5.35E
Liepaja **115 I4** 56.30N 21.00E
Lieska **115 K5** 63.20N 30.00E
Lifuka, island **84 N3**
Lightning Ridge **48 D5** 29.25S 147.59E
Lihir Group **89 F3**
Lihou Reefs **42 I7**
Likasi **126 C3** 10.58S 26.47E
Lille **116 D5** 50.39N 3.05E
Lillehammer **114 G5** 61.06N 10.27E
Lillestrom **114 G4** 59.58N 11.05E
Lilongwe **126 D3** 13.58S 33.49E
Lilydale **54 C2** 37.46S 145.21E
Lima, Peru **142 C2** 12.06S 77.03W
Lima, U.S.A. **135 F3** 40.43N 84.06W
Limbe **124 D2** 4.01N 9.12E
Limeira **143 F1** 22.34S 47.25W
Limerick **114 C3** 52.40N 8.38W
Limmen Bight **80 E7**
Limoeiro **143 G3** 7.51S 35.25W
Limoges **116 D4** 45.50N 1.15E
Limpopo River **126 D2**
Linares, Chile **144 B3** 35.47S 71.40W
Linares, Mexico **136 C3** 24.54N 99.38W
Linares, Spain **116 C2** 38.05N 3.38W

*Entry/**Page**/**Ref.**/Lat./Long.*

Lincoln **134 E3** 40.49N 96.41W
Linden **142 E4** 6.00N 58.17W
Lindi **126 D4** 10.00S 39.41E
Lindi River **126 C5**
Line Islands **84 O6**
Linfen **102 F3** 36.08N 113.40E
Lingayen **100 D4** 16.02N 120.14E
Lingen **116 E5** 52.32N 7.19E
Lingga, island **100 B2**
Linhe **102 E4** 40.45N 107.26E
Linjiang **104 C7** 41.45N 126.56E
Linkoping **114 H4** 58.25N 15.35E
Lins **143 F1** 21.40S 49.45W
Linxia **102 E3** 35.30N 102.58E
Linz **114 G2** 48.19N 14.18E
Lions, Gulf of **116 D3**
Lioppa **88 A2** 7.41S 126.01E
Lipa **100 D4** 13.57N 121.10E
Lipetsk **115 L3** 52.37N 39.36E
Liptrap, Cape **54 C1** 38.54S 145.55E
Liria **116 C2** 39.38N 0.37W
Lirik **100 B2** 0.12S 102.15E
Lisala **126 C5** 2.08N 21.37E
Lisbon **116 B2** 38.44N 9.08W
Lisburne, Cape **132 B4** 68.54N 166.18W
Lisichansk **115 L2** 48.53N 38.25E
Liski **115 L3** 51.00N 39.30E
Lismore, New South Wales **48 G5** 28.49S 153.16E
Lismore, Victoria **54 B2** 37.57S 143.21E
Litchfield, homestead **80 C8** 13.29S 130.41E
Lithgow **48 F3** 33.29S 150.10E
Lithuania, republic **115 I4**
Little Andaman, island **107 E2**
Little Cayman, island **136 E2**
Little Colorado River **134 C2**
Little Falls **132 I2** 45.58N 94.20W
Little Inagua, island **137 E3**
Little Missouri River **134 D3**
Little Rock **134 E2** 34.42N 92.17W
Liuzhou **102 E2** 24.17N 109.15E
Liverpool **114 D3** 53.25N 2.55W
Liverpool Range **48 E4**
Livingston **134 C3** 45.40N 110.33W
Livingstone **126 C3** 17.50S 25.23E
Livingstonia **126 D3** 10.36S 34.07E
Livno **117 G3** 43.50N 17.00E
Livny **115 L3** 52.25N 37.35E
Liwale **126 D4** 9.47S 38.00E
Lizard Island **42 E8**
Ljubljana **117 F4** 46.04N 14.30E
Ljusdal **114 H5** 61.50N 16.10E
Llanes **116 C3** 43.25N 4.45W
Llanos, region **142 C4**
Lloydminster **132 H3** 53.18N 110.00W
Lluchmayor **116 D2** 39.29N 2.53E
Llullaillaco, mountain **144 C4** 24.43S 68.30W
Lobatsi **126 C2** 25.11S 25.40E
Lobito **126 B3** 12.20S 13.34E
Lobo **88 B3** 3.45S 134.15E
Lock **66 E4** 33.34S 135.45E
Lockhart, Australia **48 D2** 35.14S 146.43E
Lockhart, U.S.A. **136 C4** 29.54N 97.41W
Lodar **108 C2** 13.53N 45.53E
Loddon River **54 B2**
Lodja **126 C4** 3.29S 23.26E
Lodwar **126 D5** 3.06N 35.38E
Lodz **114 H3** 51.49N 19.28E
Lofoten Islands **112 F4**
Lofty, Mount **66 G3** 34.59S 138.43E
Logan **134 C3** 41.46N 111.51W
Logan, Mount **132 D4** 60.31N 140.22W
Logansport **135 F3** 40.45N 86.25W
Loge River **126 B4**
Logone River **124 E3**
Logrono **116 C3** 42.28N 2.26W
Logstor **114 F4** 56.59N 9.16E
Loholoho **100 D2** 3.12S 121.06E
Loikaw **107 E2** 19.40N 97.17E
Loire River **116 C4**
Loja, Ecuador **142 C3** 3.59S 79.16W
Loja, Spain **116 C2** 37.10N 4.09W
Lokitaung **126 D5** 4.15N 35.45E

*Entry/**Page**/**Ref.**/Lat./Long.*

Lokoja **124 D2** 7.49N 6.44E
Lol River **125 F2**
Lom **117 H3** 43.50N 23.12E
Lomami River **126 C4**
Lombadina **72 D5** 16.31S 122.54E
Lomblen, island **100 D2**
Lombok, island **100 C2**
Lome **124 D2** 6.10N 1.21E
Lompoc **134 B2** 34.39N 120.30W
Lomza **115 I3** 53.11N 22.04E
Loncoche **144 B3** 39.23S 72.44W
London, Canada **133 J2** 42.58N 81.15W
London, U.K. **114 D3** 51.30N 0.10W
Londonderry **114 C3** 55.00N 7.19W
Londonderry, Cape **72 E6** 13.44S 126.58E
Londrina **144 D4** 23.18S 51.09W
Long Bay **135 G2**
Long Island, Bahamas **137 E3**
Long Island, P.N.G. **89 E2**
Long Island, U.S.A. **135 G3**
Long Beach **134 C2** 33.47N 118.15W
Longford **60 E3** 41.36S 147.08E
Longleju **100 C3** 2.50N 116.40E
Longmont **134 D3** 40.11N 105.07W
Longreach **42 E4** 23.27S 144.15E
Longview, Texas **134 E2** 32.30N 94.45W
Longview, Washington **134 B3** 46.06N 123.02W
Lonsdale, Lake **54 B2**
Lonsdale, Point **54 C1** 38.18S 144.37E
Lons-le-Saunier **116 E4** 46.41N 5.33E
Lookout, Cape **135 G2** 34.34N 76.34W
Loongana **72 E2** 30.57S 127.02E
Lopatka, Cape **119 P4** 50.50N 156.33E
Lop Buri **100 B4** 14.49N 100.37E
Lopez, Cape **124 D1** 0.36S 8.45E
Lop Nor, lake **102 D4**
Lorca **116 C2** 37.40N 1.41W
Lord Howe Island **84 L2**
Lordsburg **134 D2** 32.22N 108.43W
Lore **88 A2** 8.38S 127.04E
Lorengau **89 E3** 2.01S 147.15E
Lorient **116 C4** 47.45N 3.21W
Lorne **54 B1** 38.32S 143.58E
Los Alamos **134 D2** 35.52N 106.19W
Los Angeles, Chile **144 B3** 37.28S 72.23W
Los Angeles, U.S.A. **134 C2** 34.00N 118.15W
Losap, island **101 G3**
Los Palacias **136 D3** 22.35N 83.16W
Los Roques, island **137 F2**
Losuia **89 F2** 8.32S 151.04E
Lota **144 B3** 37.07S 73.10W
Lou Island **89 E3**
Loubomo **126 B4** 4.09S 12.47E
Louga **124 B3** 15.37N 16.13W
Louisiade Archipelago **89 F1**
Louisiana, state **134 E2**
Louis Trichardt **126 C2** 23.01S 29.43E
Louisville **135 F2** 38.13N 85.48W
Loukhi **118 D5** 66.04N 33.09E
Louny **114 G3** 50.22N 13.50E
Lourdes **116 D3** 43.06N 0.02W
Louth **48 C4** 30.32S 145.07E
Lovech **117 H3** 43.08N 24.45E
Loveland **134 D3** 40.24N 105.04W
Lovington **134 D2** 32.58N 103.21W
Lower California, peninsula **134 C1**
Lower Hutt **90 F4** 41.13S 174.55E
Lower Red Lake **132 I2**
Lower Tunguska River **118 J5**
Lowestoft **116 D5** 52.29N 1.45E
Lowicz **115 H3** 52.06N 19.55E
Loxton **66 H3** 34.27S 140.34E
Loyalty Islands **84 L3**
Lualaba River **126 C4**
Luama River **125 F1**
Luanda **126 B4** 8.50S 13.15E
Luang, Lake **100 B3**
Luanginga River **126 C3**
Luang Prabang **100 B4** 19.53N 102.10E
Luangwa River **126 D3**
Luanshya **126 C3** 13.09S 28.24E
Luau **126 C3** 10.44S 22.11E
Lubango **126 B3** 14.53S 13.30E

*Entry/**Page**/**Ref.**/Lat./Long.*

Lubao **126 C4** 5.19S 25.43E
Lubbock **134 D2** 33.35N 101.53W
Lubeck **114 G3** 53.52N 10.40E
Lublin **115 I3** 51.18N 22.31E
Lubny **115 K3** 50.01N 33.00E
Lubuklinggau **100 B2** 3.24S 102.56E
Lubumbashi **126 C3** 11.41S 27.29E
Lubutu **126 C4** 0.48S 26.39E
Lucena, Philippines **100 D4** 13.57N 121.36E
Lucena, Spain **116 C2** 37.25N 4.29W
Lucenec **115 H2** 48.20N 19.40E
Lucerne **116 E4** 47.03N 8.17E
Lucinda **42 F6** 18.32S 146.20E
Lucindale **66 H2** 36.58S 140.22E
Luckenwalde **114 G3** 52.05N 13.11E
Lucknow **106 D3** 26.50N 80.54E
Lucusse **126 C3** 12.38S 20.52E
Luda **103 G3** 38.53N 121.37E
Luderitz **126 B2** 26.38S 15.10E
Ludhiana **106 C4** 30.56N 75.52E
Ludvika **114 H5** 60.08N 15.14E
Ludza **115 J4** 56.30N 27.41E
Luena **126 B3** 11.48S 19.55E
Luga **115 J4** 58.42N 29.49E
Lugano **116 E4** 46.01N 8.57E
Lugansk **110 H2** 48.35N 39.20E
Lugenda River **126 D3**
Lugo **116 B3** 43.00N 7.33W
Luia River **126 C4**
Luiza **126 C4** 7.11S 22.27E
Lukenie River **126 B4**
Lukow **115 I3** 51.57N 22.21E
Lulea **118 C5** 65.35N 22.10E
Luleburgaz **108 A5** 41.25N 27.22E
Lumbala N'guimbo **126 C3** 14.08S 21.25E
Lumi **88 D3** 3.30S 142.00E
Lumsden **90 C2** 45.44S 168.26E
Lundu **100 B3** 1.40N 109.52E
Luneburg **114 G3** 53.15N 10.24E
Luohe **102 F3** 33.33N 114.00E
Luoyang **102 F3** 34.47N 112.26E
Luozi **124 E1** 4.56S 14.14E
Lusaka **126 C3** 15.26S 28.20E
Lusambo **126 C4** 4.59S 23.26E
Lut Desert **108 D4**
Lutsk **115 J3** 50.42N 25.15E
Luxembourg **114 F2** 49.37N 6.08E
Luxembourg, grand duchy **114 F3**
Luxor **125 G4** 25.41N 32.24E
Luzhou **102 E2** 28.55N 105.25E
Luzon, island **100 D4**
Luzon Strait **100 D4**
Lvov **115 I2** 49.50N 24.00E
Lyakhov Island **119 O6**
Lycksele **114 H5** 64.34N 18.40E
Lydenburg **126 D2** 25.10S 30.29E
Lynn Haven **135 F2** 30.14N 85.40W
Lynn Lake **132 H3** 56.51N 101.01W
Lyons **116 D4** 45.46N 4.50E
Lyons River **72 C4**
Lyttelton **90 D3** 43.36S 172.43E

M

Maa River **124 B2**
Ma'an **108 B4** 30.11N 35.45E
Maarianhamina **114 H5** 60.05N 19.55E
Maatsuyker Island **60 D1**
Mabber, Cape **125 I2** 9.29N 50.50E
Mabote **126 D2** 22.03S 34.09E
McAllen **134 E1** 26.13N 98.15W
Macapa **143 E4** 0.01N 51.01W
Macarthur **54 B1** 38.02S 142.00E
McArthur River **80 E6**
McArthur River, homestead **80 E6** 16.40S 135.51E
Macas **142 C3** 2.22S 78.08W
Macau **102 F2** 22.00N 113.40E
Macauley Island **84 N3**
McClintock, Mount **94 H3** 80.34S 156.09E
McClure Strait **132 G5**
McComb **136 C4** 31.13N 90.29W
McCook **134 D3** 40.13N 100.35W
MacDonald, Lake **80 B3**
Macdonnell Ranges **80 C3**

*Entry/**Page**/**Ref.**/Lat./Long.*

McDouall Peak, homestead **66 E6** 29.50S 134.54E
Macedon, Mount **54 C2** 37.27S 144.34E
Macedonia, Former Yugoslav Republic of **117 H3**
Maceio **143 G3** 9.40S 35.44W
McGill **134 C2** 39.24N 114.47W
Machala **142 C3** 3.20S 79.57W
Macintyre River **48 F5**
Mackay **42 G5** 21.09S 149.11E
Mackay, Lake **80 B3**
McKean, island **84 N5**
Mackenzie Mountains **132 E4**
Mackenzie River **132 F4**
Mackenzie King Island **132 G5**
McKinley, Mount **132 C4** 63.02N 151.01W
Maclean **48 G5** 29.28S 153.12E
Maclean Bay **60 F3**
Macleay River **48 G4**
MacLeod, Lake **72 B4**
McMurdo, antarctic base **92 I3** 77.51S 166.40E
Macon, France **116 D4** 46.18N 4.50E
Macon, U.S.A. **135 F2** 32.49N 83.37W
Macquarie Harbour **60 C2**
Macquarie Island **147 H1**
Macquarie, Lake **48 F3**
Macquarie River, New South Wales **40 E2**
Macquarie River, Tasmania **60 E3**
MacRobertson Land **94 D2**
Macumba River **66 E7**
Madagascar, republic **126 E3**
Madang **89 E2** 5.14S 145.45E
Madeira Islands **124 B5**
Madeira River **142 D3**
Madera **134 C2** 36.57N 120.04W
Madison **135 F3** 43.04N 89.22W
Madona **115 J4** 56.54N 26.10E
Madre de Dios River **142 C2**
Madrid **116 C3** 40.25N 3.43W
Madura, island **100 C2**
Madurai **106 C1** 9.55N 78.07E
Maebashi **104 I5** 36.24N 139.04E
Maffra **54 D2** 37.58S 146.59E
Mafia, island **126 D4**
Magadan **119 P4** 59.38N 150.50E
Magamo **88 B3** 0.54S 131.49E
Magangue **137 E1** 9.14N 74.46W
Magarida **89 E1** 10.18S 149.20E
Magdagachi **119 M4** 53.27N 125.44E
Magdalena, Bolivia **142 D2** 13.22S 64.07W
Magdalena, Mexico **136 A4**
Magdalena, U.S.A. **134 D2**
Magdalena River **142 C4**
Magdeburg **114 G3** 52.08N 11.37E
Magelang **100 C2** 7.28S 110.11E
Magellan, Strait of **144 B1**
Maghagha **108 B3** 28.39N 30.50E
Maghnia **116 C1** 34.50N 1.45W
Magnetic Island **42 F6**
Magnitogorsk **118 F4** 53.28N 59.06E
Maguarinho, Cape **143 F3** 0.15S 48.23W
Magwe **107 E3** 20.08N 94.55E
Mahajanga **120 H3** 15.40S 46.20E
Mahakam River **100 C3**
Mahalapye **126 C2** 23.05S 26.51E
Mahalla **108 B4** 30.59N 31.10E
Mahares **124 E5** 34.35N 10.30E
Mahdia **116 D2** 35.26N 1.40E
Mahe **106 C2** 11.41N 75.31E
Mahia Peninsula **90 G5**
Mahon **116 D2** 39.54N 4.15E
Maidan **108 C4** 34.55N 45.38E
Maiduguri **124 E3** 11.53N 13.16E
Maihar **106 D3** 24.14N 80.50E
Maimana **106 B4** 35.54N 64.43E
Mai-Ndombe, Lake **126 B4**
Maine, state **135 H3**
Maintirano **126 E3** 18.01S 44.03E
Mainz **114 F2** 50.00N 8.16E
Maipo, mountain **144 C3** 34.10S 69.52W
Maiquetia **142 D5** 10.38N 66.59W
Maitland, New South Wales **48 F3** 32.44S 151.34E
Maitland, South Australia **66 F3** 34.23S 137.40E
Maizuru **104 G4** 35.30N 135.20E
Majene **100 C2** 3.33S 118.59E
Majorca, island **116 D2**
Majuro **84 M6** 7.05N 171.08E
Majuro, island **84 M6**
Makalu, Mount **102 C2** 27.54N 87.06E
Makasar Strait **100 C2**
Makbon **88 B3** 0.44S 131.28E
Makemo, island **85 Q4**
Makeni **124 B2** 8.57N 12.02W
Makgadikgadi Pan, lake **126 C2**
Makhachkala **118 E3** 42.59N 47.30E
Maki **88 B3** 3.18S 134.13E
Makokou **126 B5** 0.38N 12.47E
Makoua **126 B4** 0.01S 15.40E
Makurdi **124 D2** 7.44N 8.35E
Malabo **124 D2** 3.45N 8.48E
Malaga **116 C2** 36.43N 4.25W
Malaimbandy **126 E2** 20.20S 45.35E
Malaita, island **89 H2**
Malakal **125 G2** 9.31N 31.40E
Malalaua **89 E2** 8.05S 146.10E
Malang **100 C2** 7.59S 112.45E
Malange **126 B4** 9.36S 16.21E
Malatya **108 B4** 38.22N 38.18E
Malawi, Lake **126 D3**
Malawi, republic **126 D3**
Malay Peninsula **98 G2**
Malaysia, federal republic **100 B3**
Malcolm, Point **72 D2** 33.48S 123.45E
Malden, island **84 P5**
Maldives, republic **96 D2**
Maldon **54 C2** 37.00S 144.04E
Maldonado **144 D3** 34.57S 54.59W
Male **96 D2** 4.00N 73.28E
Malegaon **106 C3** 20.32N 74.38E
Malekula, island **84 L4**
Maleny **42 I2** 26.46S 152.51E
Mali, republic **124 C3**
Malili **100 D2** 2.38S 121.06E
Malindi **126 E4** 3.14S 40.08E
Malita **100 D3** 6.23N 125.37E
Malkara **117 I3** 40.54N 26.54E
Mallacoota Inlet **55 E2**
Mallee, region **54 A3**
Malmesbury **126 B1** 33.28S 18.43E
Malmo **114 G4** 55.35N 13.00E
Maloelap, island **84 M6**
Maloyaroslavets **115 L3** 55.00N 36.28E
Malpelo Island **142 B4**
Malta, independent state **117 F2**
Maluku, province **100 D2**
Maluku Sea **100 D2**
Mamberamo **88 C3** 1.32S 137.53E
Mamberamo River **88 C3**
Mambere River **124 E2**
Mamburao **100 D4** 13.13N 120.39E
Mamore River **142 D2**
Mamou **124 B3** 10.24N 12.05W
Mamuju **100 C2** 2.41S 118.55E
Man **124 C2** 7.31N 7.37W
Manacor **116 D2** 39.35N 3.12E
Manado **100 D3** 1.32N 124.55E
Managua **136 D2** 12.06N 86.18W
Managua, Lake **136 D2**
Manama **108 D3** 26.13N 50.35E
Manangatang **54 B3** 35.03S 142.53E
Manapouri, Lake **90 B2**
Manar, Mount **108 C2** 14.01N 44.23E
Manassas **116 B2** 38.45N 8.00W
Manaus **142 D3** 3.06S 60.00W
Manawatu River **90 F4**
Manchester, U.K. **114 D3** 53.30N 2.15W
Manchester, U.S.A. **135 G3** 42.59N 71.28W
Manchuria, region **103 G4**
Mandab, Strait of **108 C2**
Mandal, Afghanistan **106 B4** 33.14N 61.50E
Mandal, Norway **114 F4** 58.02N 7.30E
Mandala, mountain **88 D3** 4.39S 140.18E
Mandalay **107 E3** 21.57N 96.04E
Mandalgobi **102 E4** 45.40N 106.10E
Mandali **108 C4** 33.43N 45.33E
Mandan **134 D3** 46.49N 100.54W
Mandla **106 D3** 22.35N 80.28E
Mandor **100 B3** 0.16N 109.15E
Mandurah **72 C2** 32.32S 115.43E
Manduria **117 G3** 40.24N 17.38E
Manfalut **125 G4** 27.23N 30.58E
Manfredonia **117 G3** 41.37N 15.55E
Mangaia, island **84 P3**
Mangakino **90 F5** 38.22S 175.46E
Mangalia **117 I3** 43.48N 28.36E
Mangalore **106 C2** 12.54N 74.51E
Mangareva, island **85 R3**
Manguari **38 D3** 28.58S 134.22E
Manhattan **134 E2** 39.11N 96.35W
Manicore **142 D3** 5.48S 61.16W
Manihiki, island **84 O4**
Manila **100 D4** 14.36N 120.59E
Manilla **48 F4** 30.45S 150.43E
Maningrida **80 E8** 12.03S 134.13E
Manisa **108 A4** 38.36N 27.29E
Manistee **135 F3** 44.14N 86.20W
Manitoba, Lake **132 I3**
Manitoba, province **132 I3**
Manizales **142 C4** 5.03N 75.32W
Manja **126 E2** 21.24S 44.20E
Manjimup **72 C2** 34.15S 116.09E
Mankono **124 C2** 8.01N 6.09W
Manna **100 B2** 4.29S 102.55E
Mannahill **66 G4** 32.26S 139.59E
Mannar, Gulf of **106 C1**
Mannheim **114 F2** 49.30N 8.28E
Manning River **48 F4**
Mannum **66 G3** 34.55S 139.18E
Manokwari **88 B3** 0.53S 134.05E
Manono **126 C4** 7.13S 27.24E
Manra, island **84 N5**
Manresa **116 D3** 41.43N 1.50E
Mansa **126 C3** 11.10S 28.52E
Mansel Island **133 J4**
Mansfield, Australia **54 D2** 37.03S 146.05E
Mansfield, Louisiana **136 C4** 32.02N 93.41W
Mansfield, Ohio **135 F3** 40.46N 82.31W
Manta **142 B3** 0.59S 80.44W
Mantes-la-Jolie **116 D4** 48.59N 1.43E
Manukau **90 F6** 36.59S 174.53E
Manukau Harbour **90 F6**
Manus, island **89 E3**
Manzanares **116 C2** 39.00N 3.23W
Manzanillo **136 B2** 19.00N 104.20W
Manzhouli **103 F4** 49.36N 117.28E
Manzini **126 D2** 26.30S 31.22E
Maoke Range **88 C3**
Maoming **102 F2** 21.50N 110.56E
Mapan **100 C2** 2.20S 111.10E
Mappi **88 C2** 7.04S 139.25E
Mappi River **88 C2**
Maprik **88 D3** 3.38S 143.02E
Maputo **126 D2** 25.58S 32.35E
Maquela do Zombo **126 B4** 6.06S 15.12E
Maraboon, Lake **42 F4**
Maracaibo **142 C5** 10.44N 71.37W
Maracaibo, Lake **142 C4**
Maracay **142 D5** 10.20N 67.28W
Marada **124 E4** 29.15N 19.14E
Maradi **124 D3** 13.29N 7.10E
Maragheh **108 C4** 37.23N 46.13E
Marajo Island **143 F3**
Maralal **126 D5** 1.05N 36.42E
Maralinga **66 C5** 30.10S 131.35E
Marama **66 H3** 35.11S 140.11E
Maramasike, island **89 H2**
Marambio, antarctic base **92 P3** 64.13S 56.38W
Maranboy **80 D7** 14.32S 132.47E
Maranoa River **42 G2**
Maranon River **142 C3**
Marapanim **143 F3** 0.42S 47.40W
Maras **108 B4** 37.34N 36.54E
Marble Bar **72 C4** 21.10S 119.45E
Mar Chiquita Lagoon **144 C3**
Marcus Island **84 K8**
Mardan **106 C4** 34.14N 72.05E
Mar del Plata **144 D3** 38.00S 57.32W
Mardin **108 C4** 37.19N 40.43E
Mareeba **42 E7** 17.00S 145.26E
Marek **125 H2** 3.46N 47.15E
Marganets **115 K2** 47.35N 34.37E
Margaret River **72 C2** 33.57S 115.04E
Margarita, island **142 D5**
Margate **126 D1** 30.51S 30.22E
Margherita, mountain **126 C5** 0.23N 29.54E
Margo Desert **106 B4**
Marhoum **124 C5** 34.26N 0.05W
Mari **88 D2** 9.10S 141.37E
Maria Island **60 F2**
Maria Elena **142 D1** 22.21S 69.40W
Mariana Trench **84 J7**
Marianske Lazne **114 G2** 49.59N 12.40E
Marias Islands **136 B3**
Mariato, Point **136 D1** 7.12N 80.52W
Maria van Diemen, Cape **90 E7** 34.29S 172.39E
Maribor **117 G4** 46.34N 15.38E
Marie Byrd Land **94 L3**
Marie Galante **137 F2**
Mariental **126 B2** 24.36S 17.59E
Marijampole **115 I3** 54.31N 23.20E
Marilia **143 F1** 22.13S 49.58W
Marinette **135 F3** 45.06N 87.38W
Maringa **144 D4** 23.25S 51.55W
Marion Bay **60 E2**
Marion, Lake **135 F2**
Marion Reef **42 I6**
Mariscal Estigarribia **142 D1** 22.02S 60.38W
Mariupol **115 L2** 47.05N 37.34E
Marka **125 H2** 1.42N 44.47E
Markham, Mount **94 I3** 82.45S 160.25E
Markovo **119 R5** 64.40N 170.24E
Marlo **55 E2** 37.48S 148.32E
Marmande **116 D3** 44.30N 0.10E
Marmaris **108 A4** 36.52N 28.17E
Maroa **142 D4** 2.43N 67.33W
Maromokotro, mountain **122 H3** 14.01S 49.00E
Marondera **126 D3** 18.11S 31.33E
Marotiri, island **85 Q3**
Maroua **124 E3** 10.35N 14.20E
Marquesas Islands **85 R5**
Marquette **135 F3** 46.33N 87.23W
Marrakech **124 C5** 31.49N 8.00W
Marrawah **60 B4** 40.55S 144.42E
Marree **66 G6** 29.39S 138.04E
Marrupa **126 D3** 13.10S 37.30E
Marsabit **126 D5** 2.20N 37.59E
Marsala **116 F2** 37.48N 12.27E
Marseilles **116 E3** 43.18N 5.22E
Marshall, Liberia **124 B2** 6.10N 10.23W
Marshall, U.S.A. **134 E3** 44.26N 95.46W
Marshall Bennett Islands **89 F2**
Marshall Islands, independent state **84 L7**
Marsh Harbour **136 E3** 26.31N 77.05W
Martaban, Gulf of **107 E2**
Martapura **100 B2** 4.20S 104.22E
Marte **124 E3** 12.23N 13.46E
Martinique, island **137 F2**
Marton **90 F4** 40.05S 175.23E
Maru, island **88 B2**
Marulan **48 F2** 34.43S 150.00E
Marumba, Mount **80 E8** 13.39S 134.25E
Mary **118 G2** 37.42N 61.54E
Maryborough, Queensland **42 I3** 25.32S 152.42E
Maryborough, Victoria **54 B2** 37.03S 143.44E
Maryland, state **135 G2**
Marysville **54 C2** 37.31S 145.45E
Masan **104 D4** 35.10N 128.35E
Masbate **100 D4** 12.21N 123.36E
Masbate, island **100 D4**
Masela **88 A2** 8.06S 129.50E
Masela, island **88 A2**
Maseru **126 C2** 29.19S 27.29E
Masirah, island **108 D3**
Masjed Soleyman **108 C4** 31.58N 49.18E
Mason Bay **90 B1**
Mason City **134 E3** 43.10N 93.10W
Massachusetts, state **135 H2**
Massenya **124 E3** 11.21N 16.09E
Massif Central, mountains **116 D3**

*Entry/**Page**/**Ref.**/Lat./Long.*

Masson Island **92 E3**
Masterton **90 F4** 40.57S 175.39E
Masuda **104 E4** 34.42N 131.51E
Masvingo **126 D2** 20.10S 30.49E
Maswaar, island **88 B3**
Matadi **126 B4** 5.50S 13.32E
Matagalpa **136 D2** 12.52N 85.58W
Matam **124 B3** 15.40N 13.18W
Matamoros, central Mexico **136 B3** 25.33N 103.15W
Matamoros, eastern Mexico **136 C3** 25.50N 97.31W
Matandu River **126 D4**
Matane **133 L2** 48.50N 67.31W
Matanzas **136 D3** 23.04N 81.35W
Mataram **100 C2** 8.36S 116.07E
Mataranka **80 D7** 14.56S 133.04E
Mataro **116 D3** 41.32N 2.27E
Matatiele **126 C1** 30.20S 28.49E
Mataura **90 C1** 46.12S 168.52E
Matehuala **136 B3** 23.40N 100.40W
Matera **117 G3** 40.40N 16.37E
Mateur **116 E2** 37.03N 9.40E
Mathoura **48 C2** 35.49S 144.54E
Mathura **106 C3** 27.30N 77.42E
Mati **100 D3** 6.59N 126.12E
Mato Grosso, Plateau of **143 E2**
Matrah **108 D3** 23.38N 58.34E
Matsue **104 F4** 35.29N 133.04E
Matsumae **104 J7** 41.28N 140.06E
Matsumoto **104 I5** 36.18N 137.58E
Matsuyama **104 F3** 33.50N 132.47E
Matterhorn, mountain **116 E4** 45.59N 7.39E
Matua **100 C2** 2.58S 110.46E
Maturin **142 D4** 9.45N 63.10W
Maues **142 E3** 3.22S 57.38W
Maug, island **101 F5**
Maui, island **84 P8**
Maun **126 C2** 20.00S 23.25E
Mauna Kea, mountain **84 P7** 19.50N 155.25W
Maurice, Lake **66 C6**
Mauritania, Islamic republic **124 B4**
Mauritius, independent state **146 D2**
Mavinga **126 C3** 15.44S 20.21E
Mawhai Point **90 H5** 38.10S 178.22E
Mawson, antarctic base **92 D3** 67.36S 62.52E
Mayaguana, island **137 E3**
Mayaguez **137 F2** 18.13N 67.09W
Mayamey **108 D4** 36.27N 55.40E
Maydena **60 D2** 42.45S 146.37E
Mayor Island **90 G6**
Mayor Pablo Lagerenza **142 D2** 19.55S 60.46W
Mayumba **126 B4** 3.23S 10.38E
Mazabuka **126 C3** 15.50S 27.47E
Mazar-i-Sharif **106 B4** 36.43N 67.05E
Mazatlan **136 B3** 23.11N 106.25W
Mazhan **118 F2** 32.30N 59.00E
Mbabane **126 D2** 26.20S 31.08E
Mbaiki **124 E2** 3.53N 18.01E
Mbala **126 D4** 8.49S 31.20E
Mbale **126 D5** 1.04N 34.12E
Mbambawa **88 C3** 3.42S 137.31E
Mbandaka **126 B5** 0.03N 18.28E
Mbanza-Ngungu **126 B4** 5.17S 14.51E
Mbarara **126 D4** 0.36S 30.40E
Mbeya **126 D4** 8.54S 33.29E
M'Bour **124 B3** 14.22N 16.54W
Mbuji-Mayi **126 C4** 6.10S 23.39E
Mead, Lake **134 C2**
Meander River **60 D3**
Mecca **108 B3** 21.26N 39.49E
Mechelen **114 E3** 51.02N 4.29E
Mecheria **124 C5** 33.31N 0.20W
Meda River **72 D5**
Medan **100 A3** 3.35N 98.39E
Medellin **142 C4** 6.15N 75.36W
Medford **134 B3** 42.19N 122.52W
Medias **117 H4** 46.10N 24.21E
Medicine Hat **132 G3** 50.03N 110.41W
Medina, Saudi Arabia **108 B3** 24.28N 39.36E
Medina, U.S.A. **135 G3** 43.14N 78.23W

*Entry/**Page**/**Ref.**/Lat./Long.*

Medina del Campo **116 C3** 41.18N 4.55W
Mediterranean Sea **116 E2**
Medvezhegorsk **115 K5** 62.56N 34.28E
Medvezhya, Mount **103 H4** 47.26N 137.56E
Meekatharra **72 C3** 26.36S 118.28E
Meerut **106 C3** 29.00N 77.42E
Mega **125 G2** 4.02N 38.19E
Mehakit **100 C2** 2.50S 115.56E
Meharry, Mount **72 C4** 22.59S 118.35E
Mehsana **106 C3** 23.36N 72.24E
Mejillones **144 B4** 23.06S 70.27W
Mekele **125 G3** 13.32N 39.33E
Meknes **124 C5** 33.53N 5.37W
Mekong River **102 D3**
Melaka **100 B3** 2.14N 102.14E
Melaka, Strait of **100 A3**
Melbourne, Australia **54 C2** 37.49S 144.58E
Melbourne, U.S.A. **135 F1** 28.04N 80.38W
Melfi **117 G3** 41.00N 15.33E
Melilla **124 C5** 35.20N 3.00W
Melitopol **115 L2** 46.51N 35.22E
Melo **144 D3** 32.22S 54.10W
Melrose, homestead **48 D3** 32.40S 146.55E
Melton Mowbray **60 E2** 42.28S 147.11E
Melun **116 D4** 48.32N 2.40E
Melville **132 H3** 50.57N 102.49W
Melville Bay **80 F8**
Melville, Cape **42 E8** 14.10S 144.30E
Melville Island, Australia **80 C9**
Melville Island, Canada **132 G5**
Melville Peninsula **132 J4**
Memmingen **114 G2** 47.59N 10.11E
Mempawah **100 B3** 0.23N 108.56E
Memphis **135 F2** 35.10N 90.00W
Mende **116 D3** 44.32N 3.30E
Mendi **89 D2** 6.09S 143.40E
Mendocino, Cape **134 B3** 40.26N 124.24W
Mendoza **144 C3** 32.48S 68.52W
Menihec, Lake **133 L3**
Menindee **48 B3** 32.24S 142.25E
Menindee Lake **48 B3**
Meningie **66 G3** 35.41S 139.20E
Menongue **126 B3** 14.40S 17.41E
Menyamya **89 E2** 7.15S 146.00E
Menyapa, mountain **100 C3** 0.59N 116.06E
Menzies **72 D3** 29.42S 121.02E
Menzies, Mount **94 D2** 74.41S 62.05E
Merano **116 F4** 46.41N 11.10E
Merauke **88 D2** 8.30S 140.22E
Merauke River **88 D2**
Merbein **54 B3** 34.10S 142.04E
Merced **134 B2** 37.17N 120.30W
Mercedes, northern Argentina **144 D4** 29.15S 58.05W
Mercedes, western Argentina **144 C3** 33.41S 65.28W
Mercedes, Uruguay **144 D3** 33.16S 58.05W
Mercury Islands **90 F6**
Mergui **107 E2** 12.26N 98.34E
Mergui Archipelago **107 E2**
Merida, Mexico **136 D3** 20.59N 89.39W
Merida, Spain **116 B2** 38.55N 6.20W
Merida, Venezuela **142 C4** 8.37N 71.08W
Merida Mountains **142 C4**
Meridian **135 F2** 32.21N 88.48W
Meringur **54 A3** 34.24S 141.20E
Merkus, Cape **89 E2** 6.10S 149.03E
Merowe **125 G3** 18.30N 31.49E
Merredin **72 C2** 31.29S 118.16E
Merritt **132 F3** 50.09N 120.49W
Merriwa **48 F3** 32.09S 150.21E
Mersa Fatma **108 C2** 14.57N 40.17E
Mersey River **60 D3**
Mersin **108 B4** 36.47N 34.37E
Mesewa **125 G3** 15.37N 39.28E
Meshed **108 D4** 36.16N 59.34E
Meskum **100 B3** 1.32N 102.02E
Mesolongion **117 H2** 38.21N 21.26E
Messina, Italy **117 G2** 38.13N 15.33E
Messina, South Africa **126 D2** 22.23S 30.00E
Meta River **142 C4**
Methven **90 D3** 43.38S 171.38E
Metz **116 E4** 49.07N 6.11E
Meulaboh **100 A3** 4.10N 96.09E

*Entry/**Page**/**Ref.**/Lat./Long.*

Mexia **134 E2** 31.41N 96.30W
Mexicali **134 C2** 32.36N 115.30W
Mexico, federal republic **136 B3**
Mexico, Gulf of **136 C3**
Mexico City **136 C2** 19.25N 99.10W
Mezen **118 E5** 65.50N 44.20E
Mezen River **118 E5**
Mezotur **115 I2** 47.00N 20.37E
Miami **135 F1** 25.45N 80.15W
Miami Beach **135 F1** 25.47N 80.07W
Michigan, Lake **135 F3**
Michigan, state **135 F3**
Michurinsk **115 M3** 52.54N 40.30E
Micronesia, Federated States of **84 K7**
Middle Andaman, island **107 E2**
Middlemount **42 G4** 22.50S 148.40E
Middlesbrough **114 D3** 54.35N 1.14W
Midland, Michigan **135 F3** 43.38N 84.14W
Midland, Texas **134 D2** 32.00N 102.09W
Midway, island **84 N8**
Mielec **117 H5** 50.18N 21.25E
Miena **60 D3** 41.59S 146.43E
Mieres **116 B3** 43.15N 5.46W
Mikindani **126 E3** 10.16S 40.05E
Mikkeli **115 J5** 61.44N 27.15E
Mikuni Range **104 I5**
Milagro **142 C3** 2.11S 79.36W
Milan **116 E4** 45.28N 9.12E
Milang **66 G3** 35.24S 138.57E
Milas **108 A4** 37.19N 27.48E
Milbank **134 E3** 45.15N 96.32W
Mildura **54 B3** 34.11S 142.10E
Milepa **126 D3** 11.45S 36.17E
Miles **42 H2** 26.40S 150.11E
Miles City **134 D3** 46.24N 105.48W
Milford Sound **90 B2**
Milford Haven **114 C3** 51.44N 5.02W
Milikapiti **80 C9** 11.24S 130.41E
Miling **72 C2** 30.27S 116.20E
Milingimbi **80 E8** 12.07S 134.55E
Milk River **134 C3**
Mill Island **92 E3**
Millau **116 D3** 44.06N 3.05E
Mille Lacs, lake **134 E3**
Millennium Island **85 P5**
Millicent **66 H2** 37.36S 140.22E
Millinocket **133 L2** 45.42N 68.43W
Millmerran **42 H2** 27.53S 151.16E
Millstream, homestead **72 C4** 21.36S 117.04E
Milton **90 D1** 46.07S 169.58E
Milton Keynes **114 D3** 52.02N 0.42W
Milwaukee **135 F3** 43.03N 87.56W
Minab **108 D3** 27.07N 57.06E
Minahassa Peninsula **100 D3**
Minas **144 D3** 34.20S 55.15W
Mindanao, island **100 D3**
Mindanao Sea **100 D3**
Minden **114 F3** 52.18N 8.54E
Mindona Lake **48 B3**
Mindoro, island **100 D4**
Mindoro Strait **100 C4**
Mineiros **143 E2** 17.34S 52.33W
Mingan **133 L3** 50.19N 64.02W
Mingary **66 H4** 32.08S 140.44E
Minigwal, Lake **72 D3**
Minjilang **80 D9** 11.09S 132.34E
Minlaton **66 F3** 34.46S 137.36E
Minna **124 D2** 9.39N 6.32E
Minneapolis **134 E3** 45.00N 93.15W
Minnesota River **134 E3**
Minnesota, state **134 E3**
Minnipa **66 E4** 32.51S 135.09E
Minorca, island **116 D3**
Minot **134 D3** 48.16N 101.19W
Minsk **115 J3** 53.51N 27.30E
Minto, Lake **133 K3**
Minto, Mount **94 I2** 72.00S 169.00E
Minyip **54 B2** 36.28S 142.35E
Mios Num, island **88 C3**
Miquelon, island **133 M2**
Mira **116 B3** 40.26N 8.44W
Miranda de Ebro **116 C3** 42.41N 2.57W
Mirboo North **54 D1** 38.24S 146.09E
Mirgorod **115 K2** 49.58N 33.37E

*Entry/**Page**/**Ref.**/Lat./Long.*

Miriam Vale **42 H3** 24.20S 151.34E
Mirim, Lake **144 D3**
Mirjaveh **108 E3** 29.01N 61.30E
Mirny, antarctic base **92 E3** 66.33S 93.01E
Mirnyy **119 L5** 62.30N 113.58E
Mirpur Khas **106 B3** 25.33N 69.05E
Miryang **104 D4** 35.31N 128.45E
Mirzapur **106 D3** 25.09N 82.34E
Misaha Oasis **125 F4** 22.13N 27.59E
Misawa **104 J7** 40.42N 141.26E
Mishab, Cape **108 C3** 28.12N 48.37E
Misima Island **89 F1**
Miskolc **115 I2** 48.07N 20.47E
Misool, island **88 A3**
Mississippi River **134 E3**
Mississippi, state **135 E2**
Missoula **134 C3** 46.52N 114.00W
Missouri River **134 C3**
Missouri, state **134 E2**
Mistassini **135 G3** 48.54N 72.13W
Mistassini, Lake **133 K3**
Misurata **124 E5** 32.24N 15.04E
Mita, Point **136 B3** 20.46N 105.33W
Mitchell, Australia **42 F2** 26.29S 147.58E
Mitchell, U.S.A. **134 E3** 43.40N 98.01W
Mitchell, Mount **135 F2** 35.47N 82.16W
Mitchell River, Queensland **42 D8**
Mitchell River, Victoria **54 D2**
Mitiamo **54 C2** 36.13S 144.14E
Mitiaro, island **84 P4**
Mito **104 J5** 36.22N 140.29E
Mittagong **48 F2** 34.27S 150.27E
Mitta Mitta River **54 D2**
Mitu **142 C4** 1.08N 70.03W
Miyake Island **104 I4**
Miyako **104 J6** 39.38N 141.59E
Miyakonojo **104 E2** 31.43N 131.02E
Miyazaki **104 E2** 31.56N 131.27E
Mizuho, antarctic base **92 C3** 70.42S 44.00E
Mlada Boleslav **114 G3** 50.26N 14.55E
Mlanje, Mount **126 D3** 16.00S 35.37E
Mlawa **115 I3** 53.08N 20.20E
Moa Island **42 D10**
Mobile **135 F2** 30.40N 88.05W
Mobutu Sese Seko, Lake *see* Albert, Lake
Mocha **108 C2** 13.19N 43.15E
Mochudi **126 C2** 24.28S 26.05E
Mocimboa da Praia **126 E3** 11.19S 40.19E
Mocoa **142 C4** 1.07N 76.38W
Mocorito **136 B3** 25.30N 107.53W
Modesto **134 B2** 37.37N 121.00W
Moe **54 D1** 38.11S 146.16E
Mogadishu **125 H2** 2.02N 45.21E
Mogami River **104 J6**
Mogilev **115 K3** 53.54N 30.20E
Mogilev-Podolskiy **115 J2** 48.29N 27.49E
Mogocha **119 L4** 53.44N 119.45E
Mohacs **114 H2** 46.00N 18.40E
Moisie River **133 L3**
Mojave Desert **130 H4**
Mokil, island **84 K6**
Mokoan, Lake **54 D2**
Mokpo **104 C4** 34.50N 126.25E
Molde **114 F5** 62.44N 7.08E
Moldova, republic **115 J2**
Mole Creek **60 D3** 41.34S 146.24E
Molepolole **126 C2** 24.25S 25.30E
Molfetta **117 G3** 41.12N 16.36E
Mollendo **142 C2** 17.00S 72.00W
Molndal **114 G4** 57.40N 12.00E
Molodechno **115 J3** 54.16N 26.50E
Molodezhnaya, antarctic base **92 C3** 67.40S 45.51E
Molong **48 E3** 33.06S 148.52E
Molu, island **88 B2**
Moma **126 C4** 1.25S 23.57E
Mombasa **126 D4** 4.04S 39.40E
Mombetsu **104 K9** 44.23N 143.13E
Mombo **126 D4** 4.54S 38.18E
Momchilgrad **117 I3** 41.32N 25.24E
Mompono **126 C5** 0.06N 21.50E
Mompos **142 C4** 9.15N 74.29W
Mona Passage **137 F2**
Monaco **116 E3** 43.44N 7.25E

Entry / ***Page*** / ***Ref.*** / *Lat.* / *Long.*

Monaco, principality **116 E3**
Moncao **143 F3** 3.30S 45.15W
Monchegorsk **118 D5** 67.55N 33.01E
Monclova **136 B3** 26.55N 101.25W
Moncton **133 L2** 46.04N 64.50W
Mondovi **116 E3** 44.23N 7.49E
Monforte de Lemos **116 B3** 42.32N 7.30W
Mongo **124 E3** 12.14N 18.45E
Mongolia, people's republic **102 E4**
Mongonu **124 E3** 12.43N 13.34E
Mongu **126 C3** 15.13S 23.09E
Monopoli **117 G3** 40.57N 17.18E
Monroe, Louisiana **134 E2** 32.31N 92.06W
Monroe, Utah **134 C2** 38.38N 112.06W
Monrovia **124 B2** 6.20N 10.46W
Montana, region **142 C3**
Montana, state **134 D3**
Montauban **116 D3** 44.01N 1.20E
Montbeliard **116 E4** 47.31N 6.48E
Mont-de-Marsan **116 C3** 43.54N 0.30W
Monte Azul **143 F2** 15.13S 42.53W
Monte Bello Islands **72 C4**
Monte Corno, mountain **112 F2** 46.17N 12.59E
Montego Bay **136 E2** 18.27N 77.56W
Montelimar **116 D3** 44.33N 4.45E
Montemorelos **136 C3** 25.12N 99.50W
Montenegro, province **117 G3**
Montepuez **126 D3** 13.09S 39.00E
Monterey **134 B2** 36.36N 121.53W
Monterey Bay **134 B2**
Monteria **142 C4** 8.45N 75.54W
Monterrey **136 B3** 25.40N 100.20W
Montes Claros **143 F2** 16.45S 43.52W
Montevideo **144 D3** 34.55S 56.10W
Montgomery **135 F2** 32.22N 86.20W
Mont Joli **135 H3** 48.36N 68.14W
Mont Laurier **133 K2** 46.33N 75.31W
Montlucon **116 D4** 46.20N 2.36E
Montmagny **133 K2** 46.58N 70.33W
Monto **42 H3** 24.52S 151.07E
Montpelier **135 G3** 44.16N 72.34W
Montpellier **116 D3** 43.36N 3.53E
Montreal **133 K2** 45.30N 73.36W
Montrose, U.K. **114 D4** 56.43N 2.29W
Montrose, U.S.A. **134 D2** 38.29N 107.53W
Montserrat, island **137 F2**
Monza **116 E4** 45.35N 9.16E
Monze **126 C3** 16.16S 27.28E
Moolah, homestead **48 C3** 32.39S 144.59E
Moolawatana, homestead **66 G6** 29.54S 139.44E
Moomba **66 H6** 28.07S 140.11E
Moonie **42 H2** 27.43S 150.22E
Moonie River **42 G2**
Moonta **66 F3** 34.04S 137.35E
Moor Islands **88 C3**
Moora **72 C2** 30.38S 116.00E
Moore, Lake **72 C3**
Moorhead **134 E3** 46.51N 96.44W
Moornanyah Lake **48 B3**
Mooroopna **54 C2** 36.24S 145.22E
Moose Lake **132 I3**
Moosehead Lake **135 H3**
Moose Jaw **132 H3** 50.23N 105.35W
Moosonee **133 J3** 51.18N 80.40W
Mopti **124 C3** 14.29N 4.10W
Mora **114 G5** 61.00N 14.30E
Moradabad **106 C3** 28.50N 78.45E
Moranbah **42 G4** 22.00S 148.02E
Morane, island **85 R3**
Morawa **72 C3** 29.13S 116.00E
Morden **132 I2** 49.12N 98.05W
Moree **48 E5** 29.28S 149.51E
Morehead **88 D2** 8.43S 141.38E
Morelia **136 B2** 19.40N 101.11W
Morenci **134 D2** 33.05N 109.22W
Moreton Bay **40 F3**
Moreton Island **40 F3**
Mori **104 J8** 42.07N 140.33E
Morioka **104 J6** 39.43N 141.08E
Morlaix **116 C4** 48.35N 3.50W
Mornington Island **42 B7**
Moro Gulf **100 D3**
Morobe **89 E2** 7.45S 147.37E
Morocco, kingdom **124 B4**
Morombe **126 E2** 21.47S 43.21E
Moron **136 E3** 22.08N 78.39W
Morondava **126 E2** 20.19S 44.17E
Moron de la Frontera **116 B2** 37.07N 5.27W
Moroni **126 E3** 11.40S 43.16E
Morotai, island **101 D3**
Morrinhos **143 F2** 17.45S 49.07W
Morshansk **115 M3** 53.26N 41.48E
Mortlake **54 B1** 38.05S 142.48E
Moruya **48 F2** 35.55S 150.06E
Morven **42 F2** 26.25S 147.07E
Morvi **106 C3** 22.50N 70.52E
Morwell **54 D1** 38.14S 146.24E
Moscow, Russia **115 L4** 55.45N 37.42E
Moscow, U.S.A. **134 C3** 46.44N 116.57W
Moscow University Ice Shelf **92 F3**
Mose, Cape **92 G3** 66.00S 130.12E
Mosgiel **90 D2** 45.53S 170.22E
Moshi **126 D4** 3.20S 37.21E
Mosquitos, Gulf of **136 D1**
Moss **114 G4** 59.26N 10.41E
Mosselbaai **126 C1** 34.12S 22.08E
Mossman **42 E7** 16.28S 145.23E
Moss Vale **48 F2** 34.33S 150.23E
Most **114 G3** 50.31N 13.39E
Mostaganem **124 D5** 35.54N 0.05E
Mostar **117 G3** 43.20N 17.50E
Mosul **108 C4** 36.21N 43.08E
Motueka **90 E4** 41.07S 173.01E
Moulamein **48 C2** 35.05S 144.02E
Moulins **116 D4** 46.34N 3.20E
Moulmein **107 E2** 16.30N 97.39E
Moulouya River **124 C5**
Moultrie **135 F2** 31.11N 83.47W
Moundou **124 E2** 8.36N 16.02E
Moung **100 B4** 12.47N 103.28E
Mountain Home **134 C3** 43.08N 115.43W
Mountain Valley, homestead **80 D7** 14.05S 133.49E
Mount Barker **72 C2** 34.38S 117.40E
Mount Beauty **54 D2** 36.45S 147.10E
Mount Dutton **66 E7** 27.49S 135.43E
Mount Eba, homestead **66 E5** 30.11S 135.40E
Mount Ebenezer, homestead **80 D2** 25.11S 132.41E
Mount Emu Creek **54 B1**
Mount Gambier **66 H2** 37.50S 140.47E
Mount Hagen **89 D2** 5.50S 144.13E
Mount Isa **42 B5** 20.44S 139.29E
Mount Magnet **72 C3** 28.04S 117.51E
Mount Manara, homestead **48 C3** 32.29S 143.56E
Mount Maunganui **90 G6** 37.38S 176.12E
Mount Morgan **42 H4** 23.39S 150.23E
Mount Sanford, homestead **80 C6** 16.59S 130.34E
Mount Surprise **42 E6** 18.09S 144.19E
Mount Vernon **134 B3** 48.25N 122.20W
Mount Willoughby **66 E7** 27.58S 134.09E
Moura, Australia **42 G3** 24.34S 149.59E
Moura, Brazil **142 D3** 1.32S 61.38W
Moussoro **124 E3** 13.41N 16.31E
Moyale **126 D5** 3.31N 39.04E
Mozambique **126 E3** 15.00S 40.44E
Mozambique Channel **126 D2**
Mozambique, people's republic **126 D3**
Mozhaisk **115 L4** 55.29N 36.02E
Mozyr **115 J3** 52.02N 29.10E
Mpanda **126 D4** 6.21S 31.01E
Mpika **126 D3** 11.50S 31.30E
Msaken **116 F2** 35.45N 10.33E
M'Sila **116 D2** 35.40N 4.31E
Mtwara **126 E3** 10.17S 40.11E
Mualang **100 C3** 0.44N 111.19E
Muar **100 B3** 2.01N 102.35E
Muaragusung **100 C3** 1.30N 117.30E
Muaramawai **100 C3** 0.31N 116.47E
Muarasiberut **100 A2** 1.36S 99.12E
Mubarraz **108 C3** 25.26N 49.37E
Mubrani **88 B3** 0.35S 133.30E
Muccan, homestead **72 D4** 20.38S 120.03E
Muchinga Mountains **126 D3**
Mucusso **126 C3** 18.00S 21.24E
Mudanjiang **103 G4** 44.36N 129.42E
Mudanya **117 I3** 40.23N 28.53E
Mudgee **48 E3** 32.37S 149.36E
Mueda **126 D3** 11.40S 39.31E
Mufulira **126 C3** 12.30S 28.12E
Mugla **108 A4** 37.13N 28.22E
Muhinga **126 D4** 2.50S 30.19E
Mukachevo **115 I2** 48.26N 22.45E
Mukalla **108 C2** 14.34N 49.07E
Mukomuko **100 B2** 2.30S 101.05E
Mulga Park, homestead **80 C2** 25.54S 131.40E
Mulhacen, mountain **116 C2** 37.04N 3.19W
Mulhouse **116 E4** 47.45N 7.21E
Muling River **104 E9**
Muller Mountains **100 C2**
Mullumbimby **48 G5** 28.32S 153.30E
Multan **106 C4** 30.10N 71.36E
Mulwala **48 C2** 35.59S 146.00E
Mulwala, Lake **54 D3**
Mumbai **106 C2** 18.56S 72.51E
Mumbwa **126 C3** 15.00S 27.01E
Mumeng **89 E2** 6.57S 146.34E
Muncie **135 F3** 40.11N 85.22W
Mundubbera **42 H3** 25.35S 151.18E
Mungana **42 E7** 17.06S 144.23E
Mungindi **48 E5** 28.59S 149.00E
Mungo, Lake **48 B3**
Munich **114 G2** 48.08N 11.35E
Munku Sardyk, mountain **118 K4** 51.48N 100.30E
Munro, Mount **60 F4** 40.23S 148.07E
Munster **114 F3** 51.58N 7.37E
Munte **100 C3** 0.30N 119.57E
Muntok **100 B2** 2.04S 105.12E
Murchison **54 C2** 36.37S 145.13E
Murchison River **72 C3**
Murcia **116 C2** 37.59N 1.08W
Muren **102 E4** 49.36N 100.08E
Murgon **42 H2** 26.15S 151.57E
Murmansk **118 D5** 68.59N 33.08E
Murom **115 M4** 55.34N 42.04E
Muroran **104 J8** 42.21N 140.59E
Muroto **104 G3** 33.16N 134.10E
Muroto, Cape **104 G3** 33.13N 134.11E
Murray **134 C3** 40.40N 111.53W
Murray, Lake **88 D2**
Murray, Mount **89 D2** 6.50S 144.00E
Murray River **40 E2**
Murray Bridge **66 G3** 35.07S 139.16E
Murray Downs, homestead **80 E4** 21.02S 134.40E
Murrayville **54 A3** 35.16S 141.11E
Murrumbidgee River **48 C2**
Murrumburrah **48 E2** 34.33S 148.21E
Murrurundi **48 F4** 31.46S 150.49E
Murtoa **54 B2** 36.37S 142.28E
Murupara **90 G5** 38.27S 176.42E
Mururoa Atoll **85 Q3**
Murwara **106 D3** 23.49N 80.28E
Murwillumbah **48 G5** 28.20S 153.24E
Murzuq **124 E4** 25.56N 13.57E
Murzuq Desert **124 E4**
Mus **108 C4** 38.45N 41.30E
Musa River **89 E2**
Musan **104 D8** 42.11N 129.10E
Muscat **108 D3** 23.37N 58.35E
Musgrave Ranges **80 C1**
Muskegon **135 F3** 43.13N 86.15W
Mus-Khaya, mountain **119 O5** 62.35N 140.31E
Muskogee **134 E2** 35.45N 95.21W
Musoma **126 D4** 1.31S 33.48E
Mussau Island **89 E3**
Mustafa Kemal Pasa **117 I2** 40.03N 28.25E
Musu, Cape **104 D7** 40.52N 129.45E
Muswellbrook **48 F3** 32.16S 150.54E
Mutan River **104 D9**
Mutare **126 D3** 19.00S 32.40E
Muting **88 D2** 7.23S 140.19E
Mutsu **104 J7** 41.18N 141.15E
Muttaburra **42 E4** 22.36S 144.33E
Muynak **118 F3** 43.46N 59.00E
Muzaffarnagar **102 B2** 29.28N 77.42E
Muzaffarpur **106 D3** 26.07N 85.23E
Muztagh Ata, mountain **102 B3** 38.15N 75.22E
Mwanza **126 D4** 2.30S 32.54E
Mwene-Ditu **126 C4** 7.01S 23.27E
Mweru, Lake **126 C4**
Mwinilunga **126 C3** 11.44S 24.24E
Myanmar, socialist republic **107 E3**
Myitkyina **107 E3** 25.24N 97.25E
Mymensingh **107 E3** 24.45N 90.23E
Myra, Mount **48 F4** 31.45S 151.43E
Myrtleford **54 D2** 36.34S 146.44E
Mysore **106 C2** 12.18N 76.37E
My Tho **100 B4** 10.21N 106.21E
Mytilene **117 I2** 39.06N 26.34E
Mzuzu **126 D3** 11.31S 33.50E

N

Nabadwip **102 C2** 23.24N 88.23E
Nabarlek **80 D8** 12.10S 133.19E
Nabeul **116 F2** 36.30N 10.44E
Nabire **88 C3** 3.23S 135.31E
Nacala **126 E3** 14.30S 40.37E
Nackara **66 G4** 32.48S 139.14E
Nacogdoches **134 E2** 31.36N 94.40W
Nadi **84 M4** 17.46S 177.27E
Nadiad **106 C3** 22.42N 72.55E
Nador **124 C5** 35.10N 3.00W
Nadym River **118 H5**
Naestved **114 G4** 55.14N 11.47E
Nafada **124 E3** 11.08N 11.20E
Naft-e-Shah **108 C4** 34.02N 45.30E
Naga **100 D4** 13.36N 123.12E
Naga Hills **107 E3**
Nagambie **54 C2** 36.47S 145.10E
Nagano **104 I5** 36.39N 138.10E
Nagaoka **104 I5** 37.27N 138.50E
Nagasaki **104 D3** 32.45N 129.52E
Nagercoil **106 C1** 8.11N 77.30E
Nagoya **104 H4** 35.08N 136.53E
Nagpur **106 C3** 21.10N 79.12E
Nagykanizsa **114 H2** 46.27N 17.00E
Nahuel Huapi, Lake **144 B2**
Naikliu **100 D2** 9.32S 123.48E
Nain **133 L3** 56.30N 61.45W
Nairobi **126 D4** 1.17S 36.50E
Najafabad **108 D4** 32.37N 51.21E
Najin **104 E8** 42.04N 130.32E
Najran **108 C2** 17.31N 44.19E
Nakhichevan **108 C4** 39.12N 45.24E
Nakhodka **119 N3** 42.53N 132.53E
Nakhon Ratchasima **100 B4** 15.00N 102.06E
Nakhon Sawan **100 A4** 15.42N 100.04E
Nakhon Si Thammarat **100 B3** 8.24N 99.58E
Nakhon Thai **100 B4** 17.04N 100.41E
Naktong River **104 D5**
Nakuru **126 D4** 0.16S 36.04E
Nalchik **118 E3** 43.31N 43.38E
Nalut **124 E5** 31.53N 10.59E
Nam Lake **102 D3**
Namangan **102 B4** 40.59N 71.41E
Namaripi, Cape **88 C3** 4.27S 135.10E
Namatanai **89 F3** 3.40S 152.26E
Nambour **42 I2** 26.40S 152.58E
Nambucca Heads **48 G4** 30.39S 153.00E
Nam Dinh **100 B5** 20.25N 106.12E
Namib Desert **126 B2**
Namibe **126 B3** 15.10S 12.10E
Namibia, independent republic **126 B2**
Namoi, island **84 K6**
Namoi River **48 E4**
Namoluk, island **101 G3**
Namonuito, island **101 F3**
Nampa **134 C3** 43.35N 116.34W
Nampo **104 B6** 38.47N 125.26E
Nampula **126 D3** 15.09S 39.14E
Namsos **114 G5** 64.28N 11.30E
Namu, island **84 L6**
Namwon **104 C4** 35.23N 127.23E
Nan River **107 F2**
Nanaimo **132 F2** 49.08N 123.58W
Nanango **42 I2** 26.40S 152.00E
Nanao **104 H5** 37.03N 136.58E
Nanchang **103 F2** 28.42N 115.55E
Nanchong **102 E3** 30.54N 106.06E

*Entry / **Page** / **Ref.** / Lat. / Long.*

Nancy **116 E4** 48.42N 6.12E
Nangapinoh **100 C2** 0.21S 111.44E
Nangatayap **100 C2** 1.30S 110.33E
Nangnim Range **104 C7**
Nanjing **103 F3** 32.03N 118.47E
Nanning **102 E2** 22.50N 108.19E
Nanping **103 F2** 26.40N 118.07E
Nansei Shoto Trench **103 G2**
Nantes **116 C4** 47.14N 1.35W
Nanumanga, island **84 M5**
Nanumea, island **84 M5**
Nanyang **102 F3** 33.06N 112.31E
Nanyuki **126 D5** 0.01N 37.05E
Nanzhila **126 C3** 16.04S 26.09E
Nao, Cape **116 D2** 38.44N 0.14E
Naol River **104 E10**
Naousa **117 H3** 40.38N 22.04E
Napa **134 B2** 38.20N 122.17W
Napan **88 C3** 3.08S 135.36E
Napido **88 C3** 0.39S 135.27E
Napier Broome Bay **72 E6**
Naples, Italy **117 F3** 40.50N 14.15E
Naples, U.S.A. **135 F1** 26.09N 81.48W
Naracoopa **60 B5** 39.56S 144.08E
Naracoorte **66 H2** 36.58S 140.46E
Naradhan **48 D3** 33.37S 146.20E
Narathiwat **100 B3** 6.25N 101.51E
Narbonne **116 D3** 43.11N 3.00E
Nares Strait **132 K5**
Narmada River **106 C3**
Narodnaya, Mount **118 G5** 65.02N 60.01E
Narooma **48 F1** 36.13S 150.09E
Narrabri **48 E4** 30.20S 149.47E
Narrandera **48 D2** 34.45S 146.33E
Narrogin **72 C2** 32.56S 117.11E
Narromine **48 E3** 32.14S 148.15E
Narva **115 J4** 59.28N 28.01E
Narvik **118 B5** 68.26N 17.25E
Naryan Mar **118 F5** 67.37N 53.02E
Naryilco, homestead **42 C1** 28.33S 141.55E
Nashua **135 G3** 42.44N 71.28W
Nashville **135 F2** 36.10N 86.50W
Nasik **106 C2** 20.00N 73.52E
Nassau **136 E3** 25.05N 77.20W
Nassau, island **84 O4**
Nasser, Lake **125 G4**
Natal **143 G3** 5.46S 35.15W
Natara **119 M5** 68.26N 124.02E
Natashquan **133 L3** 50.10N 61.50W
Natashquan River **133 L3**
Natchez **134 E2** 31.32N 91.24W
Natchitoches **136 C4** 31.46N 93.06W
Nathalia **54 C2** 36.03S 145.12E
Natimuk **54 A2** 36.45S 141.56E
Natitingou **124 D3** 10.17N 1.19E
Natron, Lake **126 D4**
Natuna, island **100 B3**
Naturaliste, Cape **72 B2** 33.32S 115.01E
Nauru, republic **84 L5**
Navarin, Cape **119 R5** 62.17N 179.13E
Navarre **54 B2** 36.54S 143.07E
Navojoa **136 B3** 27.04N 109.28W
Nawabshah **106 B3** 26.15N 68.26E
Naxos, island **117 I2**
Nay Band, central Iran **108 D4** 32.20N 57.34E
Nay Band, southern Iran **108 D3** 27.22N 52.38E
Nayoro **104 K9** 44.21N 142.30E
Nazilli **108 A4** 37.55N 28.20E
Ndalatando **126 B4** 9.18S 14.54E
Ndele **124 F2** 8.25N 20.38E
N'Djamena **124 E3** 12.10N 14.59E
Ndjole **126 B4** 0.07S 10.45E
Ndola **126 C3** 13.00S 28.29E
Neale, Lake **80 C2**
Neales River **66 E7**
Nebit Dag **118 F2** 39.31N 54.24E
Nebo **42 G5** 21.42S 148.42E
Nebraska, state **134 D3**
Neckarboo **48 C3** 32.04S 144.36E
Necker Island **84 O8**
Necochea **144 D3** 38.31S 58.46W
Nedroma **116 C2** 35.00N 1.44W
Nefta **124 D5** 33.53N 7.50E
Negelli **125 G2** 5.20N 39.35E

*Entry / **Page** / **Ref.** / Lat. / Long.*

Negev, desert **108 B4**
Negomano **126 D3** 11.26S 38.30E
Negra, Point **142 B3** 6.06S 81.09W
Negrais, Cape **107 E2** 15.57N 94.12E
Negro River, Argentina **144 C3**
Negro River, Brazil **142 D3**
Negros, island **100 D4**
Nehbandan **108 E4** 31.32N 60.04E
Neijiang **102 E2** 29.32N 105.03E
Neiva **142 C4** 2.58N 75.15W
Nellore **106 D2** 14.29N 80.00E
Nelson, Canada **132 G2** 49.29N 117.17W
Nelson, New Zealand **90 E4** 41.16S 173.15E
Nelson, Cape, Australia **54 A1** 38.26S 141.33E
Nelson, Cape, P.N.G. **89 E2** 9.00S 149.15E
Nelson River **132 I3**
Nelson Bay **48 G3** 32.44S 152.09E
Nelspruit **126 D2** 25.30S 30.58E
Nema **124 C3** 16.32N 7.12W
Nemuro **104 L8** 43.22N 145.36E
Nemuro Strait **104 L9**
Nen River **103 G4**
Nenjiang **103 G4** 49.10N 125.15E
Nepal, kingdom **106 D3**
Netherdale **42 G5** 21.09S 148.32E
Netherlands, kingdom **114 E3**
Netherlands Antilles, islands **137 F2**
Neumunster **114 F3** 54.05N 9.59E
Neuquen **144 C3** 38.55S 68.09W
Neustrelitz **114 G3** 53.22N 13.05E
Nevada **134 E2** 37.51N 94.22W
Nevada, state **134 C2**
Nevel **118 D4** 56.00N 29.59E
Never **119 M4** 54.00N 124.09E
Nevers **116 D4** 47.00N 3.09E
Nevis, island **137 F2**
New Amsterdam **142 E4** 6.18N 57.30W
Newark **135 G3** 40.44N 74.11W
New Britain, island **89 F2**
New Brunswick, province **133 L2**
New Caledonia **84 L3**
Newcastle **48 F3** 32.55S 151.46E
Newcastle Creek **80 D6**
Newcastle-under-Lyme **114 D3** 53.00N 2.14W
Newcastle-upon-Tyne **114 D3** 54.59N 1.35W
Newcastle Waters **80 D6** 17.20S 133.24E
Newdegate **72 C2** 33.06S 119.01E
New Delhi **106 C3** 28.37N 77.13E
Newell **134 D3** 44.44N 103.26W
Newfoundland, province **133 M3**
New Georgia Islands **89 G2**
New Georgia, island **89 G2**
New Guinea, island **88 D3**
New Hampshire, state **135 H2**
New Hanover, island **89 E3**
New Haven **135 G3** 41.18N 72.55W
New Hebrides Trench **84 L4**
New Iberia **136 C3** 30.00N 91.51W
New Ireland, island **89 F3**
New Jersey, state **135 H2**
New Liskeard **133 K2** 47.31N 79.41W
Newman **72 C4** 23.21S 119.44E
Newman, Mount **72 C4** 23.16S 119.34E
New Mexico, state **134 D2**
New Norfolk **60 E2** 42.47S 147.04E
New Orleans **135 F1** 30.00N 90.03W
New Plymouth **90 E5** 39.04S 174.04E
Newport, Oregon **134 B3** 44.39N 124.02W
Newport, Rhode Island **135 G3** 41.30N 71.19W
Newport, U.K. **114 D3** 51.35N 3.00W
Newport News **135 G2** 36.59N 76.26W
New Providence, island **136 E3**
Newry **114 C3** 54.11N 6.20W
New Siberian Islands **119 N6**
New South Wales, state **48 B4**
Newton **134 E2** 38.02N 97.22W
Newtownabbey **114 C3** 54.40N 5.57W
New York **135 G3** 40.40N 73.50W
New York, state **135 G3**
New Zealand, commonwealth **90 G4**
Neyshabur **108 D4** 36.12N 58.50E
Nezhin **115 K3** 51.03N 31.54E

*Entry / **Page** / **Ref.** / Lat. / Long.*

Ngabordamlu, Cape **88 B2** 6.58S 134.13E
Ngaiangl, island **101 E3**
Ngami, Lake **126 C2**
Ngaoundere **124 E2** 7.20N 13.35E
Ngatik, island **84 K6**
Nggatokae, island **89 G2**
Nguigmi **124 E3** 14.19N 13.06E
Nguiu **80 C9** 11.46S 130.38E
Ngukurr **80 E7** 14.44S 134.44E
Ngulu, island **101 E3**
Nguru **124 E3** 12.53N 10.30E
Nha Trang **100 B4** 12.15N 109.10E
Nhill **54 A2** 36.20S 141.39E
Nhulunbuy **80 F8** 12.11S 136.47E
Niamey **124 D3** 13.32N 2.05E
Nias, island **100 A3**
Nicaragua, Lake **136 D2**
Nicaragua, republic **136 D2**
Nicastro **117 G2** 38.59N 16.20E
Nice **116 E3** 43.42N 7.16E
Nichinan **104 E2** 31.36N 131.23E
Nicholson River **80 F6**
Nicobar Islands **107 E1**
Nicolls Town **135 G1** 25.08N 78.00W
Nicosia **125 G5** 35.09N 33.21E
Nicoya Peninsula **136 D2**
Nienburg **114 F3** 52.38N 9.13E
Niger, Mouths of the **124 D2**
Niger, republic **124 D3**
Niger River **124 D3**
Niger Delta, Inland **124 C3**
Nigeria, federal republic **124 D2**
Nihoa, island **84 O8**
Nii Island **104 I4**
Niigata **104 I5** 37.58N 139.02E
Niihama **104 F3** 33.57N 133.15E
Niitsu **104 I5** 37.40N 139.09E
Nijmegen **116 E5** 51.50N 5.52E
Nikki **124 D2** 9.55N 3.18E
Nikolayev **115 K2** 46.57N 32.00E
Nikolayevsk **119 O4** 53.10N 140.44E
Nikolsk **118 E4** 59.33N 45.30E
Nikopol **115 K2** 47.34N 34.25E
Niksic **117 G3** 42.48N 18.56E
Nikumaroro, island **84 N5**
Nila, island **88 A2**
Nile Delta **125 G5**
Nile River **125 G3**
Nilgiri Hills **106 C2**
Nimes **116 D3** 43.50N 4.21E
Nimmitabel **48 E1** 36.31S 149.17E
Ninety Mile Beach, Australia **54 D1**
Ninety Mile Beach, New Zealand **90 E7**
Ningbo **103 G2** 29.54N 121.33E
Ningjing Mountains **102 D3**
Ninigo Group **89 D3**
Ninohe **104 J7** 40.12N 141.20E
Nioaque **143 E1** 21.05S 55.50W
Niobrara River **134 D3**
Nioro du Sahel **124 C3** 15.12N 9.35W
Niort **116 C4** 46.19N 0.27W
Nipigon **132 J2** 49.02N 88.26W
Nipigon, Lake **132 J2**
Nipissing, Lake **133 J2**
Nis **117 H3** 43.20N 21.54E
Nitibe **100 D2** 9.21S 124.12E
Nitra **114 H2** 48.20N 18.05E
Niuatoputapu, island **84 N4**
Niue, independent state **84 O3**
Niut, mountain **100 B3** 0.58N 110.10E
Nive River **60 D3**
Nizamabad **106 C2** 18.40N 78.05E
Nizhni Tagil **118 F4** 58.00N 59.58E
Nizhny Novgorod **118 E4** 56.20N 44.00E
Nizmennyy, Cape **104 G8** 43.32N 135.12E
Njombe **126 D4** 9.20S 34.47E
Nkongsamba **124 D2** 4.59N 9.53E
Nobeoka **104 E3** 32.36N 131.40E
Nogales, Mexico **136 A4** 31.20N 111.00W
Nogales, U.S.A. **134 C2** 31.20N 110.56W
Noginsk **115 L4** 55.52N 38.29E
Nogoa River **42 F3**
Nojima, Cape **104 I4** 34.54N 139.54E
Nokia **115 I5** 61.29N 23.31E
Nok Kundi **106 B3** 28.49N 62.50E
Nokomis **132 H3** 51.30N 105.00W

*Entry / **Page** / **Ref.** / Lat. / Long.*

Noland Bay **60 E4**
Noma, Cape **104 E2** 31.27N 130.06E
Nomad **88 D2** 6.15S 142.13E
Nome **132 B4** 64.30N 165.30W
Nonning **66 F4** 32.31S 136.29E
Nonouti, island **84 M5**
Noosa Heads **42 I2** 26.23S 153.06E
Nordvik **119 L6** 74.01N 111.30E
Norfolk, Nebraska **134 E3** 42.01N 97.25W
Norfolk, Virginia **135 G2** 36.54N 76.18W
Norfolk Island **84 L3**
Norilsk **118 I5** 69.21N 88.02E
Norman **134 E2** 35.14N 97.27W
Norman River **42 C6**
Normanby Island **89 F1**
Normanby River **42 E8**
Normanton **42 C7** 17.40S 141.04E
Norman Wells **132 F4** 65.18N 126.42W
Norrkoping **114 H4** 58.35N 16.10E
Norrtalje **114 H4** 59.46N 18.43E
Norseman **72 D2** 32.12S 121.47E
Norte, Point **144 C2** 42.05S 63.46W
North, Cape **133 L2** 47.03N 60.24W
North Cape, New Zealand **90 E7** 34.25S 173.03E
North Cape, Norway **112 G5** 71.11N 25.40E
North Cape, P.N.G. **89 F3** 2.33S 150.48E
North Island **90 E6**
North Sea **114 E4**
Northam **72 C2** 31.39S 116.40E
Northampton **114 D3** 52.14N 0.54W
North Andaman, island **107 E2**
North Atlantic Ocean **147 M4**
North Bay **133 K2** 46.20N 79.28W
North Baykal Plateau **119 L4**
North Bruny, island **60 E1**
North Carolina, state **135 G2**
Northcliffe **72 C2** 34.38S 116.07E
North Dakota, state **134 D3**
North Dvina River **118 E5**
Northern Territory **80 D4**
Northern Marianas, commonwealth **101 F4**
North George River **60 E3**
North Korea, democratic people's republic **104 D6**
Northland, region **90 E7**
North Magnetic Pole **132 H5** 77.18N 101.48W
North Pacific Ocean **147 H4**
North Pagai, island **100 A2**
North Platte **134 D3** 41.09N 100.45W
North Platte River **134 D3**
North Saskatchewan River **132 G3**
North Siberian Lowlands **118 J6**
North Sulawesi, province **100 D3**
North Sumatra, province **100 A3**
North Taranaki Bight **90 E5**
Northumberland, Cape **66 H1** 38.04S 140.40E
North West Cape **72 B4** 21.47S 114.10E
North West Shelf **72 C5**
Northwest Territories, province **132 F4**
Norton Sound **132 B4**
Norvegia, Cape **92 R3** 71.28S 12.25W
Norway, kingdom **114 F5**
Norway House **132 I3** 53.59N 97.50W
Norwegian Sea **114 E5**
Norwich **114 E3** 52.38N 1.18E
Noshiro **104 I7** 40.13N 140.00E
Notodden **114 F4** 59.35N 9.18E
Notre Dame Bay **133 M3**
Notre Dame Mountains **135 H3**
Nottingham **114 D3** 52.58N 1.10W
Nouadhibou **124 B4** 20.54N 17.01W
Nouakchott **124 B3** 18.09N 15.58W
Noumea **84 L3** 22.16S 166.26E
Nova Iguacu **144 E4** 22.45S 43.28W
Nova Mambone **126 D2** 21.00S 35.00E
Novara **116 E4** 45.27N 8.37E
Nova Scotia, province **133 L2**
Novaya Kakhovka **115 K2** 46.45N 33.20E
Novaya Zemlya, islands **118 F6**
Novelda **116 C2** 38.24N 0.45W
Nove Zamky **114 H2** 48.00N 18.10E
Novgorod **115 K4** 58.30N 31.20E
Novi Ligure **116 E3** 44.46N 8.47E

Entry/ ***Page/Ref.*** */Lat./Long.*

Novi Pazar, Bulgaria **117 I3** 43.20N 27.12E
Novi Pazar, Yugoslavia **117 H3** 43.09N 20.29E
Novi Sad **117 G4** 45.15N 19.51E
Novograd-Volynskiy **115 J3** 50.34N 27.32E
Novogrudok **115 J3** 53.35N 25.50E
Novokazanlinsk **118 G3** 45.48N 62.06E
Novokuznetsk **118 I4** 53.45N 87.12E
Novolazarevskaya, antarctic base **92 A3** 70.46S 11.50E
Novomoskovsk, Russia **115 L3** 54.05N 38.13E
Novomoskovsk, Ukraine **115 L2** 48.38N 35.15E
Novopolotsk **115 J4** 55.32N 28.39E
Novorossisk **118 D3** 44.44N 37.46E
Novosibirsk **118 I4** 55.04N 83.05E
Novozybkov **115 K3** 52.31N 31.58E
Novy Port **118 H5** 67.38N 72.33E
Nowendoc **48 F4** 31.35S 151.45E
Nowgong **107 E3** 26.20N 92.41E
Nowingi **54 B3** 34.36S 142.15E
Nowra **48 F2** 34.53S 150.36E
Nowy Targ **115 I2** 49.28N 20.00E
Nubeena **60 E1** 43.06S 147.44E
Nubian Desert **125 G4**
Nuboai **88 C3** 2.16S 136.31E
Nueces River **134 E1**
Nueltin Lake **132 H3**
Nueva Rosita **136 B3** 27.58N 101.11W
Nuevo, Gulf of **144 C2**
Nuevo Casas Grandes **136 B4** 30.22N 107.53W
Nuevo Laredo **136 C3** 27.30N 99.30W
Nugget Point **90 C1** 46.27S 169.49E
Nukualofa **84 N3** 21.08S 175.12W
Nukuhiva, island **85 Q5**
Nukuoro, island **84 K6**
Nukus **118 F3** 42.28N 59.07E
Nullagine **72 D4** 21.53S 120.07E
Nullarbor Plain **40 C2**
Numazu **104 I4** 35.08N 138.50E
Numbulwar **80 E7** 14.17S 135.44E
Numfoor, island **88 C3**
Numurkah **54 C2** 36.06S 145.26E
Nundle **48 F4** 31.28S 151.08E
Nundooka, homestead **48 A4** 30.44S 141.41E
Nunavut, province **132 H4**
Nunivak Island **132 B3**
Nuoro **116 E3** 40.20N 9.21E
Nuremberg **114 G2** 49.27N 11.05E
Nuriootpa **66 G3** 34.28S 139.00E
Nurmes **115 J5** 63.31N 29.10E
Nurri, Mount **48 C4** 31.41S 146.02E
Nutwood Downs, homestead **80 E7** 15.49S 134.09E
Nuyts, Point **72 C1** 35.04S 116.38E
Nyabing **72 C2** 33.33S 118.09E
Nyahururu **125 G2** 0.04N 36.22E
Nyah West **54 B3** 35.11S 143.21E
Nyainqentanglha Range **102 C2**
Nyala **125 F3** 12.01N 24.50E
Nyandoma **118 E5** 61.43N 40.11E
Nyanza **126 C4** 4.19S 29.35E
Nyda **118 H5** 66.35N 72.58E
Nye Bay **60 C1**
Nyiregyhaza **115 I2** 47.57N 21.43E
Nyiru, Mount **125 G2** 2.11N 36.49E
Nykobing **114 G3** 54.47N 11.53E
Nykoping **114 H4** 58.45N 17.03E
Nylstroom **126 C2** 24.42S 28.20E
Nyngan **48 D4** 31.34S 147.12E
Nyong River **124 E2**
Nzerekore **124 C2** 7.49N 8.48W
N'zeto **126 B4** 7.13S 12.54E

O

Oahe Reservoir **134 D3**
Oahu, island **84 P8**
Oakbank, homestead **66 H4** 33.03S 140.35E
Oakey **42 H2** 27.26S 151.43E
Oak Harbor **134 B3** 48.17N 122.40W
Oakland **134 B2** 37.50N 122.15W
Oamaru **90 D2** 45.05S 170.59E
Oates Land **94 H2**
Oatlands **60 E2** 42.18S 147.22E
Oaxaca de Juarez **136 C2** 17.05N 96.41W
Ob, Gulf of **98 E15**
Ob River **118 H5**
Oban **110 D3** 56.25N 5.29W
Obbia **125 H2** 5.20N 48.30E
Obi, island **100 D2**
Obihiro **104 K8** 42.56N 143.10E
Oblachnaya, Mount **103 H4** 43.46N 134.10E
Obninsk **115 L4** 55.05N 36.37E
Obo **125 F2** 5.18N 26.28E
Obock **125 H3** 11.59N 43.16E
Obuasi **124 C2** 6.15N 1.36W
Ocala **135 F1** 29.11N 82.09W
Ocana **142 C4** 8.16N 73.21W
Oceanside **134 C2** 33.12N 117.23W
Odate **104 J7** 40.18N 140.32E
Odawara **104 I4** 35.15N 139.08E
Odendaalsrus **126 C2** 27.52S 26.42E
Odense **114 G4** 55.24N 10.25E
Oder River **114 G3**
Odessa, U.S.A. **134 D2** 31.50N 102.23W
Odessa, Ukraine **115 K2** 46.30N 30.46E
Odienne **124 C2** 9.36N 7.32W
Odorhei **117 I4** 46.18N 25.19E
Oeno, island **85 R3**
Oenpelli **80 D8** 12.20S 133.03E
Officer Creek **66 D7**
Oga **104 I6** 39.56N 139.47E
Ogaden Desert **125 H2**
Ogbomosho **124 D2** 8.05N 4.11E
Ogden **134 C3** 41.14N 111.59W
Ogilby, Mount **42 G3** 25.18S 147.55E
Ogoja **124 D2** 6.40N 8.45E
Ogooue River **124 D1**
Ohakune **90 F5** 39.25S 175.25E
Ohau, Lake **90 C2**
O'Higgins, Lake **144 B2**
Ohio River **135 F2**
Ohio, state **135 F3**
Ohopoho **126 B3** 18.03S 13.54E
Ohrid **117 H3** 41.06N 20.49E
Oita **104 E3** 33.15N 131.36E
Ojinaga **136 B3** 29.35N 104.26W
Okaba **88 C2** 8.06S 139.46E
Okapa **89 E2** 6.32S 145.36E
Okavango Basin **126 C3**
Okavango River **126 B3**
Okaya **104 I5** 36.03N 138.00E
Okayama **104 F4** 34.40N 133.54E
Okazaki **104 H4** 34.58N 137.10E
Okeechobee, Lake **135 F1**
Okene **124 D2** 7.31N 6.14E
Okha **119 O4** 53.35N 143.01E
Okhotsk **119 O4** 59.20N 143.15E
Okhotsk, Sea of **119 O4**
Oki Archipelago **104 F5**
Okinawa, island **103 G2**
Okinoerabu Island **103 G2**
Oklahoma, state **134 E2**
Oklahoma City **134 E2** 35.28N 97.33W
Ok Tedi **88 D2** 5.15S 141.06E
Oktyabriskiy **115 J3** 52.33N 28.47E
Oland, island **114 H4**
Olary **66 H4** 32.17S 140.20E
Olavarria **144 C3** 36.57S 60.20W
Olbia **116 E3** 40.56N 9.30E
Oldenburg **114 F3** 53.08N 8.13E
Old Town **135 H3** 44.55N 68.41W
Olean **135 G3** 42.05N 78.26W
Olekma River **119 M4**
Olenek River **119 L5**
Olesnica **114 H3** 51.12N 17.21E
Olga **119 N3** 43.46N 135.14E
Olga, Mount *see* Kata Tjuta
Olhao **116 B2** 37.01N 7.50W
Olimarao, island **101 F3**
Olinda **143 G3** 8.00S 34.51W
Olivenza **116 B2** 38.41N 7.06W
Ollague **142 D1** 21.14S 68.16W
Olomouc **114 H2** 49.48N 17.15E
Olsztyn **115 I3** 53.48N 20.29E
Olten **116 E4** 47.22N 7.55E
Olympia **134 B3** 47.03N 122.53W
Olympus, Mount **117 H3** 40.05N 22.21E
Olyutorsk, Cape **119 R4** 59.58N 170.25E
Oma, Cape **104 J7** 41.31N 140.55E
Omaha **134 E3** 41.15N 96.00W
Oman, Gulf of **105 A3**
Oman, sultanate **108 D3**
Omdurman **125 G3** 15.37N 32.29E
Omeo **54 D2** 37.06S 147.36E
Omiya **104 I4** 35.54N 139.39E
Omo River **125 G2**
Omsk **118 H4** 55.00N 73.22E
Omuta **104 E3** 33.02N 130.26E
Omutninsk **118 F4** 58.35N 52.28E
Ondangua **126 B3** 17.52S 15.59E
Ondjiva **126 B3** 17.03S 15.41E
Onega, Lake **115 L5**
Onega River **118 D5**
Onekotan, island **119 P3**
Ongjin **104 B5** 37.56N 125.21E
Onin Peninsula **88 B3**
Onitsha **124 D2** 6.10N 6.47E
Onslow **72 C4** 21.38S 115.07E
Onslow Bay **135 G2**
Ontario **134 C3** 44 02N 116.58W
Ontario, Lake **135 G3**
Ontario, province **133 J3**
Ontong Java Islands **89 G2**
Oodnadatta **66 E7** 27.33S 135.27E
Ooldea **66 C5** 30.28S 131.50E
Oombulgurri **72 E5** 15.15S 127.55E
Ooratippra, homestead **80 F4** 21.53S 136.05E
Ophthalmia Range **72 C4**
Opole **114 H3** 50.40N 17.56E
Oporto **116 B3** 41.09N 8.37W
Opp **135 F2** 31.16N 86.18W
Opunake **90 E5** 39.27S 173.51E
Oradea **117 H4** 47.03N 21.55E
Oran **124 C5** 35.45N 0.38W
Orange, Australia **48 E3** 33.17S 149.06E
Orange, France **116 D3** 44.08N 4.48E
Orange, Cape **143 E4** 4.25N 51.32W
Orange River **126 B2**
Orangerie Bay **89 E1**
Oranjestad **137 E2** 12.32N 70.02W
Orbetello **116 F3** 42.27N 11.07E
Orbost **55 E2** 37.43S 148.28E
Orcadas, antarctic base **92 P3** 60.45S 44.43W
Orchilla, island **142 D5**
Ord, Mount **72 E5** 17.19S 125.33E
Ord River **72 E5**
Ordos Desert **102 E3**
Ordu **108 B5** 41.00N 37.52E
Orebro **114 H4** 59.13N 15.05E
Oregon state **134 B3**
Oregon City **134 B3** 45.21N 122.36W
Orekhovo-Zuevo **115 L4** 55.49N 38.59E
Orel **115 L3** 52.58N 36.04E
Orem **134 C3** 40.20N 111.42W
Orenburg **118 F4** 51.50N 55.00E
Orense **116 B3** 42.20N 7.52W
Orford **60 E2** 42.32S 147.50E
Orgeyev **115 J2** 47.24N 28.50E
Orinoco River **142 D4**
Oristano **116 E2** 39.54N 8.36E
Oriximina **143 E3** 1.47S 55.48W
Orizaba **136 C2** 18.51N 97.08W
Orkney Islands **114 D4**
Orlando **135 F1** 28.33N 81.21W
Orleans **116 D4** 47.54N 1.54E
Ormara **106 B3** 25.12N 64.39E
Ormoc **100 D4** 11.01N 124.36E
Orocue **142 C4** 4.51N 71.21W
Oroluk, island **84 K6**
Orona, island **84 N5**
Orroroo **66 G4** 32.46S 138.37E
Orsha **115 K3** 54.30N 30.23E
Orsk **118 F4** 51.13N 58.35E
Orsova **118 C3** 44.42N 22.22E
Orsta **114 F5** 62.12N 6.00E
Oruro **142 D2** 17.59S 67.08W
Orvieto **116 F3** 42.43N 12.06E
Osaka **104 G4** 34.40N 135.30E
Oshawa **133 K2** 43.53N 78.51W
Oshkosh **135 F3** 44.01N 88.32W
Osijek **117 G4** 45.33N 18.41E
Osipovichi **115 J3** 53.19N 28.36E
Oskarshamn **114 H3** 57.16N 16.25E
Oslo **114 G4** 59.56N 10.45E
Osnabruck **116 E5** 52.17N 8.03E
Osorno **144 B2** 40.35S 73.14W
Osprey Reef **42 F9**
Ossa, Mount **60 D3** 41.52S 146.02E
Ostashkov **115 K4** 57.09N 33.10E
Ostersund **114 G5** 63.10N 14.40E
Ostrava **110 F2** 49.50N 18.15E
Ostroda **114 H3** 53.42N 19.59E
Ostrogozhsk **115 L3** 50.52N 39.03E
Ostroleka **115 I3** 53.05N 21.32E
Ostrov **115 J4** 57.22N 28.22E
Ostrowiec **115 I3** 50.58N 21.22E
Ostrow Mazowiecki **115 I3** 52.50N 21.51E
Osumi Strait **104 E2**
Otago Peninsula **90 D2**
Otaki **90 F4** 40.45S 175.08E
Otaru **104 J8** 43.14N 140.59E
Otjiwarongo **126 B2** 20.29S 16.36E
Otsu **103 H3** 35.00N 135.50E
Ottawa **133 K2** 45.25N 75.43W
Ottawa River **133 K2**
Ottumwa **134 E3** 41.02N 92.26W
Oturkpo **124 D2** 7.16N 8.16E
Otway, Cape **54 B1** 38.52S 143.31E
Otway Range **54 B1**
Ouadda **124 F2** 8.09N 22.20E
Ouagadougou **124 C3** 12.20N 1.40W
Ouahigouya **124 C3** 13.31N 2.20W
Ouanda Djale **124 F2** 8.55N 22.53E
Ouargla **124 D5** 32.00N 5.16E
Oudtshoorn **126 C1** 33.35S 22.12E
Ouenza **124 D5** 35.55N 8.08E
Ouesso **126 B5** 1.38N 16.03E
Ouezzane **116 B1** 34.52N 5.35W
Oujda **124 C5** 34.41N 1.45W
Ouled Djellal **124 D5** 34.25N 5.02E
Oulu **118 C5** 65.00N 25.26E
Ounianga Kebir **124 F3** 19.05N 20.29E
Ouro Preto **143 F1** 20.25S 43.30W
Ouse **60 D2** 42.29S 146.42E
Ouse River **60 D2**
Outer Hebrides, islands **114 C4**
Outjo **126 B2** 20.08S 16.08E
Ouyen **54 B3** 35.04S 142.19E
Ovalle **144 B3** 30.33S 71.16W
Ovar **116 B3** 40.52N 8.38W
Ovens River **54 D2**
Overland Park **134 E2** 38.57N 94.47W
Oviedo **116 B3** 43.21N 5.50W
Owase **104 H4** 34.03N 136.12E
Owen Stanley Range **89 E2**
Oxford **114 D3** 51.46N 1.15W
Oxnard **134 C2** 34.11N 119.10W
Oya **100 C3** 2.53N 111.54E
Oyo **124 D2** 7.50N 3.55E
Ozark **135 F2** 31.27N 85.40W
Ozieri **116 E3** 40.35N 9.01E

P

Paan **107 E2** 16.51N 97.37E
Paarl **126 B1** 33.45S 18.58E
Pabna **107 D3** 24.00N 89.15E
Pacaraima Mountains **142 D4**
Pachino **117 G2** 36.43N 15.06E
Pachuca **136 C3** 20.10N 98.44W
Pacific Ocean **84 L8**
Pacitan **100 C2** 8.12S 110.05E
Padang, West Kalimantan **100 B2** 1.38S 108.56E
Padang, West Sumatra **100 B2** 1.00S 100.21E
Paderborn **114 F3** 51.43N 8.44E
Padua **116 F4** 45.24N 11.53E
Pagadian **101 D3** 7.50N 123.30E
Pagan, island **101 F4**
Pagerdewa **100 B2** 3.44S 105.15E
Pagwi **88 D3** 4.03S 143.00E
Painted Desert **130 H4**

*Entry / **Page** / **Ref.** / Lat. / Long.*

Pakanbaru **100 B3** 0.33N 101.30E
Pakenham **54 C1** 38.05S 145.29E
Pakistan, Islamic republic **106 B3**
Pakokku **107 E3** 21.20N 95.05E
Pakse **100 B4** 15.07N 105.47E
Pakwach **126 D5** 2.17N 31.28E
Palamos **116 D3** 41.51N 3.07E
Palana, Australia **60 E5** 39.45S 147.55E
Palana, Russia **119 P4** 59.05N 159.59E
Palangkaraya **100 C2** 2.14S 113.53E
Palanpur **106 C3** 24.12N 72.29E
Palapye **126 C2** 22.37S 27.06E
Palatka **135 F1** 29.38N 81.40W
Palau, republic **101 E3**
Palawan, island **100 C4**
Palembang **100 B2** 2.59S 104.45E
Palencia **116 C3** 42.01N 4.32W
Palermo **116 F2** 38.08N 13.23E
Palghat **106 C2** 10.46N 76.42E
Palikir **84 K6** 6.57N 158.10E
Palime **124 D2** 6.55N 0.44E
Palk Strait **106 C1**
Palliser Bay **90 F4**
Palliser, Cape **90 F4** 41.37S 175.16E
Palm Islands **42 F6**
Palma **116 D2** 39.35N 2.39E
Palmalmal **89 F2** 5.39S 151.26E
Palmas, Cape **124 C2** 4.25N 7.50W
Palmeirinhas Point **126 B4** 9.09S 12.58E
Palmer **132 D4** 61.35N 149.10W
Palmer, antarctic base **92 O3** 64.46S 64.03W
Palmer Land **94 O2**
Palmerston, Cape **42 G5** 21.32S 149.29E
Palmerston, island **84 O4**
Palmerston North **90 F4** 40.21S 175.37E
Palmi **117 G2** 38.21N 15.51E
Palmira **142 C4** 3.33N 76.17W
Palm Springs **134 C2** 33.49N 116.34W
Palmyra **108 B4** 34.33N 38.17E
Palmyra, island **84 O6**
Paloh **100 C3** 2.26N 111.16E
Palu **100 C2** 0.54S 119.52E
Pamangkat **100 B3** 1.08N 108.55E
Pamiers **116 D3** 43.07N 1.36E
Pamirs, mountains **102 B3**
Pamlico Sound **135 G2**
Pampa **134 D2** 35.32N 100.58W
Pampas, plains **144 C3**
Pamplona **116 C3** 42.49N 1.39W
Panahan **100 C2** 1.45S 111.45E
Panama Canal **136 D1**
Panama, Gulf of **136 E1**
Panama, republic **136 D1**
Panama City, Panama **136 E1** 8.57N 79.30W
Panama City, U.S.A. **135 F2** 30.10N 85.41W
Panay, island **100 D4**
Pandan **100 D4** 14.02N 124.11E
Pandie Pandie, homestead **66 G7** 26.08S 139.23E
Panevezys **115 I4** 55.44N 24.24E
Pangkalansusu **100 A3** 4.05N 98.13E
Pangkalpinang **100 B2** 2.05S 106.09E
Panguna **89 G2** 6.19S 155.30E
Panipat **106 C3** 29.24N 76.58E
Pannawonica **72 C4** 21.39S 116.19E
Pantar, island **100 D2**
Pantonlabu **100 A3** 5.06N 97.26E
Panzi **126 B4** 7.17S 18.01E
Papeete **85 Q4** 17.32S 149.34W
Papisoi, Cape **88 B3** 4.06S 132.54E
Papua, Gulf of **89 D2**
Papua, province **88 C3**
Papua New Guinea, republic **89 E2**
Papunya **80 C3** 23.15S 131.54E
Paraburdoo **72 C4** 23.12S 117.40E
Paracel Islands **100 C4**
Parachilna **66 G5** 31.08S 138.24E
Paradise **132 F1** 39.46N 121.38W
Paradise Valley **134 D2** 31.46N 109.15W
Paragould **135 E2** 36.02N 90.30W
Paraguari **144 D4** 25.36S 57.06W
Paraguay, republic **144 C4**
Paraguay River **144 D4**
Parakou **124 D2** 9.23N 2.40E
Parakylia, homestead **66 F5** 30.24S 136.24E
Paramaribo **143 E4** 5.52N 55.14W
Paramushir, island **119 P4**
Parana **144 C3** 31.45S 60.30W
Parana River **144 C3**
Paranagua **144 E4** 25.32S 48.36W
Paranaiba **143 E2** 19.44S 51.12W
Paranaiba River **143 E2**
Paranavai **144 D4** 23.04S 52.28W
Paraparaumu **90 F4** 40.55S 175.00E
Parbhani **106 C2** 19.16N 76.51E
Parczew **115 I3** 51.39N 22.53E
Parengarenga Harbour **90 E7**
Parepare **100 C2** 4.03S 119.40E
Parinas, Point **142 B3** 4.45S 81.22W
Paris, France **116 D4** 48.52N 2.20E
Paris, Tennessee **135 F2** 36.19N 88.20W
Paris, Texas **134 E2** 33.41N 95.33W
Parkes **48 E3** 33.08S 148.11E
Parma **116 F3** 44.48N 10.19E
Parnaiba **143 F3** 2.58S 41.46W
Parnaiba River **143 F3**
Parndana **66 F3** 35.47S 137.16E
Parnu **115 I4** 58.28N 24.30E
Paroo River **42 E1**
Parral **136 B3** 26.56N 105.40W
Parry Islands **132 I5**
Parry Sound **135 G3** 45.21N 80.03W
Parthenay **116 C4** 46.39N 0.14W
Parys **126 C2** 26.55S 27.28E
Pasadena **134 C2** 34.10N 118.09W
Pascagoula **136 D4** 30.21N 88.32W
Pascani **117 I4** 47.14N 26.46E
Pasco, Mount **72 D3** 27.36S 120.33E
Paso Robles **134 B2** 35.38N 120.44W
Passau **114 G2** 48.35N 13.28E
Passo Fundo **144 D4** 28.16S 52.20W
Pasto **142 C4** 1.12N 77.17W
Patagonia, region **144 C2**
Patan **106 C3** 23.51N 72.11E
Patchewollock **54 B3** 35.23S 142.12E
Paterson, Cape **54 C1** 38.41S 145.36E
Pathiu **100 A4** 10.41N 99.18E
Patna **106 D3** 25.37N 85.12E
Patom Highlands **119 L4**
Patos, eastern Brazil **143 G3** 6.55S 37.15W
Patos, southern Brazil **143 F2** 18.35S 46.32W
Patos Lagoon **144 D3**
Patras **117 H2** 38.14N 21.44E
Pattani **100 B3** 6.50N 101.20E
Pau **116 C3** 43.18N 0.22W
Paulo Alfonso Falls **143 G3** 9.25S 38.15W
Paunairo **89 H2** 9.06S 161.09E
Pavlodar **118 H4** 52.21N 76.59E
Pavlovo **115 M4** 55.58N 43.05E
Pavlovsk **110 I3** 50.28N 40.07E
Paysandu **144 D3** 32.21S 58.05W
Peace River **132 G3**
Peace River **132 G3** 56.15N 117.18W
Peak Hill **48 E3** 32.44S 148.11E
Pearce Point **80 B7** 14.25S 129.21E
Pearl River **135 E2**
Pec **117 H3** 42.40N 20.19E
Pechenga **110 H4** 69.28N 31.04E
Pechora **118 F5** 65.14N 57.18E
Pechora River **118 F5**
Pecos **134 D2** 31.25N 103.30W
Pecos River **134 D2**
Pecs **114 H2** 46.04N 18.15E
Pedder, Lake **60 D2**
Pedra Azul **143 F2** 16.02S 41.17W
Pedra Branca, island **60 D1**
Pedro Afonso **143 F3** 8.59S 48.12W
Pedro Juan Caballero **143 E1** 22.34S 55.37W
Peebinga **66 H3** 34.56S 140.55E
Pee Dee River **135 F2**
Pegasus Bay **90 D3**
Pegu **107 E2** 17.18N 96.31E
Pegu Range **107 E2**
Peleng, island **100 D2**
Pello **118 C5** 66.47N 24.00E
Pelly River **132 E4**
Pelly Bay **132 J4** 68.38N 89.45W
Pelotas **144 D3** 31.45S 52.20W
Pematangsiantar **100 A3** 2.59N 99.01E
Pemba **126 E3** 13.00S 40.30E
Pemba Channel **126 D4**
Pemba, island **126 D4**
Pemberton **72 C2** 34.27S 116.02E
Pembroke **114 D3** 51.41N 4.55W
Penas, Gulf of **144 B2**
Pendleton **134 C3** 45.40N 118.46W
Pend Orielle Lake **134 C3**
Penguin **60 D3** 41.07S 146.05E
Peniche **116 B2** 39.21N 9.23W
Penju Islands **100 D2**
Penner River **106 C2**
Pennines, mountains **114 D3**
Pennsylvania, state **135 G3**
Penola **66 H2** 37.23S 140.50E
Penong **66 D5** 31.56S 133.01E
Penrhyn, island **84 P5**
Pensacola **135 F2** 30.26N 87.12W
Penshurst **54 B2** 37.53S 142.17E
Pensiangan **100 C3** 4.30N 116.20E
Penuguan **100 B2** 2.18S 104.30E
Penza **118 E4** 53.11N 45.00E
Penzance **114 C3** 50.07N 5.33W
Peoria, Arizona **134 C2** 33.38N 112.13W
Peoria, Illinois **135 F3** 40.43N 89.38W
Peppimenarti **80 C7** 14.08S 130.05E
Pera Head **42 C9** 12.57S 141.36E
Percival Lakes **72 D4**
Pereira **142 C4** 4.47N 75.46W
Pergamino **144 C3** 33.55S 60.32W
Perigueux **116 D4** 45.12N 0.44E
Perija Mountains **137 E1**
Perkam, Cape **88 C3** 1.29S 137.55E
Perm **118 F4** 58.01N 56.10E
Pernik **117 H3** 42.36N 23.03E
Peron Islands **80 C8**
Peron Peninsula **72 B3**
Perpignan **116 D3** 42.42N 2.54E
Perry **135 F2** 30.08N 83.36W
Perryton **134 D2** 36.23N 100.48W
Persian Gulf **108 D3**
Perth, Tasmania **60 E3** 41.34S 147.11E
Perth, U.K. **114 D4** 56.24N 3.28W
Perth, Western Australia **72 C2** 31.57S 115.51E
Peru, republic **142 C3**
Perugia **116 F3** 43.07N 12.23E
Pervomaysk **115 K2** 48.03N 30.50E
Pesaguan **100 C2** 2.02S 110.07E
Pescara **117 F3** 42.27N 14.13E
Peshawar **106 C4** 34.01N 71.40E
Pesqueira **143 G3** 8.24S 36.38W
Peterborough, Australia **66 G4** 33.00S 138.51E
Peterborough, Canada **133 K2** 44.19N 78.20W
Peterborough, U.K. **114 D3** 52.35N 0.15W
Peterhead **114 D4** 57.30N 1.46W
Petermann Ranges **72 E4**
Petersburg **132 E3** 56.49N 132.58W
Peter the Great Bay **104 E8**
Peto **136 D3** 20.08N 88.53W
Petoskey **135 F3** 45.22N 84.59W
Petrich **117 H3** 41.23N 23.10E
Petrodvorets **115 J4** 59.50N 29.57E
Petropavlovsk **118 G4** 54.53N 69.13E
Petropavlovsk-Kamchatski **119 P4** 53.03N 158.43E
Petroseni **117 H4** 45.25N 23.22E
Petrozavodsk **115 K5** 61.46N 34.19E
Pevek **119 R5** 69.41N 170.19E
Pforzheim **114 F2** 48.53N 8.41E
Phalodi **106 C3** 27.06N 72.22E
Phangan, island **100 B3**
Phan Rang **100 B4** 11.34N 109.00E
Phan Thiet **100 B4** 10.56N 108.06E
Phatthalung **100 A3** 7.38N 100.05E
Phet Buri **100 A4** 13.05N 99.58E
Philadelphia **135 G3** 40.00N 75.10W
Philippine Sea **103 G1**
Philippine Trench **84 H7**
Philippines, republic **100 D4**
Phillip Island **54 C1**
Phitsanulok **100 B4** 16.50N 100.15E
Phnom Penh **100 B4** 11.35N 104.55E
Phoenix **134 C2** 33.30N 112.03W
Phoenix, island **84 N5**
Phoenix Islands **84 N5**
Phuket **100 A3** 7.52N 98.22E
Phuket, island **100 A3**
Phu Quoc, island **100 B4**
Piacenza **116 E3** 45.03N 9.41E
Pickle Lake **132 I3** 51.30N 90.10W
Picos **143 F3** 7.05S 41.24W
Picton, Australia **48 F2** 34.11S 150.37E
Picton, New Zealand **90 E4** 41.18S 174.00E
Picton, Mount **60 D1** 43.10S 146.37E
Pidurutalagala, mountain **106 D1** 7.01N 80.45E
Piedras Negras **136 B3** 28.40N 100.32W
Pieman River **60 C3**
Pierre **134 D3** 44.23N 100.20W
Pietarsaari **115 I5** 63.41N 22.40E
Pietermaritzburg **126 D2** 29.36S 30.24E
Pietersburg **126 C2** 23.54S 29.23E
Pigeon Hole, homestead **80 C6** 16.49S 131.13E
Pigs, Bay of **136 D3**
Pikes Peak **134 D2** 38.50N 105.03W
Pila **114 H3** 53.09N 16.44E
Pilar **144 D4** 26.51S 58.20W
Pilcomayo River **144 C4**
Pillar, Cape **60 F1** 43.14S 148.00E
Pilliga **48 E4** 30.21S 148.53E
Pimba **66 F5** 31.15S 136.48E
Pinang, island **100 A3**
Pinar del Rio **136 D3** 22.24N 83.42W
Pindera Downs, homestead **48 B5** 29.25S 142.28E
Pine Bluff **134 E2** 34.13N 92.00W
Pine Creek **80 C8** 13.49S 131.50E
Pine Hill, homestead **80 D3** 22.20S 133.03E
Pine Island Bay **92 M3**
Pine Point **132 G4** 60.52N 114.30W
Pinerolo **116 E3** 44.53N 7.19E
Ping River **100 A4**
Pingdingshan **102 F3** 33.50N 113.20E
Pingelap, island **84 L6**
Pingxiang, Guanxi Zhuangzu **102 E2** 22.06N 106.44E
Pingxiang, Jiangxi **102 F2** 27.35N 113.46E
Pini, island **100 A3**
Pinjarra **72 C2** 32.38S 115.52E
Pinnaroo **66 H3** 35.16S 140.55E
Pinsk **115 J3** 52.08N 26.01E
Pioche **134 C2** 37.57N 114.30W
Piombino **116 F3** 42.56N 10.32E
Piotrkow Trybunalski **115 H3** 51.27N 19.40E
Piracicaba **143 F1** 22.45S 47.40W
Piraeus **117 H2** 37.57N 23.42E
Pirapora **143 F2** 17.20S 44.54W
Pirimapun **88 C2** 6.21S 138.24E
Pirot **117 H3** 43.10N 22.32E
Piru **88 A3** 3.01S 128.10E
Pisa **116 F3** 43.43N 10.24E
Pisagua **142 C2** 19.34S 70.14W
Pisco **142 C2** 13.46S 76.12W
Pisek **114 G2** 49.19N 14.10E
Pisz **115 I3** 53.38N 21.49E
Pita **124 B3** 11.05N 12.15W
Pitanga **144 D4** 24.45S 51.43W
Pitcairn Island **85 R3**
Pitesti **117 H3** 44.51N 24.51E
Pittsburg **134 E2** 37.25N 94.43W
Pittsburgh **135 G3** 40.26N 80.00W
Pittsfield **135 G3** 42.27N 73.15W
Pittsworth **42 H2** 27.43S 151.38E
Piura **142 B3** 5.15S 80.38W
Plainview **134 D2** 34.12N 101.43W
Plate, River **144 D3**
Platte River **134 D3**
Plauen **114 G3** 50.29N 12.08E
Pleihari **100 C2** 3.47S 114.45E
Pleiku **100 B4** 13.57N 108.01E
Plenty, Bay of **90 G6**
Pleven **117 H3** 43.25N 24.40E
Pljevlja **117 G3** 43.21N 19.21E
Plock **114 H3** 52.32N 19.40E
Ploiesti **117 I3** 44.57N 26.01E

*Entry / **Page** / **Ref.** / Lat. / Long.*

Plovdiv **117 H3** 42.08N 24.45E
Plumbago, homestead **66 G4** 32.05S 139.53E
Plumtree **126 C2** 20.30S 27.50E
Plymouth **114 D3** 50.23N 4.10W
Plzen **114 G2** 49.45N 13.25E
Po River **116 E3**
Poatina **60 D3** 41.48S 146.58E
Pobeda, mountain **119 O5** 65.10N 146.00E
Pobeda Peak **102 C4** 42.25N 80.15E
Podgorica **117 G3** 42.28N 19.17E
Pocatello **134 C3** 42.53N 112.26W
Pocklington Reef **89 F1**
Podolsk **115 L4** 55.26N 37.33E
Pofadder **126 B2** 29.09S 19.25E
Pohang **104 D5** 36.00N 129.26E
Pohnpei, island **84 K6**
Poinsett, Cape **92 F3** 65.42S 113.18E
Point-a-Pitre **137 F2** 16.14N 61.32W
Pointe Noire **126 B4** 4.46S 11.53E
Point Lookout, mountain **48 G4** 30.29S 152.25E
Point Samson **72 C4** 20.38S 117.11E
Poissonnier Point **72 C4** 19.57S 119.10E
Poitiers **116 D4** 46.35N 0.20E
Pokataroo **48 E5** 29.35S 148.42E
Poland, people's republic **115 I3**
Polillo Islands **100 D4**
Pollensa **116 D2** 39.52N 3.01E
Polnovat **118 G5** 63.47N 65.54E
Polonnoye **115 J3** 50.10N 27.30E
Polotsk **115 J4** 55.30N 28.43E
Poltava **115 K2** 49.35N 34.35E
Pombal **116 B2** 39.55N 8.38W
Pomio **89 F2** 5.31S 151.30E
Pompano Beach **135 F1** 26.14N 80.07W
Pompei **117 F3** 40.45N 14.27E
Ponca City **134 E2** 36.41N 97.04W
Ponce **137 F2** 18.01N 66.36W
Pondicherry **106 C2** 11.59N 79.50E
Pond Inlet **133 K5** 72.40N 77.59W
Pondo **89 F3** 4.33S 151.38E
Ponferrada **116 B3** 42.33N 6.35W
Pongkolaero **100 D2** 5.24S 121.56E
Pono **88 B2** 6.21S 134.37E
Ponta Grossa **144 D4** 25.07S 50.09W
Pontevedra **116 B3** 42.25N 8.39W
Pontianak **100 B2** 0.05S 109.16E
Poochera **66 E4** 32.43S 134.50E
Poona *see* Pune
Pooncarie **48 B3** 33.25S 142.34E
Poopo, Lake **142 D2**
Popayan **142 C4** 2.27N 76.22W
Popilta, Lake **48 A3**
Popocatepetl, mountain **136 C2** 19.02N 98.38W
Popondetta **89 E2** 8.45S 148.15E
Porbandar **106 B3** 21.40N 69.40E
Porcupine River **130 E7**
Porgera **88 D2** 5.30S 143.10E
Pori **114 I5** 61.28N 21.45E
Porirua **90 F4** 41.08S 174.50E
Porkkala **115 I4** 60.00N 24.25E
Porlamar **142 D5** 11.01N 63.54W
Poronaisk **119 O3** 49.13N 143.05E
Porpoise Bay **92 G3**
Porsgrunn **114 F4** 59.10N 9.40E
Portadown **114 C3** 54.26N 6.27W
Portage la Prairie **134 E4** 49.58N 98.20W
Port Alberni **132 F2** 49.11N 124.49W
Portalegre **116 B2** 39.17N 7.25W
Portales **134 D2** 34.12N 103.20W
Port Alfred **126 C1** 33.36S 26.54E
Port Antonio **137 E2** 18.10N 76.27W
Port Arthur **60 E1** 43.09S 147.51E
Port Augusta **66 F4** 32.30S 137.46E
Port-au-Prince **137 E2** 18.33N 72.20W
Port Blair **107 E2** 11.40N 92.44E
Port Broughton **66 F4** 33.36S 137.56E
Port Campbell **54 B1** 38.37S 143.00E
Port Chalmers **90 D2** 45.49S 170.37E
Port Elizabeth **126 C1** 33.58S 25.40E
Port Fairy **54 B1** 38.23S 142.14E
Port Gentil **126 A4** 0.40S 8.50E
Port Harcourt **124 D2** 4.43N 7.05E
Port Harrison **133 K3** 58.25N 78.15W
Port Hedland **72 C4** 20.18S 118.35E
Port Hope **133 K2** 43.58N 78.18W
Portimao **116 B2** 37.08N 8.32W
Portland, Maine **135 G3** 43.41N 70.13W
Portland, New South Wales **48 F3** 33.21S 149.59E
Portland, Oregon **134 B3** 45.32N 122.40W
Portland, Texas **134 E1** 27.53N 97.21W
Portland, Victoria **54 A1** 38.21S 141.36E
Portland Bay **54 A1**
Portland, Cape **60 E4** 40.45S 147.56E
Port Latta **60 C4** 40.51S 145.25E
Port Lavaca **134 E1** 28.36N 96.39W
Port Lincoln **66 E3** 34.44S 135.52E
Port Loko **124 B2** 8.50N 12.50W
Port Louis **146 D2** 20.10S 57.30E
Port MacDonnell **66 H1** 38.03S 140.42E
Port Macquarie **48 G4** 31.27S 152.54E
Port Menier **133 L2** 49.50N 64.20W
Port Moresby **89 E2** 9.28S 147.10E
Porto Alegre **144 D3** 30.03S 51.10W
Porto Amboim **126 B3** 10.47S 13.43E
Porto de Pedras **143 G3** 9.06S 35.16W
Porto Esperanca **142 D3** 19.36S 57.24W
Port of Spain **137 F2** 10.38N 61.31W
Porto Murtinho **142 E1** 21.42S 57.52W
Porto-Novo **124 D2** 6.30N 2.47E
Porto Torres **116 E3** 40.51N 8.24E
Porto Velho **142 D3** 8.45S 63.54W
Portoviejo **142 B3** 1.07S 80.28W
Port Phillip Bay **54 C2**
Port Pirie **66 F4** 33.12S 138.00E
Port Radium **132 G4** 66.04N 118.00W
Port Said **125 G5** 31.17N 32.18E
Portsmouth, New Hampshire **135 G3** 43.03N 70.47W
Portsmouth, U.K. **114 D3** 50.48N 1.05W
Portsmouth, Virginia **135 G2** 36.50N 76.20W
Port Sudan **125 G3** 19.38N 37.07E
Portugal, republic **116 B2**
Port Victoria **66 F3** 34.30S 137.29E
Posadas **144 D4** 27.25S 55.48W
Poso **100 D2** 1.23S 120.45E
Poso, Lake **100 D2**
Potchefstroom **126 C2** 26.42S 27.06E
Poteau **134 E2** 35.03N 94.36W
Potenza **117 G3** 40.38N 15.48E
Potgietersrus **126 C2** 24.15S 28.55E
Potiskum **124 E3** 11.40N 11.03E
Potomac River **135 G2**
Potosi **142 D2** 19.34S 65.45W
Potrerillos **144 C4** 26.30S 69.25W
Potsdam **114 G3** 52.24N 13.04E
Pouthisat **100 B4** 12.33N 103.55E
Povoa de Varzim **116 B3** 41.22N 8.46W
Povungnituk **133 K3** 60.10N 77.20W
Powder River **134 D3**
Powell, Lake **134 C2**
Poyang Lake **103 F2**
Pozarevac **117 H3** 44.37N 21.12E
Poza Rica de Hidalgo **136 C3** 20.34N 97.26W
Poznan **114 H3** 52.25N 16.53E
Pozoblanco **116 C2** 38.23N 4.51W
Prachuap Khiri Khan **100 A4** 11.50N 99.49E
Prague **114 G3** 50.06N 14.26E
Praia **146 B3** 15.53N 23.29W
Prato **116 F3** 43.53N 11.06E
Prerov **114 H2** 49.28N 17.27E
Prescott **134 C2** 34.30N 112.30W
Presidente Prudente **143 E1** 22.09S 51.24W
Presidio **136 B3** 29.34N 104.23W
Presque Isle **135 H3** 46.42N 68.01W
Preston **134 C3** 42.06N 111.53W
Preston, homestead **42 F2** 27.00S 147.24E
Pretoria **126 C2** 25.45S 28.12E
Preveza **117 H2** 38.58N 20.45E
Price **134 C2** 39.35N 110.49W
Prichard **135 F2** 30.46N 88.06W
Prieska **126 C2** 29.40S 22.45E
Prijedor **117 G4** 45.00N 16.41E
Priluki **115 K3** 50.35N 32.24E
Primavera, antarctic base **92 O3** 64.54S 61.01W
Prime Seal Island **60 E4**
Prince Albert **132 H3** 53.13N 105.45W
Prince Alfred, Cape **132 F5** 74.30N 124.50W
Prince Charles Island **133 K4**
Prince Edward Island **133 L2**
Prince Edward Islands **92 B2**
Prince George **132 F3** 53.55N 122.49W
Prince of Wales Island, Australia **42 D10**
Prince of Wales Island, Canada **132 H5**
Prince Patrick Island **132 F5**
Prince Rupert **132 E3** 54.18N 130.17W
Princess Charlotte Bay **42 D9**
Princess Elizabeth Land **94 E2**
Principe, island **124 D2**
Pripet River **118 C4**
Pristina **117 H3** 42.39N 21.10E
Prizren **117 H3** 42.12N 20.43E
Progreso **136 D3** 21.20N 89.40W
Prokopyevsk **118 I4** 53.55N 86.45E
Prome **107 E2** 18.50N 95.14E
Proserpine **42 G5** 20.24S 148.35E
Proston **42 H2** 26.10S 151.36E
Providence **135 G3** 41.50N 71.28W
Providencia, island **136 D2**
Provo **134 C3** 40.15N 111.40W
Przemysl **115 I2** 49.48N 22.48E
Przewalsk **115 I3** 50.04N 22.30E
Pskov **115 J4** 57.48N 28.26E
Ptuj **117 G4** 46.27N 15.51E
Pucallpa **142 C3** 8.21S 74.33W
Puebla **136 C2** 19.03N 98.10W
Pueblo **134 D2** 38.17N 104.38W
Puerto Ayacucho **142 D4** 5.39N 67.32W
Puerto Aysen **144 B2** 45.27S 72.58W
Puerto Baquerizo **142 B3** 0.51S 89.10W
Puerto Barrios **136 D2** 15.41N 88.32W
Puerto Bermudez **142 C2** 10.14S 74.56W
Puerto Berrio **142 C4** 6.28N 74.28W
Puerto Cabello **137 F2** 10.29N 68.02W
Puerto Cabezas **136 D2** 14.02N 83.24W
Puerto Carreno **142 D4** 6.08N 67.27W
Puerto Deseado **144 C2** 47.45S 65.54W
Puerto Juarez **136 D3** 21.10N 86.50W
Puerto Limon **136 D2** 10.00N 83.01W
Puertollano **116 C2** 38.41N 4.07W
Puerto Madryn **144 C2** 42.45S 65.02W
Puerto Montt **144 B2** 41.28S 73.00W
Puerto Natales **144 B1** 51.41S 72.15W
Puerto Paez **142 D4** 6.14N 67.26W
Puerto Pinasco **142 E1** 22.36S 57.53W
Puerto Plata **137 E2** 19.48N 70.41W
Puerto Princesa **100 C3** 9.46N 118.45E
Puerto Rico, island **137 F2**
Puerto Rico Trench **137 F3**
Puerto Varas **144 B2** 41.20S 73.00W
Puger **100 C2** 8.20S 113.29E
Pukaki, Lake **90 D2**
Pukapuka, island **84 O4**
Pukchong **104 D7** 40.15N 128.17E
Pukekohe **90 F6** 37.12S 174.54E
Puksubaek, Mount **104 C7** 40.40N 127.15E
Pula **116 F3** 44.52N 13.52E
Pulalak **106 B4** 30.15N 62.54E
Pulap, island **101 F3**
Pular, mountain **144 C4** 24.12S 68.05W
Pullman **134 C3** 46.44N 117.10W
Pulusuk, island **101 F3**
Puluwat, island **101 F3**
Pune **106 C2** 18.34N 73.58E
Puno **142 C2** 15.53N 70.03W
Punta Arenas **144 B1** 53.10S 70.56W
Puntarenas **136 D2** 10.00N 84.50W
Punto Fijo **142 C5** 11.42N 70.13W
Pur River **118 H5**
Puri **106 D2** 19.49N 85.54E
Purnea **106 D3** 25.47N 87.28E
Purus River **142 D3**
Pusan **104 D4** 35.05N 129.02E
Pushkin **115 K4** 59.43N 30.22E
Putaruru **90 F6** 38.03S 175.47E
Puting, Cape **100 C2** 3.35S 111.52E
Putorana Mountains **118 J5**
Putumayo River **142 C3**
Puysegur Point **90 B1** 46.10S 166.37E
Pweto **126 C4** 8.28S 28.52E
Pyatigorsk **118 E3** 44.04N 43.06E
Pyongyang **104 B6** 39.00N 125.47E
Pyramid Lake **134 C3**
Pyramid Hill **54 C2** 36.03S 144.07E
Pyrenees, mountains **116 C3**
Pyrgos **117 H2** 37.40N 21.27E

Q

Qala Nau **106 B4** 34.58N 63.04E
Qatar, sheikdom **108 D3**
Qatif **108 C3** 26.30N 50.00E
Qattara Depression **125 F5**
Qazvin **108 D4** 36.16N 50.00E
Qena **125 G4** 26.08N 32.42E
Qeshm **108 D3** 26.58N 56.16E
Qeshm, island **108 D3**
Qeys, island **108 D3**
Qila Ladgasht **106 B3** 27.55N 62.59E
Qilian Mountains **102 D3**
Qingdao **103 G3** 36.04N 120.22E
Qinghai Lake **102 D3**
Qingtongxia **102 E3** 38.03N 106.03E
Qinhuangdao **103 F3** 39.55N 119.37E
Qiqihar **103 G4** 47.23N 124.00E
Qishn **108 D2** 15.25N 51.40E
Qitaihe **103 H4** 45.47N 130.50E
Qom **108 D4** 34.39N 50.57E
Quambatook **54 B3** 35.51S 143.31E
Quambone **48 D4** 30.56S 147.52E
Quang Ngai **100 B4** 15.09N 108.50E
Quang Tri **100 B4** 16.46N 107.11E
Quan Long **100 B3** 9.27N 105.25E
Quanzhou **103 F2** 24.53N 118.36E
Qu'Appelle River **134 D4**
Quchan **108 D4** 37.04N 58.32E
Queanbeyan **48 E2** 35.21S 149.14E
Quebec **133 K2** 46.50N 71.15W
Quebec, province **133 K3**
Queen Charlotte Islands **132 E3**
Queen Charlotte Sound **132 E3**
Queen Elizabeth Islands **132 F5**
Queen Mary Land **94 F2**
Queen Maud Land **94 B2**
Queenscliff **54 C1** 38.16S 144.40E
Queensland, state **42 D5**
Queenstown, Australia **60 C2** 42.05S 145.33E
Queenstown, New Zealand **90 C2** 45.02S 168.40E
Queenstown, South Africa **126 C1** 31.54S 26.53E
Quelimane **126 D3** 17.53S 36.51E
Queretaro **136 B3** 20.38N 100.23W
Quetta **106 B4** 30.15N 67.00E
Quevedo **142 C3** 0.59S 79.27W
Quezaltenango **136 C2** 14.50N 91.30W
Quezon **100 C3** 9.14N 118.02E
Quibdo **142 C4** 5.40N 76.40W
Quillota **144 B3** 32.54S 71.16W
Quilon **106 C1** 8.53N 76.38E
Quilpie **42 E2** 26.37S 144.16E
Quimper **116 C4** 48.00N 4.06W
Quincy **134 E2** 39.55N 91.22W
Qui Nhon **100 B4** 13.47N 109.11E
Quipungo **126 B3** 14.50S 14.32E
Quirindi **48 F4** 31.30S 150.41E
Quitapa **126 B3** 10.10S 18.16E
Quito **142 C3** 0.14S 78.30W
Qujing **102 E2** 25.29N 103.47E
Quoin Mountain **60 E2** 42.33S 147.17E
Quorn **66 G4** 32.21S 138.02E
Qus **108 B3** 25.53N 32.48E
Quseir **125 G4** 26.04N 34.15E

R

Raahe **115 I5** 64.42N 24.30E
Raba **100 C2** 8.27S 118.45E
Rabaraba **89 E2** 9.58S 149.50E
Rabat **124 C5** 34.02N 6.51W
Rabaul **89 F3** 4.13S 152.11E
Rabigh **108 B3** 22.48N 39.01E

*Entry / **Page / Ref.** / Lat. / Long.*

Race, Cape **133 M2** 46.38N 53.10W
Rach Gia **100 B3** 10.02N 105.05E
Raciborz **117 G5** 50.05N 18.10E
Racine **133 J2** 42.42N 87.50W
Radhwa, Mount **108 B3** 24.36N 38.18E
Radom **115 I3** 51.26N 21.10E
Radomyshl **115 J3** 50.30N 29.13E
Radstock, Cape **66 E4** 33.11S 134.20E
Raevavae, island **85 Q3**
Rafaela **144 C3** 31.16S 61.44W
Raga **125 F2** 8.26N 25.46E
Ragusa **117 F2** 36.56N 14.44E
Raichur **102 B1** 16.15N 77.20E
Raigarh **106 D3** 21.53N 83.28E
Railton **60 D3** 41.20S 146.24E
Rainbow **54 B3** 35.54S 142.00E
Rainier, Mount **134 B3** 46.52N 121.45W
Rainy Lake **134 E3**
Raipur **106 D3** 21.16N 81.42E
Rairakhol **106 D3** 21.03N 84.23E
Rajahmundry **106 D2** 17.01N 81.52E
Rajkot **106 C3** 22.18N 70.53E
Rajshahi **102 C2** 24.24N 88.40E
Rakahanga, island **84 O5**
Rakka **108 B4** 35.57N 39.03E
Rakvere **115 J4** 59.22N 26.28E
Raleigh **135 G2** 35.46N 78.39W
Raleigh Bay **135 G2**
Ramadi **108 C4** 33.27N 43.19E
Rambutyo Island **89 E3**
Rampur **102 B2** 28.48N 79.03E
Ramu River **89 D2**
Rancagua **144 B3** 34.10S 70.45W
Ranchi **106 D3** 23.22N 85.20E
Rand **48 D2** 35.36S 146.35E
Randers **114 G4** 56.28N 10.03E
Ranfurly **90 D2** 45.08S 170.06E
Rangiora **90 E3** 43.18S 172.36E
Rangiroa, island **85 P4**
Rangitaiki River **90 G5**
Rangitata River **90 D3**
Rangpur **107 D3** 25.45N 89.21E
Rankin Inlet **132 I4** 62.52N 92.00W
Rankins Springs **48 D3** 33.51S 146.16E
Rann of Kutch, lake **106 B3**
Ranongga, island **89 G2**
Ransiki **88 B3** 1.27S 134.12E
Rantekombola, mountain **100 D2** 3.23S 120.02E
Rantepao **100 C2** 2.58S 119.58E
Rapa, island **85 Q3**
Rapallo **116 E3** 44.21N 9.13E
Rapid River **60 C3**
Rapid City **134 D3** 44.06N 103.14W
Rarotonga, island **84 P3**
Ras al-Khaimah **108 D3** 25.47N 55.57E
Ras Dashan, mountain **125 G3** 13.15N 38.27E
Ras Gharib **108 B3** 28.21N 33.06E
Rasht **108 C4** 37.18N 49.38E
Rason Lake **72 D3**
Rasskazovo **115 M3** 53.39N 41.54E
Ratlam **102 B2** 23.18N 75.06E
Ratnagiri **106 C2** 17.00N 73.20E
Raton **134 D2** 36.54N 104.27W
Raukumara Range **90 G5**
Rauma **114 I5** 61.09N 21.30E
Ravar **108 D4** 31.15N 56.51E
Ravenna **116 F3** 44.25N 12.12E
Ravensburg **114 F2** 47.47N 9.37E
Ravenshoe **42 E7** 17.36S 145.29E
Ravensthorpe **72 D2** 33.35S 120.02E
Ravi River **106 C4**
Rawalpindi **106 C4** 33.40N 73.08E
Rawlinna **72 E2** 31.01S 125.20E
Rawlins **134 D3** 41.46N 107.16W
Rawson **144 C2** 43.15S 65.06W
Raya, mountain **100 C2** 0.31S 112.37E
Rayong **100 B4** 12.38N 101.17E
Reading **135 G3** 40.20N 75.55W
Rebiana Sand Sea, desert **124 F4**
Rebun, island **104 J9**
Recherche, Archipelago of the **72 D2**
Rechitsa **115 K3** 52.21N 30.24E
Recife **143 G3** 8.06S 34.53W

*Entry / **Page / Ref.** / Lat. / Long.*

Red Basin, depression **102 E3**
Red Lake **134 E3**
Red River, China **102 E2**
Red River, Minnesota **134 E3**
Red River, Texas **134 D2**
Red Sea **125 G4**
Red Bluff **134 B3** 40.10N 122.14W
Redcliffe **42 I2** 27.14S 153.07E
Red Cliffs **54 B3** 34.18S 142.11E
Red Deer **132 G3** 52.15N 113.48W
Redding **134 B3** 40.35N 122.24W
Redford, homestead **42 F3** 25.50S 147.25E
Redmond **132 F2** 44.16N 121.10W
Reefton **90 D3** 42.07S 171.52E
Reeve, Lake **54 D1**
Regensburg **114 G2** 49.01N 12.07E
Reggane **124 D4** 26.42N 0.13E
Reggio di Calabria **117 G2** 38.06N 15.39E
Regina, Canada **132 H3** 50.30N 104.38W
Regina, French Guiana **143 E4** 4.20N 52.07W
Rehoboth **126 B2** 23.18S 17.03E
Rei **89 F3** 4.20S 153.06E
Reims **116 D4** 49.15N 4.02E
Reindeer Lake **132 H3**
Reinga, Cape **90 E7** 34.25S 172.41E
Reinosa **116 C3** 43.01N 4.09W
Relizane **116 D2** 35.44N 0.35E
Remanso **143 F3** 9.40S 42.04W
Remarkable, Mount **66 G4** 32.48S 138.10E
Rendova, island **89 G2**
Rendsburg **114 F3** 54.19N 9.39E
Renmark **66 H3** 34.10S 140.45E
Rennell, island **84 K4**
Renner Springs, homestead **80 D5** 18.19S 133.48E
Rennes **116 C4** 48.06N 1.40W
Reno **134 C2** 39.32N 119.49W
Republican River **134 D2**
Requena **116 C2** 39.29N 1.08W
Resistencia **144 D4** 27.28S 59.00W
Resita **117 H4** 45.16N 21.55E
Resolute Bay **132 I5** 74.40N 95.00W
Resolution Island, Canada **133 L4**
Resolution Island, New Zealand **90 B2**
Rethymnon **117 H2** 35.23N 24.28E
Reunion, island **146 D2**
Reus **116 D3** 41.10N 1.06E
Reutlingen **114 F2** 48.30N 9.13E
Revelstoke **132 G3** 51.02N 118.12W
Revilla Gigedo, islands **85 T7**
Rewa **106 D3** 24.32N 81.18E
Rexburg **134 C3** 43.50N 111.48W
Rey Island **136 E1**
Reykjavik **110 B4** 64.09N 21.58W
Reynosa **136 C3** 26.05N 98.18W
Rezekne **115 J4** 56.30N 27.22E
Rheine **116 E5** 52.17N 7.26E
Rhine River **114 F3**
Rhinelander **135 F3** 45.39N 89.23W
Rhode Island, state **135 H2**
Rhodes **117 I2** 36.26N 28.14E
Rhodes, island **117 I2**
Rhone River **116 D3**
Riau, province **100 B3**
Ribat Qila **106 B3** 29.48N 61.00E
Ribe **114 F4** 55.20N 8.47E
Ribeirao Preto **143 F1** 21.09S 47.48W
Riberalta **142 D2** 10.59S 66.06W
Richfield **134 C2** 38.45N 112.05W
Richland **134 C3** 46.17N 119.19W
Richmond, New Zealand **90 E4** 41.20S 173.10E
Richmond, Queensland **42 D5** 20.44S 143.08E
Richmond, Tasmania **60 E2** 42.44S 147.26E
Richmond, U.S.A. **135 G2** 37.43N 77.27W
Richmond Range **48 G5**
Ridgecrest **134 C2** 35.38N 117.36W
Rietfontein **126 C2** 26.43S 20.03E
Rieti **116 F3** 42.24N 12.51E
Riga **115 I4** 56.53N 24.08E
Rigan **108 D3** 28.40N 58.58E
Riihimaki **115 I5** 60.45N 24.45E
Rijeka **117 F4** 45.20N 14.27E

*Entry / **Page / Ref.** / Lat. / Long.*

Rimini **116 F3** 44.03N 12.34E
Rimnicu Sarat **117 I4** 45.24N 27.06E
Rimnicu Vilcea **117 H4** 45.06N 24.21E
Rimouski **133 L2** 48.26N 68.32W
Ringarooma Bay **60 E4**
Ringarooma River **60 E3**
Ringgi **89 G2** 8.08S 157.08E
Ringkobing **114 F4** 56.06N 8.15E
Ringsaker **114 G5** 60.54N 10.45E
Ringwood, homestead **80 E3** 23.50S 134.58E
Riobamba **142 C3** 1.44S 78.40W
Rio Branco **142 D3** 9.59S 67.49W
Rio Cuarto **144 C3** 33.08S 64.20W
Rio de Janeiro **144 E4** 22.53S 43.17W
Rio Grande **144 D3** 32.03S 52.08W
Rio Grande, river **134 D2**
Riohacha **142 C5** 11.34N 72.58W
Rio Pardo **144 D3** 30.00S 52.20W
Rio Verde **143 E2** 17.50S 50.55W
Ripoll **116 D3** 42.12N 2.12E
Rishiri, island **104 J9**
Riva **116 F4** 45.53N 10.50E
Rivera **144 D3** 30.55S 55.33W
Riverton **90 C1** 46.21S 168.01E
Riviere-du-Loup **135 H3** 47.49N 69.32W
Rivoli Bay **66 H2**
Riyadh **108 C3** 24.39N 46.46E
Rize **108 C5** 41.03N 40.31E
Road Town **137 F2** 18.28N 64.39W
Roanne **116 D4** 46.02N 4.05E
Roanoke **135 G2** 37.15N 79.58W
Roatan, island **136 D2**
Robat **108 D4** 30.05N 54.50E
Robbins Island **60 B4**
Robe **66 G2** 37.10S 139.45E
Robe, Mount **48 A4** 31.40S 141.20E
Robertstown **66 G3** 33.59S 139.03E
Robinson Range **72 C3**
Robinson River, homestead **80 F6** 16.46S 136.59E
Robinvale **54 B3** 34.35S 142.46E
Rob Roy, island **89 G2**
Robson, Mount **132 G3** 53.08N 119.18W
Roca, Cape **116 B2** 38.42N 9.30W
Rochefort **116 C4** 45.57N 0.58W
Rochester, Australia **54 C2** 36.22S 144.42E
Rochester, Minnesota **134 E3** 44.01N 92.27W
Rochester, New York **135 G3** 43.12N 77.37W
Rockford **135 F3** 42.16N 89.06W
Rockhampton **42 H4** 23.22S 150.32E
Rockhampton Downs, homestead **80 E5** 18.57S 135.12E
Rock Hill **135 F2** 34.55N 81.01W
Rockingham **72 C2** 32.17S 115.43E
Rocklands Reservoir **54 B2**
Rock Springs **134 D3** 41.35N 109.13W
Rockstone **142 E4** 6.00N 58.30W
Rocky Mountains **130 G6**
Rodez **116 D3** 44.21N 2.34E
Roebuck Bay **72 D5**
Rogachev **115 K3** 53.04N 30.00E
Roger River **60 C3** 41.01S 145.04E
Rokugo, Cape **104 H5** 37.30N 137.21E
Rolla **134 E2** 37.56N 91.55W
Roma **42 G2** 26.35S 148.47E
Roman **117 I4** 46.56N 26.56E
Romang, island **100 D2**
Rome, Italy **116 F3** 41.53N 12.30E
Rome, U.S.A. **135 F2** 34.01N 85.02W
Romilly **116 D4** 48.31N 3.44E
Romny **115 K3** 50.45N 33.30E
Romorantin **116 D4** 47.22N 1.44E
Roncador Reef **89 G2**
Ronda **116 B2** 36.45N 5.10W
Rongerik, island **84 L7**
Ronne **114 G4** 55.07N 14.43E
Ronne Entrance **94 O2**
Ronne Ice Shelf **92 O3**
Roon, island **88 B3**
Roosevelt Island **92 J3**
Roper River **80 D7**

*Entry / **Page / Ref.** / Lat. / Long.*

Roper Valley, homestead **80 E7** 14.56S 134.00E
Roraima, mountain **142 D4** 5.14N 60.44W
Roros **118 B5** 62.35N 11.23E
Rosario **144 C3** 33.00S 60.40W
Rosario de la Frontera **144 C4** 25.45S 65.00W
Roseau **137 F2** 15.18N 61.23W
Rosebery **60 C3** 41.46S 145.34E
Roseburg **134 B3** 43.13N 123.21W
Roslavl **115 K3** 53.55N 32.53E
Ross **60 E2** 42.01S 147.31E
Ross Ice Shelf **92 I4**
Ross Island **92 I3**
Ross Sea **92 J3**
Rossano **117 G2** 39.35N 16.38E
Rossarden **60 E3** 41.40S 147.46E
Rossel Island **89 F1**
Rossosh **115 L3** 50.12N 39.35E
Rostock **114 G3** 54.06N 12.09E
Rostov, north-western Russia **115 L4** 57.11N 39.23E
Rostov, south-western Russia **118 D3** 47.15N 39.45E
Roswell **134 D2** 33.24N 104.33W
Rota, island **101 F4**
Rothera, antarctic base **92 O3** 68.07S 68.16W
Rothschild Island **92 O3**
Roti, island **100 D1**
Roto **48 C3** 33.03S 145.28E
Rotoroa, Lake **90 E4**
Rotorua **90 G5** 38.09S 176.15E
Rotorua, Lake **90 G5**
Rotterdam **114 E3** 51.55N 4.29E
Rottnest Island **72 C2**
Rotuma, island **84 M4**
Rouen **116 D4** 49.26N 1.05E
Round Mountain **48 G4** 30.26S 152.14E
Round Rock **134 E2** 30.32N 97.42W
Rourkela **106 D3** 22.16N 85.01E
Rouyn **133 K2** 48.15N 79.00W
Rovaniemi **110 G4** 66.29N 25.40E
Rovereto **116 F4** 45.53N 11.03E
Rovieng **100 B4** 13.20N 105.06E
Rovigo **116 F4** 45.04N 11.47E
Rovno **115 J3** 50.39N 26.10E
Rowena **48 E5** 29.49S 148.54E
Roxas, Palawan **100 C4** 10.19N 119.21E
Roxas, Panay **100 D4** 11.36N 122.45E
Roxborough Downs, homestead **42 B4** 22.31S 138.50E
Roxby Downs **66 F5** 30.30S 136.55E
Roxby Downs, homestead **66 F5** 30.42S 136.46E
Royan **116 C4** 45.38N 1.02W
Ruahine Range **90 F4**
Ruapehu, mountain **90 F5** 39.17S 175.34E
Rubi River **126 C5**
Rubtsovsk **118 I4** 51.34N 81.11E
Ruchi **118 E5** 66.03N 41.20E
Rudbar **106 B4** 30.10N 62.39E
Rufiji River **122 G4**
Rukwa, Lake **126 D4**
Ruma **117 G4** 45.01N 19.50E
Rumania, socialist republic **117 H4**
Rumberpon, island **88 B3**
Rum Cay, island **137 E3**
Rumoi **104 J8** 43.57N 141.40E
Runaway, Cape **90 H6** 37.32S 177.59E
Rungwa **126 D4** 6.58S 33.31E
Rupert River **133 K3**
Rurutu, island **85 P3**
Ruse **117 I3** 43.50N 25.59E
Rushworth **54 C2** 36.35S 145.01E
Russas **143 G3** 4.56S 38.02W
Russell Islands **89 G2**
Russia, republic **118 H5**
Russkaya, antarctic base **92 L3** 74.46S 136.51W
Rustenburg **126 C2** 25.40S 27.15E
Rutba **108 C4** 33.02N 40.07E
Ruteng **100 D2** 8.35S 120.28E
Rutherglen **54 D2** 36.03S 146.28E
Ruthieres, Cape **72 E6** 13.55S 127.20E

*Entry / **Page** / **Ref.** / Lat. / Long.*

Ruvuma River **122 G3**
Ruzomberok **115 H2** 49.04N 19.15E
Rwanda, republic **126 C4**
Ryazan **115 L3** 54.37N 39.43E
Rybinsk **115 L4** 58.01N 38.52E
Rybnitsa **115 J2** 47.42N 29.00E
Rylstone **48 E3** 32.48S 149.59E
Ryukyu Islands **84 I8**
Rzeszow **115 I3** 50.04N 22.00E
Rzhev **115 K4** 56.15N 34.18E

S

Saabad **108 D4** 34.29N 57.20E
Saarbrucken **114 F2** 49.15N 6.58E
Saaremaa, island **115 I4**
Sabac **117 G3** 44.45N 19.41E
Sabadell **116 D3** 41.33N 2.07E
Sabah, state **100 C3**
Sabalan, Mount **108 C4** 38.16N 47.47E
Sabang **100 A3** 5.53N 95.17E
Sabinas Hidalgo **136 B3** 26.33N 100.10W
Sable, Cape, Canada **133 L2** 43.27N 65.33W
Sable, Cape, U.S.A. **135 F1** 25.12N 81.10W
Sabra, Cape **88 B3** 2.17S 132.18E
Sabuda, island **88 B3**
Sabzevar **108 D4** 36.13N 57.42E
Sachs Harbour **132 F5** 72.00N 124.30W
Sacramento **134 B2** 38.33N 121.30W
Sacramento River **134 B2**
Sadiya **107 E3** 27.49N 95.38E
Sado, island **104 I6**
Safi **124 C5** 32.20N 9.17W
Safonovo **115 K4** 55.08N 33.16E
Saga **104 E3** 33.16N 130.18E
Sagami Bay **104 I4**
Saginaw **135 F3** 43.25N 83.54W
Saginaw Bay **135 F3**
Sagua la Grande **136 D3** 22.48N 80.06W
Sagunto **116 C2** 39.40N 0.17W
Sahara, desert **124 D4**
Saharan Atlas, mountains **124 C5**
Saharanpur **102 B2** 29.58N 77.33E
Sahiwal **106 C4** 30.14N 73.10E
Saibai Island **88 D2**
Saida **116 D1** 34.50N 0.10E
Saidor **89 E2** 5.38S 146.28E
Saidpur **102 C2** 25.48N 89.00E
Saigo **104 F5** 36.12N 133.19E
Sailolof **88 B3** 1.15S 130.47E
Sainshanda **102 F4** 44.58N 110.10E
St Albans **133 K2** 44.49N 73.07W
St Arnaud **54 B2** 36.37S 143.16E
St Augustine **135 F1** 29.54N 81.19W
St Brieuc **116 C4** 48.31N 2.45W
St Catharines **133 K2** 43.10N 79.15W
St Christopher, island **137 F2**
St Clair, Lake **60 C3**
St Cloud **134 E3** 45.34N 94.10W
St Croix, island **137 F2**
St Dizier **116 E4** 48.38N 4.58E
Sainte-Marie, Cape **126 E2** 25.34S 45.10E
Saintes **116 C4** 45.44N 0.38W
St Etienne **116 D4** 45.26N 4.23E
St George, Australia **42 G1** 28.03S 148.35E
St George, Canada **135 H3** 45.08N 66.50W
St George, U.S.A. **134 C2** 37.05N 113.40W
St George, Cape **89 F3** 4.51S 152.53E
St George's, Canada **133 M2** 48.26N 58.31W
St George's, Grenada **137 F2** 12.04N 61.44W
St Georges Channel **89 F3**
St Georges Head **48 F2** 35.11S 150.40E
St Gregory, Mount **133 M2** 49.19N 58.13W
St Helena Bay **126 B1**
St Helena, island **122 C3**
St Helens, Australia **60 F3** 41.20S 148.15E
St Helens, U.S.A. **134 B3** 45.54N 122.50W
St Helens, Mount **134 B3** 46.12N 122.11W
St Ives **114 C3** 50.12N 5.29W
St James, Cape **132 E3** 51.58N 131.00W
St John **133 L2** 45.16N 66.03W
St Johns **137 F2** 17.07N 61.51W
St John's **133 M2** 47.34N 52.41W
St Joseph **134 E2** 39.45N 94.51W
St Kitts-Nevis, independent state **137 F2**
St Lawrence **38 E3** 22.21S 149.32E
St Lawrence, Gulf of **133 L2**
St Lawrence Island **132 A4**
St Lawrence River **133 K2**
St Lo **116 C4** 49.07N 1.05W
St Louis, Senegal **124 B3** 16.01N 16.30W
St Louis, U.S.A. **135 E2** 38.40N 90.15W
St Lucia, independent state **137 F2**
St Martin, island **137 F2**
St Mary Peak **66 G5** 31.30S 138.33E
St Marys **60 F3** 41.35S 148.11E
St Mathieu, Point **116 C4** 48.20N 4.46W
St Matthias Group **89 E3**
St Nazaire **116 C4** 47.17N 2.12W
St Paul **134 E3** 45.00N 93.10W
St Paul River **124 B2**
St Petersburg, Russia **115 K4** 59.55N 30.25E
St Petersburg, U.S.A. **135 F1** 27.45N 82.40W
St Pierre, island **133 M2**
St Quentin **116 D4** 49.51N 3.17E
St Stephen **135 H3** 45.12N 67.18W
St Veit **114 G2** 46.47N 14.22E
St Vincent, Cape **116 B2** 37.01N 8.59W
St Vincent, Gulf **66 G3**
St Vincent, independent state **137 F2**
Saipan **101 F4** 15.13 145.45E
Sajama, mountain **142 D2** 18.09S 68.52W
Sakai **104 G4** 34.35N 135.28E
Sakaka **108 C3** 29.59N 40.06E
Sakakawea, Lake **134 D3**
Sakar Island **89 E2**
Sakaraha **126 E2** 22.54S 44.31E
Sakata **104 I6** 38.55N 139.51E
Sakhalin, island **119 O4**
Sakon Nakhon **100 B4** 17.10N 104.08E
Sakrivier **126 C1** 30.50S 20.26E
Sala **114 H4** 59.55N 16.38E
Salado River, central Argentina **144 C3**
Salado River, northern Argentina **144 C4**
Salalah **108 D2** 17.00N 54.04E
Salamanca **116 B3** 40.58N 5.40W
Salamat River **124 E2**
Salavat **118 F4** 53.22N 55.50E
Salawati, island **88 B3**
Salayar, island **100 D2**
Sala y Gomez, island **85 U3**
Sale **54 D1** 38.07S 147.04E
Salekhard **118 G5** 66.33N 66.35E
Salem, India **106 C2** 11.38N 78.08E
Salem, Oregon **134 B3** 44.57N 123 01W
Salem, Virginia **135 F2** 37.17N 80.04W
Salgotarjan **115 H2** 48.05N 19.47E
Salihli **117 I2** 38.29N 28.08E
Salimbatu **100 C3** 2.59N 117.23E
Salina **134 E2** 38.53N 97.36W
Salina Cruz **136 C2** 16.11N 95.12W
Salinas **134 B2** 36.39N 121.40W
Salinas Bay **136 D2**
Salinas, Point **126 B3** 12.50S 12.54E
Salinas Grandes, lake **144 C3**
Salisbury **114 D3** 51.05N 1.48W
Salkhad **108 B4** 32.29N 36.42E
Salmon River **134 C3**
Salo **115 I5** 60.23N 23.10E
Salonika **117 H3** 40.38N 22.58E
Salonta **117 H4** 46.49N 21.40E
Salta **144 C4** 24.46S 65.28W
Salt Creek **66 G2** 36.08S 139.39E
Saltillo **136 B3** 25.30N 101.00W
Salt Lake City **134 C3** 40.45N 111.55W
Salto **144 D3** 31.27S 57.50W
Salum, Gulf of **125 F5**
Salvador **143 G2** 12.58S 38.29W
Salween River **107 E3**
Salzburg **114 G2** 47.54N 13.03E
Salzgitter **114 G3** 52.13N 10.20E
Samar, island **100 D4**
Samara **118 F4** 53.10N 50.10E
Samarai **89 F1** 10.37S 150.40E
Samarinda **100 C2** 0.30S 117.09E
Samarkand **118 G2** 39.40N 66.57E
Samarra **108 C4** 34.13N 43.52E
Samate **88 B3** 0 59S 131.03E
Samawa **108 C4** 31.18N 45.18E
Sambalpur **106 D3** 21.28N 84.04E
Samboja **100 C2** 1.03S 117.04E
Sambor **115 I2** 49.31N 23.10E
Samchok **104 D5** 37.30N 129.10E
Samoa, independent state **84 N4**
Samsun **108 B5** 41.17N 36.22E
Samuel, Mount **80 E5** 19.41S 134.09E
Samui, island **100 B3**
Samut Prakan **100 B4** 13.32N 100.35E
Samut Songkhram **100 A4** 13.25N 100.01E
San **124 C3** 13.21N 4.57W
San'a **108 C2** 15.23N 44.14E
Sanae, antarctic base **92 R3** 70.18S 2.24W
Sanaga River **124 E2**
San Ambrosio, island **85 W3**
Sanana **100 D2** 2.03S 125.59E
Sanana, island **100 D2**
Sanandaj **108 C4** 35.18N 47.01E
San Andres, island **136 D2**
San Andres Tuxtla **136 C2** 18.28N 95.15W
San Angelo **134 D2** 31.28N 100.28W
San Antonio, Chile **144 B3** 33.35S 71.39W
San Antonio, U.S.A. **134 E1** 29.25N 98.30W
San Antonio, Cape **136 D3** 21.50N 84.57W
San Antonio de los Cobres **144 C4** 24.10S 66.17W
San Bernardino **134 C2** 34.07N 117.18W
San Blas, Cape **136 D3** 29.40N 85.23W
San Carlos, Chile **144 B3** 36.25S 71.58W
San Carlos, Philippines **100 D4** 15.56N 120.21E
San Cristobal **142 C4** 7.46N 72.15W
San Cristobal, island, Ecuador **142 B3**
San Cristobal, island, Solomon Islands **84 L4**
Sandakan **100 C3** 5.52N 118.04E
Sandefjord **114 G4** 59.12N 10.10E
Sanderson **134 D2** 30.08N 102.25W
Sandia **142 D2** 14.18S 69.25W
San Diego **134 C2** 32.45N 117.10W
San Diego, Cape **144 C1** 54.43S 65.07W
Sandnes **114 F4** 58.51N 5.45E
Sandomierz **115 I3** 50.40N 21.45E
Sandover River **80 E3**
Sandstone **72 C3** 27.59S 119.18E
Sandviken **114 H5** 60.38N 16.50E
Sandy Cape, Queensland **42 I3** 24.42S 153.16E
Sandy Cape, Tasmania **60 B3** 41.25S 144.45E
San Felipe, Chile **144 B3** 32.45S 70.30W
San Felipe, Mexico **134 C2** 31.03N 114.52W
San Felix, island **85 W3**
San Fernando, Chile **144 B3** 34.40S 71.00W
San Fernando, Philippines **100 D4** 16.39N 120.19E
San Fernando, Trinidad and Tobago **137 F2** 10.16N 61.28W
San Fernando de Apure **142 D4** 7.53N 67.15W
San Fernando de Atabapo **142 D4** 4.03N 67.45W
Sanford **135 G3** 43.25N 70.46W
Sanford, Mount **132 D4** 62.14N 144.08W
San Francisco, Argentina **144 C3** 31.20S 62.06W
San Francisco, U.S.A. **134 B2** 37.45N 122.27W
Sangan, Mount **106 B4** 33.33N 64.56E
Sangar **119 M5** 64.02N 127.30E
Sangha River **126 B5**
Sangihe Islands **100 D3**
San Giovanni **117 G2** 39.15N 16.42E
Sangju **104 D5** 36.25N 128.08E
Sangli **106 C2** 16.55N 74.37E
San Ignacio, Mexico **134 C1** 27.22N 113.00W
San Ignacio, Paraguay **144 D4** 26.51S 57.00W
San Javier **142 D2** 16.22S 62.38W
Sanjo **104 I5** 37.38N 138.59E
San Jorge, Gulf of **144 C2**
San Jorge, island **89 G2**
San Jose, Costa Rica **136 D1** 9.59N 84.04W
San Jose, U.S.A. **134 B2** 37.20N 121.55W
San Jose Island **134 C1**
San Jose de Amacuro **142 D4** 8.30N 60.30W
San Jose del Cabo **134 D1** 23.01N 109.40W
San Juan, Argentina **144 C3** 31.33S 68.31W
San Juan, U.S.A. **137 F2** 18.29N 66.08W
San Juan River **136 D2**
San Juan Bautista **144 D4** 26.37S 57.06W
San Juan del Norte **136 D2** 10.55N 83.40W
San Juan de los Morros **142 D4** 9.53N 67.23W
San Juan del Sur **136 D2** 11.16N 85.51W
San Julian **144 C2** 49.17S 67.45W
San Lorenzo **144 C3** 32.45S 60.45W
San Lorenzo, mountain **144 B2** 47.40S 72.18W
Sanlucar de Barrameda **116 B2** 36.46N 6.21W
San Lucas, Cape **134 C1** 22.50N 109.55W
San Luis, Argentina **144 C3** 33.20S 66.23W
San Luis, Mexico **134 C2** 32.26N 114.48W
San Luis Obispo **134 B2** 35.18N 120.40W
San Luis Potosi **136 B3** 22.10N 101.00W
San Marino, republic **116 F3**
San Martin **144 C3** 33.05S 68.28W
San Martin, antarctic base **92 O3** 68.07S 67.08W
San Matias, Gulf of **144 C2**
San Miguel **136 D2** 13.28N 88.10W
San Miguel de Tucuman **144 C4** 26.47S 65.15W
San Miguelito **136 E1** 9.02N 79.30W
Sanming **103 F2** 26.16N 117.35E
Sannicandro Garganico **117 G3** 41.50N 15.34E
San Nicolas **144 C3** 33.25S 60.15W
Sanok **115 I2** 49.32N 22.12E
San Pablo **100 D4** 14.03N 121.20E
San Pedro, Argentina **144 D3** 33.43S 59.45W
San Pedro, Paraguay **144 D4** 24.07S 56.59W
San Pedro de Macoris **137 F2** 18.30N 69.30W
San Pedro Sula **136 D2** 15.26N 88.01W
San Rafael, Argentina **144 C3** 34.35S 68.24W
San Rafael, U.S.A. **134 B2** 37.58N 122.30W
San Ramon **142 D2** 13.19S 64.41W
San Remo **116 E3** 43.48N 7.47E
San Salvador **136 D2** 13.40N 89.10W
San Salvador, island, Bahamas **137 E3**
San Salvador, island, Ecuador **142 A3**
San Sebastian **116 C3** 43.19N 1.59W
Santa **142 C3** 8.57S 78.37W
Santa Ana, Ecuador **142 B3** 1.16S 80.24W
Santa Ana, El Salvador **136 D2** 14.00N 89.31W
Santa Ana, U.S.A. **134 C2** 33.44N 117.54W
Santa Ana, mountain **136 D2** 13.51N 89.37W
Santa Barbara, Mexico **136 B3** 26.48N 105.50W
Santa Barbara, U.S.A. **134 C2** 34.26N 119.43W
Santa Catalina Island **85 T9**
Santa Clara **136 E3** 22.25N 79.58W
Santa Cruz, Argentina **144 C1** 50.01S 68.31W
Santa Cruz, Bolivia **142 D2** 17.45S 63.14W
Santa Cruz, Spain **124 B4** 28.28N 16.15W
Santa Cruz, U.S.A. **134 B2** 36.58N 122.03W
Santa Cruz, island **142 A3**
Santa Cruz Islands **84 L5**
Santa Cruz de la Palma **124 B4** 28.41N 17.46W
Santa Fe, Argentina **144 C3** 31.38S 60.43W
Santa Fe, U.S.A. **134 D2** 35.41N 105.57W
Santa Isabel, island **89 G2**
Santa Maria, Brazil **144 D4** 29.40S 53.47W

Entry/***Page***/***Ref.***/*Lat.*/*Long.*

Santa Maria, Malawi **126 C3** 11.09S 29.59E
Santa Maria, U.S.A. **134 B2** 34.57N 120.28W
Santa Marta **142 C5** 11.18N 74.10W
Santa Marta, Cape **126 B3** 13.52S 12.24E
Santana do Livramento **144 D3** 30.52S 55.30W
Santander **116 C3** 43.28N 3.48W
Sant' Antioco **116 E2** 39.04N 8.27E
Santarem, Brazil **143 E3** 2.26S 54.41W
Santarem, Portugal **116 B2** 39.14N 8.40W
Santa Rosa, Argentina **144 C3** 36.37S 64.17W
Santa Rosa, Ecuador **142 C3** 3.29S 79.59W
Santa Rosa, U.S.A. **134 B2** 38.27N 122.42W
Santa Rosalia **134 C1** 27.20N 112.20W
Santa Teresa **80 E2** 24.08S 134.22E
Santiago, Brazil **144 D4** 29.11S 54.52W
Santiago, Chile **144 B3** 33.30S 70.40W
Santiago de Compostela **116 B3** 42.52N 8.33W
Santiago de Cuba **137 E2** 20.00N 75.49W
Santiago del Estero **144 C4** 27.48S 64.15W
Santiago de los Caballeros **137 E2** 19.30N 70.42W
Santiaguillo, Lake **136 B3**
Santo Amaro **143 G2** 12.35S 38.41W
Santo Andre **144 E4** 23.39S 46.29W
Santo Angelo **144 D4** 28.17S 54.14W
Santo Domingo **137 F2** 18.30N 69.57W
Santos **144 E4** 23.56S 46.22W
Santo Tome de Guayana **142 D4** 8.22N 62.37W
San Valentin, Peak **144 B2** 46.33S 73.20W
Sao Bento **143 F3** 2.45S 44.48W
Sao Borja **144 D4** 28.35S 56.01W
Sao Carlos **143 F1** 22.01S 47.54W
Sao Francisco **143 F2** 15.51S 44.52W
Sao Francisco River **143 F2**
Sao Francisco do Sul **144 E4** 26.17S 48.39W
Sao Gabriel **144 D3** 30.24S 54.20W
Sao Jose do Rio Preto **143 F1** 20.50S 49.20W
Sao Leopoldo **144 D4** 29.45S 51.11W
Sao Luis **143 F3** 2.34S 44.16W
Sao Luis Gonzaga **144 D4** 28.25S 55.00W
Sao Paulo **144 E4** 23.33S 46.39W
Sao Roque, Cape **143 G3** 5.28S 35.17W
Sao Sebastiao, Cape **126 D2** 22.05S 35.32E
Sao Tome **124 D2** 0.19N 6.43E
Sao Tome, island **124 D1**
Sao Tome and Principe, democratic republic **124 D2**
Sapporo **104 J8** 43.05N 141.21E
Saragossa **116 C3** 41.39N 0.54W
Sarajevo **117 G3** 43.52N 18.26E
Saransk **118 E4** 54.12N 45.10E
Sarapul **118 F4** 56.30N 53.49E
Sarasota **135 F1** 27.20N 82.32W
Saratov **118 E4** 51.30N 45.55E
Saravan **108 E3** 27.25N 62.17E
Saravane **100 B4** 15.43N 106.24E
Sarawak, state **100 C3**
Sardinia, island **116 E3**
Sargodha **106 C4** 32.01N 72.40E
Sarh **124 E2** 9.08N 18.22E
Sari **108 D4** 36.33N 53.06E
Sarigan, island **101 F4**
Sarina **42 G5** 21.26S 149.13E
Sariwon **104 B6** 38.30N 125.45E
Sariyer **117 I3** 41.11N 29.03E
Sarkoy **117 I3** 40.37N 27.07E
Sarmi **88 C3** 1.51S 138.45E
Sarny **115 J3** 51.21N 26.31E
Sarpsborg **114 G4** 59.17N 11.06E
Sarria **116 B3** 42.47N 7.25W
Saruhanli **117 I2** 38.43N 27.35E
Sasak **100 A2** 0.00N 99.42E
Sasamungga **89 G2** 7.03S 156.47E
Sasaram **106 D3** 24.58N 84.01E
Sasebo **104 D3** 33.10N 129.42E
Saskatchewan, province **132 H3**
Saskatchewan River **132 H3**
Saskatoon **132 H3** 52.10N 106.40W
Sassandra **124 C2** 4.58N 6.08W
Sassari **116 E3** 40.43N 8.34E
Sassnitz **114 G3** 54.32N 13.40E
Satara **106 C2** 17.43N 74.05E
Satna **106 D3** 24.33N 80.50E
Satui **100 C2** 3.49S 115.29E
Satun **100 B3** 6.38N 100.02E
Saudharkrokur **110 C4** 65.45N 19.39W
Saudi Arabia, kingdom **108 C3**
Saugor **106 C3** 23.50N 78.44E
Saul **143 E4** 3.32N 53.15W
Sault Ste Marie **133 J2** 46.32N 84.20W
Saumarez Reef **42 I5**
Saumlakki **88 B2** 7.59S 131.22E
Saumur **116 D4** 47.16N 0.05W
Saunders, Cape **90 D2** 45.53S 170.44E
Saurimo **126 C4** 9.39S 20.24E
Sausapor **88 B3** 0.31S 132.07E
Savage River **60 C3**
Savage River **60 C3** 41.28S 145.15E
Savaii, island **84 N4**
Savalou **124 D2** 7.59N 2.03E
Savannah, Georgia **135 F2** 32.04N 81.07W
Savannah, Tennessee **136 D4** 35.12N 88.15W
Savannah River **135 F2**
Savannakhet **100 B4** 16.34N 104.45E
Save River **126 D2**
Savona **116 E3** 44.18N 8.28E
Savonlinna **115 J5** 61.54N 28.55E
Savu, island **100 D1**
Savu Sea **100 D2**
Sawahlunto **100 B2** 0.41S 100.52E
Saweda, Cape **88 B3** 0.44S 133.57E
Saxby Downs, homestead **42 D5** 20.02S 142.29E
Sayaboury **100 B4** 19.18N 101.46E
Scamander **60 F3** 41.28S 148.16E
Scandinavian Peninsula **114 G5**
Scarborough, Trinidad and Tobago **137 F2** 11.11N 60.45W
Scarborough, U.K. **114 D3** 54.17N 0.24W
Sceale Bay **66 E4**
Schanck, Cape **54 C1** 38.30S 144.53E
Schefferville **133 L3** 54.50N 67.00W
Schleswig **114 F3** 54.32N 9.34E
Schouten Island **60 F2**
Schwedt **114 G3** 53.04N 14.17E
Schwerin **114 G3** 53.38N 11.25E
Sciacca **116 F2** 37.31N 13.05E
Scicli **117 F2** 36.48N 14.43E
Scone **48 F3** 32.03S 150.52E
Scotia Sea **92 P2**
Scott Base, antarctic base **92 I3** 77.51S 166.45E
Scottsbluff **134 D3** 41.52N 103.40W
Scottsdale **60 E3** 41.09S 147.31E
Scranton **135 G3** 41.25N 75.40W
Seal Islands **54 D1**
Sea Lake **54 B3** 35.30S 142.51E
Searcy **134 E2** 35.14N 91.43W
Seaspray **54 D1** 38.23S 147.11E
Seattle **134 B3** 47.35N 122.20W
Seba **100 D1** 10.29S 121.54E
Sebanga **100 B3** 1.15N 101.20E
Sebes **117 H4** 45.58N 23.34E
Sebha **124 E4** 27.02N 14.26E
Secretary Island **90 B2**
Seelig, Mount **94 M3** 82.30S 103.00W
Sefadu **124 B2** 8.41N 10.55W
Sefrou **116 C1** 33.50N 4.50W
Segezha **115 K5** 63.24N 34.19E
Segou **124 C3** 13.28N 6.18W
Segovia **116 C3** 40.57N 4.07W
Seguntur **100 C3** 1.56N 117.50E
Seinajoki **115 I5** 62.45N 22.55E
Seine River **116 D4**
Seiyun **108 C2** 16.00N 49.00E
Sekondi-Takoradi **124 C2** 4.59N 1.43W
Selaru, island **88 B2**
Selatan, Cape **100 C2** 4.09S 114.36E
Selatpampang **100 B2** 0.15S 109.11E
Sele **88 B3** 1.22S 131.06E
Selenga River **102 E4**
Selkirk **132 I3** 50.10N 96.52W
Selma **135 F2** 32.24N 87.01W
Selu, island **88 B2**
Selvas, region **142 D3**
Selwyn Range **42 B5**
Semarang **100 C2** 6.58S 110.29E
Semipalatinsk **118 I4** 50.26N 80.16E
Semnan **108 D4** 35.30N 53.25E
Sena **126 D3** 17.26S 35.01E
Senanga **126 C3** 16.09S 23.16E
Sendai **104 J6** 38.16N 140.52E
Senegal, republic **124 B3**
Senegal River **124 B3**
Sens **116 D4** 48.12N 3.18E
Seoul **104 C5** 37.30N 127.00E
Sepang **100 C2** 1.24S 113.54E
Sepasu **100 C3** 0.42N 117.40E
Sepik River **88 D3**
Sept Iles **133 L3** 50.13N 66.22W
Seram, island **88 A3**
Seram Sea **88 A3**
Serang **100 B2** 6.07S 106.09E
Serbia, province **117 H3**
Serbia and Montenegro, republic **117 H3**
Seremban **100 B3** 2.42N 101.54E
Sergiev-Posad **115 L4** 56.20N 38.10E
Serian **100 C3** 1.10N 110.35E
Serov **118 G4** 59.42N 60.32E
Serowe **126 C2** 22.25S 26.44E
Serpa **116 B2** 37.56N 7.36W
Serpentine Lakes **66 B6**
Serpukhov **115 L3** 54.53N 37.25E
Serrai **117 H3** 41.03N 23.33E
Serua, island **88 B2**
Serui **88 C3** 1.53S 136.15E
Sesheke **126 C3** 17.30S 24.30E
Sete Lagoas **143 F2** 19.27S 44.14W
Setif **124 D5** 36.11N 5.24E
Settat **124 C5** 33.04N 7.37W
Setubal **116 B2** 38.31N 8.54W
Seul, Lake **132 I3**
Sevan, Lake **108 C5**
Sevastopol **115 K1** 44.36N 33.31E
Severn River **132 I3**
Severnaya Zemlya, islands **118 J6**
Severodonetsk **115 L2** 48.58N 38.29E
Severodvinsk **118 D5** 64.35N 39.50E
Sevilla de Niefang **124 E2** 1.55N 10.08E
Seville **116 B2** 37.24N 5.59W
Seward **132 D4** 60.05N 149.34W
Seward Peninsula **132 B4**
Seychelles, republic **96 C1**
Seydhisfjordhur **110 C4** 65.16N 14.02W
Seyhan River **108 B4**
Seymour **54 C2** 37.02S 145.09E
Sfantu Gheorghe **117 I4** 45.51N 25.48E
Sfax **124 E5** 34.45N 10.43E
Shache **102 B3** 38.27N 77.16E
Shackleton Ice Shelf **92 E3**
Shah Fuladi, mountain **106 B4** 34.39N 67.39E
Shahjahanpur **106 C3** 27.53N 79.55E
Shahr Kord **108 D4** 32.20N 50.52E
Shaki **124 D2** 8.39N 3.25E
Shala, Lake **125 G2**
Sham, Mount **108 D3** 23.13N 57.16E
Shan Plateau **107 E3**
Shanghai **103 G3** 31.06N 121.22E
Shangrao **103 F2** 28.28N 117.54E
Shannon **90 F4** 40.33S 175.25E
Shantar Islands **119 N4**
Shantou **103 F2** 23.23N 116.39E
Shaoguan **102 F2** 25.54N 113.33E
Shaoxing **103 G2** 30.02N 120.35E
Shaoyang **103 F2** 27.10N 111.25E
Shaqra **108 C3** 25.15N 45.15E
Sharjah **108 D3** 25.22N 55.23E
Shark Bay **72 B3**
Sharpe, Lake **134 D3**
Sharya **118 E4** 58.22N 45.20E
Shashi **102 F3** 30.16N 112.20E
Shasta, Mount **134 B3** 41.25N 122.12W
Shawinigan **133 K2** 46.33N 72.45W
Shay Gap **72 D4** 20.30S 120.10E
Shaykh Uthman **108 C2** 12.54N 45.00E
Shchekino **115 L3** 54.00N 37.34E
Shebelle River **125 H2**
Sheboygan **135 F3** 43.46N 87.44W
Shebshi Mountains **124 E2**
Sheffield, Australia **60 D3** 41.23S 146.20E
Sheffield, U.K. **114 D3** 53.23N 1.30W
Shelagski, Cape **119 R6** 70.08N 170.20E
Shelbourne **54 C2** 36.51S 144.01E
Shelburne Bay **42 D10**
Shelikhov, Gulf of **119 P5**
Shendi **125 G3** 16.41N 33.22E
Shenyang **103 G4** 41.50N 123.26E
Shenzhen **103 F2** 22.31N 114.08E
Shepetovka **115 J3** 50.12N 27.01E
Shepparton **54 C2** 36.23S 145.24E
Sherbro Island **124 B2**
Sherbrooke **133 K2** 45.24N 71.54W
Sheridan **134 D3** 44.48N 106.57W
Sheringa **66 E4** 33.51S 135.14E
Sherman **134 E2** 33.39N 96.35W
Shetland Islands **114 D5**
Shibarghan **106 B4** 36.40N 65.42E
Shibetsu **104 K9** 44.10N 142.10E
Shibotsu, island **103 I4**
Shield, Cape **80 F8** 13.19S 136.20E
Shihezi **102 C4** 44.19N 86.10E
Shihr **108 C2** 14.44N 49.35E
Shijiazhuang **102 F3** 38.04N 114.28E
Shikarpur **106 B3** 27.58N 68.42E
Shikoku, island **104 F3**
Shikotan, island **103 I4**
Shikotsu, Lake **104 J8**
Shilka River **119 L4**
Shillong **107 E3** 25.34N 91.53E
Shimla **106 C4** 31.07N 77.09E
Shimoga **106 C2** 13.56N 75.31E
Shimonoseki **104 E3** 33.59N 130.58E
Shindand **106 B4** 33.16N 62.05E
Shinyanga **126 D4** 3.40S 33.20E
Shiogama **104 J6** 38.19N 141.00E
Shiono, Cape **104 G3** 33.28N 135.47E
Shirakawa **104 J5** 37.07N 140.11E
Shirane, Mount **104 I5** 36.49N 139.23E
Shiraz **108 D3** 29.38N 52.34E
Shiretoko, Cape **104 L9** 44.24N 145.20E
Shiriya, Cape **104 J7** 41.24N 141.30E
Shiyan **102 F3** 32.31N 110.45E
Shizuishan **102 E3** 39.04N 106.22E
Shizunai **104 K8** 42.20N 142.23E
Shizuoka **104 I4** 34.59N 138.24E
Shkoder **117 G3** 42.03N 19.30E
Shoalhaven River **48 E2**
Shoalwater Bay **42 H4**
Sholapur **106 C2** 17.43N 75.56E
Shorawak **106 B4** 31.48N 64.16E
Shortland Islands **89 G2**
Shoshone Mountains **134 C2**
Shostka **115 K3** 51.53N 33.30E
Shreveport **134 E2** 32.30N 93.46W
Shuangyashan **103 H4** 46.42N 131.20E
Shumen **117 I3** 43.17N 26.55E
Shuqra **108 C2** 13.23N 45.44E
Shushong **126 C2** 22.59S 26.30E
Shushtar **108 C4** 32.03N 48.51E
Shuya **115 M4** 56.49N 41.23E
Sialkot **106 C4** 32.29N 74.35E
Siargao, island **100 D3**
Siassi **89 E2** 5.37S 147.49E
Siauliai **115 I4** 55.51N 23.20E
Sibenik **117 G3** 43.45N 15.55E
Siberia, region **118 I5**
Siberut, island **100 A2**
Sibigo **100 A3** 2.50N 95.58E
Sibiu **117 H4** 45.46N 24.09E
Sibolga **100 A3** 1.42N 98.48E
Sibu **100 C3** 2.18N 111.49E
Sicily, island **117 F2**
Sideia Island **89 F1**
Sidi-bel-Abbes **124 C5** 35.15N 0.39W
Sidi Ifni **124 B4** 29.24N 10.12W
Sidi Kacem **116 B1** 34.15N 5.49W
Sidra, Gulf of **124 E5**
Siena **116 F3** 43.19N 11.19E
Sierra Leone, republic **124 B2**
Sierra Madre, mountains **136 B3**
Sierra Nevada, mountains **134 C2**
Sierra Vista **134 C2** 31.34N 110.21W
Sigep **100 A2** 0.59S 98.50E

*Entry / **Page** / **Ref.** / Lat. / Long.*

Sighet **117 H4** 47.56N 23.53E
Sighisoara **117 H4** 46.12N 24.48E
Sigli **100 A3** 5.21N 95.56E
Sigmaringen **114 F2** 48.05N 9.13E
Signy, antarctic base **92 P3** 60.54S 45.56W
Sigoisooinan **100 A2** 2.05S 99.35E
Siguiri **124 C3** 11.28N 9.07W
Siirt **108 C4** 37.56N 41.56E
Sikar **106 C3** 27.33N 75.12E
Sikasso **124 C3** 11.18N 5.38W
Sikhote Mountains **103 H4**
Sikuati **100 C3** 6.54N 116.39E
Silet **124 D4** 22.40N 4.34E
Siliguri **106 D3** 26.42N 88.30E
Silistra **117 I3** 44.06N 27.17E
Silivri **117 I3** 41.05N 28.15E
Silkeborg **114 F4** 56.10N 9.39E
Silute **115 I4** 55.18N 21.30E
Silver City **134 D2** 32.47N 108.16W
Simanggang **100 C3** 1.10N 111.32E
Simav **117 I2** 39.05N 28.59E
Simbo, island **89 G2**
Simcoe, Lake **135 G3**
Simeulue, island **100 A3**
Simferopol **115 K1** 44.57N 34.05E
Simla *see* Shimla
Simonstown **126 B1** 34.12S 18.26E
Simpang **100 B2** 1.15S 104.05E
Simpson Desert **40 D3**
Sinai, Mount **125 G4** 28.32N 33.59E
Sinai Peninsula **125 G4**
Sincelejo **142 C4** 9.17N 75.23W
Sinchu **103 G2** 24.48N 120.59E
Sindangan **100 D3** 8.13N 123.01E
Sindangbarang **100 B2** 7.26S 107.01E
Sines **116 B2** 37.58N 8.52W
Sinfra **124 C2** 6.35N 5.56W
Singa **125 G3** 13.11N 33.55E
Singapore **100 B3** 1.20N 103.50E
Singapore, republic **100 B3**
Singaraja **100 C2** 8.06S 115.07E
Singida **126 D4** 4.45S 34.45E
Singkarak **100 B2** 0.41S 100.38E
Singkep, island **100 B2**
Singkil **100 A3** 2.16N 97.47E
Singleton **48 F3** 32.34S 151.10E
Singleton, Mount, Northern Territory **80 C4** 21.53S 130.39E
Singleton, Mount, Western Australia **72 C3** 29.28S 117.19E
Sinnamary **143 E4** 5.28N 53.00W
Sintra **116 B2** 38.48N 9.22W
Sinuiju **104 B7** 40.04N 124.25E
Sioux City **134 E3** 42.30N 96.28W
Sioux Falls **134 E3** 43.34N 96.42W
Siping **103 G4** 43.15N 124.25E
Siple, antarctic base **92 N3** 75.55S 83.55W
Siple Island **92 L3**
Sipura, island **100 A2**
Siracusa **117 G2** 37.04N 15.18E
Sir Edward Pellew Group **80 F7**
Sirte **124 E5** 31.10N 16.39E
Sirte Desert **124 E5**
Sisak **117 G4** 45.30N 16.22E
Sitapur **102 B2** 27.33N 80.40E
Sitka **132 E3** 57.05N 135.20W
Sittwe **107 E3** 20.09N 92.55E
Sivas **108 B4** 39.44N 37.01E
Siwa **125 F4** 29.12N 25.31E
Siwalik Range **106 D3**
Skagen **114 G4** 57.44N 10.37E
Skagerrak, passage **114 F4**
Skagway **132 E3** 59.23N 135.20W
Skedsmo **114 G5** 60.00N 10.55E
Skeena River **132 F3**
Skelleftea **114 I5** 64.47N 20.59E
Skien **114 F4** 59.14N 9.37E
Skikda **124 D5** 36.53N 6.54E
Skipton **54 B2** 37.41S 143.22E
Skive **114 F4** 56.34N 9.02E
Skopje **117 H3** 42.00N 21.28E
Slamet, mountain **100 B2** 7.14S 109.10E
Slantsy **115 J4** 59.09N 28.09E
Slatina **117 H3** 44.26N 24.22E
Slave River **132 G4**
Slavyansk **115 L2** 48.51N 37.36E
Sligo **114 C3** 54.17N 8.28W
Sliven **117 I3** 42.40N 26.19E
Slovakia, republic **115 H2**
Slovenia, republic **117 F4**
Slupsk **114 H3** 54.28N 17.00E
Slutsk **115 J3** 53.02N 27.31E
Smallwood Reservoir **133 L3**
Smara **124 B4** 26.44N 11.41W
Smithton **60 C4** 40.52S 145.07E
Smoky Bay **66 D4** 32.23S 133.56E
Smolensk **115 K3** 54.49N 32.04E
Smolyan **117 H3** 41.34N 24.42E
Smyley Island **92 O3**
Snake Island **54 D1**
Snake River **134 C3**
Snasa **118 B5** 64.15N 12.23E
Snowtown **66 G4** 33.47S 138.13E
Snowy Mountains **48 E1**
Snowy River **55 E2**
Snug **60 E1** 43.04S 147.15E
Snyder **134 D2** 32.43N 100.54W
Sobger River **88 D3**
Sobral **143 F3** 3.45S 40.20W
Sochi **118 D3** 43.35N 39.46E
Society Islands **85 P4**
Socorro, Colombia **142 C4** 6.30N 73.16W
Socorro, U.S.A. **134 D2** 34.05N 106.55W
Socorro, island **85 T7**
Socotra, island **108 D2**
Sodankyla **118 C5** 67.26N 26.35E
Soddu **125 G2** 6.49N 37.41E
Soderhamn **114 H5** 61.19N 17.10E
Sodertalje **114 H4** 59.11N 17.39E
Soe **100 D2** 9.51S 124.16E
Sofia **117 H3** 42.40N 23.18E
Sogamoso **142 C4** 5.43N 72.56W
Sogeri **89 E2** 9.25S 147.25E
Sohag **125 G4** 26.33N 31.42E
Sohano **89 F2** 5.26S 154.40E
Soke **117 I2** 37.45N 27.26E
Sokode **124 D2** 8.59N 1.11E
Sokolka **115 I3** 53.26N 23.30E
Sokota **125 G3** 12.40N 39.02E
Sokoto **124 D3** 13.02N 5.15E
Solander Island **90 B1**
Solomon Sea **89 F2**
Solomon Islands, independent state **89 G2**
Solothurn **116 E4** 47.13N 7.32E
Soltau **114 F3** 52.59N 9.50E
Somalia, democratic republic **125 H2**
Sombor **117 G4** 45.46N 19.09E
Somerset Island **132 I5**
Somerset East **126 C1** 32.44S 25.35E
Sondrio **116 E4** 46.11N 9.52E
Songea **126 D3** 10.42S 35.39E
Songhua Lake **104 C8**
Songhua River **103 G4**
Songjin **104 D7** 40.38N 129.10E
Songkhla **100 B3** 7.12N 100.35E
Songnim **104 B6** 38.45N 125.39E
Sonhat **106 D3** 23.28N 82.35E
Sonsonate **136 D2** 13.43N 89.44W
Sonsorol, island **101 E3**
Sopron **114 H2** 47.40N 16.38E
Sora **116 F3** 41.43N 13.37E
Sorel **133 K2** 46.03N 73.06W
Sorell **60 E2** 42.47S 147.34E
Sorell, Cape **60 C2** 42.12S 145.12E
Sorell, Lake **60 E2**
Soria **116 C3** 41.46N 2.28W
Sorocaba **144 E4** 23.30S 47.32W
Soroki **115 J2** 48.08N 28.12E
Sorol, island **101 F3**
Sorong **88 B3** 0.50S 131.17E
Soroti **126 D5** 1.46N 33.40E
Sorrento **54 C1** 38.21S 144.44E
Sortavala **115 K5** 61.40N 30.40E
Souk Ahras **116 E2** 36.14N 8.00E
Souk el Arba du Rharb **124 C5** 34.43N 6.01W
Sousse **124 E5** 35.50N 10.38E
South Island **90 C3**
South Africa, republic **126 C1**
South Alligator River **80 D8**
Southampton **114 D3** 50.55N 1.25W
Southampton Island **133 J4**
South Andaman, island **107 E2**
South Atlantic Ocean **147 M1**
South Australia, state **66 D6**
South Bend **135 F3** 41.40N 86.15W
South Bruny, island **60 E1**
South Carolina, state **135 F2**
South China Sea **100 C4**
South Dakota, state **134 D3**
South East Cape **60 D1** 43.38S 146.48E
South-east Sulawesi, province **100 D2**
Southern Alps **90 C2**
Southern Cross **72 C2** 31.14S 119.19E
Southern Indian Lake **132 I3**
Southern Ocean **92 K3**
Southern Sierra Madre, mountains **136 B2**
South Esk River **60 E3**
South Georgia, island **92 Q2**
South Kalimantan, province **100 C2**
South Korea, republic **104 D5**
South Natuna, island **100 B3**
South Orkney Islands **92 P2**
South Pacific Ocean **147 J1**
South Pagai, island **100 B2**
South Platte River **134 D2**
South Sandwich Islands **92 Q2**
South Saskatchewan River **132 G3**
South Sulawesi, province **100 C2**
South Sumatra, province **100 B2**
South Taranaki Bight **90 E5**
South West Cape **60 D1** 43.32S 146.03E
South West Rocks **48 G4** 30.53S 153.02E
Sovetsk **115 I4** 55.02N 21.50E
Sovetskaya Gavan **119 O3** 48.57N 140.16E
Sowek **88 C3** 0.50S 135.50E
Soya, Cape **104 J9** 45.33N 141.58E
Spain, kingdom **116 C3**
Spalding **66 G4** 33.30S 138.37E
Sparks **132 G1** 39.34N 119.45W
Sparta **117 H2** 37.05N 22.25E
Spassk-Dalniy **103 H4** 44.37N 132.37E
Spenard **132 D4** 61.05N 150.00W
Spencer **134 E3** 43.08N 95.08W
Spencer, Cape **66 F3** 35.18S 136.53E
Spencer Gulf **66 F3**
Spenser Mountains **90 E3**
Spittal **114 G2** 46.48N 13.30E
Split **117 G3** 43.31N 16.28E
Spokane **134 C3** 47.40N 117.25W
Spoleto **116 F3** 42.44N 12.44E
Spratly Islands **100 C3**
Springbok **126 B2** 29.43S 17.55E
Springdale **134 E2** 36.10N 94.09W
Springerville **134 D2** 34.09N 109.18W
Springfield, Illinois **135 F2** 39.49N 89.39W
Springfield, Massachusetts **135 G3** 42.07N 72.35W
Springfield, Missouri **134 E2** 37.11N 93.19W
Springfield, Oregon **134 B3** 44.01N 123.02W
Springhill **134 E2** 33.00N 93.29W
Springs **126 C2** 26 15S 28.26E
Springsure **42 G3** 24.07S 148.05E
Squires, Mount **72 E3** 26.14S 127.28E
Sredinnyy Range **119 Q4**
Sre Umbell **100 B4** 11.08N 103.46E
Sri Lanka, independent state **106 D1**
Srinagar **106 C4** 34.08N 74.50E
Stacks Bluff, mountain **60 E3** 41.37S 147.42E
Stafford **114 D3** 52 48N 2.07W
Stanger **126 D2** 29 20S 31.18E
Stanhope **54 C2** 36.27S 144.59E
Stanley, Australia **60 C4** 40.46S 145.18E
Stanley, U.K. **144 D1** 51.45S 57.56W
Stann Creek **136 D2** 16.59N 88.15W
Stanovoy Range **119 M4**
Stanthorpe **42 H1** 28.40S 151.56E
Staraya Russa **115 K4** 58.00N 31.22E
Stara Zagora **117 I3** 42.25N 25.37E
Starbuck, island **84 P5**
Stargard Szczecinski **114 H3** 53.2[illegible]N 15.01E
Starokonstantinov **115 J2** 49.48N 27.10E
Stary Oskol **115 L3** 51.20N 37.50E
Statesboro **135 F2** 32.28N 81.47W
Stavanger **114 F4** 58.58N 5.45E
Stavropol **118 E3** 45.03N 41.59E
Stawell **54 B2** 37.03S 142.47E
Steep Point **72 B3** 26.08S 113.10E
Stefanie, Lake **125 G2**
Steffen Strait **89 F3**
Steinbach **132 I2** 49.32N 96.40W
Steinkjer **114 G5** 64.00N 11.30E
Stellenbosch **126 B1** 33.56S 18.51E
Stenhouse Bay **66 F3** 35.17S 136.57E
Stepanakert **108 C4** 39.48N 46.45E
Stephens, Cape **90 E4** 40.42S 173.57E
Stephens, Port **48 G3**
Stephenville, Canada **133 M2** 48.33N 58.34W
Stephenville, U.S.A. **134 E2** 32.12N 98.13W
Sterling **134 D3** 40.38N 103.14W
Sterlitamak **118 F4** 53.40N 55.59E
Stevenson Creek **66 E7**
Stewart, Cape **80 E9** 11.57S 134.45E
Stewart Island **90 C1**
Steyr **114 G2** 48.04N 14.25E
Stip **117 H3** 41.44N 22.12E
Stirling **114 D4** 56.07N 3.57W
Stirling, Mount **80 D4** 21.40S 133.33E
St-Jean, Lake **133 K2**
Stjordal **114 G5** 63.27N 10.57E
Stockholm **114 H4** 59.20N 18.05E
Stockton **134 B2** 37.59N 121.20W
Stockton-on-Tees **114 D3** 54.34N 1.19W
Stoke-on-Trent **114 D3** 53.00N 2.10W
Stonehenge **42 D3** 24.21S 143.17E
Stony Tunguska River **118 J5**
Storm Bay **60 E1**
Strahan **60 C2** 42.09S 145.15E
Stralsund **114 G3** 54.18N 13.06E
Strangnas **114 H4** 59.22N 17.02E
Stranraer **114 C3** 54.54N 5.02W
Strasbourg **116 E4** 48.35N 7.45E
Stratford **54 D2** 37.58S 147.05E
Strathalbyn **66 G3** 35.16S 138.54E
Strathgordon **60 D2** 42.46S 146.03E
Streaky Bay **66 E4**
Streaky Bay **66 E4** 32.48S 134.13E
Strzelecki Creek **66 G6**
Strzelecki, Mount **80 D4** 21.08S 133.52E
Stuart Range **66 E6**
Stuart Creek, homestead **66 F6** 29.42S 137.02E
Stung Treng **100 B4** 13.31N 105.59E
Sturt Creek **72 E5**
Sturt Desert **66 H7**
Sturt Plain **80 B6**
Stutterheim **126 C1** 32.35S 27.26E
Stuttgart **114 F2** 48.47N 9.12E
Subotica **117 G4** 46.04N 19.41E
Suceava **117 I4** 47.37N 26.18E
Sucre **142 D2** 19.05S 65.15W
Sudan, democratic republic **125 F3**
Sudbury **133 J2** 46.30N 81.01W
Sudd, region **125 F2**
Sudetic Mountains **114 H3**
Sudirman Range **88 C3**
Suez **125 G4** 29.59N 32.33E
Suez Canal **122 G8**
Suez, Gulf of **108 B3**
Sugarloaf Point **48 G3** 32.25S 152.30E
Suhar **108 D3** 24.22N 56.45E
Suihua **103 G4** 46.40N 127.00E
Sukabumi **100 B2** 6.55S 106.50E
Sukhebator **102 E5** 50.10N 106.14E
Sukhumi **118 E3** 43.01N 41.01E
Sukkur **106 B3** 27.42N 68.54E
Sula Islands **100 D2**
Sulaiman Range **106 B3**
Sulaimaniya **108 C4** 35.32N 45.27E
Sulawesi, island **100 D2**
Sulawesi Sea **100 D3**
Sulechow **114 H3** 52.05N 15.37E
Sulina **118 C3** 45.07N 29.40E
Sullana **142 B3** 4.52S 80.39W
Sulmona **117 F3** 42.03N 13.56E

*Entry / **Page** / **Ref.** / Lat. / Long.*

Sulu Sea **100 D3**
Sumatra, island **100 B3**
Sumba, island **100 C1**
Sumbawa, island **100 C2**
Sumbawa Besar **100 C2** 8.30S 117.25E
Sumbe **126 B3** 11.13S 13.50E
Sumgait **108 C5** 40.35N 49.38E
Sumpangbinangae **100 C2** 4.25S 119.39E
Sumy **115 K3** 50.55N 34.49E
Sunbury **54 C2** 37.35S 144.44E
Sunchon **104 C4** 34.56N 127.28E
Sundarbans, region **102 C2**
Sundargarh **106 D3** 22.04N 84.08E
Sundsvall **114 H5** 62.22N 17.20E
Sungaibatu **100 C3** 0.56N 110.44E
Sungaisalak **100 B2** 0.29S 103.00E
Sungei Patani **100 B3** 5.34N 100.29E
Sungsang **100 B2** 2.22S 104.50E
Sunyani **124 C2** 7.22N 2.18W
Suo Sea **104 F3**
Suomussalmi **118 C5** 64.52N 29.10E
Superior **135 E3** 46.42N 92.05W
Superior, Lake **135 F3**
Suphan, Mount **108 C4** 38.55N 42.49E
Supiori, island **88 C3**
Sur **108 D3** 22.35N 59.31E
Surabaya **100 C2** 7.14S 112.45E
Surakarta **100 C2** 7.32S 110.50E
Surat, Australia **42 G2** 27.09S 149.04E
Surat, India **106 C3** 21.10N 72.54E
Surat Thani **100 A3** 9.09N 99.20E
Surgut **118 H5** 61.13N 73.20E
Surigao **100 D3** 9.47N 125.29E
Surin **100 B4** 14.53N 103.29E
Surinam, republic **143 E4**
Susanville **134 B3** 40.26N 120.40W
Susurluk **117 I2** 39.55N 28.10E
Sutlej River **106 C4**
Suva **84 N4** 18.08S 178.25E
Suwalki **115 I3** 54.06N 22.56E
Suwarrow, island **84 O4**
Suwon **104 C5** 37.16N 126.59E
Suzhou **103 G3** 31.21N 120.40E
Svalbard, islands **98 B15**
Svendborg **114 G4** 55.04N 10.38E
Sverdrup Islands **132 I5**
Svetlogorsk **115 J3** 53.08N 29.43E
Svetlovodsk **115 K2** 49.05N 33.15E
Svishtov **117 I3** 43.36N 25.22E
Svobodny **119 M4** 51.24N 128.05E
Swain Reefs **42 I5**
Swakopmund **126 B2** 22.40S 14.34E
Swan River **72 C2**
Swan Hill **54 B3** 35.20S 143.33E
Swansea, Australia **60 F2** 42.08S 148.04E
Swansea, U.K. **114 D3** 51.38N 3.57W
Swaziland, kingdom **126 D2**
Sweden, kingdom **114 G5**
Sweetwater **134 D2** 32.27N 100.25W
Swift Current **132 H3** 50.17N 107.49W
Swinoujscie **114 G3** 53.55N 14.16E
Switzerland, confederation **116 E4**
Sydney, Australia **48 F3** 33.55S 151.13E
Sydney, Canada **133 L2** 46.10N 60.10W
Syktyvkar **118 F5** 61.42N 50.45E
Sylhet **107 E3** 24.53N 91.51E
Syowa, antarctic base **92 C3** 69.00S 39.35E
Syr River **118 G3**
Syracuse **135 G3** 43.03N 76.10W
Syria, Arab republic **108 B4**
Syrian Desert **108 B4**
Syzran **118 E4** 53.10N 48.29E
Szczecin **114 G3** 53.25N 14.32E
Szczecinek **114 H3** 53.42N 16.41E
Szeged **115 I2** 46.15N 20.09E
Szekesfehervar **114 H2** 47.11N 18.22E
Szekszard **114 H2** 46.21N 18.41E
Szombathely **117 G4** 47.12N 16.38E

T

Tabaco **100 D4** 13.21N 123.44E
Tabar Islands **89 F3**
Tabas **108 D4** 33.37N 56.54E
Taber **134 C3** 49.48N 112.09W
Tabiteuea, island **84 M5**
Tablas, island **100 D4**
Table Cape **60 C4** 40.56S 145.45E
Tabor **114 G2** 49.25N 14.40E
Tabora **126 D4** 5.02S 32.50E
Tabou **124 C2** 4.28N 7.20W
Tabriz **108 C4** 38.05N 46.18E
Tabuk **108 B3** 28.22N 36.35E
Tacheng **102 C4** 46.42N 83.00E
Tacloban **100 D4** 11.15N 125.01E
Tacna **142 C2** 18.00S 70.15W
Tacoma **134 B3** 47.16N 122.30W
Tacuarembo **144 D3** 31.42S 56.00W
Tademait Plateau **124 D4**
Tadzhikistan, republic **118 G2**
Taebaek Range **104 C6**
Taedong River **104 C6**
Taegu **104 D4** 35.52N 128.36E
Taejon **104 C5** 36.20N 127.26E
Tafalla **116 C3** 42.32N 1.41W
Tafemaar **88 B2** 6.49S 134.12E
Taftan, Mount **108 E3** 28.37N 61.08E
Tagab **106 B4** 34.35N 69.20E
Tagrifet **124 E4** 29.12N 17.22E
Tagula **89 F1** 11.20S 153.10E
Tagula Island **89 F1**
Tagus River **116 B2**
Tahan, Mount **100 B3** 4.34N 102.17E
Tahat, mountain **124 D4** 23.18N 5.33E
Tahiti, island **85 Q4**
Tahoe, Lake **132 G1**
Tahoua **124 D3** 14.57N 5.19E
Tahta **125 G4** 26.47N 31.31E
Taichung **103 G2** 24.09N 120.40E
Taif **108 C3** 21.15N 40.21E
Taihape **90 F5** 39.41S 175.48E
Tailem Bend **66 G3** 35.15S 139.28E
Taimyr, Lake **118 K6**
Taimyr Peninsula **118 I6**
Tainan **103 G2** 23.01N 120.14E
Tainaron, Cape **117 H2** 36.23N 22.29E
Taipei **103 G2** 25.05N 121.32E
Taiping **100 B3** 4.54N 100.42E
Tais **100 B2** 4.05S 102.29E
Taiwan, independent state **103 G2**
Taiwara **106 B4** 33.21N 64.25E
Taiyuan **102 F3** 37.50N 112.30E
Taizz **108 C2** 13.35N 44.02E
Tajumulco, mountain **136 C2** 15.03N 91.54W
Takabba **126 E5** 3.27N 40.14E
Takamatsu **104 G4** 34.20N 134.01E
Takapuna **90 F6** 36.48S 174.47E
Takayama **104 H5** 36.09N 137.16E
Takefu **104 H4** 35.54N 136.10E
Takikawa **104 J8** 43.35N 141.55E
Takwa **89 H2** 8.20S 160.46E
Talangbatu **100 B2** 3.59S 105.29E
Talara **142 B3** 4.38S 81.18W
Talasea **89 E2** 5.19S 150.02E
Talaud Islands **100 D3**
Talavera de la Reina **116 C2** 39.58N 4.50W
Talca **144 B3** 35.28S 71.40W
Talcahuano **144 B3** 36.40S 73.10W
Taliabu, island **100 D2**
Tallahassee **135 F2** 30.26N 84.19W
Tallangatta **54 D2** 36.12S 147.10E
Tallebung **48 D3** 32.45S 146.33E
Tallering Peak **72 C3** 28.06S 115.38E
Tallinn **115 I4** 59.22N 24.48E
Talok **100 C3** 1.00N 118.53E
Talsi **115 I4** 57.18N 22.39E
Taltal **144 B4** 25.26S 70.33W
Tamale **124 C2** 9.26N 0.49W
Tamanrasset **124 D4** 22.50N 5.28E
Tamar, River **60 E3**
Tambo **42 F3** 24.53S 146.15E
Tambo River **55 D2**
Tamboritha, Mount **54 D2** 37.29S 146.42E
Tambov **118 E4** 52.44N 41.28E
Tambura **125 F2** 5.38N 27.30E
Tampa **135 F1** 27.58N 82.28W
Tampere **115 I5** 61.32N 23.45E
Tampico **136 C3** 22.18N 97.52W
Tamsag Bulag **102 E4** 40.45N 103.33E
Tamworth **48 F4** 31.07S 150.57E
Tana **118 C6** 70.26N 28.14E
Tana, Lake **125 G3**
Tana River **125 H1**
Tanabe **104 G3** 33.43N 135.22E
Tanahmerah **88 D2** 6.08S 140.18E
Tanami **80 B5** 19.59S 129.43E
Tanami Desert **80 C5**
Tanana River **132 C4**
Tanchon **104 D7** 40.29N 128.50E
Tanderra, homestead **42 F3** 24.36S 147.39E
Tandil **144 D3** 37.18S 59.10W
Tandou Lake **48 B3**
Tanega Island **104 E2**
Tanga **126 D4** 5.07S 39.05E
Tanga Islands **89 F3**
Tanganyika, Lake **126 D4**
Tangarare **89 G2** 9.34S 159.39E
Tangier **124 C5** 35.48N 5.50W
Tangjin **104 C5** 36.55N 126.34E
Tangla **107 E3** 26.40N 91.57E
Tangshan **103 F3** 39.37N 118.05E
Tanimbar Islands **88 B2**
Tanjay **100 D3** 9.31N 123.10E
Tanjungbaliha **100 D2** 2.25S 126.03E
Tanjungbatu **100 C3** 2.15N 118.06E
Tanjungkarang **100 B2** 5.28S 105.16E
Tanna, island **84 M4**
Tanout **124 D3** 15.05N 8.50E
Tanta **125 G5** 30.48N 31.00E
Tanumbirini, homestead **80 E6** 16.27S 134.13E
Tanunda **66 G3** 34.34S 138.56E
Tanzania, united republic **126 D4**
Taongi, island **84 L7**
Taormina **117 G2** 37.51N 15.17E
Taoudenni **124 C4** 22.38N 3.58W
Tapachula **136 C2** 14.54N 92.15W
Tapajos River **142 E3**
Tapan **100 B2** 2.14S 101.02E
Tapolca **114 H2** 46.54N 17.29E
Tapti River **106 C3**
Tapuaenuku, mountain **90 E3** 42.00S 173.40E
Tapul **100 D3** 5.44N 120.55E
Tapul, island **100 D3**
Tapurucuara **142 D3** 0.23S 65.01W
Tara, Australia **42 H2** 27.17S 150.28E
Tara, Russia **118 H4** 56.55N 74.24E
Tarakan **100 C3** 3.20N 117.38E
Tarancon **116 C3** 40.01N 3.01W
Taranto **117 G3** 40.28N 17.15E
Tararua Range **90 F4**
Tarata **142 D2** 17.35S 66.04W
Tarawa, island **84 M6**
Tarazona **116 C3** 41.54N 1.44W
Tarbes **116 D3** 43.14N 0.05E
Tarcoola **66 E5** 30.43S 134.34E
Taree **48 G4** 31.54S 152.26E
Tarfawi Oasis **125 F4** 22.57N 28.55E
Tarfaya **124 B4** 27.58N 12.55W
Tarhunah **124 E5** 32.26N 13.38E
Tari **88 D2** 5.50S 142.55E
Tarifa **116 B2** 36.01N 5.36W
Tarija **142 D1** 21.33S 64.45W
Tariku River **88 C3**
Tarim **108 C2** 16.08N 48.58E
Tarim Basin **102 C3**
Tarim River **102 C4**
Taringo Downs, homestead **48 C3** 32.14S 145.33E
Tarlac **100 D4** 15.29N 120.35E
Tarnow **115 I3** 50.01N 20.59E
Taroom **42 G3** 25.39S 149.48E
Taroudant **124 C5** 30.31N 8.55W
Tarragona **116 D3** 41.07N 1.15E
Tartu **115 J4** 58.20N 26.44E
Tartus **108 B4** 34.55N 35.52E
Tashauz **118 F3** 41.49N 59.58E
Tashkent **118 G3** 41.16N 69.13E
Taskul **89 F3** 2.34S 150.25E
Tasman Bay **90 E4**
Tasman, Mount **90 D3** 43.33S 170.10E
Tasman Mountains **90 E4**
Tasman Peninsula **60 F1**
Tasman Point **80 F7** 14.15S 136.20E
Tasman Sea **84 L2**
Tasmania, state **60 D3**
Tatabanya **114 H2** 47.31N 18.25E
Tatar Strait **119 O4**
Tathra Head **48 F1** 36.44S 149.58E
Tatura **54 C2** 36.26S 145.14E
Taumarunui **90 F5** 38.52S 175.15E
Taung **126 C2** 27.32S 24.48E
Taunggyi **107 E3** 20.55N 96.53E
Taunton **114 D3** 51.01N 3.06W
Taupo **90 G5** 38.41S 176.05E
Taupo, Lake **90 F5**
Taurage **115 I4** 55.12N 22.16E
Tauranga **90 G6** 37.42S 176.10E
Tauroa Point **90 E7** 35.10S 173.04E
Taurus Mountains **108 B4**
Tauy Bay **119 O4**
Tavani **132 I4** 62.00N 93.20W
Tavas **117 I2** 37.33N 29.04E
Tavira **116 B2** 37.07N 7.39W
Tavoy **107 E2** 14.02N 98.12E
Tavsanli **117 I2** 39.34N 29.28E
Tawau **100 C3** 4.16N 117.54E
Tawitawi, island **100 D3**
Tawurgha **124 E5** 32.02N 15.08E
Taylor, Mount **134 D2** 35.14N 107.36W
Tayshet **118 J4** 55.56N 98.01E
Taytay **100 C4** 10.47N 119.32E
Taz River **118 I5**
Taza **124 C5** 34.16N 4.01W
Tazerbo Oasis **124 F4** 25.44N 21.55E
Tazovskiy **118 H5** 67.28N 78.43E
Tbilisi **108 C5** 41.43N 44.48E
Tchepone **100 B4** 16.41N 106.14E
Tchibanga **126 B4** 2.49S 11.00E
Te Anau **90 B2** 45.25S 167.43E
Te Anau, Lake **90 B2**
Te Awamutu **90 F6** 38.00S 175.19E
Tebessa **124 D5** 35.21N 8.06E
Tecuci **117 I4** 45.50N 27.27E
Tedzhen **108 E4** 37.26N 60.30E
Tegal **100 B2** 6.52S 109.07E
Tegucigalpa **136 D2** 14.05N 87.14W
Tehran **108 D4** 35.40N 51.26E
Tehuacan **136 C2** 18.30N 97.26W
Tehuantepec **136 C2** 16.21N 95.13W
Tehuantepec, Gulf of **136 C2**
Teixeira Pinto **124 B3** 12.03N 16.00W
Tekapo, Lake **90 D3**
Tekirdag **108 A5** 40.59N 27.31E
Te Kuiti **90 F5** 38.20S 175.10E
Telanaipura **100 B2** 1.34S 103.37E
Tel Aviv-Jaffa **108 B4** 32.05N 34.46E
Telefomin **88 D2** 5.08S 141.38E
Teles Pires River **142 E3**
Telfer **72 D4** 21.42S 122.13E
Telsiai **115 I4** 55.58N 22.16E
Telukdalam **100 A3** 0.33N 97.48E
Tema **124 D2** 5.41N 0.01W
Tembilahan **100 B2** 0.16S 103.14E
Teminabuan **88 B3** 1.30S 131.59E
Temirtau **118 H4** 50.05N 72.55E
Temora **48 D2** 34.27S 147.32E
Tempio Pausania **116 E3** 40.54N 9.07E
Temple **134 E2** 31.06N 97.22W
Tempue **126 B3** 13.27S 18.53E
Temuco **144 B3** 38.45S 72.40W
Temuka **90 D2** 44.15S 171.16E
Tenali **106 D2** 16.13N 80.36E
Tenerife, island **124 B4**
Tenes **116 D2** 36.35N 1.18E
Tengchong **107 E3** 25.02N 98.28E
Teniente Rodolfo Marsh, antarctic base **92 P3** 62.12S 58.54W
Tennant Creek **80 E5** 19.39S 134.11E
Tennessee River **135 F2**
Tennessee, state **135 F2**
Tenterfield **48 G5** 29.01S 152.04E
Tepa **88 A2** 7.52S 129.35E
Tepic **136 B3** 21.30N 104.51W
Te Puke **90 G6** 37.47S 176.20E
Teraina, island **84 O6**
Terang **54 B1** 38.14S 142.55E
Teresina **143 F3** 5.09S 42.46W

Entry/***Page***/***Ref.***/*Lat.*/*Long.*

Termez **106 B4** 37.15N 67.15E
Terminos Lagoon **136 C2**
Termoli **117 G3** 42.00N 15.00E
Terni **116 F3** 42.34N 12.39E
Ternopol **115 J2** 49.35N 25.39E
Terpeniya, Cape **119 O3** 48.05N 144.43E
Terracina **116 F3** 41.17N 13.15E
Terrigal **48 F3** 33.27S 151.27E
Teruel **116 C3** 40.21N 1.06W
Tervola **118 C5** 66.05N 24.50E
Teshio **104 J9** 44.53N 141.46E
Teshio River **104 K9**
Tessalit **124 D4** 20.11N 1.02E
Tessaoua **124 D3** 13.46N 7.55E
Tessenei **125 G3** 15.05N 36.41E
Tete **126 D3** 16.10S 33.35E
Tetepare, island **89 G2**
Tetouan **124 C5** 35.34N 5.22W
Tetovo **117 H3** 42.00N 20.59E
Teulada, Cape **116 E2** 38.52N 8.38E
Te Waewae Bay **90 B1**
Texarkana **134 E2** 33.28N 94.02W
Texas **42 H1** 28.51S 151.10E
Texas, state **134 D2**
Texas City **134 E1** 29.23N 94.55W
Texoma, Lake **134 E2**
Thabana Ntlenyana, mountain **122 F2** 29.28S 29.16E
Thailand, Gulf of **100 B4**
Thailand, kingdom **100 B4**
Thamar, Mount **108 C2** 13.53N 45.15E
Thames **90 F6** 37.08S 175.33E
Thames River **114 D3**
Thana **106 C2** 19.14N 73.02E
Thangool **42 H3** 24.29S 150.35E
Thar Desert **106 C3**
Thargomindah **42 D1** 27.59S 143.49E
Thasos, island **117 H3**
Thebes **117 H2** 38.19N 23.19E
The Entrance **48 F3** 33.21S 151.30E
The Hague **114 E3** 52.05N 4.16E
Theodore **42 H3** 24.57S 150.05E
The Pas **132 H3** 53.50N 101.15W
Thief River Falls **134 E3** 48.12N 96.48W
Thiers **114 E2** 45.51N 3.33E
Thies **124 B3** 14.49N 16.52W
Thika **126 D4** 1.03S 37.05E
Thimphu **107 D3** 27.32N 89.43E
Thionville **116 E4** 49.22N 6.11E
Thisted **114 F4** 56.58N 8.42E
Thistle Island **66 F3**
Thompson **132 I3** 55.45N 97.54W
Thomson River **42 D3**
Thonburi **100 B4** 13.43N 100.27E
Thorshavn **114 C5** 62.02N 6.47W
Thouars **116 C4** 46.59N 0.13W
Three Forks **134 C3** 45.54N 111.32W
Three Hummock Island **60 C4**
Three Kings Islands **84 M2**
Three Points, Cape **124 C2** 4.43N 2.09W
Thule **133 L5** 77.30N 69.29W
Thunder Bay **132 J2** 48.27N 89.12W
Thurloo Downs, homestead **48 B5** 29.17S 143.29E
Thursday Island **42 D10**
Thurso **114 D4** 58.35N 3.32W
Thurston Island **92 N3**
Tianjin **103 F3** 39.08N 117.12E
Tianshui **102 E3** 34.25N 105.58E
Tiaret **116 D2** 35.20N 1.20E
Tiber River **116 F3**
Tiberias, Lake **60 E2**
Tibesti, mountains **124 E4**
Tibet, Plateau of **102 C3**
Tibooburra **48 A5** 29.26S 142.01E
Tidjikja **124 B3** 18.29N 11.31W
Tieling **103 G4** 42.19N 123.52E
Tien Mountains **118 H3**
Tieri **42 G4** 23.02S 148.21E
Tierra del Fuego, islands **144 C1**
Tigard **134 B3** 45.25N 122.46W
Tigris River **108 C4**
Tijuana **134 C2** 32.29N 117.10W
Tikhoretsk **110 I2** 45.52N 40.07E
Tiksi **119 M6** 71.40N 128.45E
Tilaiya **106 D3** 24.20N 85.38E
Tillabery **124 D3** 14.28N 1.27E
Timan Ridge, mountains **118 F5**
Timaru **90 D2** 44.24S 171.15E
Timber Creek **80 C7** 15.53S 130.31E
Timboon **54 B1** 38.29S 142.59E
Timbuktu **124 C3** 16.49N 2.59W
Timimoun **124 D4** 29.15N 0.14E
Timisoara **117 H4** 45.45N 21.15E
Timmins **133 J2** 48.30N 81.20W
Timor, island **100 D2**
Timor Sea **80 A9**
Timpah **100 C2** 1.39S 114.26E
Tinaca Point **100 D3** 5.34N 125.20E
Tindouf **124 C4** 27.42N 8.10W
Tineo **116 B3** 43.20N 6.24W
Tingwon Island **89 E3**
Tinian, island **101 F4**
Tinompo **100 D3** 0.25N 120.16E
Tintinara **66 H3** 35.52S 140.04E
Tioman, island **100 B3**
Tioor, island **88 B3**
Tirane **117 G3** 41.20N 19.49E
Tiraspol **115 J2** 46.50N 29.38E
Tire **117 I2** 38.04N 27.45E
Tirgu-Jiu **117 H4** 45.03N 23.18E
Tirgu-Mures **117 H4** 46.33N 24.34E
Tirgu Ocna **117 I4** 46.18N 26.44E
Tirich Mir, mountain **106 C4** 36.18N 71.55E
Tirua Point **90 F5** 38.23S 174.38E
Tiruchchirappalli **106 C2** 10.50N 78.43E
Tiruvananthapuram **106 C1** 8.30N 76.57E
Tisza River **115 I2**
Tit-Ary **119 M6** 71.55N 127.02E
Titicaca, Lake **142 D2**
Tititira Head **90 C3** 43.38S 169.26E
Titova Mitrovica **117 H3** 42.54N 20.52E
Titovo Uzice **117 G3** 43.52N 19.50E
Titov Veles **117 H3** 41.43N 21.49E
Ti-Tree **80 D3** 22.08S 133.16E
Tivaouane **124 B3** 14.57N 16.45W
Tizimin **136 D3** 21.10N 88.09W
Tizi Ouzou **116 D2** 36.44N 4.05E
Tiznit **124 C4** 29.43N 9.44W
Tlemcen **124 C5** 34.53N 1.21W
Toamasina **120 H3** 18.10S 49.23E
Toba, Lake **100 A3**
Tobago, island **137 F2**
Tobo **88 B3** 3.33S 130.10E
Toboali **100 B2** 3.00S 106.30E
Tobol River **118 G4**
Tobolsk **118 G4** 58.15N 68.12E
Tobruk **124 F5** 32.06N 23.56E
Tocantins River **143 F2**
Tocopilla **142 C1** 22.05S 70.10W
Tocumwal **48 C2** 35.49S 145.34E
Todd River **80 E2**
Todd River, homestead **80 E3** 23.51S 134.30E
Todeli **100 D2** 1.40S 124.30E
Tofua, island **84 N4**
Togo, republic **124 D2**
Toi, Cape **104 E2** 31.22N 131.21E
Tokar **125 G3** 18.27N 37.41E
Tokara Group **104 D1**
Tokara Strait **104 D2**
Tokelau Islands **84 N5**
Tokoroa **90 F5** 38.13S 175.52E
Tokuno Island **103 G2**
Tokushima **104 G4** 34.03N 134.34E
Tokyo **104 I4** 35.40N 139.45E
Tokzar **106 B4** 35.57N 66.21E
Tolbukhin **117 I3** 43.34N 27.51E
Toledo, Spain **116 C2** 39.52N 4.02W
Toledo, U.S.A. **135 F3** 41.40N 83.35W
Toliara **126 E2** 23.20S 43.41E
Tolima, mountain **142 C4** 4.39N 75.22W
Tolo Bay **100 D2**
Tolosa **116 C3** 43.09N 2.04W
Toluca **136 C2** 19.20N 99.40W
Tolyatti **118 E4** 53.32N 49.24E
Tomakomai **104 J8** 42.39N 141.33E
Tombigbee River **135 F2**
Tombstone **134 C2** 31.44N 110.04W
Tombua **126 B3** 15.50S 11.51E
Tome **144 B3** 36.38S 72.57W
Tomelloso **116 C2** 39.09N 3.01W
Tomini Bay **100 D2**
Tom Price **72 C4** 22.40S 117.48E
Tomsk **118 I4** 56 30N 85.05E
Tondano **100 D3** 1.19N 125.56E
Tonga, homestead **48 B4** 30.30S 143.47E
Tonga, kingdom **84 N3**
Tonga Trench **84 N3**
Tongatapu, island **84 N3**
Tongchuan **102 E3** 35.05N 109.02E
Tonghua **104 B7** 41.45N 125.50E
Tongliao **103 G4** 43.37N 122.15E
Tonkin, Gulf of **100 B4**
Tooms Lake **60 E2**
Toowoomba **42 H2** 27.34S 151.57E
Topeka **134 E2** 39.02N 95.41W
Topolobampo **136 B3** 25.36N 109.04W
Tops, Mount **80 D4** 21.50S 133.59E
Top Springs, homestead **80 C6** 16.32S 131.48E
Torbay **114 D3** 50.28N 3.30W
Torgau **114 G3** 51.34N 13.01E
Tornio **110 G4** 65.52N 24.10E
Toronto **133 K2** 43.42N 79.25W
Tororo **126 D5** 0.42N 34.12E
Torquay **54 C1** 38.20S 144.20E
Torrelavega **116 C3** 43.21N 4.03W
Torrens, Lake **66 F5**
Torreon **136 B3** 25.34N 103.25W
Torres Strait **88 D2**
Torres Vedras **116 B2** 39.05N 9.15W
Tortosa **116 D3** 40.49N 0.31E
Tortuga, island **142 D5**
Torun **114 H3** 53.01N 18.35E
Torzhok **115 K4** 57.02N 34.51E
Tottenham **48 D3** 32.14S 147.21E
Tottori **104 G4** 35.32N 134.12E
Toubkal, mountain **124 C5** 31.03N 7.57W
Toudao River **104 C8**
Touggourt **124 D5** 33.08N 6.04E
Tougue **124 B3** 11.29N 11.48W
Toukmatine **124 D4** 24.48N 7.08E
Toulon **116 E3** 43.08N 5.55E
Toulouse **116 D3** 43.37N 1.27E
Tours **116 D4** 47.23N 0.42E
Townshend, Cape **42 H4** 22.13S 150.29E
Townsville **42 F6** 19.16S 146.49E
Towuti Lake **100 D2**
Toyama **104 H5** 36.42N 137.14E
Toyama Bay **104 H5**
Trabzon **108 B5** 41.00N 39.43E
Trafalgar **54 D1** 38.13S 146.09E
Trafonomby, mountain **126 E2** 24.38S 46.44E
Trail **132 G2** 49.04N 117.39W
Tralee **114 C3** 52.16N 9.42W
Trangan, island **88 B2**
Trangie **48 D3** 32.03S 147.59E
Transantarctic Mountains **94 H2**
Transylvanian Alps **117 H4**
Trappes **116 D4** 48.47N 2.00E
Traralgon **54 D1** 38.12S 146.32E
Trat **100 B4** 12.16N 102.30E
Traverse City **135 F3** 44.46N 85.38W
Treasury Islands **89 G2**
Treinta y Tres **144 D3** 33.16S 54.17W
Trelew **144 C2** 43.13S 65.15W
Trelleborg **114 G4** 55.22N 13.10E
Trencin **114 H2** 48.53N 18.00E
Trent **116 F4** 46 04N 11.08E
Tres Arroyos **144 C3** 38.26S 60.17W
Tres Cruces, mountain **144 C4** 27.05S 68.46W
Tres Puntas, Cape **144 C2** 47.05S 65.55W
Triabunna **60 E2** 42.30S 147.55E
Trial Bay **48 G4**
Trichur **106 C2** 10.32N 76.14E
Trieste **116 F4** 45.39N 13.47E
Triglav, mountain **116 F4** 46.21N 13.59E
Trikkala **117 H2** 39.33N 21.46E
Trikora, Peak **88 C3** 4.15S 138.41E
Trincomalee **106 D1** 8.34N 81.13E
Trinidad, Bolivia **142 D2** 14.46S 64.50W
Trinidad, U.S.A. **134 D2** 37.11N 104.31W
Trinidad, island **137 F2**
Trinidad and Tobago, independent state **137 F2**
Trinity Bay **133 M2**
Trinity River **134 E2**
Tripoli, Lebanon **108 B4** 34.27N 35.50E
Tripoli, Libya **124 E5** 32.58N 13.12E
Tripolis **117 H2** 37.31N 22.22E
Tristan da Cunha, island **146 B1**
Trivandrum *see* Tiruvananthapuram
Trnava **114 H2** 48.23N 17.35E
Trobriand Islands **89 F2**
Trois Rivieres **133 K2** 46.21N 72.34W
Trollhattan **114 G4** 58.17N 12.20E
Tromso **118 B5** 69.42N 19.00E
Tronador, mountain **144 B2** 41.10S 71.50W
Trondheim **114 G5** 63.25N 10.23E
Troudaninna **40 D3** 29.28S 139.22E
Troyes **116 D4** 48.18N 4.05E
Trujillo, Peru **142 C3** 8.06S 79.00W
Trujillo, Venezuela **142 C4** 9.20N 70.26W
Truk, island **101 G3**
Tsabong **126 C2** 26.03S 22.27E
Tsane **126 C2** 24.05S 21.54E
Tsetserlig **102 E4** 47.26N 101.22E
Tshikapa **126 C4** 6.28S 20.48E
Tshuapa River **126 C4**
Tsiafajavona, mountain **122 H3** 19.20S 47.15E
Tsimlyansk Reservoir **118 E3**
Tsugaru Strait **104 I7**
Tsumeb **126 B3** 19.13S 17.42E
Tsuruga **104 H4** 35.40N 136.05E
Tsuruoka **104 I6** 38.42N 139.50E
Tsushima, island **104 D4**
Tsuyama **104 G4** 35.04N 134.01E
Tuakau **90 F6** 37.15S 174.58E
Tual **88 B2** 5.38S 132.40E
Tuamotu Archipelago **85 Q4**
Tuapse **118 D3** 44.06N 39.05E
Tubau **100 C3** 3.10N 113.41E
Tubuai, island **85 Q3**
Tucano **143 G2** 10.57S 38.46W
Tucson **134 C2** 32.15N 110.57W
Tucumcari **134 D2** 35.11N 103.43W
Tucupita **142 D4** 9.02N 62.04W
Tucurui **143 F3** 3.42S 49.44W
Tudela **116 C3** 42.04N 1.37W
Tufi **89 E2** 9.05S 149.19E
Tuguegarao **100 D4** 17.36N 121.44E
Tukums **115 I4** 56.58N 23.10E
Tukzar **118 G2** 35.57N 66.21E
Tula, Mexico **136 C3** 23.00N 99.41W
Tula, Russia **115 L3** 54.11N 37.38E
Tulaghi **89 H2** 9.08S 160.09E
Tulare **134 C2** 36.12N 119.22W
Tulcea **117 I4** 45.10N 28.50E
Tulchin **115 J2** 48.40N 28.49E
Tullahoma **135 F2** 35.21N 86.12W
Tulle **116 D4** 45.16N 1.46E
Tully **42 E7** 17.56S 145.56E
Tulsa **134 E2** 36.07N 95.58W
Tulua **142 C4** 4.05N 76.12W
Tulun **118 K4** 54.32N 100.35E
Tumaco **142 C4** 1.51N 78.46W
Tumbarumba **48 E2** 35.46S 148.01E
Tumbes **142 B3** 3.37S 80.27W
Tumby Bay **66 F3** 34.22S 136.08E
Tumkur **106 C2** 13.20N 77.06E
Tumut **48 E2** 35.18S 148.13E
Tunbridge **60 E2** 42.08S 147.26E
Tungku **100 C3** 5.05N 119.00E
Tunis **124 E5** 36.50N 10.13E
Tunisia, republic **124 D5**
Tunja **142 C4** 5.33N 73.23W
Tupa **143 E1** 21.56S 50.30W
Tupelo **135 F2** 34.15N 88.43W
Tupiza **142 D1** 21.27S 65.45W
Tupungato, Mount **144 C3** 33.20S 69.50W
Tura **118 K5** 64.20N 100.17E
Turangi **90 F5** 38.59S 175.48E
Turayf **108 B4** 31.43N 38.35E
Turbat **106 B3** 26.00N 63.06E
Turbo **142 C4** 8.06N 76.44W
Turda **117 H4** 46.35N 23.50E

Entry/***Page***/***Ref.***/*Lat.*/*Long.*

Tureia, island **85 R3**
Turfan **102 C4** 42.55N 89.06E
Turfan Depression **102 C4**
Turin **116 E4** 45.04N 7.40E
Turkana, Lake **126 D5**
Turkestan **118 G3** 43.17N 68.16E
Turkey, republic **108 B4**
Turkmenistan, republic **118 F2**
Turks Islands **137 E3**
Turku **115 I5** 60.27N 22.15E
Turkwel River **125 G2**
Turlock **134 B2** 37.30N 120.51W
Turnagain, Cape **90 G4** 40.30S 176.37E
Turnovo **117 I3** 43.04N 25.39E
Turnu Magurelle **117 H3** 43.44N 24.53E
Turnu Severin **117 H3** 44.37N 22.39E
Turtle Point **80 B7** 14.51S 129.13E
Tuscaloosa **135 F2** 33.12N 87.33W
Tutoia **143 F3** 2.45S 42.16W
Tutrakan **117 I3** 44.02N 26.40E
Tutuala **100 D2** 8.25S 127.15E
Tutuila **84 O4**
Tuvalu, independent state **84 M5**
Tuxpan **136 B3** 21.58N 105.20W
Tuxtla Gutierrez **136 C2** 16.45N 93.09W
Tuy **116 B3** 42.03N 8.39W
Tuy Hoa **100 B4** 13.08N 109.10E
Tuz, Lake **108 B4**
Tuzla **117 G3** 44.33N 18.41E
Tver **115 L4** 56.49N 35.57E
Tweed Range **48 G5**
Twin Falls **134 C3** 42.33N 114.29W
Twizel **90 D2** 44.15S 170.06E
Twofold Bay **48 F1**
Tyers, Lake **55 E2**
Tyler **134 E2** 32.22N 95.18W
Tyre **108 B4** 33.16N 35.12E
Tyrnavos **117 H2** 39.45N 22.18E
Tyrrell Creek **54 B3**
Tyrrell, Lake **54 B3**
Tyrrhenian Sea **116 F2**
Tzaneen **126 D2** 23.50S 30.09E

U

Uahuka, island **85 R5**
Uaupes **142 D4** 0.07S 67.05W
Uba **143 F1** 21.07S 42.56W
Ubangi River **124 E2**
Ubari **124 E4** 26.35N 12.47E
Ubari Desert **124 E4**
Ube **104 E3** 33.57N 131.16E
Uberaba **143 F2** 19.47S 47.57W
Uberlandia **143 F2** 18.57S 48.17W
Ubsu Lake **102 D5**
Ubundu **126 C4** 0.24S 25.30E
Ucayali River **142 C3**
Uchiura Bay **104 J8**
Uda River **118 J4**
Udaipur **106 C3** 24.36N 73.47E
Uddevalla **114 G4** 58.20N 11.56E
Udialla, homestead **72 D5** 17.57S 123.44E
Udine **116 F4** 46.04N 13.14E
Udon Thani **100 B4** 17.25N 102.45E
Udsk Bay **119 N4**
Uele River **125 F2**
Uelzen **114 G3** 52.58N 10.34E
Ufa **118 F4** 54.45N 55.58E
Uganda, republic **126 D5**
Uige **126 B4** 7.45S 15.09E
Uitenhage **126 C1** 33.46S 25.25E
Ujelang, island **84 L6**
Ujjain **102 B2** 23.11N 75.50E
Ujung Pandang **100 C2** 5.09S 119.28E
Ukhta **118 F5** 63.33N 53.44E
Uki, island **89 H1**
Ukiah **134 B2** 39.09N 123.12W
Ukmerge **115 I4** 55.14N 24.49E
Ukraine, republic **115 K2**
Ulamona **89 F3** 5.00S 151.15E
Ulan Bator **102 E4** 47.54N 106.52E
Ulangom **102 D4** 49.59N 92.00E
Ulanhot **103 G4** 46.06N 122.05E
Ulan-Ude **119 K4** 51.55N 107.40E

Entry/***Page***/***Ref.***/*Lat.*/*Long.*

Ulawa, island **89 H2**
Ulegei **102 D4** 48.54N 90.00E
Ulhasnagar **106 C2** 19.15N 73.08E
Uliastay **102 D4** 47.42N 96.52E
Ulingan **89 E3** 4.32S 145.23E
Ulithi, island **101 E4**
Ulladulla **48 F2** 35.21S 150.29E
Ullung Island **104 E5**
Ulm **114 G2** 48.24N 10.00E
Ulsan **104 D4** 35.32N 129.21E
Ultima **54 B3** 35.28S 143.16E
Ulugh Muztagh, mountain **102 C3** 36.28N 87.29E
Uluru **80 C2** 25.21S 131.05E
Ulverstone **60 D3** 41.09S 146.10E
Ul'yanovsk **118 K2** 48.45N 30.10E
Uman **115 K2** 48.45N 30.10E
Umboi Island **89 E2**
Umea **118 E4** 54.19N 48.22E
Umm al-Quaiwain **108 D3** 25.35N 55.34E
Umm Lajj **108 B3** 25.03N 37.17E
Umm Said **108 D3** 24.59N 51.32E
Umtata **126 C1** 31.35S 28.47E
Umuarama **144 D4** 23.45S 53.20W
Unalaska Island **132 B3**
Underbool **54 A3** 35.10S 141.49E
Undurkhan **102 F4** 47.20N 110.35E
Ungava Bay **133 L3**
Ungava Peninsula **133 K4**
Ungerem **88 D2** 6.30S 141.07E
Unimak Island **132 B3**
United Kingdom **114 D4**
United Arab Emirates, republic **108 D3**
United States of America, republic **134 C2**
Upington **126 C2** 28.28S 21.14E
Upolu, island **84 N4**
Upper Klamath Lake **134 B3**
Upper Seal Lake **133 K3**
Upper Yarra Reservoir **54 C2**
Uppsala **114 H4** 59.55N 17.38E
Ural Mountains **118 F5**
Ural River **118 F3**
Uralla **48 F4** 30.38S 151.30E
Uralsk **118 F4** 51.19N 51.20E
Urana **48 D2** 35.20S 146.16E
Urandangi **80 G4** 21.36S 138.19E
Uranium City **132 H3** 59.32N 108.43W
Urapunga, homestead **80 E7** 14.44S 134.33E
Urbino **116 F3** 43.43N 12.38E
Urfa **108 B4** 37.08N 38.45E
Urgench **118 G3** 41.35N 60.41E
Urla **108 A4** 38.19N 26.47E
Urmia **108 C4** 37.32N 45.02E
Urmia, Lake **108 C4**
Uruapan **136 B2** 19.26N 102.04W
Uruguaiana **144 D4** 29.45S 57.05W
Uruguay, republic **144 D3**
Uruguay River **144 D3**
Urumqi **102 C4** 43.43N 87.38E
Urup, island **119 P3**
Urzhum **118 F4** 57.08N 50.00E
Urziceni **117 I3** 44.34N 26.39E
Usak **108 A4** 38.42N 29.25E
Ushuaia **144 C1** 54.48S 68.19W
Usoke **126 D4** 5.05S 32.21E
Ussuri River **119 N3**
Ussurisk **119 N3** 43.48N 131.59E
Ust-Kamchatsk **119 Q4** 56.14N 162.28E
Ust-Kamenogorsk **118 I3** 49.58N 82.36E
Ust-Kut **118 K4** 56.48N 105.42E
Ust Tsilma **118 F5** 65.28N 52.09E
Ust Urt Plateau **118 F3**
Usumacinta River **136 C2**
Utah, state **134 C2**
Utena **115 J4** 55.30N 25.35E
Uthai Thani **100 B4** 15.22N 100.03E
Utica **133 K2** 43.06N 75.15W
Utiel **116 C2** 39.33N 1.13W
Utopia, homestead **80 E3** 22.14S 134.34E
Utrecht **114 F3** 52.06N 5.07E
Utrera **116 B2** 37.10N 5.47W
Utsunomiya **104 I5** 36.33N 139.52E
Uuldza **102 F4** 49.03N 112.35E

Entry/***Page***/***Ref.***/*Lat.*/*Long.*

Uusikaupunki **114 I5** 60.48N 21.25E
Uvira **126 C4** 3.22S 29.06E
Uwajima **104 F3** 33.13N 132.32E
Uyuni **142 D1** 20.28S 66.50W
Uzbekistan, republic **118 G3**
Uzhgorod **115 I2** 48.38N 22.15E
Uzlovaya **115 L3** 53.55N 38.11E
Uzunkopru **117 I3** 41.16N 26.42E

V

Vaal River **126 C2**
Vaasa **114 I5** 63.06N 21.36E
Vac **115 H2** 47.46N 19.08E
Vadodara **106 C3** 22.19N 73.14E
Vadso **118 C6** 70.05N 29.47E
Vaghena, island **89 G2**
Vaigach Island **118 F6**
Vaitupu, island **84 M5**
Valdepenas **116 C2** 38.46N 3.24W
Valdes Peninsula **144 C2**
Valdez **132 D4** 61.07N 146.17W
Valdivia **144 B3** 39.46S 73.15W
Val d'Or **133 K2** 48.07N 77.47W
Valdosta **135 F2** 30.51N 83.20W
Valenca **143 G2** 13.23S 39.06W
Valence **116 E3** 44.56N 4.54E
Valencia, Spain **116 C2** 39.29N 0.24W
Valencia, Venezuela **142 D5** 10.14N 67.59W
Valentine **134 D3** 42.53N 100.31W
Valera **142 C4** 9.21N 70.38W
Valga **115 J4** 57.44N 26.00E
Valjevo **117 G3** 44.16N 19.56E
Valkeakoski **115 I5** 61.17N 24.05E
Valladolid **116 C3** 41.39N 4.45W
Valle de la Pascua **142 D4** 9.15N 66.00W
Valledupar **142 C5** 10.31N 73.16W
Vallenar **144 B4** 28.36S 70.45W
Valletta **117 F2** 35.54N 14.32E
Valleyfield **135 G3** 45.15N 74.08W
Val Marie **132 H2** 49.15N 107.44W
Valmiera **115 J4** 57.32N 25.29E
Valparaiso **144 B3** 33.05S 71.40W
Van **108 C4** 38.28N 43.20E
Van, Lake **108 C4**
Vancouver, Canada **132 F2** 49.13N 123.06W
Vancouver, U.S.A. **132 F2** 45.37N 122.40W
Vancouver Island **132 F2**
Vanderlin Island **80 F7**
Van Diemen, Cape **80 C9** 11.10S 130.22E
Van Diemen Gulf **80 C9**
Vanern, Lake **114 G4**
Vanersborg **114 G4** 58.23N 12.19E
Vangunu, island **89 G3**
Vanimo **88 D3** 2.40S 141.17E
Vannes **116 C4** 47.40N 2.44W
Van Rees Range **88 C3**
Vanrhynsdorp **126 B1** 31.36S 18.45E
Vanua Lava, island **84 L4**
Vanua Levu, island **84 N4**
Vanuatu, republic **84 K4**
Varanasi **106 D3** 25.20N 83.00E
Varazdin **117 G4** 46.18N 16.21E
Varberg **114 G4** 57.06N 12.15E
Varese **116 E4** 45.49N 8.49E
Varkaus **115 J5** 62.20N 27.50E
Varna **117 I3** 43.12N 27.57E
Varnamo **114 G4** 57.11N 14.03E
Vasilkov **115 K3** 50.12N 30.15E
Vaslui **117 I4** 46.37N 27.46E
Vasteras **114 H4** 59.36N 16.32E
Vastervik **114 H4** 57.45N 16.40E
Vasto **117 F3** 42.07N 14.43E
Vattern, Lake **114 G4**
Vaughn **134 D2** 34.35N 105.14W
Vaupes River **142 C4**
Vava'u, island **84 N4**
Vaxjo **114 G4** 56.52N 14.50E
Vegreville **132 G3** 53.30N 112.02W
Velebit Mountains **117 G4**
Veliki Ustyug **118 E5** 60.48N 46.15E
Velikiye Luki **115 K4** 56.19N 30.31E
Vella Lavella, island **89 G2**

Entry/***Page***/***Ref.***/*Lat.*/*Long.*

Velletri **116 F3** 41.41N 12.47E
Vellore **106 C2** 12.56N 79.09E
Velsk **118 E5** 61.05N 42.06E
Venable Ice Shelf **92 N3**
Vendome **116 D4** 47.48N 1.04E
Venezuela, Gulf of **142 C5**
Venezuela, republic **142 D4**
Venice **116 F4** 45.26N 12.20E
Ventspils **115 I4** 57.22N 21.31E
Ventura **134 C2** 34.18N 119.18W
Veracruz **136 C2** 19.11N 96.10W
Verdal **114 G5** 63.47N 11.23E
Verde, Cape **124 B3** 14.43N 17.33W
Verdun **116 E4** 49.10N 5.24E
Verin **116 B3** 41.55N 7.26W
Verkhoyansk **119 N5** 67.35N 133.25E
Verkhoyansk Range **119 M5**
Vermillion **134 E3** 42.48N 96.55W
Vermont, state **135 H2**
Vernon, Canada **134 C4** 50.17N 119.19W
Vernon, U.S.A. **134 E2** 34.10N 99.19W
Vero Beach **135 F1** 27.39N 80.24W
Veroia **117 H3** 40.32N 22.11E
Verona **116 F4** 45.26N 11.00E
Versailles **116 D4** 48.48N 2.08E
Vesuvius, mountain **117 F3** 40.49N 14.26E
Veszprem **114 H2** 47.06N 17.54E
Vetlanda **114 H4** 57.26N 15.05E
Viana do Castelo **116 B3** 41.41N 8.50W
Vibo Valentia **117 G2** 38.40N 16.06E
Vicenza **116 F4** 45.33N 11.32E
Vich **116 D3** 41.56N 2.16E
Vichy **116 D4** 46.08N 3.25E
Victor Harbor **66 G3** 35.33S 138.37E
Victoria, Canada **132 F2** 48.25N 123.22W
Victoria, Seychelles **146 D2** 4.38S 55.28E
Victoria, U.S.A. **134 E1** 28.49N 97.01W
Victoria Falls **126 C3** 17.58S 25.45E
Victoria Island **132 H5**
Victoria, Lake, New South Wales **48 A3**
Victoria, Lake, Uganda **126 D4**
Victoria, Lake, Victoria **55 D1**
Victoria Land **94 H2**
Victoria, Mount **89 E2** 8.52S 147.32E
Victoria River **80 C7**
Victoria, state **54 C2**
Victoria River Downs, homestead **80 C6** 16.24S 131.01E
Victoria West **126 C1** 31.25S 23.08E
Viedma **144 C2** 40.45S 63.00W
Viedma, Lake **144 B2**
Vienna **114 H2** 48.13N 16.22E
Vienne **114 F2** 45.32N 4.54E
Vientiane **100 B4** 17.59N 102.38E
Vierzon **116 D4** 47.14N 2.03E
Vieste **117 G3** 41.53N 16.11E
Vietnam, socialist republic **100 B5**
Vigan **103 G1** 17.35N 120.23E
Vigo **116 B3** 42.15N 8.44W
Vijayawada **106 D2** 16.34N 80.40E
Vila **84 L4** 17.45S 168.18E
Vila Real **116 B3** 41.17N 7.45W
Vilcanota, mountain **142 C2** 14.24S 71.00W
Vilhelmina **118 B5** 64.38N 16.40E
Viljandi **115 J4** 58.22N 25.30E
Vilkitski Strait **118 J6**
Villach **114 G2** 46.37N 13.51E
Villa Dolores **144 C3** 31.58S 65.15W
Villahermosa **136 C2** 18.00N 92.53W
Villa Maria **144 C3** 32.25S 63.15W
Villanueva y Geltru **116 D3** 41.13N 1.43E
Villarrica **144 D4** 25.45S 56.28W
Villavicencio **142 C4** 4.09N 73.38W
Villefranche-de-Rouergue **116 D3** 44.21N 2.02E
Villena **116 C2** 38.39N 0.52W
Villeneuve-sur-Lot **116 D3** 44.25N 0.43E
Vilnius **115 J3** 54.40N 25.19E
Vilyui River **118 K5**
Vina del Mar **144 B3** 33.02S 71.35W
Vinaroz **116 D3** 40.29N 0.28E
Vincennes Bay **92 F3**
Vindhya Mountains **106 C3**

*Entry/**Page**/**Ref.**/Lat./Long.*

Vineland **135 G2** 39.29N 75.02W
Vinh **100 B4** 18.42N 105.41E
Vinh Loi **100 B3** 9.17N 105.44E
Vinnitsa **115 J2** 49.11N 28.30E
Vinson Massif, mountain **94 N2** 79.04S 86.21W
Virac **100 D4** 13.35N 124.14E
Virgin Islands **137 F2**
Virginia **134 E3** 47.30N 92.28W
Virginia, state **135 G2**
Virovitica **117 G4** 45.50N 17.25E
Visby **114 H4** 57.37N 18.20E
Viseu **116 B3** 40.40N 7.55W
Vishakhapatnam **106 D2** 17.42N 83.24E
Vistula River **114 H3**
Vitebsk **115 K4** 55.10N 30.14E
Viterbo **116 F3** 42.24N 12.06E
Vitiaz Strait **89 E2**
Viti Levu, island **84 M4**
Vitim Plateau **119 L4**
Vitim River **119 L4**
Vito **89 G2** 6.05S 155.25E
Vitoria, Brazil **143 F1** 20.19S 40.21W
Vitoria, Spain **116 C3** 42.51N 2.40W
Vivero **116 B3** 43.39N 7.35W
Vivonne Bay **66 F2**
Vizianagaram **106 D2** 18.07N 83.30E
Vladikavkaz **118 E3** 43.02N 44.43E
Vladimir **115 M4** 56.08N 40.25E
Vladivostok **119 N3** 43.09N 131.53E
Vlore **117 G3** 40.29N 19.29E
Voiron **116 E4** 45.22N 5.35E
Vojvodina, province **117 G4**
Volcano Islands **84 J8**
Volchansk **115 L3** 50.19N 36.55E
Volga River **118 E3**
Volgograd **118 E3** 48.45N 44.30E
Volkovysk **115 I3** 53.10N 24.28E
Volksrust **126 C2** 27.22S 29.54E
Volochanka **118 J6** 70.59N 94.48E
Vologda **115 L4** 59.10N 39.55E
Volos **117 H2** 39.22N 22.57E
Volta, Lake **124 C2**
Voltaire, Cape **72 E6** 14.16S 125.35E
Vorkuta **118 G5** 67.27N 64.00E
Voronezh **115 L3** 51.40N 39.13E
Voru **115 J4** 57.48N 26.54E
Voss **114 F5** 60.38N 6.25E
Vostok, antarctic base **92 F3** 78.27S 106.51E
Vostok, island **85 P4**
Voyeykov Ice Shelf **92 F3**
Voznesensk **115 K2** 47.34N 31.21E
Vranje **117 H3** 42.33N 21.54E
Vratsa **117 H3** 43.12N 23.32E
Vrsac **117 H4** 45.07N 21.19E
Vryburg **126 C2** 26.57S 24.44E
Vryheid **126 D2** 27.45S 30.48E
Vung Tau **100 B4** 10.21N 107.04E
Vyatka River **118 E4**
Vyazma **115 K4** 55.12N 34.17E
Vyborg **115 J5** 60.45N 28.41E
Vyksa **115 M4** 55.19N 42.11E
Vyshni Volochek **115 K4** 57.34N 34.23E
Vysokaya, Mount **103 H4** 45.57N 136.37E

W

Wa **124 C3** 10.07N 2.28W
Wabag **89 D2** 5.30S 143.44E
Wabash River **130 K4**
Wabuda Island **89 D2**
Waco **134 E2** 31.33N 97.10W
Waddamana **60 D2** 42.08S 146.45E
Waddan **124 E4** 29.10N 16.10E
Waddington, Mount **132 F3** 51.22N 125.14W
Wadeye **80 B7** 14.15S 129.31E
Wadi Halfa **125 G4** 21.46N 31.20E
Wad Medani **125 G3** 14.24N 33.30E
Wagga Wagga **48 D2** 35.07S 147.22E
Wagin **72 C2** 33.19S 117.20E
Wahpeton **132 I2** 46.16N 96.36W
Waiau **90 E3** 42.39S 173.03E
Waiau River **90 E3**
Waibeem **88 B3** 0.29S 132.59E
Waigeo **88 B3** 0.14S 130.30E
Waigeo, island **88 B3**
Waihi Beach **90 G6** 37.24S 175.56E
Waikare, Lake **90 F6**
Waikaremoana, Lake **90 G5**
Waikato River **90 F6**
Waikerie **66 H3** 34.11S 139.59E
Waimate **90 D2** 44.45S 171.03E
Waingapu **100 D2** 9.40S 120.16E
Waipapa Point **90 C1** 46.40S 168.51E
Waipukurau **90 G5** 40.00S 176.33E
Wairarapa, Lake **90 F4**
Wairau River **90 E4**
Wairoa **90 G5** 39.03S 177.26E
Waitaki River **90 D2**
Waitangi **90 F7** 35.18S 174.06E
Waitemata **90 F6** 36.50S 174.42E
Wakasa Bay **104 G4**
Wakatipu, Lake **90 C2**
Wakayama **104 G4** 34.12N 135.10E
Wake Island **84 L7**
Wakkanai **104 J9** 45.26N 141.43E
Wakool **48 C2** 35.28S 144.24E
Waku Kungo **126 B3** 11.23S 15.08E
Wakunai **89 G2** 5.52S 155.13E
Walbrzych **114 H3** 50.48N 16.19E
Walcha **48 F4** 30.59S 151.35E
Walgett **48 E4** 30.01S 148.07E
Walir, island **88 B2**
Walker Island **60 B4**
Wallaroo **66 F4** 33.56S 137.38E
Walla Walla **134 C3** 46.03N 118.20W
Wallis Islands **84 M4**
Walpole **72 C2** 34.59S 116.44E
Walungurru **80 B3** 23.17S 129.23E
Walvis Bay **126 B2** 22.59S 14.31E
Wamba River **126 B4**
Wamena **88 C3** 3.50S 138.38E
Wami **88 B3** 2.45S 133.18E
Wanaka **90 C2** 44.42S 169.08E
Wanaka, Lake **90 C2**
Wanapiri **88 C3** 4.31S 135.59E
Wanda Mountains **104 F9**
Wandingzhen **107 E3** 24.01N 98.00E
Wang River **100 A4**
Wanganella **48 C2** 35.13S 144.49E
Wanganui **90 F5** 39.56S 175.02E
Wanganui River **90 F5**
Wangaratta **54 D2** 36.21S 146.19E
Wangary **66 E3** 34.33S 135.29E
Wanneroo **72 C2** 31.46S 115.48E
Wannon River **54 A2**
Wanxian **102 E3** 30.54N 108.20E
Waranga Basin, lake **54 C2**
Warangal **106 C2** 18.00N 79.35E
Waratah **60 C3** 41.27S 145.32E
Waratah Bay **54 D1**
Warburton, Victoria **54 C2** 37.45S 145.42E
Warburton, Western Australia **72 E3** 26.08S 126.35E
Warburton Creek **66 G7**
Warburton Range **72 E3**
Wardang Island **66 F3**
Ward Hunt, Cape **89 E2** 8.04S 148.08E
Waren **114 G3** 53.32N 12.42E
Warialda **48 F5** 29.32S 150.35E
Warmbad, Namibia **126 B2** 28.29S 18.41E
Warmbad, South Africa **126 C2** 24.55S 28.15E
Warm Springs **134 C2** 38.12N 116.24W
Warracknabeal **54 B2** 36.15S 142.24E
Warragul **54 C1** 38.10S 145.56E
Warrego River **42 E2**
Warren **48 D4** 31.42S 147.50E
Warrenton **126 C2** 28.07S 24.51E
Warri **124 D2** 5.36N 5.46E
Warrnambool **54 B1** 38.23S 142.30E
Warrumbungle Range **48 E4**
Warruwi **80 D9** 11.36S 133.20E
Warsaw **115 I3** 52.15N 21.00E
Waru **88 B3** 3.26S 130.37E
Warwick **42 I1** 28.13S 152.02E
Washington **135 G2** 38.55N 77.00W
Washington, Cape **92 I3** 75.00S 165.00E
Washington, state **134 B3**
Wasian **88 B3** 1.51S 133.21E
Waram **89 D3** 3.55S 144.30E
Waterbag, homestead **48 B4** 31.28S 142.17E
Waterbury **135 G3** 41.33N 73.03W
Waterford **114 C3** 52.15N 7.06W
Waterhouse Island **60 E4**
Waterloo **134 E3** 42.30N 92.20W
Watertown **134 E3** 44.54N 97.08W
Waterville **135 H3** 44.34N 69.41W
Watsa **126 C5** 3.02N 29.33E
Watson **66 C5** 30.29S 131.30E
Watson Lake **132 F4** 60.02N 128.47W
Wau, P.N.G. **89 E2** 7.21S 146.43E
Wau, Sudan **125 F2** 7.40N 28.04E
Wauchope, New South Wales **48 G4** 31.28S 152.44E
Wauchope, Northern Territory **80 E4** 20.39S 134.13E
Waukarlycarly, Lake **72 D4**
Wausau **135 F3** 44.58N 89.40W
Wave Hill, homestead, New South Wales **48 B4** 31.55S 143.30E
Wave Hill, homestead, Northern Territory **80 C6** 17.23S 131.07E
Waverley **90 F5** 39.46S 174.38E
Waverley Downs, homestead **48 C5** 29.06S 143.55E
Wayakuba **88 A3** 0.50S 127.48E
Wayamli **101 D3** 0.59N 128.30E
Wayatinah **60 D2** 42.23S 146.30E
Waycross **135 F2** 31.12N 82.22W
Weam **88 D2** 8.37S 141.08E
Weda Bay **88 A4**
Weddell Sea **92 Q3**
Wedderburn **54 B2** 36.26S 143.39E
Wednesday Island **88 D1**
Wee Waa **48 E4** 30.13S 149.27E
Wei River **102 E3**
Weiden **114 G2** 49.40N 12.10E
Weifang **103 F3** 36.44N 119.10E
Weihai **103 G3** 37.30N 122.04E
Weipa **42 C9** 12.37S 141.52E
Weiser **134 C3** 44.15N 116.59W
Weishan Lake **103 F3**
Wejh **108 B3** 26.13N 36.28E
Weld Range **72 C3**
Welkom **126 C2** 27.59S 26.44E
Wellesley Islands **42 B7**
Wellington, Australia **48 E3** 32.33S 148.59E
Wellington, New Zealand **90 F4** 41.17S 174.47E
Wellington Island **144 B2**
Wellington, Lake **54 D1**
Wellington, Mount **60 E2** 42.54S 147.14E
Wells, Lake **72 D3**
Wells, Mount **72 E5** 17.26S 127.14E
Wellsford **90 F6** 36.18S 174.31E
Wels **118 B3** 48.10N 14.02E
Welshpool **54 D1** 38.40S 146.27E
Wenatchee **132 F2** 47.24N 120.18W
Wenlock River **42 D9**
Wentworth **48 A2** 34.06S 141.55E
Wenzhou **103 G2** 28.02N 120.40E
Werribee **54 C2** 37.55S 144.40E
Werribee River **54 C2**
Werris Creek **48 F4** 31.21S 150.39E
Wesel **114 F3** 51.39N 6.37E
Wessel, Cape **80 F9** 11.00S 136.46E
Wessel Islands **80 F9**
West Ice Shelf **92 E3**
West Island **80 F7**
West Point, South Australia **66 E3** 35.00S 135.57E
West Point, Tasmania **60 B4** 40.56S 144.37E
Westall, Point **66 E4** 32.55S 134.04E
West Barney Peak **42 I1** 28.17S 152.42E
Westbury **60 D3** 41.32S 146.47E
West Dvina River **115 J4**
Western Port **54 C1**
Western Australia, state **72 D4**
Western Ghats, mountains **106 C2**
Western Sayan, mountains **118 I4**
Western Sierra Madre, mountains **136 B3**
West Falkland, island **144 C1**
West Indies, islands **130 L3**
West Java, province **100 B2**
West Kalimantan, province **100 B2**
West Korea Bay **104 B6**
West Nusa Tenggara, province **100 C2**
West Palm Beach **135 F1** 26.42N 80.05W
Westport **90 D4** 41.45S 171.36E
West Siberian Plain **118 H4**
West Sister Island **60 E5**
West Sumatra, province **100 A2**
West Virginia, state **135 F2**
West Wyalong **48 D3** 33.55S 147.12E
Wetar, island **88 A2**
Wetaskiwin **132 G3** 52.57N 113.20W
Wewak **89 D2** 3.35S 143.35E
Weyburn **132 H2** 49.31N 103.50W
Whakatane **90 G6** 37.58S 176.59E
Whale River **133 K3**
Whangarei **90 F7** 35.43S 174.19E
Wheeling **135 F3** 40.05N 80.43W
White Island, Antarctica **92 C3**
White Island, New Zealand **90 G6**
White Island, Russia **118 G6**
White, Lake **80 B4**
White River **134 D3**
White Sea **118 D5**
White Cliffs **48 B4** 30.50S 143.08E
Whitehorse **132 E4** 60.41N 135.08W
Whitemark **60 E4** 40.07S 148.00E
White Nile, river **125 G2**
White Nile Dam **125 G3** 15.18N 32.30E
White Volta, river **124 C2**
Whitney, Mount **134 C2** 36.35N 118.17W
Whitsunday Island **42 G5**
Whittlesea **54 C2** 37.31S 145.07E
Whyalla **66 F4** 33.02S 137.35E
Wichita **134 E2** 37.43N 97.20W
Wichita Falls **134 E2** 33.55N 98.30W
Wickenburg **134 C2** 33.59N 112.44W
Wickham, Cape **60 A5** 39.35S 143.59E
Wide Bay **89 F2**
Widgiemooltha **72 D2** 31.30S 121.34E
Wiener Neustadt **114 H2** 47.49N 16.15E
Wilcannia **48 B4** 31.33S 143.22E
Wilhelm, Mount **89 D2** 5.46S 144.59E
Wilhelm II Land **94 F2**
Wilhelmshaven **114 F3** 53.32N 8.07E
Wilkes Land **94 G2**
Willandra Lakes **48 B3**
Willemstad **142 D5** 12.12N 68.56W
Willeroo, homestead **80 C7** 15.17S 131.35E
William, Mount, Queensland **42 G5** 21.02S 148.38E
William, Mount, Victoria **54 B2** 37.18S 142.36E
William Creek **66 F6** 28.55S 136.20E
Williams Lake **132 F3** 52.08N 122.09W
Willis Group **42 H7**
Williston **134 D3** 48.09N 103.39W
Willmar **134 E3** 45.06N 95.03W
Willowra, homestead **80 D4** 21.15S 132.35E
Wills, Lake **72 E4**
Wilmington, Australia **66 G4** 32.39S 138.06E
Wilmington, U.S.A. **135 G2** 34.14N 77.55W
Wilpena, homestead **66 G5** 31.31S 138.37E
Wilson Bluff **66 B5** 31.41S 129.01E
Wilson, Mount **134 D2** 37.51N 107.59W
Wilsons Promontory **54 D1**
Wiluna **72 D3** 26.35S 120.14E
Wimmera River **54 B2**
Winchelsea **54 B1** 38.15S 143.59E
Windhoek **126 B2** 22.34S 17.06E
Windorah **42 D3** 25.25S 142.39E
Windsor, Australia **48 F3** 33.37S 150.49E

*Entry/**Page**/**Ref.**/Lat./Long.*

Windsor, Canada **133 J2** 42.18N 83.00W
Windward Islands **137 F2**
Windward Passage **137 E2**
Winisk **133 J3** 55.20N 85.15W
Winisk River **130 K6**
Winkler **134 E3** 49.12N 97.55W
Winnemucca **134 C3** 40.58N 117.45W
Winnipeg **132 I3** 49.53N 97.10W
Winnipeg, Lake **132 I3**
Winnipegosis **132 H3** 51.40N 99.56W
Winnipegosis, Lake **132 H3**
Winona **134 E3** 44.02N 91.37W
Winston-Salem **135 F2** 36.05N 80.18W
Winton, Australia **42 D4** 22.23S 143.02E
Winton, New Zealand **90 C1** 46.09S 168.20E
Wirrulla **66 E4** 32.24S 134.32E
Wisconsin, state **134 E3**
Witchelina, homestead **66 G5** 30.01S 138.03E
Wittenberg **114 G3** 51.53N 12.39E
Wittenberge **114 G3** 52.59N 11.45E
Wittenoom **72 C4** 22.14S 118.20E
Wloclawek **114 H3** 52.39N 19.04E
Wlodawa **117 H5** 51.33N 23.31E
Wodonga **54 D2** 36.07S 146.53E
Wokam, island **88 B2**
Woleai, island **101 F3**
Wolfsburg **114 G3** 52.30N 10.37E
Wollaston Lake **132 H3**
Wollogorang **80 F6** 17.13S 137.57E
Wollongong **48 F2** 34.26S 150.52E
Wondai **42 H2** 26.19S 151.53E
Wonju **104 C5** 37.24N 127.52E
Wonsan **104 C6** 39.07N 127.26E
Wonthaggi **54 C1** 38.37S 145.36E
Woocalla **66 F5** 31.41S 137.11E
Woodenbong **48 G5** 28.23S 152.37E
Woodend **54 C2** 37.22S 144.32E
Woodlark Island **89 F2**
Woodroffe, Mount **66 C7** 26.19S 131.45E
Woods, Lake **80 D6**
Woods, Lake of the **132 I2**
Woods Point **54 D2** 37.34S 146.15E
Woodstock **135 H3** 46.10N 67.36W
Woodstock, homestead **42 D6** 19.22S 142.45E
Woodward **134 E2** 36.26N 99.25W
Woody Head **48 G5** 29.22S 153.22E
Woolgoolga **48 G4** 30.07S 153.12E
Woomelang **54 B3** 35.41S 142.40E
Woomera **66 F5** 31.12S 136.49E
Wooramel River **72 B3**
Worcester **116 C5** 52.11N 2.13W
Workington **114 D3** 54.39N 3.33W
Worland **134 D3** 44.02N 107.56W
Worthington **134 E3** 43.37N 95.36W
Wosi **88 A3** 0.10S 127.57E
Wotje, island **84 M6**
Wowoni, island **100 D2**
Wrangel Island **119 R6**
Wrangell **132 E3** 56.28N 132.23W
Wright Island **92 M3**
Wroclaw **114 H3** 51.05N 17.00E
Wu River **102 E2**
Wuhai **102 E3** 39.40N 106.40E
Wuhan **102 F3** 30.35N 114.19E
Wuhu **103 F3** 31.23N 118.25E
Wuliang Mountains **102 E2**
Würzburg **114 F2** 49.48N 9.57E
Wuwei **102 E3** 38.00N 102.54E
Wuxi **103 G3** 31.35N 120.19E
Wuyi Mountains **103 F2**
Wuyuan **102 E4** 41.07N 108.30E
Wuzhong **102 E3** 37.55N 106.11E
Wuzhou **100 C5** 23.30N 111.21E
Wycheproof **54 B2** 36.05S 143.14E
Wynbring **66 D5** 30.34S 133.32E
Wyndham **72 E5** 15.28S 128.06E
Wynyard **60 C3** 41.00S 145.44E
Wyoming, state **134 D3**
Wyong **48 F3** 33.17S 151.26E
Wyszkow **115 I3** 52.36N 21.25E

*Entry/**Page**/**Ref.**/Lat./Long.*

X

Xai-Xai **126 D2** 25.25S 33.38E
Xangongo **126 B3** 16.44S 15.00E
Xanthe **117 H3** 41.07N 24.56E
Xi River **103 F2**
Xiaguan **102 E2** 25.33N 100.09E
Xiamen **103 F2** 24.28N 118.05E
Xi'an **102 E3** 34.16N 108.54E
Xiangfan **102 F3** 32.05N 112.03E
Xiangtan **102 F2** 27.55N 112.47E
Xianyang **103 G4** 44.35N 127.30E
Xichang **102 E2** 27.52N 102.16E
Xigaze **102 C2** 29.18N 88.50E
Xilin Hot **102 F4** 43.56N 116.11E
Xinbin **104 B7** 41.41N 124.55E
Xingtai **102 F3** 37.08N 114.29E
Xingu River **143 E3**
Xining **102 E3** 36.35N 101.55E
Xinxiang **102 F3** 35.16N 113.51E
Xinyang **102 F3** 32.10N 114.05E
Xique-Xique **143 F2** 10.47S 42.44W
Xizang Plateau **106 D4**
Xuanhua **102 F4** 40.36N 115.01E
Xuchang **102 F3** 34.03N 113.48E
Xuzhou **103 F3** 34.17N 117.18E

Y

Ya'an **102 E3** 30.00N 103.02E
Yaapeet **54 B3** 35.46S 142.03E
Yablonovy Range **119 L4**
Yacheng **102 E1** 18.30N 109.12E
Yafran **124 E5** 32.04N 12.31E
Yakeshi **103 G4** 49.17N 120.48E
Yakima **134 B3** 46.37N 120.30W
Yaku Island **104 E2**
Yakumo **104 J8** 42.14N 140.19E
Yakutsk **119 M5** 62.10N 129.50E
Yalata **66 C5** 31.29S 131.50E
Yalgoo **72 C3** 28.21S 116.41E
Yallourn **54 D1** 38.12S 146.21E
Yalta **115 K1** 44.30N 34.09E
Yalu River **104 B7**
Yamagata **104 J6** 38.16N 140.19E
Yamaguchi **104 E4** 34.10N 131.28E
Yamal Peninsula **118 G6**
Yambol **117 I3** 42.28N 26.30E
Yamdena, island **88 B2**
Yamin, Peak **88 C3** 4.30S 139.55E
Yamma Yamma, Lake **42 C2**
Yamoussoukro **124 C2** 6.49N 5.17W
Yamtup, Cape **88 B3** 1.40S 130.23E
Yamuna River **106 C3**
Yana Bay **119 N6**
Yana River **119 N5**
Yanac **54 A2** 36.08S 141.26E
Yanan **102 E3** 36.41N 109.20E
Yancannia, homestead **48 B4** 30.16S 142.47E
Yanco Creek **48 C2**
Yandina **89 G2** 9.05S 159.13E
Yangambi **126 C5** 0.47N 24.24E
Yanggu **104 C6** 38.05N 128.00E
Yangon **107 E2** 16.47N 96.10E
Yangquan **102 F3** 37.52N 113.29E
Yangtze River **102 D3**
Yanji **103 G4** 42.52N 129.32E
Yankton **134 E3** 42.53N 97.24W
Yantai **103 G3** 37.30N 121.22E
Yaounde **124 E2** 3.51N 11.31E
Yap, island **101 E3**
Yapekopra **88 C3** 4.23S 135.05E
Yapen, island **88 C3**
Yapen Strait **88 C3**
Yaraka **42 E3** 24.53S 144.05E
Yaren **84 K5** 0.33S 166.56E
Yarkant River **102 B3**
Yaroslavl **115 L4** 57.34N 39.52E
Yarra River **54 C2**
Yarram **54 D1** 38.34S 146.41E
Yarrawonga **54 C2** 36.01S 146.00E
Yartsevo **115 K4** 55.06N 32.43E
Yasawa Group **84 M4**
Yass **48 E2** 34.51S 148.55E

*Entry/**Page**/**Ref.**/Lat./Long.*

Yasuj **108 D4** 30.40N 51.34E
Yathkyed Lake **132 I4**
Yatsushiro **104 E3** 32.32N 130.35E
Ya Xian **102 E1** 18.20N 109.31E
Yazd **108 D4** 31.55N 54.22E
Yazdan **108 E4** 33.30N 60.55E
Ye **107 E2** 15.15N 97.50E
Yea **54 C2** 37.13S 145.26E
Yeelanna **66 E3** 34.09S 135.44E
Yefremov **115 L3** 53.08N 38.08E
Yegorova, Cape **103 H4** 44.44N 136.18E
Yekaterinburg **118 G4** 56.52N 60.35E
Yelets **115 L3** 52.35N 38.30E
Yelizavety, Cape **119 O4** 54.30N 142.42E
Yellow Sea **103 G3**
Yellowknife **132 G4** 62.30N 114.29W
Yellowstone Lake **134 C3**
Yellowstone River **134 D3**
Yelnya **118 D4** 54.36N 33.13E
Yemen, Arab republic **108 C2**
Yenangyaung **107 E3** 20.28N 94.54E
Yenbo **108 B3** 24.05N 38.03E
Yendi **124 C2** 9.30N 0.00W
Yenisey River **118 I5**
Yeoval **48 E3** 32.45S 148.39E
Yeppoon **42 H4** 23.08S 150.44E
Yerevan **108 C5** 40.10N 44.31E
Yevlakh **118 E3** 40.35N 47.08E
Yevpatoriya **115 K2** 45.12N 33.20E
Yeysk **115 L2** 46.43N 38.17E
Yibin **102 E2** 28.50N 104.35E
Yichang **102 F3** 30.43N 111.22E
Yichun **103 G4** 47.41N 129.10E
Yinchuan **102 E3** 38.30N 106.19E
Yingkou **103 G4** 40.40N 122.17E
Yining **102 C4** 44.00N 81.20E
Ylivieska **115 I5** 64.05N 24.30E
Yobi **88 C3** 1.48S 138.04E
Yogyakarta **100 C2** 7.48S 110.24E
Yokohama **104 I4** 35.28N 139.38E
Yokosuka **104 I4** 35.18N 139.39E
Yonezawa **104 J5** 37.56N 140.06E
Yonghung **104 C6** 39.31N 127.18E
Yonkers **135 G3** 40.56N 73.54W
Yopal **137 E1** 5.20N 72.19W
York, Australia **72 C2** 31.53S 116.46E
York, U.K. **114 D3** 53.58N 1.05W
York, Cape **40 E4** 10.42S 142.32E
Yorke Peninsula **66 F3**
Yorketown **66 F3** 35.01S 137.37E
York Factory **132 I3** 57.30N 92.30W
Yorkton **132 H3** 51.12N 102.29W
Yoro **136 D2** 15.08N 87.10W
Yoshkar Ola **118 E4** 56.38N 47.52E
Yosu **104 C4** 34.46N 127.45E
Young **48 E2** 34.19S 148.18E
Young Nicks Head **90 H5** 38.45S 177.58E
Youngstown **135 F3** 41.05N 80.40W
Youth, Isle of **136 D3**
Ysabel Channel **89 E3**
Ystad **114 G4** 55.25N 13.50E
Yuba City **134 B2** 39.08N 121.38W
Yubari **104 K8** 43.04N 141.59E
Yucatan Channel **136 D3**
Yucatan Peninsula **136 D3**
Yuci **102 F3** 37.40N 112.44E
Yudnapinna Hill **66 F4** 32.06S 137.08E
Yuendumu **80 C3** 22.15S 131.48E
Yukon Plateau **132 E4**
Yukon, province **132 E4**
Yukon River **132 C4**
Yulara **80 C2** 25.10S 130.55E
Yuma **134 C2** 32.40N 114.39W
Yumen **102 D3** 39.54N 97.43E
Yunta **66 G4** 32.35S 139.34E
Yurimaguas **142 C3** 5.54S 76.07W
Yushu **102 D3** 33.06N 96.48E
Yuty **144 D4** 26.31S 56.20W
Yuxi **102 E2** 24.24N 102.29E
Yuzhno-Sakhalinsk **119 O3** 46.58N 142.45E

*Entry/**Page**/**Ref.**/Lat./Long.*

Z

Zabid **108 C2** 14.10N 43.18E
Zabol **108 E4** 31.00N 61.32E
Zacatecas **136 B3** 22.48N 102.33W
Zadar **117 G3** 44.07N 15.14E
Zafra **116 B2** 38.25N 6.25W
Zagan **114 H3** 51.37N 15.20E
Zagreb **117 G4** 45.48N 15.58E
Zagros Mountains **108 C4**
Zahedan **108 E3** 29.32N 60.54E
Zahle **125 G5** 33.50N 35.55E
Zaire *see* Congo, Democratic Republic of
Zaire River *see* Congo River
Zaisan, Lake **102 C4**
Zajecar **117 H3** 43.55N 22.16E
Zalantun **103 G4** 48.02N 122.45E
Zalau **117 H4** 47.14N 22.57E
Zambezi River **126 D3**
Zambia, republic **126 C3**
Zamboanga **100 D3** 6.55N 122.05E
Zamora, Ecuador **142 C3** 4.05S 79.01W
Zamora, Mexico **136 B3** 20.00N 102.18W
Zamora, Spain **116 B3** 41.30N 5.45W
Zamosc **117 H5** 50.43N 23.15E
Zanjan **108 C4** 36.40N 48.30E
Zanzibar **126 D4** 6.10S 39.12E
Zanzibar, island **126 D4**
Zaouatene **124 D4** 19.58N 2.58E
Zaozhuang **103 F3** 34.53N 117.38E
Zaporozhye **115 L2** 47.50N 35.10E
Zaranj **106 B4** 30.58N 61.53E
Zarate **144 D3** 34.07S 59.00W
Zaraza **142 D4** 9.23N 65.20W
Zard, Mount **108 D4** 32.22N 50.04E
Zaria **124 D3** 11.01N 7.44E
Zarzis **124 E5** 33.34N 11.04E
Zawia **124 E5** 32.49N 12.43E
Zeehan **60 C3** 41.53S 145.21E
Zeil, Mount **80 D3** 23.24S 132.23E
Zelenogradsk **115 I3** 54.57N 20.28E
Zemio **125 F2** 5.00N 25.09E
Zenica **117 G3** 44.11N 17.53E
Zeya River **119 M4**
Zhangguangcai Mountains **104 C8**
Zhangjiakou **102 F4** 40.51N 114.59E
Zhangye **102 E3** 38.58N 100.30E
Zhangzhou **103 F2** 24.31N 117.40E
Zhanjiang **102 F2** 21.10N 110.20E
Zhaodong **103 G4** 46.03N 125.58E
Zhaotong **102 E2** 27.20N 103.39E
Zhelaniya, Cape **118 G6** 77.00N 68.31E
Zheleznogorsk **115 L3** 52.20N 35.26E
Zheltyye Vody **115 K2** 48.21N 33.31E
Zhengzhou **102 F3** 34.45N 113.38E
Zhigansk **119 M5** 66.48N 123.27E
Zhitomir **115 J3** 50.18N 28.40E
Zhlobin **115 K3** 52.50N 30.00E
Zhmerinka **115 J2** 49.00N 28.02E
Zhob **106 B4** 31.21N 69.31E
Zhongshan **102 F2** 22.30N 113.20E
Zhuzhou **102 F2** 27.53N 113.07E
Zibo **103 F3** 36.51N 118.01E
Zielona Gora **114 H3** 51.57N 15.30E
Zigong **102 E2** 29.25N 104.47E
Ziguinchor **124 B3** 12.35N 16.20W
Zilina **114 H2** 49.14N 18.43E
Zimbabwe, republic **126 C3**
Zinder **124 D3** 13.46N 8.58E
Znamenka **115 K2** 48.42N 32.40E
Znojmo **114 H2** 48.52N 16.05E
Zodino **115 J3** 54.07N 28.27E
Zomba **126 D3** 15.22S 35.22E
Zonguldak **108 B5** 41.26N 31.47E
Zouar **124 E4** 20.30N 16.30E
Zrenjanin **117 H4** 45.22N 20.23E
Zunyi **102 E2** 27.41N 106.50E
Zurich **116 E4** 47.23N 8.33E
Zvenigorodka **115 K2** 49.05N 30.58E
Zvishavane **126 D2** 20.20S 30.05E
Zwai, Lake **125 G2**
Zwickau **114 G3** 50.43N 12.30E

First published 1996 by
John Wiley & Sons Australia, Ltd
42 McDougall Street, Milton, Qld 4064

Offices also in Sydney and Melbourne

Typeset in Times and Gill Sans

Reprinted in Hong Kong 1997 (twice), 1998
Reprinted in Singapore by
Craft Print International Ltd 1999, 2000, 2001, 2002, 2003, 2004, 2005, 2006

National Library of Australia
Card Number and ISBN-10 0 7016 3302 6
ISBN-13 978 0 7016 3302 8

Cartography by MAPgraphics Pty Ltd, Brisbane

Illustrated by Craig Jackson, Paul Lennon, Peter Reardon, Stephen Francis and Ikon Computergraphics, Brisbane

Designer: Kate Barry
Compositor: Jeff Hofmann

Cover photograph: © Ken Duncan

Acknowledgements
The publisher and editorial board wish to thank the following people and institutions for their assistance in the production of this atlas.

Illustrative material
Geoff Higgins: page 1; Sporting Pix/Stuart Milligan: page 2 (*top*); VicImage: pages 2 (*bottom*), 8; Viking O'Neil *Explore Australia*, Penguin Books Australia: page 4; illustration by permission of Sea World, courtesy of Village Roadshow Exhibition: pages 6–7; Open Spaces Photography/Glenn Tempest: page 10 (*photo*); courtesy of Bob Lowres: page 10 (*top map*); Copyright © Commonwealth of Australia, AUSLIG, Australia's national mapping agency. 1985. All rights reserved: Contents and page 15, pages 16–17 (© 1981), 68 (*bottom*) (© 1992), 79 (© 1991); Crown copyright — Aerial photography supplied by the Sydney Map Shop, 22–33 Bridge St, Sydney NSW 2000. Telephone (02) 228 6111, (02) 221 5980: page 18 (top); Stock Photos Pty Ltd/Lance Nelson: page 18 (*bottom*), Bruce Peebles: pages 44 (*bottom*), 50 (*bottom*); Sunmap/Department of Lands: page 19 (*top and bottom*); Photo Library-Sydney/Hulton Deutsch: pages 22 (*top*), 186 (poster), Earth Satellite Corp.: pages 20–21, Acres, Auslig: pages 46, 52, 58, 64, 70, 76, 163, Robert McFarlane: page 55, Geoff Higgins: page 63 (*bottom*), Ed Simpson: page 83, Science Photo Library: pages 99, 113, 123, 131, 141, 153, (Photo Library Sydney): pages 105, 184, Hiroshi Higuchi: page 111, Rick Strange: pages 117, 121, Peter Poulides: page 129, Robin Smith: pages 37, 139, 155 (*bottom*), John Reader/SPL: page 168 (*bottom right*), SNS/Sipa Press: page 169 (*top*), Claire Alix: page 171 (*bottom left*), Brian Alexander: page 172 (*top*), P. H. Cornut: page 173 (*top*), Alain Evrard: page 183; Courtesy of Dr Robert D. Ballard: page 22 (*bottom*), page 26 (*left*); paintings by Ken Marschall from *Exploring the Lusitania* © 1995 by Robert D. Ballard, a Madison Press Book: pages 23, 26–27; Courtesy of Mark Reynolds from *Exploring the Lusitania*, © 1995 by Robert D. Ballard, a Madison Press book: page 24 (*bottom*); Eric Sauder Collection: page 24 (*top*); Australian Picture Library/Reuters/Bettmann: page 25; *Age*: page 28; Department of Lands, Queensland: page 29; Archives Office of Tasmania: page 34 (*top*), page 35 (*left*); Nick Mooney: page 35 (*bottom right*); Tasmap: pages 36, 62 (*bottom*); Coo-ee Picture Library: page 39 (*left*), page 69 (*bottom*), page 151 (*top*), 155 (*top*); Australian Picture Library/Reuters/Bettmann: page 25, Woodfin/Grossma: page 168 (*left*), C. Bowman: page 170 (*top*), (Australian Picture Library): page 170 (*bottom left*), Steve Vidler: page 171 (*top right*), Westlight: page 181 (*left*); Neil McLeod: page 41; Jim McEwan, *Age*, page 43; John Oxley Library, Brisbane: Contents and pages 44 (*top*), 186 (*top*); *Central station 7 a.m (The morning rush) (1931)* by Vida Lahey, 1882–1968. Collection Brisbane City Hall Art Gallery and Museum — bequest of the artist. Reproduced by kind permission of Shirley Lahey, Brisbane: page 45 (*top*); Fred Adler/Kino Studios: page 45 (*bottom*); A.N.T. Photo Library/Otto Rogge: page 48 (*bottom*), Martin Harvey: pages 127, 154, 164 (*top right*), P. & M. Walton: page 151 (*bottom*), Natural Images: page 157 (*top*), J. P. & E. S. Baker: page 157 (*bottom*), Gerard Lacz: page 164 (*bottom left*), page 165 (*top* and *second top*), Jonathan Chester: page 164 (*bottom centre*), 165 (*bottom*), M. Cermak: page 165 (*second bottom*), Dennis & Theresa O'Byrne: page 167, Grant Dixon: page 169 (*bottom*); Juno Gemes: page 49 (*top*); National Library of Australia: pages 50 (*top*), 56 (*top*), 62 (*top*); Alex MacCartney: page 51 (*top*); Mark Heriot: page 51 (*bottom*); Ian Smith & Associates Pty Ltd: page 56 (*bottom*); Coo-ee Historical Picture Library: pages 57 (*top*), 61, 63 (top), 69 (top); Sporting Pix/John Daniels: page 57 (*bottom*); Luise Hercus: page 67; City of Adelaide Civic Collection: page 68 (*top*); Courtesy of Waringarri Aboriginal Art and kind permission of Rover Thomas: page 73; Department of Land Administration: page 74 (*top*); Central Mapping Perth: page 74 (*bottom*); Photo Index: page 75 (*bottom*); W. B. Griffin 1913 Plan of Canberra/National Library of Australia — map section: page 78 (*top*); *Canberra Times*/Capital Pictures: page 81; Map courtesy of NT Department of Lands, Planning and Environment: page 82; Panos Pictures — Penny Tweedie: page 85, Hamish Wilson: page 125, Paul Smith: page 137, Martin Adler: page 170 (*bottom right*); Betty Press: page 172 (*bottom*), Nancy Durell McKenna: page 182; AAP Photo Library/AP: page 87 (*bottom*), Liu Heung Shing: page 101, AAP Photo Library: page 173 (centre); Courtesy of BHP Ok Tedi: page 89; Fotopress News Picture Agency: page 91; Australian Antarctic Division: page 93; Robert Harding Picture Library: page 97; The Image Bank: page 103; Topham Picture Source: pages 107, 119, 143; Gunther Schörlitz, Fotozentrum der Friedrich-Schiller-Universität Jena: page 109; Austral International/Ken Welsh: page 112, Camera Press/John Watney: page 115, Camera Press: page 135; International Photographic Library: pages 133, 145, 164 (*bottom right*); *Courier-Mail*: page 160; Horizon: page 164 (*top left*); Marie Loi: page 165 (*illustration*); Nature Focus/Babs and Bert Wells: page 167 (*bottom photograph*), Rod Wells: page 171 (*bottom centre*); Jonathan N. Tubb: page 168 (*top right*), page 175; South American Pictures/Tony Morrison: page 171 (*bottom right*), Bill Leimbach: page 173 (*bottom*); Museum and Art Gallery of the Northern Territory/George Chaloupka (curator of rock art): page 174; Bruce Petty: page 176 (*cartoon*); David Witts: page 176; General Motors — Holden's Automotive Limited: pages 180–81, 181 (*right*); John Oxley Library: page 186 (*top*).

Other material
Australian Bureau of Statistics: pages 47, 53, 59, 65, 71, 77

ARCTIC
Svalbard (Nor.)
Greenland (Den.)
Arctic Circle
ICELAND
Reykjavik
Faroe Is (Den.)
RUSSIA
Moscow
SWEDEN
NORWAY
FINLAND
Oslo
Stockholm
Helsinki
Tallinn
ESTONIA
LATVIA
Riga
LITHUANIA
Vilnius
Minsk
BELARUS
DENMARK
Copenhagen
UNITED KINGDOM
London
IRELAND
Dublin
POLAND
Warsaw
Berlin
GERMANY
Kiev
UKRAINE
FRANCE
Paris
ITALY
Rome
RUMANIA
Bucharest
BULGARIA
Sofia
GREECE
Athens
SPAIN
Madrid
PORTUGAL
Lisbon
1 NETHERLANDS Amsterdam
2 BELGIUM Brussels
3 LUXEMBOURG Luxembourg
4 SWITZERLAND Bern
5 AUSTRIA Vienna
6 CZECH Prague
7 SLOVAKIA Bratislava
8 HUNGARY Budapest
9 SLOVENIA Ljubljana
10 CROATIA Zagreb
11 BOSNIA and HERZEGOVINA Sarajevo
12 SERBIA and MONTENEGRO Belgrade
13 MOLDOVA Kishinev
14 KOSOVO (UN) Pristina
15 FORMER YUGOSLAV REPUBLIC OF MACEDONIA Skopje
16 ALBANIA Tirane
17 MONACO Monaco
GEORGIA
Tbilisi
ARMENIA
Yerevan
Baku
AZERBAIJAN
TURKEY
Ankara
CYPRUS
Nicosia
SYRIA
Damascus
IRAQ
Baghdad
Amman
JORDAN
IRAN
Tehran
KAZAKHSTAN
Astana
UZBEKISTAN
Tashkent
TURKMENISTAN
Ashkhabad
KIRGIZIA
Bishkek
Dushanbe
TADZHIKISTAN
AFGHANISTAN
Kabul
Islamabad
PAKISTAN
MONGOLIA
Ulan Bator
CHINA
Beijing
NORTH KOREA
Pyongyang
Seoul
SOUTH KOREA
JAPAN
Tokyo
NEPAL
Kathmandu
BHUTAN
Thimphu
New Delhi
INDIA
BANGLADESH
Dacca
MYANMAR
Yangon
LAOS
Vientiane
Hanoi
THAILAND
Bangkok
VIETNAM
CAMBODIA
Phnom Penh
Taipei
TAIWAN
Manila
PHILIPPINES
NORTHERN MARIANAS
Saipan
Guam (U.S.A.)
MICRONES
PALAU
Koror
BRUNEI
Bandar Seri Begawan
Kuala Lumpur
MALAYSIA
SINGAPORE
Jakarta
INDONESIA
Dili
EAST TIMOR
PAPUA NEW GUI
Port Moresby
Christmas I (Aust.)
Cocos Is (Aust.)
AUSTRALIA
Canberra
SRI LANKA
Colombo
Male
MALDIVES
Chagos Arch (U.K.)
Madeira Is (Port.)
Rabat
MOROCCO
Algiers
Tunis
TUNISIA
Tripoli
ALGERIA
LIBYA
EGYPT
Cairo
Canary Is (Sp.)
Western Sahara
Tropic of Cancer
MAURITANIA
Nouakchott
MALI
NIGER
CHAD
SUDAN
Khartoum
ERITREA
Asmara
SAUDI ARABIA
Riyadh
YEMEN
San'a
Aden
OMAN
Muscat
18 LEBANON Beirut
19 ISRAEL Jerusalem
20 KUWAIT Kuwait
21 BAHRAIN Manama
22 QATAR Doha
23 UNITED ARAB EMIRATES Abu Dhabi
CAPE VERDE ISLANDS
Praia
Dakar
SENEGAL
GAMBIA
Banjul
Bissau
GUINEA BISSAU
GUINEA
Conakry
Freetown
SIERRA LEONE
Monrovia
LIBERIA
BURKINA FASO
Bamako
Niamey
Ouagadougou
NIGERIA
Abuja
N'Djamena
TOGO
BENIN
Porto-Novo
GHANA
IVORY COAST
Accra
Lome
Yamoussoukro
CAMEROON
Yaounde
Malabo
EQUATORIAL GUINEA
SAO TOME AND PRINCIPE
Libreville
GABON
CENTRAL AFRICAN REPUBLIC
Bangui
DJIBOUTI
Djibouti
Addis Ababa
ETHIOPIA
SOMALIA
Mogadishu
Equator
DEMOCRATIC REPUBLIC OF CONGO
UGANDA
Kampala
KENYA
Nairobi
CONGO
Brazzaville
Kinshasa
RWANDA
Kigali
Bujumbura
BURUNDI
Dodoma
TANZANIA
SEYCHELLES
Victoria
Luanda
ANGOLA
ZAMBIA
Lusaka
MALAWI
Lilongwe
COMOROS
Moroni
St Helena (U.K.)
Harare
ZIMBABWE
MOZAMBIQUE
MADAGASCAR
Antananarivo
MAURITIUS
Port Louis
Reunion (Fr.)
NAMIBIA
Windhoek
BOTSWANA
Gaborone
Pretoria
Maputo
Mbabane
SWAZILAND
Bloemfontein
Maseru
LESOTHO
SOUTH AFRICA
Cape Town
Tropic of Capricorn
SOUTH ATLANTIC OCEAN
INDIAN OCEAN
Tristan da Cunha (U.K.)
Gough I (U.K.)
Kerguelen Is (Fr.)
Heard I (Aust.)
Antarctic Circle
ANTARCTICA
N
Scale 1:68 000 000 at 45°N and 45°S
0 1000 2000 3000 4000 km
Modified Cylindrical Projection
60°W 30°W 0° 30°E 60°E 90°E 120°E
60°N 30°N 0° 30°S 60°S